Aims and Methods
of Vegetation Ecology

Dieter Mueller-Dombois
Heinz Ellenberg

Aims and Methods
of Vegetation Ecology

JOHN WILEY & SONS

New York   London   Sydney   Toronto

Copyright © 1974, by John Wiley & Sons, Inc.

All rights reserved. Published simultaneously in Canada.

No part of this book may be reproduced by any means,
nor transmitted, nor translated into a machine language
without the written permission of the publisher.

Library of Congress Cataloging in Publication Data:

Mueller-Dombois, Dieter, 1925–
     Aims and methods of vegetation ecology.

     Bibliography: p.
     1. Botany—Ecology.   2. Botany—Methodology.
I. Ellenberg, Heinz, 1913–   joint author.
II. Title.

QK901.M8   581.5   74-5492
ISBN 0-471-62290-7

Printed in the United States of America

10 9 8 7 6 5 4 3 2 1

Foreword

Vegetation is one of the major geographical features of almost all parts of the earth's surface. Only the polar ice-sheets are devoid of self-reproducing plants, the aggregate of which make up vegetation. In some regions there may be only microscopic planktonic algae, soil-fungi, or bacteria, but there is a plant component to all major, and most minor ecosystems. In major ecosystems, at least, it is an essential component, as it includes the primary production apparatus that fuels the system by capture of solar energy.

Practically every terrestrial view that man has of his environment, outside his own constructions—cities, houses, excavations, and vehicles —is almost sure to be primarily of vegetation. It is the most obvious surface feature of the land. On it depends the existence of all animal life. Most human activities deal in some respect with vegetation or its products. In other words, vegetation is an inescapable fact of life. As such, it is one of the most important of all subjects for investigation and study. Such attention may be directed toward its beauty, intrinsic interest, or practical importance and uses.

There are many books and innumerable papers on vegetation. Only one who has compiled bibliographies on the subject can realize how many. Most of these include descriptions of vegetation, its environmental relations, its importance and usefulness to man, its management and control, and interpretation of its significance and interrelations of any or all of these aspects. Very few of the works on vegetation deal in any comprehensive way with how one studies vegetation. TANSLEY and CHIPP, in 1926, published a book that was a landmark in this respect, called *Aims and Methods in the Study of Vegetation*. It was a general discussion of vegetation study and an exposition of ideas on effective methods of vegetation description and study as seen at that time, especially through English eyes. Because of the then far-flung extent of British interests and the British Empire the English viewpoint was at that time a peculiarly appropriate one for a general text.

Whether because of the TANSLEY and CHIPP book, or because of the forceful and convincing writing of a group of American plant ecologists, or, perhaps more likely, because of the linguistic provincialism of English-speaking people even among scientists, an "Anglo-American" school of plant ecology developed, and for nearly a half-century has practically dominated vegetation-science in English-speaking parts of the world. The remarkable developments in continental and Soviet geobotany and phytosociology have remained little-known and scarcely understood by most English-speaking ecologists and their disciples and students. Even FULLER and CONARD's English translation of BRAUN-BLANQUET's work (1932) produced scarcely a ripple in the self-contained complacency of the Clementsian and Tansleyan schools of thought in this field.

In recent years, with the great upsurge of ecological interest and concern, has come an increasing awareness of the ecological work and thought of other parts of the world. A demand has arisen for English translations of the outstanding German and Russian books on vegetation-science, inaccessible to most English-speaking scientists because of the inadequacy of our educational systems in foreign language training.

Good translation of scientific writing is rare, because there are so few truly bilingual scientists and because scientists are usually too busy with their own work to have time to translate that of others. Hence most vegetation texts have remained untranslated.

One of the best of the continental vegetation books available has been *Aufgaben und Methoden der Vegetationskunde* (1956) by HEINZ ELLENBERG, one of Europe's truly outstanding plant ecologists and phytosociologists. Written in German, it has remained essentially unavailable to most English-speaking ecologists. With the rapid development of the field of vegetation-science, a new edition became desirable. We are most fortunate that a new and completely revised edition of this book has been prepared by DIETER MUELLER-DOMBOIS in close cooperation with HEINZ ELLENBERG. German-born, MUELLER-DOMBOIS has, in over 20 years of work and residence in America and Ceylon, acquired perfect fluency in English while at the same time developing into one of the ablest and best-balanced of American plant ecologists. Eleven years in Central Canada and the Pacific Northwest plus almost a decade of experience in the tropics has given him a breadth particularly appropriate for the task of making the results of ELLENBERG's vast experience, ability, and soundness available to the non-German-speaking scientific public. To ELLENBERG's profound knowledge of European ecology, MUELLER-DOMBOIS has added a broad spectrum of American thought and methodology, both theoretical and practical. In addition, he has been able to extend the scope of the work so it is eminently suited to guide the increasing and vitally important study of tropical vegetation.

As a result, one will find in the present volume a discussion of vegetation theory, an explanation of the interest and uses of vegetation study, and a vegetation-science cookbook. Perusal of the table of contents will show what a comprehensive approach has been adopted. The first part attempts to give the student and nonspecialized reader a general background and appreciation of vegetation ecology with some of its underlying philosophy. PARTS II, III, and IV combine a detailed explanation of vegetation phenomena with a thorough exposition of most of the methods in use for investigating them. Nowhere else, to the best of my knowledge, has there been brought together such a careful and detailed, but at the same time well integrated, view of the essential methodology for the scientific study of the plant-cover of the earth.

PART V synthesizes aims and methods in vegetation ecology. The essentially uniform viewpoints of TANSLEY and CHIPP and the Anglo-American school have been reconciled with the several diverse continental and Russian approaches with a considerable degree of success. At least the disregard of European phytosociology by American workers must now be from choice rather than from ignorance.

From the student's standpoint, the lazy ones may start their investigations with sound methodology without the irksome necessity for much thought. The mentally alert and aggressive ones may carry out their probing into understanding of nature and investigation along the frontiers of knowledge from a secure and reliable base of the present "state-of-the-art."

How well the authors have attained their objective of providing a basis for research into the patterns and nature of vegetation may only be judged after reading the book. I have no reservations about recommending it as by far the best work of its scope that I know.

F. R. Fosberg

Preface

This book, we believe, is the first integrated synthesis of European and Anglo-American approaches to plant synecology or vegetation science. The book combines classical and current concepts with detailed prescriptions for vegetation analysis and data processing by well-proven methods appropriate to specified objectives.

The book is based on an earlier vegetation methods book of H. ELLENBERG written in German and entitled, "Aufgaben und Methoden der Vegetationskunde" (Verlag Eugen Ulmer, Stuttgart 1956, 136 p.), which has long been out of print. It received wide popularity beyond the borders of Europe because of the detailed and critical explanation of the relevé (i.e., vegetation abstract or sample) method introduced by BRAUN-BLANQUET. The potential of this method has rarely been appreciated in the English speaking countries, most probably because of a lack of understanding due to the language barrier. The BRAUN-BLAN-QUET method is often merely identified with a crude estimate scale of species abundance and cover, with a subjective placement of sample areas and with a subsequent classification of vegetation units. A closer understanding will show that there is really very much more to it, that discriminate selection of sample stands has its place in vegetation sampling and that species abundance can be assessed with differing degrees of sophistication depending on the objectives. Moreover, the BRAUN-BLANQUET method of relevé synthesis has become a basic technique underlying modern methods of computer synthesis and classification. ELLENBERG's original book contained a number of additional vegetation analysis methods and concepts, for example, the "ecological group concept," which are as yet relatively unfamiliar to English speaking ecologists.

The present book is the outcome of a decision by the two authors to prepare a basic translation of the 1956 book, to update the earlier information and to enlarge the scope to include the Anglo-American methods and concepts. Vegetation analysis methods of Soviet investigators are not specifically treated in this text since their major aim is directed to a rapid and relatively simple overview of large areas, and since such overviews have long been available for both western Europe and North America. However, some important Russian and other eastern European conceptual contributions are included. As in the first book, the idea was to accept only the well-proven and widely applied techniques and not those that represent various minor modifications that are still in the research stage or those that are only rarely applied. The methods and concepts newly incorporated in this book are treated in the same detail as those in the 1956 book.

The synthesis of European and American approaches was attained not by amalgamating these different methods and concepts, but by evaluating them with regard to their limitations and specified objectives. As will be seen, the approaches that originated in the different language camps are only competitive to a limited extent. For the most part they are truly complementary. By bringing the proven concepts and methods together in one volume, the subject area of vegetation science can be more adequately treated as a discipline in its own right. However, from an ecological viewpoint, this discipline is not considered an end in itself, but rather an introduction to other related areas, in particular to functional ecosystem ecology.

A very helpful and critical reading of the developing book manuscript was done by F. RAYMOND FOSBERG, who provided many useful suggestions. We are also grateful for the constructive criticism of MICHAEL

G. BARBOUR, who as professional ecologist, reviewed the manuscript for the publisher, JOHN WILEY and SONS, Inc. Furthermore, we wish to acknowledge with many thanks the help of Dr. H. HELLER and Dr. ERIKA GEYGER (University of Göttingen) in the search of the earlier literature. Finally, our thanks go to students in vegetation ecology (spring semesters 1971 and 1972) at the University of Hawaii, who helped indirectly in shaping the manuscript and directly in providing some of the sample data used in the book. Thanks are due also to Mr. T. NAKATA for drafting most of the line drawings and to HARRIET MATSUMOTO, BARBARA MYERS and LYNNETTE ARAKI, for typing sections of the manuscript.

Honolulu, Hawaii
Göttingen, West Germany

Dieter Mueller-Dombois
Heinz Ellenberg

Contents

**PART
III**

**CLASSIFYING AND ORDINATING
VEGETATION DATA**

Aims and Methods
of Vegetation Ecology

PART
I

General

Introduction–Clarification of Concepts

1

1.1 VEGETATION ECOLOGY

Vegetation ecology (a term coined for the first time by this book) combines both European and Anglo-American terms to define an area of research and a body of knowledge and techniques. This area of research is common to both Europe and America, but differences in terminology and methodology in the past have obscured the similarities and hindered cooperation.

This book presents a modern summary of vegetation science (as it is known in Europe), or plant synecology (as it is called in England and America); a synthesis which we hope will both inform the new student and stimulate the professional.

3

Vegetation ecology will be defined more carefully later in this chapter, but for now we will broadly describe it as the study of plant communities, or vegetation. Plants typically occur together in repeating groups of associated plants. These groups are called communities, and they are best described by noting the identity and growth form of the most abundant species, the largest species, or the most characteristic species of the particular community. That is, the community is not described by simply listing all the species which compose it, as though each were of equal importance. Instead, a community is characterized by detailing those species which most contribute to its unique structure and composition.

Notice, then, that a community (or vegetation in general) is not equivalent to the flora of an area. In its simplest form, flora refers to a list of species or to the plant taxa occurring in that area. A flora in document form may range from a floristic checklist to a complete taxonomic treatment with keys—morphological and nomenclatural information. Floras, as a rule, do not give information on the combinations (communities) in which the species occur in nature, nor do they comment on their abundance, importance, or uniqueness—all species have equal weight.

Vegetation ecology is concerned not only with identifying the plant communities (the vegetation) on an area, but also with determining how they are related to one another and to the environmental factors.

1.2   HOW TO RECOGNIZE PLANT COMMUNITIES

Because groups of plants form communities, many vegetation ecologists assume that the plants in a community have some influence upon one another and/or that they have something in common with their environment. This implies interdependence, that the whole is greater than the sum of the parts, and that communities are integrated entities.

For practical purposes, plant communities may be considered subdivisions of a vegetation cover. Wherever the cover shows more or less obvious spatial changes, one may distinguish a different community.

These changes may be caused by spatial changes in species composition, changes in spacing and height of plants, changes in growth form or life form of plants, or seasonal plant responses of other vegetation properties, which in turn may correspond to spatial changes in environment. Whatever vegetation parameters are involved in causing such changes, they form part of the definition, description, and interpretation of a community. This will be discussed later in detail.

Spatial changes in species combinations are often obvious even to the casual observer. He may notice that certain species combination changes are related to changes in environment. For example, changes that are related either to increases in soil moisture in local depressions in a grass covered area, or to spatial changes in climate are fairly obvious. Undergrowth vegetation in forests also reacts to the intensity and seasonal rhythms of available light beneath a deciduous forest canopy. In more extreme habitats, such as bogs, coastal dunes, and salt marshes, plant life forms and species combinations are often so different from the adjacent plant cover that they are self-evident as different communities.

Depending on the properties of the vegetation and the area, changes vary from relatively abrupt to transitional or very gradual and diffuse. Consequently, plant communities may be self-evident or nonevident to the investigator on first inspection. Communities that are nonevident to the unacquainted may be self-evident to the experienced investigator. Moreover, certain patterns may not be noted on first inspection even by an experienced investigator; they become evident only through increased familiarity with the vegetation of a given area. Recognition and definition of plant communities is a skill that can be learned. On the other hand, certain variations in the plant cover that may be very obvious on first inspection may turn out to be only transitory phases of the same community type. This may apply, for example, to a juvenile or older stages of a forest community type or to the summer or winter aspects of a semidesert community type, characterized by short-lived herbaceous vegetation. The ability to recognize temporary variations as mere short-time phases of a more basic community type comes from frequent analyses and comparisons and experience with a regional vegetation.

Not only the larger plants (including ferns, mosses, and lichens) that are more obvious to the eye are members of the plant communities; there are also fungi, algae, and numerous microorganisms. The latter are often of special importance to the entire community for the circulation of matter. The same applies to the many animals that live on organic matter in and above the soil.

However, since the study of these different groups of organisms requires specialization and the application of special techniques, it is customary to limit the study of vegetation to higher plants, *cryptogamic tracheophytes*, mosses, and lichens. These plants are the dominant producers in terrestrial communities.

We do not intend to overlook the opinion of some plant ecologists (for example, WHITTAKER 1970) that communities are often less than discrete units and may be thought of then as hardly more than the chance meeting of several species whose tolerance ranges to the environment overlap. In this view, interdependence is not assumed or invoked, and the whole is not greater than the sum of the parts. Indeed, a great debate exists today as to the utility of the community concept, and we will come back to it indirectly in later chapters.

1.3 EUROPEAN AND ANGLO-AMERICAN TERMINOLOGY

We said that the term vegetation ecology is a new one, coined for this book, and that its synthesis from European and American components symbolizes our attempt in this work to bring together European and American viewpoints on plant ecology. One hurdle in this objective is the confusion that exists between terminology used by Anglo-American and European researchers.

In Europe, field botany is called *geobotany* (GRISEBACH 1866, RÜBEL 1922, ELLENBERG 1968a). At first, geobotany consisted only of traditional plant taxonomy, but, around 1800, ALEXANDER VON HUMBOLDT expanded it into the discipline of plant geography. By 1900, the study of communities had added another dimension to geobotany. The present areas of specialization within geobotany are shown in TABLE 1.1.

In the context of botany as a whole, geobotany is concerned with the higher levels of organization (DANSEREAU 1963). This means that the scientific interest begins with the structural, systematic, and functional aspects of individual plants in the field and extends to the same aspects of the plant population, the community, the ecosystem, the biogeoclimatic zone (KRAJINA 1965), and the floristic province or phytogeographic region.

There is no Anglo-American equivalent to the term geobotany. However, each of the subdivisions of geobotany is rife with European and Anglo-American synonyms or near equivalents. As the purpose of this section is more or less to make each of us comfortable in the presence of others, we have chosen to condense the terminology in tabular form and to emphasize similarities rather than split hairs on fine differences. This means that our TABLE 1.1 gives only approximate equivalents; it is not meant to serve as a short form for rigorous definitions of the terms.

It is worthwhile to point out some of the discrepancies in terminology here, in addition to what the reader may see for himself in TABLE 1.1. First, of the four general areas of specialization, the subject matter that the term ecology subjectively brings to mind appears in only two: sociologic and ecological geobotany. This brings us to the first discrepancy, the term ecology itself.

TABLE 1.1   *Areas of Specialization within the Field of Geobotany; their Synonyms and Anglo-American Equivalents. More Specific Definitions Appear in the Text.*

AREA OF SPECIALIZATION (AND SYNONYMS, EUROPEAN TERMS)	SUBJECT MATTER	ANGLO-AMERICAN EQUIVALENTS (AND SYNONYMS)
Floristic geobotany	Study of geographic distribution of plant taxa and their evolutionary relationships	Plant geography (phytogeography)
Sociologic geobotany[a] (vegetation science, plant sociology, phytosociology, phytocoenology)	Study of composition, development, geographic distribution, and environmental relationships of plant communities	Synecology[a] (community ecology, plant ecology in part)
Ecological geobotany (plant ecology)		
Autecology (ecophysiology)	Study of physiological functions of individual organisms in field environments and communities; life history studies of species or ecotype	Autecology (physiological ecology, population ecology in part)
Demecology (population ecology)	Study of structure and function of populations	Population ecology
	Study of genetic variation in populations	Genecology
Synecology (habitat science; ecosystem research)	Study of habitat factors and the physiological response of species and species groups to these factors; study of community functioning, and niche functions of plant populations in an ecosystem context	Ecosystem ecology (community process ecology, functional ecology, systems ecology)
Historical geobotany	Study of historical origin and development of populations and communities	Paleobotany (paleoecology)

[a] Equivalent to the term vegetation ecology, as used in this book.

7

Ecology was coined by the German zoologist ERNST HAECKEL in 1866. It refers to the science (logos) concerned with the house (oikos) or household or organism. In America, ecology includes the study of both groups of organisms (communities) *and* individual organisms or of populations in relation to environment. Thus, it covers such endeavors as the description and mapping of plant communities, measuring the productivity of an annual grassland, and experimenting with the effect of temperature on the growth and reproduction of a particular species. In continental European understanding, plant ecology has a more restricted meaning. It is equivalent to the ecological geobotany area and thus does not include vegetation mapping.

A second discrepancy is with the terms aut- and synecology (SCHRÖTER and KIRCHNER 1896, 1902). The Anglo-American sense of the term synecology is equivalent to sociologic geobotany, while the European sense of the term includes community- process- and habitat-ecology. Autecology does appear to have the same meaning in both camps, but the synonyms differ. Autecology refers to the ecological study of individual organisms, as opposed to the study of communities. Between the individual and the community levels is that of the population, and SCHWERDTFEGER (1963) has distinguished the ecological study of populations as *demecology* (from Greek demos, people; also see the deme-terminology advocated by GILMOUR and HESLOP-HARRISON, 1954, for use in population studies). In Anglo-American usage this field is known as: population ecology (primarily among zoologists); or as part of autecology when the study is concerned with the life history and physiological ecology of a species or ecotype; or as genecology when the object is investigation of the genetic variation of a species or other kind of population. DAUBENMIRE (1962) has used the term autecology even more generally as the study of environmental factors in relation to plants, but in Europe the study of environmental factors is part of both aut- and synecology (BRAUN-BLANQUET 1965, WALTER 1971).

1.4   THE SCOPE OF THIS BOOK

The scope of this book is limited to sociologic geobotany (TABLE 1.1) in the European sense. In the Anglo-American literature this area has been equated with plant ecology (GREIG-SMITH 1964, KERSHAW 1964), with community ecology (OOSTING 1956), or with plant synecology (DAUBENMIRE 1968). Since the word ecology is so widely accepted in the English-speaking literature, it seems appropriate to use the new term vegetation ecology for this book. The term implies equivalence with vegetation science (sociologic geobotany) and synecology (in the American sense), for the term *vegetation* in contrast to *plant* signifies synecology and excludes autecology.

The emphasis in vegetation science (sociologic geobotany) has been on vegetation systematics, that is, classification of typical communities (*sensu* DAUBENMIRE 1966). But vegetation systematics is no longer considered an end in itself. Instead, environmental investigation of plant communities and the sociological interaction of plant species have become the main concerns. Therefore, it seems appropriate to convey the modern meaning of vegetation science in a new term, vegetation ecology.

Vegetation ecology is the study of both the structure of vegetation and vegetation systematics. This includes the investigation of species composition and the sociological interaction of species in communities. It further includes the study of community variation in the spatial or geographic sense, and the study of community development, change, and stability in the time sense. Vegetation ecology is concerned with all geographic levels of plant communities, from broad physiognomic formations in the sense of biomes (tundra, grassland, desert, etc.) to the very fine floristic patterns occurring on an area less than a square meter in size. Vegetation ecology is very much concerned with correlations between environment and vegetation, and with the causes of community formation.

We would like to emphasize that vegetation ecology, as covered in this book, is merely one part of the study of plant communities. As seen in TABLE 1.1, the portion described by WALTER (1960) as habitat science (Standortslehre), or by ODUM (1959) as functional ecology, will be omitted from this book. We also will omit discussion of the increasingly large body of autecological material, and refer the reader to treatments by WALTER (1960), ODUM (1959), and DAUBENMIRE (1962).

Background and Current Trends
in Vegetation Ecology

2

2.1 THE GEOGRAPHIC-DESCRIPTIVE ORIENTATION

Vegetation science has a tradition of almost three centuries. Earlier work was primarily concerned with the description of unusual landscapes and their vegetation. For example, European naturalists gave much attention to the step-wise changes in pattern from forest to the vegetation near permanent snow fields in the Alps and to the colorful vegetation of distant lands. The character of the landscape is strongly influenced by such outstanding vegetation types as tropical rain forest, savanna, prairie, cactus-deserts and others. Landscape character is also influenced by such distinct life-form differences as shown between

deciduous and evergreen forests, or between forest stands with broad
crowns as contrasted with those having narrow crowns, or between
small-leaved and large-leaved plant communities. Because such vegetation features are obvious, they received much attention by travelling
naturalists and geographers. But systematic descriptions of recurring
vegetation patterns began only with A. VON HUMBOLDT (1806) who
classified repeatedly recurring growth forms into types. He used the
currently applied term "associations" for plants growing in communities (1805). However, the Danish naturalist SCHOUW (1832) deserves
the credit for presenting the first clearly defined systematic description
of numerous plant communities. VON HUMBOLDT's system of classifying vegetation by growth or life form types was later enlarged, particularly by GRISEBACH (1872), who described the vegetation of the
world in macroclimate-related categories.

In the twentieth century efforts were directed to simplifying vegetation description in order to increase its accuracy and to find a standard
basis for quantitative evaluation. Different vegetation-analysis methods
were developed, which allowed presentation of data in great detail
through coding and tabulation. At the same time these were in a form
easily understood and comparable to studies of other areas. Particularly the methods of RAUNKIAER (1913, 1918), CLEMENTS (1905,
1916), DU RIETZ (1921, 1930), BRAUN (1915), and BRAUN-BLANQUET
(1928), found favorable acceptance. These methods simplified scientific
communication among specialists, but because of limited descriptions,
they often had a discouraging effect on people interested in vegetation
from different viewpoints.

The pioneers in vegetation science did not restrict their efforts to
mere description and field analysis of plant communities. The innumerable variations in plant forms and combinations required reasonably
clear systems for presentation and discussion of plant communities
from specific viewpoints. At the same time it became desirable to explain the described communities in terms of their causality and function and to elucidate their environmental and successional relationships. Emphases on these aspects of vegetation research have varied in
the course of time. The important current descriptive concepts will be
discussed in detail in PARTS III and IV.

General descriptions of vegetation and plant communities through
plant life forms and dominant species are by no means a thing of the
past. In fact, they are urgently needed for many undescribed areas of
the world. This has become obvious through the International Biological Program (IBP). Recent guidelines are given in the IBP Handbook
Number 4 by PETERKEN (1967) including FOSBERG's (1961) descrip-

tion scheme for general purposes, and in the 1964 UNESCO discussion draft written by ELLENBERG and MUELLER-DOMBOIS (1967a) for an inventory of physiognomic-ecological vegetation units of the earth.

Extensive, qualitative descriptions are not automatically worse than intensive, quantitative descriptions. The measure stick for the value of a description is the adequacy in relation to the stated objectives, not the accuracy per se. For example, a very detailed, quantitative description may not satisfy the objectives of general orientation. However, qualitative descriptions, though still very important, are no longer pursued as an end in themselves, but as the beginning of more to come.

2.2 THE SYSTEMATIC-TYPOLOGICAL ORIENTATION

Systematic classification of plant communities received a special impetus through BRAUN-BLANQUET, who combined many initiations of his predecessors into a convincing program. His system of "phytosociology" (1928, 1932) has found much favorable acceptance throughout the world, and it has influenced the development of European vegetation science for decades. His hierarchy of systematical units was purposely patterned after the LINNAEAN system of plant taxonomy. It is, therefore, an artificial system that works almost exclusively with floristic criteria. These criteria relate to character or differential species, i.e., to species with restricted ecological amplitudes that at the same time show a high degree of presence within the area of study.

His system was by no means accepted by all vegetation scientists, especially not by American and English ecologists, such as CLEMENTS, TANSLEY and their collaborators. These authors also used mainly floristic criteria, but classified communities into types on the basis of dominant species or dominant species groups (WHITTAKER 1962). Dominant species usually have wider ecological amplitudes. Consequently, the communities were larger in size and had a more heterogenous environment. Application of the two different floristic criteria precluded comparison of the communities of Anglo-American and European authors.

Among most of BRAUN-BLANQUET's students the development of a hierarchical system of vegetation units received prime emphasis. This gave the impression that this school of vegetation science would find an end in establishing, naming, and applying such integrated classcategories. This trend was hardly in accord with BRAUN-BLANQUET's original intentions. In the first edition of his well-known book, he devoted more than half the space to ecology in the proper sense, i.e., to the functional and causal relations.

2.3   THE ENVIRONMENTAL ORIENTATION

The relations of plant communities to their habitats were emphasized by many authors, such as UNGER (1836), SENDTNER (1854), DRUDE (1896), WARMING (1909), SCHIMPER (1898), SCHRÖTER (1904), and many others. A few of these, particularly WARMING and DRUDE (1913), made environmental relations the basis of their vegetation studies. They also developed systems for classifying vegetation by environment. Vegetation science became an important branch of ecology around the turn of the century. This understanding led to the definition of ecology as known and interpreted in all English speaking countries.

Today, the trend seems to lead in the direction of environmental relation studies. The use of environmental criteria for characterizing plant communities is more common, and efforts are made to arrive at a synthesis of floristic and ecological vegetation treatments (KRAJINA 1969, DAUBENMIRE 1968). Particular emphasis is given to detailed ecological investigations of single communities (ODUM 1959, OVINGTON 1962). Along with this there occurred an important change in methods during the last decades. The earlier authors worked mainly deductively and drew their ecological and even physiological conclusions from comparative observations in the field, without attempting to support them by experimentation. In contrast to this, modern investigators work inductively and use experiments to increase the accuracy of their conclusions.

2.4   THE EXPERIMENTAL ORIENTATION

Investigation of causal relations in plant communities is not yet fully accomplished with careful studies of their environmental conditions in separated localities. For example, competition among plants is known to play an important role in conditioning their grouping patterns. The influences of competition, however, can only be clarified through experimentation, whereby the interaction of important plants of the community are studied under controlled conditions. Such experiments were done by CLEMENTS, WEAVER and HANSON (1929) and later by KLAPP (1951) and other grassland ecologists. ELLENBERG (1953, 1954), KNAPP (1954), MUELLER-DOMBOIS and SIMS (1966), among others used experiments in interspecific competition for elucidating causal relations in vegetation studies. A summary of results is given by ELLENBERG (1963). Experimental studies in the genetic adaptation of species to communities and ecosystems are reviewed by McMILLAN (1960, 1969).

2.5   THE DYNAMIC ORIENTATION

Experimental work with simplified plant communities can also lead to a more exact basis for studies in vegetation dynamics. This subject received its strongest attention through North American ecologists, such as COWLES (1899), COOPER (1913) and CLEMENTS (1916). Chronological sequences in vegetation seemed better noticeable in the more continuously distributed and less anthropogenically influenced plant communities of North America than in the strongly man-modified fragmented vegetation segments of the Old World. CLEMENTS (1916) developed a system of classification based on successional relations, i.e., on changes in time that he deduced from spatial similarities and differences in dominant species and their community environments. The key position formed a regional climatic climax community to which all other communities were related in a chrono-sequence. This approach was accepted by many ecologists and applied in similar or modified form, for example, by TANSLEY (1920, 1939) in his description of the vegetation of the British Isles, but also by BRAUN-BLANQUET and his students, by LÜDI (1930) and others.

WHITTAKER (1951, 1953, 1957) strongly criticized CLEMENTS' orderly phylogenetic classifying scheme based on succession, and de-emphasized environment as an aid in recognizing successional trends. But, WHITTAKER emphatically restates the dynamic view of vegetation and sees different rates and kinds of temporal population changes as the only means of finding an order in the study of succession.

The dynamic viewpoint has always been maintained in North American ecology, although in variously modified forms. One form has given rise to the continuum concept in Eastern North America, which views vegetation as a continually changing phenomenon in space and time (GLEASON 1926, CURTIS 1959). Another form, represented by DAU-BENMIRE (1952, 1968), emphasizes dynamics through secondary succession. This emphasis is related to the importance of fire in Western North America. DAUBENMIRE establishes community types only from relatively undisturbed stands, while modified vegetations are traced through habitat similarities as successionally related stands. Such modified or disturbed vegetation types are not separately classified. As a result vegetation patterns are not equally evaluated. Because of the temporal variations in vegetation patterns, environment is built into the classification of communities. This concept of treating vegetation by giving more weight to some than to other patterns has certain advantages, but also several disadvantages, which will be explained in PARTS III and IV.

2.6 THE MATHEMATICAL-STATISTICAL ORIENTATION

The problem of devising an objective approach to recognizing and defining plant communities has concerned investigators, particularly in those regions where self-evident communities are either rare, absent, or seem to occur only as broad physiognomic units (e.g., as forests, grasslands, etc.). CURTIS in Wisconsin (CURTIS and McINTOSH 1950, 1951, CURTIS 1955, 1959), COTTAM (1949), WHITTAKER (1951, 1953, 1954, 1970, 1972), GOODALL (1953a, 1953b, 1954a), and others were impressed by the absence of absolute boundaries between plant communities. Plant communities were studied through systematic and randomized sampling of individual plant populations along selected distribution gradients. Upon recomposing the studied populations to communities it was found that no two populations have exactly parallel distributions and thus communities become extremely difficult to define on an exact mathematical basis. Instead, these studies emphasized the continuity of distribution not only of populations but also of communities.

Another mathematical-statistical orientation, developed in England, particularly by GREIG-SMITH (1964), KERSHAW (1964) and others, is concerned primarily with the quantitative investigation of small vegetation patterns. Larger self-evident communities are more or less accepted as providing objective entities for more detailed studies. Questions of major interest include the mathematical analysis of positive and negative associations between species and the detection of variations in departure from randomness in small-area plant assemblages (see also PIELOU 1969).

Both orientations, the North American, working with relatively broad vegetation segments, and the English with relatively small ones, consider ordination of communities on the basis of graduated similarity an effective alternative to classification. This is based on the universal observation that no two communities are exactly alike. This fact was most emphatically stressed by GLEASON (1926, 1939) in his individualistic concept of the plant community. Interest in classification has been revived through the development of new mathematical analysis methods (see CHAP. 10). A summary review of the entire subject area was made by GOODALL (1970) .

2.7 THE AREAL-GEOGRAPHIC AND HISTORICAL ORIENTATION

Studies of the ecology and dynamics of vegetation are meaningful only in relation to well defined, limited geographic areas. The results are applicable only to these same areas. If vegetation is studied over broad geographic segments of the earth's surface, one can recognize, in addition to environmental and successional variations among plant communities, still other variations whose cause lies in the different historical development of the flora, in particular, in non-overlapping regional distribution ranges of species. This is the object of study in floristic and historic geobotany as summarized, for example, by WALTER and STRAKA (1970).

Floristic-historical analyses of plant distribution are important for a deeper understanding of variation in vegetation.

An interesting attempt was made by SCHMID (1954, 1963) to combine taxonomic phytogeographical viewpoints with those of vegetation analysis in the classification of landscapes. He demonstrated his system with Europe as an example, specifically with Switzerland. He distinguishes eight so-called "vegetation belts" (Vegetationsgurtel), which largely correspond to the altitudinal vegetation belts in the mountains. SCHMID's "vegetation belts" are not recognized by certain vegetation types, but by floristic criteria. As floristic criteria he uses all plant species whose center of distribution lies within a specific vegetation belt. For further details see SECTION 14.8.

This ordering principle requires many subjective decisions, and renders any finer subdivisioning of the rather broadly defined "belts" very difficult. In spite of the broad geographic definition, the method requires very detailed floristic knowledge. For this reason, SCHMID's method is applied only to a very limited extent in Europe and not at all in other parts of the world. Nevertheless, SCHMID's vegetation map of Switzerland (1:200,000) offers a remarkably good overview.

2.8 THE CARTOGRAPHIC ORIENTATION

Along with increased information on plant communities rose the desirability of geographic representation. Vegetation mapping developed from different viewpoints. The greatest impetus came from the applied fields of forestry, agriculture, and watershed management.

The great utility of vegetation maps is being appreciated more and more in nearly all parts of the world, and as a result, the preparation of vegetation maps is now rapidly expanding. An international bibliography of vegetation maps has been compiled KÜCHLER and McCORMICK 1965, KÜCHLER 1966, 1968, 1970). KÜCHLER (1967) has also written an authoritative manual on vegetation mapping.

2.9   THE APPLIED ORIENTATION

It goes beyond the scope of this methodological introduction to discuss the many practical applications of vegetation research. Vegetation ecology has a tradition of bridging the gap between the basic and applied approach to research. The interest in the behavior of botanical organisms in their actual field environments forms a natural link to the applied sciences of forestry, agriculture, and wildlife management. However, the awareness of the often complex interdependencies of organisms among each other and with their environment, and the knowledge that there are limitations in the responsiveness of biota and communities to technological modifications have preserved a basic or fundamental outlook towards problem solving in vegetation ecology.

2.10   THE ECOSYSTEM ORIENTATION:
A SYNTHESIS OF CONCEPTS

It cannot be emphasized too strongly that a thorough approach to vegetation ecology contains an element of all the orientations mentioned, so far. The basis for such a synthesis is given through the ecosystem concept. This is well documented through the Symposium on Forest Ecosystems held at the Ninth International Botanical Congress in Montreal in 1959 and published in SILVA FENNICA (1960). The historic development of the ecosystem concept was reviewed by MAJOR (1969). One of his main points is that this concept has always been part of men of different cultures. MAJOR cites a number of landscape terms from different languages, such as tundra, taiga, paramo, chaparral, that denote specific kinds of ecosystems with regard to location, habitat factors and organismic life forms.

The term "ecological system" was introduced for aquatic habitats with their specific biota and processes by WOLTERECK (1928). TANSLEY (1935) introduced the term "ecosystem" in the same sense for terrestrial communities with their habitats. He expressed the view that the organisms cannot be considered separately from their specific environments in any fundamental treatment, as both organisms and environment form a functional system in nature, an ecosystem. As shown before, this viewpoint has been an important underlying theme already in earlier vegetation studies. However, it was only since about the Second World War that the significance and potential of this concept has been more fully realized in ecology.

A very similar concept was introduced by the Russian ecologist

SUKACHEV (1945, SUKACHEV and DYLIS 1964), who considered a phytocoenosis (plant community) together with its environment as a "biogeocoenosis." In contrast to the ecosystem concept, the biogeocoenosis concept prescribes the boundary as set by the phytocoenosis. From a vegetation systematic viewpoint the biogeocoenosis concept has the advantage of defining a specific unit of plant synecological interest, while the ecosystem concept has the more universal advantage of suggesting only the essential focus of synecological research, leaving open specific boundary definitions in relation to different research inquiries.

In this sense, an ecosystem can be defined in various ways. The only requirement is that the major components, living organisms and an amenable environment, are present and that they operate together in some sort of functional stability (ODUM 1959). We may now look at the ecosystem from the standpoint of its general structural definition, and from its functional and typological aspects.

2.10.1 Structural Aspects. In a typical forest ecosystem, one can use the component strata as a general structural definition. These strata consist of tree layer, shrub layer, herb layer, and often moss and lichen layers. The latter may form a ground stratum, but commonly mosses and lichens occur also as epiphytes and they are often associated with specific microsubstrates in a habitat such as outcropping rocks or decaying logs. Beneath the vegetation are the litter and humus layers. In the soil one can usually distinguish three main soil strata (A, B, C) plus the rooting strata and the geological substratum. Other important components are the microfauna and flora in the soil, litter, humus, and those occupying the foliage, branches, and bark. Also there are larger animals that live in or visit the forest ecosystem periodically. Superimposed is the climatic complex of the habitat, consisting of the regional macroclimate, the local topographical climate, and the various microclimates within the ecosystem itself.

These different components can be studied as to their structural composition and functional activity and interaction.

Most individual ecosystem components have given rise to special disciplines. The vegetation component receives the particular attention of the vegetation ecologist, but he is equally concerned with those aspects of the environment that influence the behavior of the vegetation. Yet, the climatic component is the particular field of study of the climatologist, the geological substrate that of the geologist, the landform and physiography that of the geomorphologist, the soil that of the soil scientist, the soil organisms that of the soil microbiologist, and the animal populations that of the animal ecologist. Even within the vegetation component itself, different plant life forms receive specialists' attention; the trees that of the forester, the mosses that of the bryologist, the algae that of the phycologist, and the fungi that of the mycologist. Similar specializations occur within the environmental components. The list can be expanded and diversified, particularly in ecosystem-metabolism studies. The point is: the ecosystem concept provides a sound theoretical basis for bringing specialists together in the field for the study of more complex problems requiring team work. Moreover, it encourages the synthesis of findings relating to the separate components that, if viewed alone, apply only to part-aspects of the same ecosystem.

2.10.2 Functional Aspects. From a functional quantitative viewpoint, OVINGTON (1962) lists fourteen parameters worthy of analysis in the forest ecosystem. Input parameters are solar energy, rain, dust, rock, weathering, and soil weathering. Output parameters are logs, etc., animal migration, and outward drainage. Within parameters are the use of solar energy by the green plants, leaching from the plants, litter fall, deposition of other plant remains and animal residue, litter breakdown, and root breakdown in the soil.

OVINGTON's problem analysis brings out another important point, namely that an ecosystem is not a closed but an open system that takes in, discharges, and shows a specific within-set of activities. This openness has been used as a criticism of the concept. However, the same applies to an organism which responds nevertheless as a unit, and few people will doubt that an organism has a certain individuality of its own.

Less specific, but no less elucidative, is ODUM's (1959) emphasis of the ecosystem parts. He recognizes two organic components, an autotrophic and a heterotrophic component. These are classified into producers (the green plants), consumers (chiefly animals), and decomposers (mostly saprophytic microorganisms). The ecosystem is completed with the abiotic component containing the basic inorganic and organic substances.

This process-oriented viewpoint is designed to emphasize functional similarity in different ecosystems. This is very important in the formulation of more general activity principles. Yet, it is important also to recognize the diversity of ecosystems. The more specific the studies of processes and interactions become, the more important will be the limitation of the extent to which results can be extrapolated in nature. Therefore, the size and kind of ecosystems present typological problems worthy of consideration.

2.10.3 Size and Kind. As EVANS (1956) pointed out, there should be no limitation in size and kind. One may recognize as ecosystems very small units, such as the phyllosphere of microorganisms on a leaf of a plant, to the very largest, the earth or ecosphere. One may also recognize very different kinds, for example, a forest, a lake, an island, or an urban ecosystem. In vegetation ecology an ecosystem may be delimited by one plant community, as in the biogeocoenosis concept, or by several interrelated communities. For example, one may speak of a delta ecosystem, which typically includes several habitats and communities that are functionally and closely interrelated through the river system (FOSBERG 1965a).

From this it is quite apparent that the ecosystem concept cannot replace established vegetation and plant community concepts. These are needed to define particular ecosystems in space (i.e., geographically) and in time. The ecosystem concept, however, has led to the emphasis of giving equal attention to all of the ecosystem's major components in field studies.

2.10.4 Ecosystem Classification.

Three different approaches to ecosystem classification can be distinguished. They are: the combined approach, the independent approach, and the functional approach.

The combined approach aims at a synthesis of vegetation and environment from the start. It is represented in the works and concepts of SUKACHEV (1928, 1945), SUKACHEV and DYLIS (1964), DAUBEN-MIRE (1952, 1968), KRAJINA and his students (1969), HILLS (1960, 1961), MARR (1967), SCHLENKER (1951), KOPP and HURTIG (1965), EBERHARDT, KOPP, and PASSARGE (1967), and others. Depending on the emphasis, ecosystem boundaries may either be determined by plant community boundaries (SUKACHEV), by soil or landform boundaries (HILLS), or by a combination of vegetation and environmental characteristics. The latter is considered least artificial and truly in line with the ecosystems concept (ROWE et al. 1961).

The independent approach considers the individual components of the ecosystem as separate entities and evaluates them independently. Subsequently they are combined on the basis of maps and profiles. In this way the ecosystems are established (FOSBERG 1961, MUELLER-DOMBOIS 1966).

The first approach has been found useful in forest and site evaluation studies, particularly for basic ecological research serving applied ends, where the ecosystem components can be employed as indicators of the more hidden site factors. For example, they may be used to define factors such as soil moisture and nutrients that are critical for tree growth and productivity differences (MUELLER-DOMBOIS 1964, 1965a). In this case the correlation of plant communities to their environment is built into the established units and the components cannot be evaluated independently. The second approach allows for an objective correlation of the various ecosystem components since their variations are established independently, namely from criteria that were found within each component. The approach is particularly useful in basic vegetation studies that start from the broader or more general aspects and then extend to the more specific ones by a process of successive approximations (*sensu* POORE 1962).

Recently, ELLENBERG (1973) proposed a classification of the ecosystems of the earth, the categories of which are based on functional relationships and differences. This classification will be explained in more detail in SECTION 8.82.

Plant Community Hypotheses

3   Before discussing the methods of vegetation analysis, it is important to give some considerations to theoretical viewpoints on the nature of the plant community. This is important because the viewpoints influence the basic objectives in vegetation science and these in turn have a strong bearing on the methods applied in field research.

3.1   ORGANISM ANALOGIES AND THEIR CONSEQUENCES

3.11   The Holistic Viewpoint.   To view the plant community as a unit or an entity, certain analogies to an organism were proposed. CLEM-ENTS (1916, 1928) compared the successional development of a community from its pioneer stage to the relatively stable climax stage, e.g., from aquatic vegetation to bogforest, with the life history of an organism. He held that the community, like an organism, is born, grows, matures, reproduces, and dies, and that these different developmental stages or successionally related communities can be interpreted as an organic entity. He believed the reproduction aspect of this analogy was particularly valid. In his view, each climax community can reproduce itself at any time by repeating the developmental stages in essentially the same pattern.

However, important differences were not mentioned. The disappearance of a climax community or its "death" cannot be compared to the natural death of an organism imposed by the loss of function of its organs. The populations of plant species, which were considered the organs in this analogy, do not disappear because of senility. They are replaced in part or wholly by other populations through a catastrophic event, a gradual change in environment, or through competitive replacement. Likewise, the "growing" and "maturing" processes in plant communities do not occur in the same organs that develop from young to old, but rather in an exchange of populations. Unlike organisms, plant communities cannot reproduce in environmentally different habitats or different climates without losing their identity. TÜXEN (1947) has even shown in his phytosociological garden that forest communities and other complicated communities can be reconstructed from their parts in a relatively short time. Furthermore, the individual community members are not, as a rule, structurally connected as are the organs of an organism or the tissues of an organ.

TANSLEY (1920) considered CLEMENTS' organism idea as too hypothetical. But TANSLEY believed that plant communities can be described as organic entities by using the more appropriate term, quasi-organism. In comparison to the tight structure of an organism, he pointed out that in plant communities certain populations are independent as they can establish themselves quite well in other communities, while in contrast, others are strongly dependent. However, he emphasized that communities behave in many respects as wholes, and thus should be studied as wholes. This later led to his ecosystem concept (see SECTION 2.10).

3.12   The Systematic Viewpoint.   A third organism analogy was presented by BRAUN-BLANQUET (1928, 1932) and others (e.g. NICHOLS, 1917; WARMING 1909) for quite different reasons than inferred by either CLEMENTS or TANSLEY. Particularly, BRAUN-BLANQUET had in mind the aspect of classifying communities similarly as organisms are classified into taxonomic groups. For this reason, he compared the plant community to a species. BRAUN-BLANQUET regarded the plant community as the basic unit of vegetation classification similarly as the species is considered the basic unit in the taxonomic classification system of organisms. He overlooked in this analogy that individual organisms of a species are members of genetically related populations, whereas the individual plant community is not genetically related to other similar communities that may be grouped into a type or class. Here, the relationship is based merely on certain features of structural or compositional similarity. For community classification, it seems reasonable to relate an individual community to a community type, similarly as is done, for example, in soil classification, where the individual soil is related to certain class concepts (such as the great soil group, a soil series, or a soil type), or in petrological classification, where the individual rock is related to certain rock type concepts (such as granite, basalt, etc.).

3.13 The Individualistic Viewpoint. GLEASON (1926, 1939) considered all three organism analogies to be misleading. In their place he proposed his "individualistic" concept of the plant community. GLEASON claimed that while the plant community depends for its existence on the selective forces of its particular environment and the surrounding vegetation, the environment changes constantly in space and time. Therefore, in his view, no two communities can be considered alike or closely related to one another.

Truly, every square meter of habitat and plant cover shows certain differences when compared on an absolute basis. However, when considered on a relative basis, one can distinguish between greater and lesser similarities and differences. This principle of relative similarities and differences underlies all systems of biological and environmental classifications, including the taxonomic system of organism classification. No two individual members of a species or other biological population are absolutely identical. It is hard to understand why this principle should not be applicable to plant communities.

These four plant community concepts have had important consequences in vegetation research.

3.14 Emphasis on Dynamic Relations. CLEMENTS' approach emphasized the dynamic relationships of plant communities seen as an organic development from youthful to mature. GLEASON's ideas emphasized the dynamic nature of plant communities in time, but also in space. However, in contrast to CLEMENTS, who elaborated much on the stages in succession, GLEASON did not offer any tangible concept with regard to the change in time. While the flow of time is constant, it is not also true that the rate of population change is equally continuous in all communities (WHITTAKER 1953, 1957). It is important to make at least a relative distinction between temporary and more or less stable communities (DAUBENMIRE 1952, 1966) or between fast changing and slowly changing communities, while the rate and direction of change as such remains an important subject of vegetation research.

By emphasizing the continuity in space, GLEASON pointed to the absence of absolute boundaries between adjacent communities. With this, he gave the underlying concept for the phytosociological research inquiry of the Wisconsin school, also known as the continuum school. The continuum research approach was effectively started by CURTIS (CURTIS and McINTOSH 1951, CURTIS 1955). CURTIS and his students contributed to vegetation science largely by formalizing the methods of gradient analysis and ordination (WHITTAKER 1967).

A modification of the continuum concept came from McINTOSH (1967a). He admits that vegetation discontinua are found in nature. However, he reemphasizes that continuity is always found when a comparison is made between similar vegetation stands. This is an important clarification of concept that will be referred to later.

3.15 Emphasis on Order. BRAUN-BLANQUET worked to find the conceptual basis for community classification. The species analogy of the community was to provide for a hierarchical system of community classification on a worldwide scale. This research resulted in a wealth of community-ecological information. Yet, increasing geographic coverage brought forth the frustration that an integrated system of classification of small, strictly floristically defined communities with "character" species cannot be extended to the same worldwide application as the taxonomic system of organism classification.

"Character" species may be interpreted as key species by which individual communities in the field can be identified as members of a particular community type. The reason for a certain disenchantment with the "character" species concept is that the key species may lose their diagnostic value when the community study is enlarged beyond its original regional limits. The same key species may show differing ecological and sociological relationships when followed over a wider geographic range. This will be further explained.

3.2 THE COMMUNITY—A COMBINATION OF INDIVIDUALS

As explained above, the authors of the organism analogies had particular community attributes in mind that they thought could best be explained by such an analogy. It is unfortunate that their explanatory attempts caused much confusion that still persists today.

Plant communities have many attributes that are very unlike those of organisms. Only in exceptional situations are the individual plants of a community structurally interconnected. This applies, for example, to certain *Populus tremuloides* (ZAHNER and CRAWFORD 1965) and *Acacia koa* stands (MUELLER-DOMBOIS 1967), where individual stems form suckers arising from a common root system. Such a clonal stand is certainly an organism, if one ignores the associated undergrowth plants.

However, a plant community consists of a group of plants that retain their individuality. This is not a new concept. In 1900 FLAHAULT (FLAHAULT and SCHRÖTER 1910) pointed this out by saying that "the plant association does not imply a harmonious concurrence of diverse activities working towards a common end, as in every society founded on the division of labor. It is applied to the coexistence of forms, which specifically and morphologically are foreign to one another, each having its object, its own exclusive profit. They live side by side according to the similarity or diversity of their environment or determined by the presence of other organisms."

This stress of the plant's individuality does not preclude different kinds of relationships among the plants of a community. WALTER (1964, 1971) has classified these as:

1.   Direct competitors that compete for the same environmental resources by occupying the same strata above and below the ground.

2.   Dependent species that can only exist in their particular niche because of the presence of another plant, such as, for example, certain mosses that can only grow within the specific microclimate produced by a stand of trees.

3.   Complementary species that do not compete with one another, because their requirements are satisfied by occupying different strata above and below the ground or by having a different seasonal rhythm.

It should be noted, however, that dependent species are not, as a rule, dependent on a particular dominant species or species group. This is commonly the case only in parasitic, symbiotic, and, even less so, in epiphytic relationships. Instead, the dependence relates only to the general environmental conditions created by the dominant species. These conditions may be recreated by other means or artificially in the laboratory. Moreover, the three kinds of interdependencies are by no means clear cut, and various transitory forms occur. For example, two species may be complementary in their rooting patterns and requirements, but may compete strongly for light.

This brief explanation is an oversimplification merely for the purpose of ready comprehension.

Beyond their interacting influences the plants of a community share a common area, habitat, or environment. Therefore, integration in plant communities is an established phenomenon (POORE 1964), which brings about differing degrees of organization. Well-integrated communities may show a certain resistance to fluctuations in their environment (homeostasis), and specific environmental changes may bring about predictable responses in the community. Yet, integration should not be made a requirement for calling a plant assemblage a community, because the degrees and kinds of integration are also important subjects of ecological research. At best, the community can be described as a "spatial and temporal organization of organisms" with differing degrees of integration, and clearly, the community occupies a higher level of organization than the organism itself (DANSEREAU 1963).

Therefore, a plant community can be understood as a combination of plants that are dependent on their environment and influence one another and modify their own environment. They form, together with their common habitat and other associated organisms, an ecosystem (*sensu* TANSLEY 1935), which is also related to neighboring ecosystems and to the macroclimate of the region.

However, in spite of their close interrelationships the individual members retain their individuality, because each species can exist also outside the community. Therefore, the ultimate unit of vegetation is not the plant community but the individual plant type. By this, we mean genetically related populations of any taxonomic rank (such as species, subspecies, variety, races, or ecotypes), whose representatives show a simliar ecological behavior.

3.3 IDENTIFICATION OF COMMUNITIES

An attempt to answer the question, "What is a plant community?" was given in part by the previous discussion. However, the question is by no means fully answered. For example, how should one identify a plant community in the field? ALECHIN (1926) provided a solution for this when he wrote a paper on the subject. According to ALECHIN, plant assemblages may form, (a) open groups or (b) closed groupings. Among the latter one may distinguish: settlements without integration; pure stands, that may either be temporary or permanent; or communities, i.e., mixed population stands.

ALECHIN's outline restricts the meaning of community to integrated, mixed population stands that occur as closed groupings.

Integration is considered by many authors a prerequisite for recognizing a plant grouping as a community (POORE 1964). It is our opinion that population integration in communities is a subject of vegetation research. It cannot always be judged in advance, nor made a criterion for community identification. Likewise, pure population stands, if not composed entirely of artificially seeded or planted crop plants, are groups of plants of interest to phytosociological research. There appears to be little justification to exclude such plant groups from consideration as communities. Nor should plant-to-plant spacing be a limitation. Even groups of desert plants whose structure is open above the ground can form highly integrated communities (WALTER, 1964, 1971). They may be closed below the ground by an interlocking lateral root system.

LIPPMAA (1939) restricted the community concept even further, namely to the individual horizontal strata, such as shrub, herb or near-ground layers, which are component parts of many communities.

However, any such structural restriction, while bringing out interesting community attributes, imposes an a priori limitation on vegetation research. Therefore, for the purposes of identification, the community concept should be as unrestrictive as possible. It appears quite sufficient to identify communities (at all levels of geographic scale) through the variations in the homogeneity or uniformity of the vegetation cover of an area, where these variations are obvious to the eye. Whether these can be called formations, consociations, associations, synusiae, etc. is another question that will be discussed in CHAPTER 8.

3.4 VEGETATION PARAMETERS FOR DEFINING COMMUNITIES

Various vegetation parameters have been proposed by which communities can be defined geographically or in space. These include life form or growth form, species dominance, and the presence or absence of certain diagnostic species. Different schools have evolved around the more exclusive use of one or another of these parameters. WHITTAKER (1962) presented a detailed review on the subject, wherein he emphasized the approach through dominance in English-speaking countries and the application of species presence and absence criteria in the continental European approaches. While this separation is justified on the basis of finding the historical background of differing approaches in vegetation ecology, it would oversimplify methods now being used in these and other parts of the world. The traditions by country are disappearing wherever methods best suited for particular objectives and vegetations are applied. Also the use of more than one vegetation parameter is increasingly practiced as it leads to a less artificial system of classification (POORE 1962).

Since variations in the vegetation cover are caused by a number of plant attributes other than species dominance or composition, such attributes may also be useful in the definition of patterns, even if they are difficult to measure. Other such attributes include flowering, canopy coloration, seasonal changes in foliage or shoots, differences in height and stature, etc. These can be recognized at all geographic levels.

3.5 CONCRETE AND ABSTRACT COMMUNITIES

So far we have discussed the plant community concept only with reference to stands of plants as they may be seen in the field, that is, in the concrete sense. The absence of absolute boundaries between adjacent communities in the field is no greater a problem in community identification than is the absence of absolute boundaries between two adjacent but different soils in their identification. Sharp boundaries are rare between any related natural phenomena. Yet, soil classifications have received little objection on this account, while the concept of classifying vegetation is generally less well understood (WHITTAKER 1962). DAUBENMIRE (1966) has clearly explained a number of concrete discontinua that may occur in any vegetation cover over a gradually changing substrate, and McINTOSH (1967a) has reduced the continuum problem as relating only to the abstract community concept. Therefore, there seems general agreement now that a concrete, regional vegetation cover may show both, discontinua or sharp boundaries and gradually changing patterns or continua. Both kinds of distribution patterns may occur mosaicly within the same region. Such a mosaic can be studied by sample stands and classified or it can be evaluated by ordination.

A classification groups similar sample stands into types; an ordination interprets sample stands in relation to one another according to their similarities and dissimilarities. Both methods are abstractions. The choice of either form of representation is primarily a matter of objectives and to a lesser extent one related to the nature of the vegetation pattern (WHITTAKER 1972). This will be further discussed in CHAPTER 4.

Yet, certain problems in understanding vegetation classification have been attributed to a confusion between the individual, concrete class member and the abstract class unit. We wish to emphasize that the abstract vegetation units have no absolute reality in nature. They are somewhat like arithmetic means and ranges that may give a lesser or closer fit to a population of numbers. But they show no absolute identity with the population members, which, in the abstract vegetation unit, are the individual, concrete communities.

To avoid confusion, we believe it is useful to clearly distinguish between the concrete plant community and the abstract community or vegetation type. Terms like plant association or sociation should be used only in the abstract sense, that is, they should not be used for a particular concrete community or sample stand somewhere in the field. If one refers to the latter, one should use such terms as plant community or phytocoenosis, just as one employs the term biocoenoses for life communities in the concrete sense. POORE (1955) has suggested the term *nodum* to refer to all abstract communities regardless of rank.

Considerations in Vegetation Sampling

4

4.1 ESSENTIAL STEPS IN SAMPLING

Any detailed vegetation study is based on the description and investigation of plant communities or vegetation segments that must first be recognized in the field. Then, the vegetation segments must be sampled through analysis of representative subareas or stands within the recognized segments. A 100 percent survey of even one vegetation segment would for most purposes be much too time consuming. Therefore, the description must be based on samples. Also, one must decide what parameters or statistics of the vegetation one should record or measure and what size and shape the samples should take.

These four steps have to be considered in any sampling of vegetation: (*a*) *segmentation* of the vegetation cover or recognition of entities (i.e., entitation), (*b*) *selection of samples* in the recognized segments, (*c*) the decision of what *size and shape* the sample should take, and (*d*) the decision of *what to record* once the samples or plots are established.

The four steps may not always be followed in this sequence and they may differ very much in detail with the viewpoint and vegetation concept of the investigator, with the character of the vegetation itself, with the objectives of the investigation and with the time available for the study.

We intend to discuss only the approaches most commonly used, since a complete treatment would go beyond the scope of this book. Moreover, we wish to emphasize that, however important careful investigation of plant communities or sampling is, it is not the aim of vegetation research. It only provides the necessary raw material for documentation of the study. The importance of vegetation sampling is that all subsequent treatment of the data and the conclusions one may draw depend on the initial selection and the characteristics of the samples.

The first step, the entitation, segmentation or subdividing of the vegetation cover is always a subjective one. This is so, whether or not one chooses objective methods for the subsequent selection of sample areas within the segments.

4.2   SUBJECTIVE VERSUS OBJECTIVE SELECTION OF SAMPLE STANDS

Three approaches can be used for placing samples into initially recognized vegetation segments. They are:

1.   Subjective with preconceived bias.
2.   Subjective without preconceived bias.
3.   Objective, according to chance.

The first approach has often been used in vegetation sampling in continental Europe. It implies that the investigator may consciously overlook certain unconformities or deviations in the vegetation cover to present the image he wishes to convey. This has led to erroneous conclusions and well-deserved criticism. Unfortunately, few critics have made a distinction between the two kinds of subjective sampling. As a result, many American ecologists consider all subjective sampling as unscientific.

However, the second form of subjective sampling has great validity.

In fact this approach has led to the most rapid advances in science. It differs from the first kind of subjectivity in a major aspect, namely, that the investigator approaches his study object with a negative hypothesis in mind. This means that he applies his mind to the fullest in the process of entitation; furthermore, that he takes nothing for granted and that he maintains a flexible program during the course of his investigation. Therefore, he is ready to accept a new working hypothesis as soon as further knowledge indicates a modification or change. PLATT (1964) described this approach as "strong inference," which in essence implies that it is expedient even in science to use one's imagination to the fullest, and to discriminate wherever common sense and experience makes a distinction obligatory. A similar idea was presented by POORE (1962), i.e., that knowledge progresses most rapidly through a process of successive approximations.

The objective approach requires sampling according to chance. This implies systematic or random sampling and the elimination of any further choice in locating the samples after the initial entities have been recognized. This approach often suffers from inflexibility. But objective or predetermined locating of sample stands is necessary, where vegetation patterns are nondistinct or unclear to the investigator. Moreover, random sampling is necessary when the ecologist wants to use probability statistics to back up his conclusions.

4.3   RECOGNITION OF ENTITIES OR COMMUNITIES

4.31   Level of Entitation—A Question of Viewpoint and Purpose.   The viewpoints differ primarily with regard to the degree of cover segmentation one may use, i.e., with regard to the detail of observing community boundaries. This in turn depends on the community concept, on the thoroughness of the preceding reconnaissance, on the purposes of the study, on the required geographic scale, and on the vegetation itself.

For example, if one holds with GLEASON that recurring similar plant assemblages cannot be grouped or classified into an abstract community type (because they are not alike in all detail), one will use only a general degree of entitation. This would apply to the recognition only of such physiognomic entities as forest, scrub or grassland, or within forest vegetation to significant groupings of dominant tree species. CURTIS (1955, 1959) used further subdivisions of broad landscape units in the Wisconsin prairie. Such a general degree of entitation does not require much reconnaissance and familiarization with the vegetation before sampling. Unless one is very careful, subjective sampling in such large units will easily lead to biased conclusions, because of the greater within-variation usually associated with broadly defined units. In such broad units it is difficult to locate samples subjectively in such a way that they represent the whole unit. Instead, many samples located either systematically or randomly may give a less biased description. However, detailed recurring variations of ecological significance may never be revealed through this approach in spite of the great physical effort in sampling.

In contrast to GLEASON's individualistic concept stands the philosophy of BRAUN-BLANQUET and many others that recurring similar plant assemblages can be usefully grouped into types regardless of the fact that no two similar plant assemblages are exactly alike. However, it is accepted in this concept that only a certain number of species of a community recur frequently in like combinations, while other community members do not. The frequently recurring plant combinations form a basis for community distinction. It is this contrast in community concept, i.e., recurrence versus nonrecurrence of similar plant combinations, that focuses attention on a different approach to sampling. It is not the idea, cited earlier, that communities form entities like organisms and that they can be viewed as discrete units. Where the purpose is to describe vegetation through classification with a view to establishing meaningful environmental relationships to the vegetation units and then, to search for causal explanations of the recurrence of similar and different plant assemblages, more emphasis lies on familiarization before sampling. This requires repeated and detailed reconnaissance. The better the initial knowledge of the area as a whole from reconnaissance, the better can be the subsequent sampling.

CAIN and CASTRO (1959) describe the intensification in familiarization as (a) reconnaissance, (b) primary survey, and (c) intensive survey. The intensive survey is the sampling phase in which quantitative or semi-quantitative data are obtained. Both the reconnaissance and primary surveys are qualitative, resulting largely in entitation, stratification, or general mapping. The detailed and intensive reconnaissance required before sample stands can be located subjectively without preconceived bias, is more or less equivalent to CAIN and CASTRO's primary survey.

The kind of sampling that follows the initial stratification is a question of purpose. If the purpose is to get an idea of the floristic composition that differentiates the entities, some form of centralized sampling will be sufficient. That is, one will look for areas within each segment that appear to represent the total segment.

Where the segments are defined only by the dominant tree species, as is the case in most Eastern North American studies (WHITTAKER 1962), sampling is more difficult than the initial stratifying. Communi-

ties defined by dominant species may be called "dominance communities" for convenience. In searching for a representative sample in
such dominance communities, one may discover all kinds of variations
that seem to obscure the representativeness. Broad vegetation segments
usually contain many finer floristic and environmental variations and
are, therefore, rarely homogeneous.

Where, in contrast, the segments are defined not only by the dominant tree species, but also by the associated undergrowth species, as is
the case in most European and several Western North American studies
(KRAJINA 1969, DAUBENMIRE 1968, MARR 1967), the entities are
much smaller and, therefore, more homogeneous. This makes it easier
to find sample locations representative of a segment. However, the
initial reconnaissance and primary survey to arrive at such a segmentation requires a greater effort.

Where a floristic classification of finer recurring plant assemblages is
not sought, as in the continuum analysis, replicate sampling is not considered. Instead sampling is done along floristic or environmental
gradients in such a way that the samples are spread out either randomly
or systematically. This means that each sample includes a lesser or
greater variation in plant combinations. Any replicate sampling would
thus be purely accidental.

Both approaches are initially subjective, but the approach in search
for recurring patterns carries the subjective element much further. Its
emphasis lies on entitation and the development of qualitative analyses
or working hypotheses before investing in any quantitation. The first
aim in this approach is to observe similarities and differences in the
vegetation cover and to document their importance with sampling data.
Because of the much finer degree of entitation in this approach, the
within-variation of the segments is also much reduced, and the major
variations are brought out through the segments themselves. The
internal variation of these finer and more homogeneous segments becomes a secondary aim of investigation. A substantial part of the
internal variation is assessed by placing several samples into what
appear to be recurring plant assemblages of the same kind. Since each
one differs in detail, the samples represent a spectrum of variation
usually showing a continuum of graded similarities.

4.32 Limitations in Objectifying Entitation. It took GOODALL
(1953a) five days to objectively divide an Australian *Eucalyptus* stand
of only 640 m² size into four entities, which he called subassociations.
The 640 m² area was an ordinary sample plot within a subjectively
chosen broader segment of the forest. He sampled this plot by placing
256 quadrats each of 5 m² size in the form of a restricted randomiza-

tion, i.e., almost systematically over the plot surface. This represents a sampling intensity of 200 percent. The objective subentitation achieved was based on frequency of species grouping. Therefore, cover and size of plants, which are equally valid for subentitation, were not considered. The four entities that GOODALL detected by an objective method of sampling were quite evident. Application of BRAUN-BLANQUET's relevé* method would have clarified the four entities in a few hours through floristic comparison.

GOODALL's work was a method-study. The dense network of samples brought out the fact that the types were connected by many transitions. Therefore, vegetation represents a "varying continuum" and distinction into units cannot be accomplished without abstraction. To apply GOODALL's sampling intensity of 200 percent would be impractical. On a practical sampling basis, samples are scattered. Equal weight is given to each sample, whether its location is centralized or decentralized with respect to the vegetation segment. Thus, boundaries disappear through scattered random or systematic sampling, even though they may be quite obvious to the eye.

For example, BESCHEL and WEBBER (1962) sampled the tree layer of a wetland forest in Ontario at 50 m intervals along transects by the point-centered quarter method (see CHAP. 7) of COTTAM and CURTIS (1956) and found no sufficient breaks in species combinations to warrant recognition of separate communities, although the authors recognized beforehand distinct belts of dominants from the air. These distinct belts had been described by other authors (GATES 1942, CONWAY 1962) as separate communities. They can be mapped from air photos. But the objective analysis of BESCHEL and WEBBER failed to reveal them.

The examples show that subdivisions can be discovered by objective methods only through an extremely dense network of samples, which is impractical for most purposes. An alternative would be to weight the samples by the vegetation parameters that show the boundaries. This, however, would destroy most of the objectiveness of such an approach. A second objective approach has also been used. It deals with individual species instead of groups of species (WHITTAKER 1951, DE VRIES 1952, BRAY and CURTIS 1957). In the species approach the amplitude and magnitude (usually density, i.e., number of individuals) of a few selected individual species are assessed, and their combination with associated species is tested. Community boun-

---

* The term relevé (French for "abstract") is the European equivalent to sample stand or vegetation sample. Wherever used in this book, the term relevé refers to the relatively small sample stands based on the minimal area concept. For relevé method see CHAPTER 5.

daries are drawn subsequently where there is the least overlap in amplitude and magnitude. This approach suffers from the selection of only a few abundant species that can be analyzed for statistical reasons and from the limitations to equidistant or random sampling. The sampling positions are not modified or weighted with regard to visible boundaries, transitions, or ecotones. Therefore, the resulting recombinations of associated species groups give only a vague indication of community formation or none at all. This difficulty was expressed by McINTOSH (1967a), when he wrote that "quantitation in the sense of analytic methods is well developed; entitation in the sense of objective and quantitative synthetic methods is poorly developed."

Since entitation of the vegetation cover into communities through objective methods is impractical, it seems expedient to stratify by using one's head. This means that all obvious variations in the vegetation cover including those caused through variations in the undergrowth should be noted, and subsequent samples should be distributed accordingly.

The viewpoint that we like to emphasize in vegetation analysis has very adequately been stated by GERARD (1965) as follows: "Before measurements can be meaningful they must be directed to the right things and, even in science, finding these things is a major achievement; entitation is more important than quantitation."

4.4   SELECTION OF SAMPLE STANDS IN RECOGNIZED VEGETATION SEGMENTS

4.41   Centralized Replicate Sampling. Only through good reconnaissance and a subsequent knowledge of the area as a whole will it be possible to recognize frequently recurring plant combinations and to determine the locations at which these show optimal development.

A perfect fulfillment of this requirement was exemplified by, among others, STEBLER and SCHRÖTER (1892) in their investigation of grassland types in Switzerland. Their complicated and time-consuming method of analysis, which is not used anymore today, and the small size of their sample areas (one square foot), forced them to select their sample areas with great care. Because of their thorough knowledge of the grasslands in Switzerland they defined types with relatively few samples. These are still valid today, and they are supported by modern phytosociological studies.

A good reconnaissance knowledge is particularly important in the study of plant communities that recur many times over a wide geographic area, for example, the European beech forests or the *Arrhenatherum elatius* grasslands. Ideally, such communities should be investigated first in this distribution center; that is, in areas where they are most common, where they occupy large tracts, and where they show their greatest within-community variation and maximum numbers of species.

Unless defined by purpose, there is no absolute criterion by which one can judge in advance which variation and recurrence of plant assemblages are ecologically significant. However, repeated walking and searching through the area from one environmental extreme to the other will usually give an indication. If not, one will have to use more samples.

After recognizing such relatively small-area variations in the vegetation cover that repeat themselves in more or less closely similar patterns throughout the area, one has, in effect, achieved an initial entitation. Its validity is then tested by sample stands or plots. These plots are established in the vegetation segment in locations wherever the segment appears to be most typically represented. If the community is very homogeneous, sample stands can be established anywhere within its range. But communities usually vary from their centers outward. In those cases, the sample stands should be placed as centrally as possible. Therefore, one may speak of centralized or nonrandom sampling.

The approach has been described by CAIN (1943, CAIN and CASTRO 1959) and others as the single plot method. This emphasis is somewhat misleading as a single plot per entity or community is rarely a satisfactory sample. Instead, several sample stands (usually a minimum of between 5 to 10) are established particularly in what has initially been recognized as similar recurring plant combinations. This will permit us to see whether these recombinations, which seem useful for classifying into types, really are similar. This is what may be called sampling of noda (POORE 1962). If the stand data supports the recognized similarity, one may consider this quite an achievement in entitation, since no two plant assemblages resemble each other in all detail. However, the variability among replicate stand samples is not a criterion of the sampling efficiency. It is a reflection of the variability in recurring patterns of a vegetation region, its geographic extent, the number of replicate samples, and the degree of entitation.

A lower limit of subdivisions has to be set arbitrarily, since too many units would defeat the value of a classification. A classification should result in a meaningful generalization. This lower limit requires additional judgment on the part of the investigator. This judgment should be oriented on the objectives for which the classification is prepared.

4.42 Random, Systematic, and Stratified Random Sampling. These types of sampling are necessary, where a measurement is required of variation across a vegetation segment too large for a 100 percent sampling. This applies particularly to the continuum analysis, which aims at evaluating the variation throughout broadly defined dominance communities. But it applies also to the releve' analysis, if one is interested in measuring rather than estimating species quantities within the sample stand or releve'. Measuring has particular application where releve's are used as permanent plots.

Random placement of samples requires locating them entirely through chance within the segment. This is best accomplished by establishing two coordinates with equidistant intervals at the fringes of the sample area. Then, a pair of random numbers may be obtained from a random table (published in most statistical handbooks) or by drawing two numbers out of a hat. With these two numbers any random plot position can be fixed in the segment through the coordinates. Random sampling has the advantage that a statistical error term can be assigned to the mean value. Therefore, the variation about the mean can be accurately assessed, but only for the more common species.

With increasing knowledge of plant population distributions, it was found that truly random dispersal is hardly, if ever, found in nature. Instead, plant populations depart more or less from randomness, tending to be more clumped or contagiously distributed (ASHBY 1948, GREIG-SMITH 1964). Unless the clumps themselves are randomly distributed, a grid of regularly distributed samples across the vegetation segment tends to give a better coverage of the range of variation.

The advantages of systematic and random sampling are often combined by a system called stratified random sampling. In this form of sampling the vegetation segment is arbitrarily subdivided (stratified) into several more or less even-sized subsegments. Within these subsegments the samples are laid out at random. Sampling in each subdivision insures coverage across the whole vegetation segment, while random locating of sampling sites within the subdivisions allows for establishment of statistical error terms. Yet, these are meaningful only for the more abundant plant populations and only if these are randomly distributed in each subdivision. However, arbitrary subdivisioning of a large vegetation segment has the added advantage that the homogeneity between the subsegments can be tested. For this, systematic sampling within the subsegments may also be used.

4.5   TWO DIFFERENT SAMPLING APPROACHES

The reasons for using a particular sampling method are—like the degree of entitation—primarily a matter of viewpoint and purpose. In aiming for a detailed floristic vegetation description, two major approaches can be distinguished. These are, the relevé analysis for classification and the continuum analysis for ordination.

A classification aims at grouping individual stands into categories. The stands that are closely similar with one another form one class, which is separated from other classes that also consist of similar stands. The properties common to a group of similar stands in a class are then abstracted to serve as a description of that class. Therefore, the abstracted class properties may be compared to the average or mean of a set of values when combined with a measure of range. However, the nature of variation within a class unit is not always indicated. This depends on the form of presentation. For practical and scientific validity, the abstracted class features should adequately describe the individual members of each class.

In contrast, an ordination aims at portraying the individuality of each stand. This is done by demonstrating the similarity or dissimilarity of all stands to one another in the form of geometric models.

The essential differences with regard to community sampling are:

1.   *The Classification Method*

a.   Evolved from the plant-list (relevé) method (CHAPS. 5 and 9) which aims at obtaining a complete check list of plants in a series of relatively small, environmentally uniform habitats.

b.   Works with communities that are considered homogeneous with regard to both dominant and subdominant species and layers of vegetation.

c.   Aims at description through classification by sampling for recurring plant assemblages, which are visually distinguished.

d.   As a rule sampling intensity is high, because sampling for recurring patterns involves replicate sampling.

e.   Species presence or absence is considered more important than minor variations in quantity. The method is therefore primarily qualitative.

2.   *The Ordination Method*

a.   Evolved primarily from timber survey methods (CHAP. 7) and usually ignores details in pattern and composition of undergrowth species during the sampling phase.

b.   Works with communities that are homogeneous only with regard to dominant species or the dominant vegetation layer. Therefore, the communities are usually rather large and thus heterogeneous in undergrowth and habitat.

c.   Aims at description through ordination (i.e., arranging sample stands in order of similarity, see CHAP. 10) by sampling either randomly, systematically, or subjectively along floristic continua or environmental gradients.

   d.   Sampling intensity may be high or low. Each sample stand may contain a more or less strongly differing plant assemblage. Replicate sampling is not emphasized, and the demonstration of recurring patterns is therefore largely a matter of chance.

   e.   Species presence and absence is considered less important than minor variations in quantity. The method is, therefore, primarily quantitative.

   In spite of their differences, both methods compete in obtaining a thorough description of a restricted vegetation region or a widely distributed vegetation formation, such as a forest or grass cover.

   Yet, both methods are also complementary in their treatment of sample data. Stands sampled for a classification can be subjected to ordination, while stands sampled for an ordination can be classified, provided they are floristically complete enough.

   With an equal intensity of sampling applied in the relevé and the continuum analyses, the difference in approach lies in the degree of entitation and the sample distribution.

   The generally broader degree of entitation applied in continuum analysis and the subsequent distribution of samples throughout such a broader vegetation-entity will result in member-stands that usually differ by more or less continuous increments of similarity or dissimilarity. Such a continuum can be broken up into classes.

   The generally finer degree of entitation applied in relevé analysis and the subsequent more or less central distribution of samples in each of the smaller entities will result in member-stands that are more similar to one another or more clearly clustered into groups or classes. In this method recurring plant assemblages are emphasized. In the continuum method transitions between recurring patterns are equally emphasized. This also implies that with the same sampling intensity in both methods, recurring patterns are deemphasized in the continuum analysis.

   The two methods are truly complementary only where the degree of entitation is the same and where the sampling intensity in continuum analysis is increased over that of the relevé analysis.

   Under such conditions, the relevé analysis for classification can be considered as supplying the first order of information. This would be to portray the primarily qualitative differences of the abstracted recurring patterns or classes. These classes can also be interpreted as convenient stations along a floristic continuum or environmental gradient.

Under such conditions, the continuum analysis can be applied for supplying the second order of information. This would be to portray the "fine structure" within each class and among the classes by sampling the variation throughout each entity.

PART
II

Vegetation Analysis in the Field

Community Sampling:
The Relevé Method

5   Once the entitation or subdivisioning of the vegetation cover has been clarified, the communities are essentially established. This is the reason why a thorough reconnaissance and familiarization before sampling is so important. Subsequent sampling and data collection will merely derive more detailed information on these communities, irrespective of one's choice for semiquantitative or quantitative methods for description.

However, certain changes in entitation may become necessary with increasing knowledge during the investigation. The approach should be flexible enough to allow for such changes.

5.1   THREE REQUIREMENTS OF A SAMPLE STAND—
PARTICULARLY THE HOMOGENEITY REQUIREMENT

Regardless of the method used for field analysis, a sample stand (releve) should fulfill the following requirements.

1.   It should be large enough to contain all species belonging to the plant community.

2.   The habitat should be uniform within the stand area, as far as one can determine this.

3.   The plant cover should be as homogeneous as possible. For example, it should not show large openings or should not be dominated by one species in one half of the sample area and by a second species in the other half.

This requirement of relative homogeneity throughout the sample stand area rests on the premise that the vegetation parameters or statistics recorded for the stand should result in a meaningful average. An average obtained from a sample area reaching partly into an opening or across an obvious community boundary is an artifact. However, if openings are a characteristic feature of a community, they must be sampled as well.

Depending on the degree of segmentation and geographic scale desired, openings may be recognized as separate entities, or they may be included in larger samples. Where stand openings are stratified as separate entities, they may either be ignored or sampled as separate communities.

An important difference in the homogeneity concept has been practiced in the European releve analysis as opposed to the American continuum analysis.

In the releve analysis of a forest stand the requirement for homogeneity applies to both the tree layer and the undergrowth vegetation. DAUBENMIRE (1968) has used the same concept in Western North America. In the continuum analysis of eastern American forests the homogeneity requirement so far has primarily been applied to the tree layer only. In this homogeneity concept, therefore, the undergrowth vegetation and habitat can be quite heterogeneous.

CURTIS (1959) devised a test of homogeneity by quartering a sample stand and by comparing the variation in tree composition among the four quarters in a chi-square test. However, it was found that homogeneity could easily be judged subjectively. Many other attempts have been made to quantify homogeneity (RAABE 1952, DAHL 1960, MORAVEC 1971) of the plant cover. However, there is no satisfactory objective method, and the degree of plant-cover variation one considers integratable into a meaningful expression of average is still a matter of judgment.

At first one is guided by physiognomy and structure of the vegetation, and then by the prevalence or dominance of certain species that must be relatively uniformly present across the sample stand. Less obvious or rarer species may occur dispersed among the more obvious or dominant ones, and the dispersal pattern of these less obvious species may be considered of secondary importance. But as a major qualitative difference in the homogeneity concept we may reemphasize the European tradition as considering the field or herb and low-shrub layer as the guide while the eastern North American tradition pays less attention to the undergrowth in forests and considers the cover composition of the uppermost layer as the guide to homogeneity. These important qualitative differences result in much larger sample stands to be recognized for traditional North American vegetation studies as compared to the traditional European ones.

5.2   MINIMAL SAMPLE AREA—THE SIZE REQUIREMENT

In all within-segment or community sampling it is important that the species are as fully represented as possible.

Where the emphasis is on sampling for recurring plant assemblages, it is common practice to determine the so-called "minimal area" of the community. This is defined as the smallest area on which the species composition of the community in question is adequately represented. This smallest area gives an indication of the releve′ or quadrat size that should be used.

In contrast, where the emphasis is on sampling the quantitative variation of species within large communities defined only by dominant species, the idea of representative species composition is commonly reduced to certain quantitative measures of the more abundant species only. Instead of using a few large "minimal area" quadrats or plots, the sample is spread over each vegetation segment as much as possible in the form of small quadrats or point samples. Individual quadrat sizes are adjusted primarily to the convenience of assessing the quantitative parameters selected. The size of the sample as a whole is either determined arbitrarily or adjusted to the quantitative variation of the vegetation segment.

A selection prior to sampling of the "more important" (dominant) from the "less important" (quantitatively less well represented) species is considered inadequate by many ecologists (e.g., FLAHAULT and

SCHRÖTER 1910, BRAUN-BLANQUET 1928, KRAJINA 1960, HANSON and CHURCHILL 1961, DAUBENMIRE 1968, among others).

However, both viewpoints can be combined and the number of small quadrats can be determined to include the minimal area of the community (CAIN 1938, OOSTING 1956, DAUBENMIRE 1968).

The minimal area depends on the kind of community and varies within wide limits. For temperate-zone vegetations, the following empirical values can be given:

Forests (including tree stratum)	200–500 m²
(undergrowth vegetation only)	50–200 m²
Dry-grassland	50–100 m²
Dwarf-shrub heath	10– 25 m²
Hay meadow	10– 25 m²
Fertilized pasture	5– 10 m²
Agricultural weed communities	25–100 m²
Moss communities	1– 4 m²
Lichen communities	0.1– 1 m²

5.21 The Nested Plot Technique. The minimal area can only be determined in a community that is relatively homogeneous and not fragmentary. A community can be called fragmentary if it lacks species that are usually present in the recurring plant assemblages of this kind. Such fragmentation may be caused by selective destruction of certain species, for example through grazing, or simply through fragmentation of the surface area into too small segments. The total number of species (richness) is in itself an important characteristic of a community type (McINTOSH 1967b).

The minimal area is determined by initially lining out a small area, for example 0.5 × 0.5 m (0.25 m²) and by recording all species that occur within this small area. Then the sample area is enlarged to twice the size, then to four and eight times the size, etc. The additionally occurring species are listed separately for each enlarged area (TABLE 5.1). The sample area is increased until the species added to the list become very few. FIGURE 5.1 shows the arrangement of the sample quadrats in the form of nested plots.

5.22 Criteria for Size of Releve. The species number is then plotted over size of sample area. This results in a species/area curve (FIG. 5.2). The minimal area is the sample area at which the initially steeply increasing curve becomes almost horizontal. This is not an exact definition. It is, therefore, advisable to decide on a somewhat larger area

TABLE 5.1. Example Data for Determining the Minimal Area of a Pasture (Lolieto–Cynosuretum typicum in Northwest Germany)

SUBPLOT NUMBER	SIZE (m²)	SPECIES	CUMULATIVE TOTAL NUMBER OF SPECIES
1	0.25	Lolium perenne	
		Poa pratensis	
		Poa trivialis	
		Festuca pratensis	
		Trifolium repens	
		Crysanthemum leucanthemum	
		Rumex acetosella	
		Plantago lanceolata	
		Bellis perennis	
		Cirsium arvense	10
2	0.5	Cynosurus cristatus	
		Trifolium pratense	
		Cerastium fontanum	
		Centaurea jacea	14
3	1	Leontodon autumnalis	
		Achillea millefolium	16
4	2	Holcus lanatus	
		Vicia cracca	
		Prunella vulgaris	19
5	4	Plantago major	
		Festuca rubra var. genuina	21
6	8	Anthoxanthum odoratum	22
7	16	Trifolium dubium	
		Taraxacum officinale	24
8	32	Rumex crispus	25
9	64	Lathyrus pratensis	26

for an adequate size of sample plot or quadrat. This was done in the empirical values cited in SECTION 5.2.

CAIN (1938) pointed out that the minimal area in such a graphical presentation is influenced by the ratio of the ordinate (y) to the abscissa (x). A contraction of the ordinate in proportion to the abscissa results in a decrease also of the apparent minimal area, since the curve flattens at a faster rate. To overcome this deficiency, he developed a criterion for minimal area delineation that is unaffected by the y/x ratio. He suggested using the point along the curve at which an increase in 10 per-

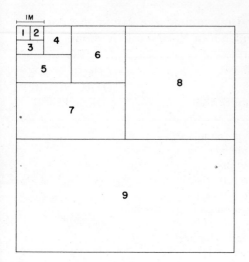

FIGURE 5.1. A system of nested plots for establishing minimal area. Each subplot numbered consecutively to 1 includes the area of the previous subplot. Thus, uneven numbered subplots are square, even numbered ones are rectangular. The plot layout relates to the data shown in TABLE 5.1.

cent of the total sample area yields only 10 percent more species of the total number recorded.

The example in TABLE 5.1 shows that 10 percent of the total area equals 6.4 m². Ten percent of the total species number is 2.6. This point ($x = 6.4$ and $y = 2.6$) is plotted on the graph. Then a line is drawn through this point and the intersection of the y and x axes (the origin). However, the same line can be drawn more accurately from the origin through a point defined by the 100 percent values of x and y (see FIG. 5.2, line a_1). Next, a line parallel to the first is drawn that touches the species/area curve tangentially (FIG. 5.2, line a_2). The tangential intersection point on the curve is then protracted to the x axis (FIG. 5.2, line a_3), on which the minimal area is indicated.

A more conservative estimate of minimal area is obtained where a 10 percent increase in area yields only 5 percent more species. This point is found by drawing a line through the origin and a point defined by 50 percent of the total species count and 100 percent of the area sampled (FIG. 5.2, line b_1).

In the two examples shown on FIG. 5.2, the minimal areas defined in this way are:

Pasture at **10 percent species increase**	**5 m²**	
at **5 percent species increase**	**12 m²**	
Hay meadow at **10 percent species increase**	**8 m²**	
at **5 percent species increase**	**10 m²**	

The minimal area values are still only approximations and are therefore rounded off to the nearest square meter.

However, CAIN's method is not an absolute guide, because it can be

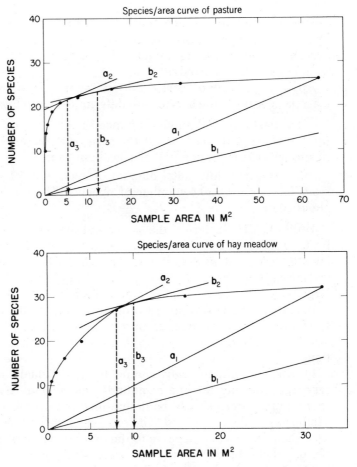

FIGURE 5.2. *Species/area curves of a pasture (see TABLE 5.1) and hay meadow. Legend: a_1=10 percent line; a_2=tangent parallel to 10 percent line; a_3= protraction to minimal area based on 10 percent species increase; b_1=5 percent line; b_2=tangent parallel to 5 percent line; b_3=protraction to minimal area based on 5 percent species increase. Further explanation in text.*

applied to any shape of curve, whether levelling off or not. Yet, only curves that level off can be used for determining minimal areas. RICE and KELTING (1955) have drawn attention to the fact that the 10 percent point shifts to the right on the species/area curve with any increase in size of sample area. This is so even if the curve remains horizontal.

However, for a relevé analysis there is no advantage in increasing the sample size if it yields no further species.

The objection of RICE and KELTING can be overcome by setting an objective standard through the total number of species sampled. For example, one may require that the sample plot contain at least 90 or 95 percent of the maximum number of species encountered in the largest sample unit of the nested plot.

Applied to the two curves, the minimum releve´ sizes containing at least 95 percent of the species are determined as follows:

1.   Pasture; total number of species is 26. Subtract 5 percent from total number (1.3) and read off the area from the curve at 24.7 number of species. The minimum plot size is about 25 m².

2.   Hay meadow; total number of species is 32. Read off the area from the curve at 30.4 number of species. The minimum plot size is about 20 m².

Unlike CAIN's criterion, the 95 percent species requirement is affected by the y/x ratio. But since the 95 percent point lies to the right of the levelling-off point, as determined through CAIN's criterion, the effect of the curve shape becomes insignificant. The plot size should always be larger than the minimal area obtained through CAIN's criterion.

In spite of the lack of an absolute criterion for the minimal area, the species/area curve remains an important practical guide to stand or plot size in studies aiming at portraying a representative species composition.

Ideally, the minimal area should be established for the community type and not only for one community member of a type. This means that minimal areas should be determined in several recurring plant assemblages of the same kind. The one that indicates the largest minimal area should be used as a guide to the minimum size of a vegetation sample or releve´. MORAVEC (1973) suggested another method for determining the minimal area, which involves enlarging separate (in contrast to nested) plots in a vegetation segment until the floristic similarity (CHAP. 10) between the plots reaches a maximum value. This method appears sound theoretically but it requires much more time in field sampling and subsequent computation, so that its practicability seems rather doubtful.

5.23   Evaluation of the Minimal Area Concept.   The species/area curve is clearly quantitative in terms of species number, but it gives little information on the number of individuals per species.

For example, the pasture showed 24 species in an area of 16 m², the hay meadow 30 species on the same unit area. The greater number of species in the hay meadow for the same unit area may indicate a smaller size of individuals, or a smaller number of individuals per species, or a combination of these parameters. Therefore, the species/area curve is often considered less consequential in quantitative studies (McINTOSH 1967b). But the problem of counting individuals per species is an entirely different one that requires usually smaller sizes of quadrats or plots. The number of small quadrats depends on the variability or homogeneity of the vegetation segments caused by the more dominant or abundant species in the community. Yet, there is no absolutely objective guide either for determining the sample size or number of small quadrats for quantitative analyses. The problem involves the same sort of judgment as is required for deciding the minimal area from a species/area curve (see discussion under sample size, SECTION 6.4).

The species/area curve has its application also to quantitative community analyses. A quantitative vegetation sample, to be representative of the species composition of a community, should never be smaller than the minimal area.

A species/area curve is of interest also for its own information value, because it indicates an important community property, namely, species diversity in relation to increasing size of area. Any community can be compared on that basis. FIGURE 5.3 shows six species/area curves from tropical rain forests in Borneo that were assembled by ASHTON (1965). Only the lowest curve from Belalong shows a clear trend of levelling-off. In this community a 0.5 acre relevé could be considered satisfactory. In the other communities, even a 5 acre (or 2 hectare)

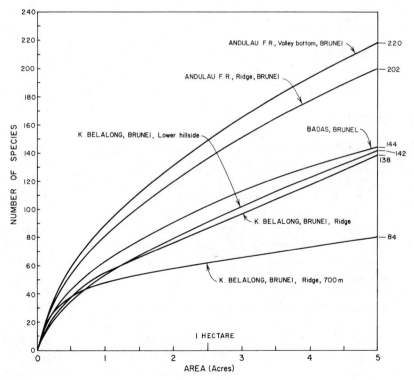

FIGURE 5.3. *Species/area curves for sites in Brunei. (Redrafted with modifications from ASHTON 1965:236.)*

releve´seems too small. The minimal area of these communities is probably around 5 hectares. In these communities a compromise has to be
made. Instead of one large releve´with a near-complete species assemblage, one must choose to work with a releve´ size that is practical and
manageable, even if it contains only a fraction of the number of species
of the community type. But then, several such releve´s should be used
in place of one to recompose a near-complete species composition of
the whole community. Of course, in such rain forest stands, this is a
formidable task.

5.3   EXAMPLE OF A RELEVÉ ANALYSIS

After establishing how large a releve´ or vegetation sample should be,
we then list all species present. A good floristic knowledge and careful
collection of all plant species (that cannot be identified immediately) is a
necessary prerequisite for such investigations. Depending on the purpose of the study and the questions asked, the species list can be completed by more or less accurate information on the quantitative relations, the current status of development, or other species properties
that are readily determinable in the field.

Furthermore, a description of the habitat and the soil profile can be
added. The date of analysis and an accurate description of the location
should not be missing. A sketch map will facilitate relocation of the
sample plot if one intends repeated analyses.

As an example, we will describe the analysis of a forest stand
(TABLE 5.2).

The analysis begins with noting down the releve´ number (which is
used also for marking the releve´ location on a map; scale 1:24,000 or
similar), the date, a record of the geographic area, general location, detailed location, etc., size, and position of the sample area. The community should be given a structural name and a tentative floristic name describing the most obvious species composition. A few general remarks
on the condition of the plant cover, particularly on its vertical stratification into layers of different heights or into life form groups and their
percent cover should be recorded in the introductory part of the field
note sheet.

5.31   Species Quantities and Stratification.   A species list alone is
usually not satisfactory for portraying the special characteristics of a
plant cover segment. Therefore, the list should be augmented with
notes on the quantity of each species. Other aspects often worth recording are certain characteristics of species dispersion and morpho

logical structure. Also, phenological information may be noted with relative ease (see CHAP. 8). In all cases the community structure should be described with respect to its layering into height strata (if evident) and the cover extent of each layer should be estimated in percent of the sample area. Preferably the species should be recorded in height strata right away. This allows for a simple order of recording. Any further preordering of the species list is likely to complicate the analysis.

No definite height limits can be set in advance for any community layer, because this depends on the community structure itself. Nevertheless, certain height limits that are commonly found convenient in forest communities are:

1. T = Tree layer, any plant taller than 5 m. In taller forests, this layer is usually subdivided into 2, 3, or even 4 layers of decreasing heights ($T1$, $T2$, etc.).

2. S = Shrub layer, plants between 50 cm and 5 m tall, or 30 cm and 5 m tall. A subdivision is commonly indicated at 2 or 3 m height. For example, an $S1$ layer may be indicated from 2 to 5 m, and an $S2$ layer from 30 (or 50) cm to 2 m height.

3. H = Herb layer, from < 30 cm (or 50 cm) to 1 m height. Subdivisions analogous to the T and S layers are often useful. For example, $H1$ as tall herb layer > 30 cm height, $H2$, as medium-tall herb layer from 10 to 30 cm height, $H3$ as low herb layer < 10 cm height.

4. M = Moss and lichen layer. This refers usually to a ground-appressed low carpet of less than 5 or 10 cm height. Where necessary it should be subdivided into an M layer on bare mineral soil, an M layer on humus, an M layer on decaying wood, an M layer on stones or exposed rocks. This leads over also to a recognition of one or several epiphyte layers (E), the plants that occur attached (but not parasitically) to other plants, for example, mosses on tree stumps or trunks.

A compromise must often be made between emphasizing height stratification or life form. For example, a decision is required in certain stands with a distinct moss layer (M) whether to record small herbs and seedlings in the same layer (M) or to recognize a separate low vascular plant layer (H) of < 10 cm or so in height. This may initially cause some difficulties. But the problem can always be solved by a decision that, of course, should be closely adapted to the prevailing conditions.

The species are recorded within these height strata or in life form groups in the order as they are noticed. It is possible also to record the

TABLE 5.2. *Example of a Releve′ Analysis in a Tropical Insular Rain Forest Near Honolulu, Hawaii.*

Releve′ #11 March 4, 1972

LOCATION: Tantalus Mountain in the rain forest, Honolulu, Hawaii. Along trail leading to Pauoa Flats at 450 m (1480 feet) elevation, 160 m before planted *Eucalyptus* stand on Pauoa Flats.

Below the trail on a gently (21 percent) sloping convex ridge, where *Acacia koa* is present as an emergent tree. Location mapped on 1:24,000 topographic sheet of Honolulu area.

COMMUNITY: A low-stature submontane rain forest with scattered emergent *Acacia koa* trees up to 10 m tall.

Acacia koa–Metrosideros–Psidium guajava forest with few tree ferns (*Cibotium splendens*) and *Oplismenus* grass undergrowth.

POSITION   Aspect N 50 W, slope 21 percent, elevation 420 m (1380 feet)

SIZE OF RELEVE: 10 × 20 m (200 m²); the 20 m part upslope

STRATIFICATION: T   Scattered *Acacia koa* 5–10 m tall, 40 percent cover

S1 Woody plants 2–5 m tall, 80 percent cover

S2 Woody plants 0.3–2 m tall, 10 percent cover

H   Herbaceous plants, most of them up to 30 cm tall, 95 percent cover

REMARKS: This stand is composed of native trees (*Acacia, Metrosideros*) and partly of nonwillfully introduced, exotic trees (*Psidium, Citharexylum*). All *Metrosideros* and *Psidium* trees are multistem trees, forked or branched near the base. *Psidium* regeneration is entirely vegetative from root sprouts near the base of the older trees. *Citharexylum* regeneration is from seed.

---

species right away in a certain order, for example, in order of decreasing abundance. However, this usually takes more time and increases the danger of overlooking certain species. It must be emphasized that the primary requirement in the releve′ method is a list of all species present. The consideration of species quantity takes a secondary rank.

It is useful to leave some space on the list in front of the species names for notes on their abundance. By listing each species on a separate line, the names are easily spotted afterwards. The methods for estimating their quantities will be discussed in SECTIONS 5.4–5.43.

Notes on the habitat and the description of the soil profile are often conveniently written on the backside of the note sheet. This is practical, if one uses a loose-leaf data-holder. Loose-leaves facilitate subsequent sorting of field notes.

Other plant communities, such as heath, grassland, bogs, etc. are analyzed in a similar way. However, the general remarks and notes on

TABLE 5.2 (Continued) LIST OF PLANTS BY STRATA WITH
BRAUN-BLANQUET RATINGS IN FRONT OF NAME:

Stratum Symbol	BRAUN-BLANQUET Rating Symbol[a]	Plant Species
T	3	*Acacia koa*[b]
	1	*Metrosideros collina* subsp. *polymorpha*[b]
	+	*Psidium guajava*
	1	*Metrosideros tremuloides*[b]
S1	+	*Metrosideros collina* subsp. *polymorpha*[b]
	+	*Citharexylum caudatum*
	4	*Psidium guajava*
S2	+	*Cordyline terminalis*
	1	*Cibotium splendens*[b]
	1	*Citharexylum caudatum*
	+	*Psidium guajava*
	r	*Sadleria* sp.[b]
	1	*Rubus rosaefolius*
H	4	*Oplismenus hirtellus*
	1	*Commelina diffusa*
	1	*Setaria palmifolia*
	+	*Nephrolepis hirsutula*
	1	*Microlepia setosa*[b]
	+	*Athyrium proliferum*
	r	*Asplenium falcatum*

SOIL: Shallow (50–75 cm deep) rain forest soil (Hydrol Humic Latosol or Oxisol)
with few rocks outcropping at surface.

[a] For explanation see SECTION 5.42.1.
[b] Native plants.

the habitat features are modified according to the different situation
and purposes of the analyses.

5.32 Essential Information Parts. The relevé record may be summarized as having to show always three basic items of information,
(a) geographic and physiographic, (b) the species list with information
on the quantity of each species and their membership in different layers, and (c) soil information and specific remarks. Of course, the relevé
record may be much further extended and systematized. It can go into
great detail of quantitative measurements, into an assessment of plant
life form and size-structure of populations, and into a study of their
functions. Moreover, it may be accompanied or followed by an in-depth
analysis of the environment. However, a list of at least all the higher
plant species found in the sample stand is mandatory.

CURTIS (1955, CURTIS and GREENE 1949) called this type of analysis the species-presence method but preferred to prepare the species list from random locations of as many sample stands as possible in very broad, at least 6 hectare large forest vegetation segments or in broad landscape segments for prairie vegetation. However, such a large-area sample is necessarily heterogeneous and thus cannot be used for the same purpose as the above described relevé method. Many such relevés (as described in TABLE 5.2) are placed in all the differentiated vegetation segments and repeatedly into recurring plant assemblages of the same kind.

5.33 Further Detail for Complete Investigations. A good example of a systematic extension of important ecological information are the plot or relevé record forms published by KRAJINA, ORLOCI and BROOKE (1962). One plot record form provides space for specific information on geographic and topographic location and space for details on vertical stratification with space for separate percent cover estimates for each layer. Included is an estimate of the ground surface in percent cover of humus, exposed rocks and mineral soil, and decaying wood. A separate form deals with important habitat characteristics, such as position within landform, soil type identification, ground water, and drainage characteristics. Another form provides an outline for a very detailed soil profile description. The plant list form, extending over several sheets with strata subdivisions, provides columns for recording of several phenological, life form, and functional features. Further forms are provided for detailed tree-mensurational data and for subsequent laboratory analyses of the essential physical and chemical soil properties and for micro- or ecoclimatological investigations.

Such an outline suggests a very complete relevé analysis. But this sort of information may be obtained in several stages. The relevé example in TABLE 5.2 is the first completed stage of an investigation that yields a thorough description. Many more such relevés must be obtained to arrive at a well-documented regional classification of vegetation. The relevé method was similarly interpreted by BENNINGHOFF (1966).

5.4 ESTIMATING SPECIES QUANTITIES

A relatively crude estimate of species quantities is satisfactory for many purposes of vegetation description. In earlier times, each plant geographer and ecologist used more or less his own method. Therefore, the results of earlier authors were usually hard to compare (see, for example, the summarized comparisons of DU RIETZ 1921, RÜBEL 1922, CLEMENTS 1928, BRAUN-BLANQUET 1932, and BECKING 1957).

5.41   Relative Magnitude Terms.   Relative magnitude terms ascribed to each species are useful on a very general basis. A division into five classes was often applied by earlier ecologists (KERSHAW 1964). These were in order of decreasing magnitude: dominant, abundant, frequent, occasional, and rare. The first three terms have received specified meanings in the course of time. Today, the term dominant is meaningful only if supplied with a definition. It may mean dominant in height (the usual understanding of foresters), dominant in crown cover (the usual understanding in combined estimate scales), dominant in basal area (the usual understanding of the Wisconsin School), or dominant in number. The term abundance implies the same meaning as number of individuals per species or density. But when speaking of abundance, one usually refers to a number estimate; when speaking of density, to an actual count of individuals. Frequency has a special definition. It refers to the number of times a species is found in a given total number of frame placings or quadrats, or at a given total number of sample points.

Now, the magnitude values mentioned above can be replaced by more meaningful terms (e.g., very rare, occasional, abundant, and very abundant) or by code values. But these remain relative as they are only related to each other and not to a specified reference area.

5.42   Absolute Scale Values.   BRAUN-BLANQUET made a major contribution in selecting, simplifying, and modifying a system for analysis that is convincingly simple and yet not superficial. It was favorably and rapidly accepted by many investigators and proved to be useful in areas with large numbers of species. The method requires relatively little time per relevé analysis and can be used for all plant communities composed of higher plants and for most moss and lichen communities. Therefore, it is useful not only for monographical work of vegetation zones, but also for a single community type studied throughout its range of distribution.

For reasons of comparability one should not unnecessarily deviate from this method, even if one does not intend to process the data further according to BRAUN-BLANQUET's synthesis table technique. (CHAPTER 9).

5.42.1   The BRAUN-BLANQUET Cover-Abundance Scale. The following scale values of BRAUN-BLANQUET are absolute in so far as the values relate to a reference area, which is fixed by the size of the relevé.

5 Any number, with cover more than ¾ of the reference area ($> 75\%$)

4 Any number, with $\frac{1}{2}$–$\frac{3}{4}$ cover (50–75%)

3 Any number, with $\frac{1}{4}$–$\frac{1}{2}$ cover (25–50%)

2 Any number, with $\frac{1}{20}$–$\frac{1}{4}$ cover (5–25%)

1 Numerous, but less than $\frac{1}{20}$ cover, or scattered, with cover up to $\frac{1}{20}$ (5%)

+ (Pronounced cross) few, with small cover

r Solitary, with small cover

An idea of the application of the scale values is given in FIGURE 5.4. An estimate is assigned independently to each species.

The upper four scale values (5, 4, 3, 2) refer only to cover, which is understood as the vertical crown or shoot-area projection per species in the plot. The lower three scale values are primarily estimates of abundance, that is, number of individuals per species. Therefore, the scale is often called a combined estimate scale. This combination of the two parameters into one scale was an improvement over a separate estimation of abundance and cover. Abundance can be estimated with some precision only for herb and shrub layer species with little or insignificant crown or shoot cover. Cover can be estimated more accurately only for species that contribute significantly to the biomass of the community. Cover can be expressed in fractions or percentages of the reference area. It is usually less than the total leaf area of a species, since leaves do often overlap each other. Earlier, BRAUN-BLANQUET used only five values. Subsequently the scale was expanded to seven values by extending the scale in the lower classes.

The combined estimate of abundance and cover was later called "species magnitude" (Artmächtigkeit) by BRAUN-BLANQUET (1965) after SCHWICKERATH (1940). KRAJINA (1960) calls it "species significance." The estimating method can be learned quite easily if one recalls and clarifies the range and meaning of each scale value.

The method is often referred to as semiquantitative in the literature, because of the almost qualitative character of the large intervals among the scale values.

One may object that this method is rather crude and does not require measurements. However, estimates cannot be made in many vegetations on a much more refined scale. Measurements always require much more time. Since plant communities vary so much from place to place, it seems more useful for community classification or ordination to analyze several sample stands of the same type of community by estimating species quantities rather than to analyze only one relevé in greater quantitative detail.

During establishment of the plant list, the investigator walks several times through the relevé. The general quantitative relations of the spe-

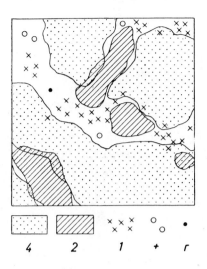

FIGURE 5.4. Schematic presentation of combined estimate values for cover and abundance (4, 2, 1, + and r) after the BRAUN-BLANQUET scale.

| 4 | 2 | 1 | + | r |

cies become apparent during the listing of the species. After all species are recorded, the investigator walks into the middle of the relevé or to a location where he can once more overview the entire plant assemblage of the relevé. From this point he assigns a cover-abundance rating to each species, taking into account his experience during the listing. The rating must apply to the entire area of the relevé. It is convenient to have the scale values on the notebook cover, because it is important to clarify their meaning each time they are used. The procedure for each species on the list is to first ask, does the species cover more than 50 percent of the area or less than 50 percent? If it covers more than 50 percent of the area, the next question is, does it cover more than 75 percent or less than 75 percent? If the latter applies, the rating is 4. Similarly, if a species covers less than 50 percent of the area, the next question is, does it cover more than 25 percent or less than 25 percent? If less than 25 percent, there are again two alternatives, namely, a cover of more than 5 percent or less than 5 percent. This area can be quickly visualized in a 200 m² relevé as a 10 m² area into which one can mentally contract the species that may be scattered over the 200 m² relevé. A species with less than 5 percent cover can either receive the value 1 or + (cross). If the cover seems to be less than 1 percent (or 2 m² in a 200 m² relevé) the species receives the value of +. If really only seen once in the relevé the species receives an r for rare or solitary in the relevé. Species occurring outside the relevé but that were not seen inside, should also be recorded if noticed. In that case one assigns a scale value in parentheses.

The rating is quantitatively crude, but if properly applied it is not superficial. At the limit of a range of a scale value, even an experienced investigator may choose a higher or a lower rating symbol. This means, that an error by one rating value can easily occur in border cases. However, since the releve' method aims at describing the spatial floristic variation of a regional vegetation cover, the emphasis is put on more releve's with semiquantitative estimates rather than fewer vegetation samples with exact measurements of species quantities.

Perhaps a more accurate, but also more time-consuming, modification of this method is to estimate the cover of the herbaceous plants in a number of small frames (DAUBENMIRE 1968) and that of shrubs and trees in larger subplots lined out accurately with string. A square meter unit is convenient for many kinds of herbaceous layers. In this case one should estimate the cover directly in percent for each frame placement. This will permit calculation of an average. This method can be used to check the estimate ratings made for the total plot area. Further refinement of this method approaches cover measurement (SECTION 6.5) rather than cover estimation.

5.42.2 *The DOMIN-KRAJINA Cover-Abundance Scale.* KRAJINA (1933) introduced a somewhat more detailed scale (originally developed by DOMIN) that was applied successfully in many studies of forest communities in British Columbia within the last 20 years (KRAJINA 1969), and also, for example, by KERSHAW (1968). It gives an advantage in forest communities where differences in abundance among rarer species are often quite noticeable. The scale values are shown in the table. In species-rich herbaceous communities, estimative errors are

BRAUN-BLANQUET		DOMIN–KRAJINA	COVER %
5	10	any number, with complete cover	≈100
	9	any number, with more than 3/4 but less than complete cover	> 75
4	8	any number, with 1/2–3/4 cover	50–75
3	7	any number, with 1/3–1/2 cover	33–50
	6	any number, with 1/4–1/3 cover	25–33
2	5	any number, with 1/10–1/4 cover	10–25
	4	any number, with 1/20–1/10 cover	5–10
1	3	scattered, with cover under 1/20	1–5
	2	very scattered, with small cover	< 1
+	1	seldom, with insignificant cover	
r	+	solitary, with insignificant cover	

more likely with the finer scale intervals than with the cruder scale intervals of BRAUN-BLANQUET. The DOMIN-KRAJINA scale can readily be converted to the latter, but the reverse is never possible, as the correspondence of the two scales shows.

5.42.3   The DAUBENMIRE Cover Scale.   DAUBENMIRE (1959, 1968) prefers to estimate cover alone according to a similar rating scale as follows:

COVER CLASS	RANGE OF COVER (%)	CLASS MIDPOINTS (%)
6	95–100	97.5
5	75– 95	85
4	50– 75	62.5
3	25– 50	37.5
2	5– 25	15
1	0– 5	2.5

Values 1 to 4 are exactly the same as the BRAUN-BLANQUET values, except that r and + species would be included in cover class 1 of DAUBENMIRE. The two upper values in DAUBENMIRE's cover scale are the exact equivalents of the two upper values in the DOMIN-KRAJINA scale.

A recent modification of the DAUBENMIRE scale by BAILEY and POULTON (1968) separates the 0–5 percent cover class into two classes, one for a range of 0–1 percent (midpoint 0.5 percent), a second for 1–5 percent (3 percent). This finer breakdown is analogous to the + and 1 ratings of the BRAUN-BLANQUET scale.

One may wonder why most scales show unequal class intervals. The main reason for this is a practical one. The chosen scale values allow for an easier estimation of species-cover-to-area relationship than do equal intervals of cover. Also, the less abundant species or species with small cover may sometimes have an important diagnostic significance, which requires a finer breakdown in the lower scale values as compared to the larger scale values.

5.43   Estimating Biomass by Partitioning in Grass Communities.   The estimate method of KLAPP (1929:714, compare also WACKER 1943) can be applied to grass communities, and it is based on shoot dry weight proportions as estimated from fresh weight proportions under field conditions. The method allows for a more complete quantitative evaluation than is possible with the BRAUN-BLANQUET scale since it is concerned with biomass rather than with cover. It has been applied widely in investigating pastures and tall-grass stands for their feed value. The method is also useful for basic studies, since it provides a good indication of the competitive relations among species. The method is based on an estimate of the weight proportions of each species in percent of total biomass. It replaces an earlier technique of determining the biomass weight for each species separately.

However, a very refined accuracy can never be achieved by mere estimation. One has to allow always for a personal error that may be as high as one fourth of the estimated percentage in this method. But sufficient accuracy for practical purposes can be attained by using the following approach.

One estimates initially the total standing biomass per m² (with weight-checking). Then the two more obvious main groups are partitioned, namely, the grasses (including the grass-like species) and the forbs (including the legumes). These are estimated in relation to one another (also initially by weight-checking). In the example shown in TABLE 9.1 their ratio was probably larger than 75:25, but smaller than 90:10 and is probably quite correctly estimated with 85:15. Next, the two main groups are further subdivided in a similar manner. In this case the legumes are estimated to make up 1/10 at the most of the nongrass species, therefore, only about 1 percent of the total plant material. Sedges form only a small proportion of the graminoids. Their mass appears about double that of the legumes and can be estimated as 2 percent. Therefore, the following percentages can be derived (the extreme values, shown in brackets, were estimates of several unexperienced persons):*

Grasses	83 (75–90 percent)
Sedges	2 (1– 5 percent)
Legumes	1 (+ – 2 percent)
Other herbs	14 (10–22 percent)

The different species that are always recorded separately on the lists, are then estimated within each group on the basis of the remaining percentages. Here, the "rare" species that probably contribute less than 1 percent are singled out first by designating them with +. The remaining species are then evaluated in order of their magnitude. The begin-

---

* A check of air-dry weights resulted in the following values 85.4 : 1.8 : 1.3 : 11.5 percent.

ner will usually have to change the figures several times before they portray the proper dry weight proportions. Subsequent corrections of initial estimates are sometimes necessary.

Of course, this method requires much more experience than the estimation of species magnitudes after BRAUN-BLANQUET and about four times the time. Moreover, the method can only be learned if the estimates are frequently checked by weighing or by experienced investigators. In spite of this, the method has gained considerable popularity in grassland investigations in Central Europe. We, therefore, discussed it in some detail here.

5.44 Estimating Vitality or Vigor.   When estimating cover-abundance or biomass percentages, one sometimes notices species with extremely poor growth or others with extremely good growth. (Poor and good growth, that is, as compared with the growth in the same species on other sites.) Such observations should be recorded, even if one does not wish to assign finer categories of growth-capacities as suggested by ZOLLER (1954) and others. Such a notation may give some information on the competitive status of a species in the plant community. It also may indicate the developmental trend of species in a community in comparison to other communities.

Following a suggestion of BRAUN-BLANQUET, vitality is conveniently shown by adding an index to the magnitude value, provided that the vigor deviates substantially from that normally shown. The following symbols are recommended:

°° very feeble and never fruiting (e.g., +°° or 2°°)

° feeble (e.g., 1°)

no index: normal

• exceptionally vigorous (e.g., 3•)

5.45 Estimating Sociability and Dispersion.   Two species with the same cover-abundance value may be quite differently distributed. For example, *Crepis biennis* usually grows singly, *Deschampsia caespitosa* in clumps and *Phragmites australis* in large stands. Following BRAUN-BLANQUET, the degrees of dispersion or sociability can be assigned by a scale with five values. These are added to the cover-abundance value (usually separated by a dot, e.g., 3 • 1 meaning species covers 25–50 percent of the area, but plants grow singly or solitarily):

5 = growing in large, almost pure population stands

4 = growing in small colonies or forming larger carpets

3 = forming small patches or cushions

2 = forming clumps or dense groups

1 = growing solitarily

However, it has been found that sociability is a characteristic property of a species in most instances. Therefore, it does not need to be recorded in normal cases. Sociability ratings have been omitted recently by many authors mainly for this reason, but also to make the tables more easily readable and to save space. We consider it satisfactory to record sociability only in cases of abnormal dispersion of a species. However, even in such cases it is often not necessary, since the species magnitude is then also low. Therefore, species magnitude and sociability are often designated by the same scale value.

Also, the distance between individuals of a species is often a characteristic property of that species. For example, *Androsace helvetica* always forms dense cushions. In contrast, *Phragmites* usually occurs in loose stands, which permit other species to grow between their shoots at least during spring. If one wants to take note of particularly dense growth, one can show this by underlining the sociability value (e.g., 3). Unusually widely spaced distributions can be designated by a dotted line (e.g., 3).

Measuring Species Quantities

6   **6.1   QUANTITATIVE VEGETATION PARAMETERS**

The more important measurable quantities in community sampling are:

1. Number of individuals or density (= abundance).
2. Frequency, the number of times a species is recorded in a given number of small quadrats or at a given number of sample points.
3. Cover, either of crown and shoot area or of basal area.

In addition, there are several other measurable quantities, such as height, stem diameter, and biomass. The latter is measured in volume (e.g., timber surveys) or obtained through cropping, and is usually expressed in fresh weight, dry weight, or gram calories per unit area.

Many other structural life form criteria are measurable, for example, leaf size, bark thickness, or current year's twig diameters. Also, functional parameters, such as leaf persistence, vegetative reproduction, and shade-tolerance can be subjected to quantitative analyses as has been demonstrated, for example, by KNIGHT and LOUCKS (1969).

Of great importance are the physiological parameters, such as transpiration rate, water potential, net assimilation rates, or other productivity parameters, such as litter production, seed production, annual diameter increment, etc. A widely useful measure related to cover and productivity is the leaf-area index (WALTER 1971). This is the ratio of the total leaf area in square meters (or other area unit) of a plant individual, species, or stand to the ground surface expressed in square meters (or other area unit). Only one side of the leaves is considered in the leaf-area index.

When speaking of quantitative ecology, these parameters definitely belong in the discussion. However, they are measured primarily for experimental rather than descriptive purposes. For the latter, only the first three parameters, listed above, are usually applied.

6.2   DENSITY MEASUREMENTS IN QUADRATS

This parameter relates to the counting of individuals per unit area. Counting is usually done in small quadrats placed several times into the community. Afterwards, the sum of the individuals per species is calculated for the total area sampled by the small quadrats, and the result is expressed in terms of species density per convenient area unit, such as a square meter, an acre (approx. 4000 m²)* or a hectare (10,000 m²).

Counting is perhaps the easiest analytical concept to grasp, but it often causes difficulties in application. One difficulty is the recognition of individuals. Trees and single-stemmed annuals present little difficulties, but nearly all other plant life forms do. In spreading shrubs, particularly where standing close together, it is often difficult to decide where one individual begins and another ends. This problem is aggravated in creeping shrubs or krummholz. Bunch-grasses or tufted fern fronds and caespitose or single-stemmed herbs are usually countable, particularly if their individual outlines are well shown. But many other perennial herbs, such as rhizomatous or stoloniferous forms, can

---

* Exactly 4046.85m².

hardly be counted accurately. In grasses it sometimes helps to cut off the shoots near ground level. One may then count individual stems more easily, but still may have difficulties in deciding whether a stem represents an individual plant or just an upright stem or branch coming from a rhizome. It is usually impossible to count individual mosses in a moss carpet.

In those cases, a decision must be made whether one can really count individuals or just parts of individuals. The latter may have little meaning. Although for certain experimental purposes, the number of shoots may be of greater significance than the number of individuals (STEBLER and SCHRÖTER 1887, SPATZ and MUELLER-DOMBOIS 1973). A count of some sort can often be established. It then gives the impression of great accuracy. But unless backed by a proper definition, it may be less accurate than a visual estimate of abundance. Therefore, application of counting as a sampling tool is limited by the plant life forms and their spacing, and accurate counting in all slightly difficult situations requires a good knowledge of plant life forms (see CHAP. 8).

A second difficulty is the marginal effect of the quadrat. The quadrat boundary may go through an individual and a decision has to be made whether to count it or to exclude it. This problem becomes aggravated by denser vegetation and smaller quadrats. The smaller the quadrat the greater the boundary in relation to the area, and the more frequently will a decision be required whether to count a marginal individual or not. Also here, an arbitrary definition helps to ease the problem, if one decides, for example, to include only those individuals that are rooted within the quadrat area; but even this may be difficult to decide.

The boundary problem in counting has been overcome to some extent by the plotless distance measures, such as the random pairs method and the point-centered quarter method (CHAP. 7). However, these are primarily applicable only to woody plants that are randomly distributed.

A third major difficulty is the time it takes to count herbaceous and shrubby individuals. Before investing time in counting, the purpose of the study must be very clear. Counting has particular value in assessing changes in studies of succession or changes caused by treatment in experiments. Also, for comparison of closely similar communities, counting of individuals may reveal important insights. For ordinary descriptive purposes, however, the time factor is often prohibitive, because the result conveys little more meaning than a less time-consuming abundance estimate.

6.21 Size of Density Quadrats. Quadrat size must be related to the size and spacing of the individuals, because counting of numerous individuals per species cannot be done accurately in large plots unless they are subdivided, or the individuals are marked off after each is enumerated. How many individuals (regardless of species) one may count accurately within a given quadrat is almost entirely a matter of judgment. Therefore, the quadrat size is not very important. However, for statistical analysis, a certain limitation is indicated (see discussion under sample size, SECTION 6.4).

In spite of the personal judgment involved in determining suitable sizes for density quadrats, the sizes usually vary within limits for each height stratum. Commonly used sizes are for the tree layer, 10×10 m quadrats; for all woody undergrowth up to 3 m height, 4×4 m quadrats; and for the herb layer 1×1 m quadrats (OOSTING 1956).

This decreasing range shows resemblance to the minimal area sizes stated in SECTION 5.2. But this is only because smaller plants usually occupy less space than larger plants. Otherwise, the two kinds of quadrats are for entirely different purposes: the minimal area quadrats for obtaining a representative combination of species, and the density quadrats for obtaining conveniently an accurate estimate of number of individuals per unit area.

6.22   Shape of Density Quadrats.   CLAPHAM (1932) and others (e.g., BORMANN 1953) have demonstrated that the shape of density quadrats also has an effect on the accuracy of the count. Rectangular shapes are more efficient than square or circular shapes, because of the general tendency of clumping in vegetation (GREIG-SMITH 1964).

BORMANN (1953) further qualified this phenomenon. He found that the reduction in variance associated with rectangular as opposed to square sampling units applies only if the long axis of the rectangular plot cuts across any banding in vegetation pattern. Such a perpendicular alignment apparently increases the variation within the sample unit, but it decreases the variation between them. Therefore, the sampling intensity per sample unit is increased and the variance between them is reduced.

This observation does not apply to minimal area plots. Their purpose lies only in sampling a near-complete species composition of the vegetation segment. Therefore, their shape is inconsequential. They may be square, circular, rectangular, or even irregular, if the segment demands such a shape.

6.3   FREQUENCY DETERMINATION

Frequency relates to the number of times a species occurs in a given number of repeatedly placed small sample plots or sample points. It is expressed as a fraction of the total, usually in percent. No counting is involved, just a record of species presence. Frequency is a much more readily established quantitative measure than either the counting of individuals or the measurement of cover.

Small plots or points may either be distributed randomly, for example, by throwing a metal ring, or systematically, by following a regular pattern (FIG. 6.1). In each placement, the species are recorded without regard to their quantity or number of individuals. For comparing different communities, frequency is best expressed as a percentage of the total number of placements, i.e., the so-called "frequency percentage" or "frequency index" (GLEASON 1920) is determined.

In our example, the frequency percentage of the species represented by dots is 84 percent (Fig. 6.1a) and 72 percent (Fig. 6.1b), and the frequency for the species shown by crosses is 24 and 20 percent, respectively. This difference can be decreased only by using a larger number of placements (e.g., 100), which requires much more time.

The determination of frequency was originally developed by RAUN-KIAER (1913). It was applied and further developed particularly by Scandinavian and Anglo-American investigators. BRAUN-BLANQUET and his students use frequency determinations only in special cases, as it requires much time in species-rich communities. The method proved to be very useful, however, in the species-poor communities of

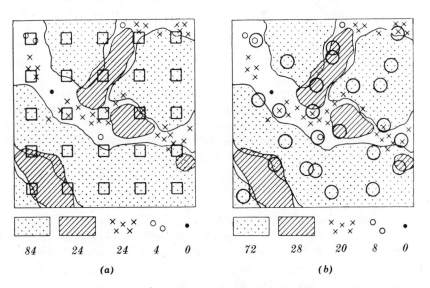

FIGURE 6.1. *Frequency determination referring to the sample quadrat or plot shown in FIGURE 5.4:* (a) *regular distribution of 25 frequency quadrats;* (b) *random distribution of 25 circular frequency plots.*

the North and in alpine areas, on intensively grazed meadows in the Netherlands (DE VRIES 1949), and in investigations of aquatic and marsh plant communities in Central Europe (TÜXEN and PREISING 1942). Frequency is the most commonly applied quantitative parameter for the analysis of forest undergrowth and herbaceous communities in North American descriptive studies.

6.31 Frequency in Quadrats—a Nonabsolute Measure. Frequency provides for an objective assessment similar to density and cover measurements, but, in contrast, it is a nonabsolute measure. This means that the result is in part a function of the size and shape of the quadrat frame. Depending on the species richness per unit area, a slight increase in frequency-frame size usually results in quite different frequency results for species of intermediate abundance. Since individuals of a species normally show concentrations within an otherwise homogeneous vegetation cover, a rectangular frame is likely to assess a somewhat different frequency than an equally sized square or circular frame. Because the results depend on the frame size and shape, they have meaning only in relation to a particular frame size and shape selected for a determination.

Frequency is often considered a measure of abundance. Therefore it should be related to density. However, GREIG-SMITH (1964) has clarified that frequency rarely gives an indication of number of individuals per species, because, for this to be true, individual plants must be regularly or randomly distributed. Instead, plants are usually contagiously distributed. Therefore, sociability or dispersion, i.e., local or small-area patterns enter into the measure of frequency. A species with a large number of individuals may show low frequency values simply because the individuals are concentrated in patches, whereas a species with the same number of individuals spread out evenly over the sample area may show 100 percent frequency. Therefore, frequency gives a certain indication of uniformity of distribution rather than of density. Of course, a species with few individuals can never show high frequency values, even if they are uniformly distributed, unless the quadrat size is very much enlarged. Therefore, frequency confounds the two parameters of density and dispersion.

Frequency gives little or no indication of cover when determined in frames or quadrats. A species with very small individuals evenly spread out over the sample area will give high frequency values, even though its cover may be insignificant. A species with few individuals but large crown or basal areas that cover a considerable portion of the sample area, will give low frequencies.

However, this depends on the criteria set for what plant part to include or exclude in the frequency count. RAUNKIAER's concept was to count a species, if the perennating bud of an individual of that species was inside the frame. This concept is difficult to apply in the humid tropics, where plants lack seasonal shoot reduction. In North America, the usual criterion is that the plant must be rooted inside the frame (CAIN and CASTRO 1959). In a creeping or matted life form the latter criterion results in lower frequency. Therefore, the perennating bud criterion takes cover somewhat into consideration. A stoloniferous matted grass may be included wherever it is rooted at the nodes. In this case, cover is included in the frequency count, rather than number of individuals. This shows that a number of subjective decisions must be made before one can assess frequency objectively.

Frequency can be made an absolute measure by eliminating the effect of quadrat or frame size. This is accomplished by reducing the quadrat to a point. The point may be a needle, sharpened rod, or a sighting cross made of wires or hairs (such as in rifle telescopes, for example). A needle lowered at predetermined points over a herbaceous cover will either miss or intercept a plant part at each lowering. This technique gives a record of presence or absence for the more abundant species composing the cover. Therefore, in this case frequency is used to measure cover. This method, generally known as the "point-quadrat" method, will be discussed among the methods of measuring cover (see SECTION 6.54).

Another way of determining frequency that does not involve quadrats or boundaries is provided by the Wisconsin distance methods (SECTION 7.6).

6.32 Frame Size for Frequency Determinations. Frame size is primarily a function of plant size and species richness per unit area. If, for example, 20 to 30 species occur in 1 m² of a low (10± cm) herbaceous cover, a subdivision of that frame into one hundred 10×10 cm (0.01 m²) frequency subquadrats appears suitable. A number of 3 to 8 species per frequency frame can be counted conveniently. Any larger number of species slows down the progress considerably.

Such a 100-square quadrat is shown in FIGURE 6.4 (SECTION 6.53). A record of the presence of species in such a grid of quadrats is known as "local frequency" (GREIG-SMITH 1964). Local frequency can also be analyzed in larger frames as long as they are placed into a contiguous matrix.

Since species counting depends on the ease of recognition at any one stage of development, quadrat sizes may vary within limits according to individual preferences of the investigator. In general they can be somewhat larger than quadrats for counting individuals, but the same size may be used for both purposes. Enlarging the frequency quadrat can have the effect of showing a 100 percent frequency for a sparsely represented species, which then would have the same frequency value as the abundantly represented species. DAUBENMIRE (1968) recommends, as an empirical rule, a reduction of the frequency frame size when more than one or two species show a frequency of 100 percent.

CAIN and CASTRO (1959) suggest the following empirical sizes:

Moss layer	0.01–0.1 m²
Herb layer	1– 2 m²
Low shrubs and tall herbs	4 m²
Tall shrubs	16 m²
Trees	100 m²

RAUNKIAER used a ring of 0.1 m² size for herbaceous and low undergrowth forest vegetation. Another commonly used unit is 1 m² of either 1×1 m or 0.5×2 m sides. Since the results depend on the frame size, the size must be uniform for comparisons. In the larger quadrats suggested by CAIN and CASTRO for tall shrubs and trees, frequency is usually only a by-product to the counting of individuals or the assessment of basal area or cover. To line out such large quadrats simply for frequency would, for most purposes, be a poor time investment.

6.33 Frequency and Minimal Area. This topic is discussed because a certain confusion exists in the literature, that is, how can one establish the minimal area from frequency data?

If one is interested in sampling a representative species composition in addition to a reasonably accurate quantitative assessment of the more dominant species, the species/area curve should not be ignored. To obtain an adequate species complement of a stand, the area covered by a number of small quadrats for quantitative parameters should approximately equal or exceed the same area in square meters as indicated by the minimal area (RICE and KELTING 1955).

CAIN (1943) experimented with frequency quadrats in an alpine fell-field vegetation and found that the accumulated minimal area obtained from placing several small, even-sized quadrats in a scattered arrangement was very much smaller than the minimal area established from successively enlarged nested plots. He found a minimal area of 32 m² through a successively enlarged, nested plot, while only four randomly placed 0.1 m² plots indicated a minimal area for that community. CAIN considered this small (0.4 m²) minimal area insufficient for a sample of frequency, because a frequency determination from only four quadrats can hardly be expected to yield adequate results. He suggested using twenty 0.1 m² plots, a total sample of 2 m². (The number twenty was an arbitrary choice.) CAIN concluded from his data that many small plots take less time and give a more adequate description of a stand than a single, large plot.

An inspection of CAIN's (1943) species/area curves shows that the 32 m² minimal area obtained from the nested plots contained 22 species, while the 0.4 m² minimal area obtained from the four 0.1 m² frames contained only 15 species. Therefore, CAIN's conclusion of a more "adequate" description can only relate to his satisfaction obtained by an objective and quantitative determination of frequency. But even 20 frame placements are considered too few for almost any community evaluation (GREIG-SMITH 1964).

An adequate description should include nearly all species found in the community. If a sampling scheme ignores one-third of the species composition of the stand, it cannot be used to define the minimal area. Therefore, CAIN's study shows that the minimal area of a community cannot be defined through a species/area curve obtained from random placements of small plots.

By using small quadrats in scattered formation, there is always a chance that species represented by only a few individuals (i.e., the rare ones) are not included in the sample. The inclusion of a rare species influences the shape of the species/area curve just as much as does the inclusion of an abundant species. Therefore, only a system of con- tiguous or nested plots will define the minimal area adequately. There is no theoretical reason why this cannot be done with small, even-sized quadrats. This would permit the determination of frequency at the same time. But these quadrats must then be placed in a contiguous matrix as is done in the nested plot technique. Thus, the minimal area can be established from frequency data only through what is known as a "local frequency" analysis. In CAIN's example, it would require 320 side-by-side placements of the 0.1 m² frame to cover the minimal area of 32 m². This would obviously be too time-consuming for the objectives at hand. The same area could be covered with thirty-two 1 m² frames. This would permit enumeration of an adequate (i.e., repre- sentative or near-complete) species sample of the community combined with a frequency determination of even greater accuracy than sug- gested by CAIN.

CURTIS and GREENE (1949) referred to CAIN's (1943) findings and restated that frequency samples should be larger than the minimal area. This statement becomes understandable only in view of a mini- mal area concept that ignores the less frequent (rare) species. How- ever, this interpretation is in disagreement with the original intent of the minimal area concept. It is not surprising, therefore, that some confusion has arisen about the concept and the meaning of the species/ area curve.

In trying to find an alternative to the species/area curve, RICE (1967) suggested eliminating altogether the dependency on this curve for finding a suitable sample size. His reason, also held by GOODALL and others, is that the ecologically most important aspect of plant distribution would be the distribution of the quantity of plant material (probably meaning biomass), rather than the distribution of individuals. This is a clear shift in objective, which has little to do with the value of the species/area curve as an indication of the representative species composition of a community. From a different perspective it can just as well be argued that the most important aspect of plant distribution is species diversity. For obtaining the smallest sample area with a maximum number of species of a community, the species/area curve is still the best tool. However, it cannot serve to define the sample size needed to evaluate adequately the number of individuals per species, their cover, or their frequency.

In conclusion it may therefore be well to restate that there are basically two types of sample quadrats: the large minimal area quadrat, which is used for the purpose of sampling a representative species composition in recurring plant assemblages; and the small quadrat adapted in size to height and spacing of species individuals, which is used for the quantitative analysis of individuals per species (or cover or frequency). The density quadrat should be small for the convenience of accurate counting, but not so small as to cause proportionately too high a personal error through the edge effect.

6.4 HOW TO DETERMINE SAMPLE SIZE

The sample size, which relates to the number of times a given density or frequency quadrat should be repeated, is often arbitrarily delimited. In timber volume surveys, it is common practice to set a percentage limit. For example, one may set a standard of 5 or 10 percent sampling intensity. This refers to the area that is covered by the vegetation segment. If the latter covers 6 hectares (15 acres), a 5 percent sample would extend over an area of $0.05 \times 60,000$ m^2 = 3000 m^2. This area could be sampled, for example, by thirty 10×10 m plots.

GREIG-SMITH (1965) emphasized that the accuracy of the count is not a function of the area sampled, but a function of the number of enumerations. This is related to the factor of spacing. Where the individuals are widely apart, far fewer are counted in the same size of plot than where the individuals are close together. Therefore, in stands with great differences in spacing, it seems advisable to use as many plots as are necessary to count a given number of individuals. As a result, the actual area sampled may vary considerably. Of course, the number of plots to be counted is a function of the variation of individuals between plots. The greater this variation, the more plots are needed.

6.41 A Statistical Approach. In probability-statistics one may use the ratio of the standard error of the mean to the mean as a measure of sample size (GREIG-SMITH 1964). For a POISSON distribution, this can be expressed as

$$\frac{\sqrt{x}}{\underline{n}} : \frac{x}{n} = \frac{\sqrt{x}}{\underline{n}} \times \frac{n}{\cancel{n}} = \frac{\sqrt{x}}{x} = \frac{1}{\sqrt{x}}$$

where x is the sum of the enumerations and n the number of quadrats. The ratio of standard error of the mean to the mean reads $1/\sqrt{x}$, and n is cancelled out. Where species individuals are randomly distributed, the accuracy of the density estimate is not affected by the size of the quadrats, only by the number of individuals counted. Such random distributions may be found in tree communities if one considers all tree individuals together as a group regardless of species. However, the ratio applies only where large numbers of sample quadrats or counts of individuals are involved. The larger the total count of individuals, the smaller this error term (FIG. 6.2). However, GREIG-SMITH (1964) points out that such statistical error terms cannot usually be applied to the individuals of single species, because the individuals of a species are rarely randomly distributed. Therefore, the curve is only a guide. The standard error of the mean as $\sqrt{x}/n$ is applicable only to a POISSON distribution, where the mean equals the variance. Where one cannot assume this, calculate the standard error of the mean (SEM) as the variance (s^2) divided by the number of samples (n), i.e., SEM= s^2/n, where s = standard deviation. An example is given in the next section.)

6.42 Plotting the Running Mean.

A practical guide to estimating adequacy of sample size for small density quadrats is to stop sampling at the point at which additional quadrats do not significantly affect the mean of the more important (or abundant) species. This can be tested by calculating and plotting a cumulative or running mean during the quadrat analysis (KERSHAW 1964).

In practice it is often satisfactory to set an arbitrary standard of sampling size by requiring that a sample be within 5 or 10 percent of a more time-consuming maximum sample.

The two ideas on sample size are closely related, and one may interpret the 5 percent limit as a nonsignificant variation from such a sample size curve (FIG. 6.3). For example, the density count of a small single-stem lichen, *Stereocaulon vulcani*, occurring on recent lava rock in Hawaii was made in eighteen 1 cm² quadrats. The cumulative or running mean was calculated for always two quadrats at a time, giving the results shown in TABLE 6.1.

The running mean values are plotted over the number of quadrats in FIGURE 6.3. After an initial greater variation, the curve becomes less variable already after the sixth's quadrat.

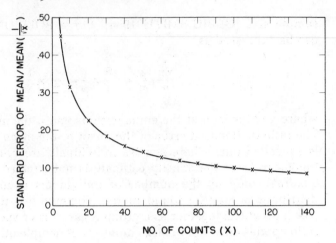

FIGURE 6.2. *Relation of standard error of mean/mean* $(1/\sqrt{x})$ *to number of counts among randomly dispersed individuals. Like the species/area curve this relationship is one of decreasing returns with increasing sample size.*

If we intend to stop sampling when the running mean shows insignificant variation, we might choose to stop after 8 quadrats were sampled. Here the mean was identical to that of 18 samples. In other words, a count of 104 individuals gave the same mean as a count of 233 individuals. In applying the criterion to stop sampling, when the running mean comes to within 5 percent of a more time-consuming maximum sample, we might stop counting after 6 quadrats. Five percent of the mean of the arbitrary maximum sample of 18 quadrats is $0.05 \times 13 = 0.65$. Thus, the mean of 6 quadrats of 12.7 lies within 13 ± 0.65. In the latter case, only 76 individuals were counted.

If the *Stereocaulon* individuals were randomly distributed, the error term $(1/\sqrt{x})$ for 72 individuals would be $1/\sqrt{72} = 0.12$ (FIG. 6.2). We may not wish to make this assumption, and instead calculate the error term (SEM/MEAN) by using the formula SEM $= s^2/n$ (where s = standard deviation). Applied to the first 6 quadrats of the *Stereocaulon* example, the calculation would be as follows:

Quadrat no. = n	Count x	x^2
1	13	169
2	15	225
3	11	121
4	9	81
5	15	225
6	13	169
Sum 6	76	990

$$\text{MEAN} = \frac{\text{sum } x}{\text{sum } n} = \frac{76}{6} = 12.67$$

$$\underbrace{\text{VARIANCE} = s^2} = \frac{\text{sum } x^2 - \dfrac{(\text{sum } x)^2}{n}}{(n-1)}$$

$$s^2 = \left(990 - \frac{76^2}{(6-1)}\right) = \frac{(990 - 962.67)}{5}$$

$$s^2 = \frac{27.33}{5} = 5.47$$

SEM = standard error of the mean

$$\text{SEM} = \frac{s^2}{n} = \frac{5.47}{6} = 0.91$$

$$\text{Ratio SEM/MEAN} = \frac{0.91}{12.67} = 0.07$$

The result shows that the error term (SEM/MEAN) for the count of 72 *Stereocaulon* individuals is even less than that expected for a random distribution. Therefore, the *Stereocaulon* individuals seem to approach a regular distribution.

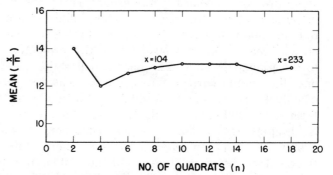

FIGURE 6.3.   *Plotting of running mean.*

TABLE 6.1   *Running Mean Number of* Stereocaulon *Stems from Each of Two Quadrats for 18 Random Quadrats of 1 cm².*

NUMBER OF *STEREOCAULON* STEMS PER CM²	CUMULATIVE TOTAL NUMBER	RUNNING MEAN
13, 15	28	14.0
11, 9	48	12.0
15, 13	76	12.7
13, 15	104	13.0
13, 15	132	13.2
13, 14	159	13.2
13, 12	184	13.1
11, 11	206	12.9
14, 13	233	13.0

It may be noted that the determination of an adequate sample size for quantitative analysis is handicapped by the same limitation as the determination of the minimal area through the species/area curve. In both cases there is no strictly objective criterion, and a decision has to be made by the investigator. A decision is always subjective, but it should be based on good judgment.

Moreover, the decision of what constitutes an adequate sample size is relatively easy for one-population stands as in the *Stereocaulon* example. A problem is introduced when there are more than one species to be quantitatively assessed. This is the case in almost all plant communities.

It is not always appreciated that in quantitative analyses it is possible only to evaluate the more abundant species with reasonable accuracy.

6.5   COVER MEASUREMENT

Usually cover is defined as the vertical projection of the crown or shoot area of a species to the ground surface expressed as a fraction or percent of a reference area. This definition applies to the estimation methods of cover as well as to the measurement. Instead of crown area, cover may also imply the projection of the basal area to the ground surface. The basal area is the area outline of a plant near the ground surface. In trees it is measured through the diameter, usually at breast height (dbh), i.e., 1.5 m above the ground, by the formula πr^2, where r equals ½ dbh. The breast height measurement is acceptable as a basal area measurement in most temperate tree stands. However, in tropical forests, where many species have distinct butt-flares or in stands with multistem trees, it is necessary to measure the diameters at the tree base, if one claims to have measured the basal area. Even in temperate forests breast height area is smaller than real basal area. The basal area concept is sometimes applied to caespitose life forms such as bunch grasses. Here it relates to the space occupied by the shoot system at ground level.

6.51   Ecological Significance of Cover. Cover as a measure of plant distribution has been emphasized as being of greater ecological significance than density (RICE 1967, DAUBENMIRE 1968). This idea is based on the observation that cover gives a better measure of plant biomass than does the number of individuals.

Plant biomass is the first and second order criterion for the structural classifications developed by FOSBERG (1961) and by UNESCO (ELLENBERG and MUELLER-DOMBOIS 1967a). The first structural divisions are based on spacing and height of the plant biomass. Plant biomass is an indication of the capacity of a vegetation to accumulate

organic material if something is known about the developmental status
of the community and its use as food supply for animals. Plant biomass
has a major influence on the stand climate in terms of light and temperature relations. It influences the water relations through rainfall
interception and transpiration rate per unit area, and it is closely related
to the volume of circulating nutrients in the ecosystem. Moreover, the
amount and characteristics of the plant biomass are of direct importance to the animals associated with the vegetation, because the plant
biomass provides their shelter and food.

Plant biomass is evaluated through cover only in conjunction with
a measure of depth or height. For descriptive purposes, this is accomplished by the stratification of a community into the various height
layers as discussed in CHAPTER 5. Therefore, cover must be evaluated
separately for each height layer or vegetation stratum.

Another great advantage of cover as a quantitative measure is that
nearly all plant life forms, from trees to mosses, can be evaluated by
the same parameter and thereby in comparable terms. This does not
apply to density or frequency. However, cover can be measured in
several ways, depending on the kind of vegetation and the objectives
of the study.

6.52   The Crown-Diameter Method.   A method for trees or bushes
analogous to the basal area measurement is as follows. A meter tape
is laid out on the ground from one side of the crown perimeter of the
tree or bush across the center to the other side of the crown perimeter.
This results in one diameter reading. Since crowns do not usually form
a perfect circle, it is necessary to run at least a second crown-diameter
measurement more or less perpendicular to the first one. The crown
cover (cc) is then obtained from the formula

$$cc = \left(\frac{D_1 + D_2}{4}\right)^2 \pi$$

where D_1 equals the first measured crown diameter and D_2 equals the
second measurement. The result can be expressed in square meters of
crown cover. To relate the crown-cover measurement to a unit of
ground area, it is necessary to measure such a unit of ground area as
well. However, this method is not practical where one is interested in
the cover by species over a larger area to obtain a more widely representative sample. In that case one can often use the line-intercept
method (SECTION 6.55).

6.53   The Quadrat-Charting Method.   In low herbaceous vegetation
(for example, in pastures) cover can sometimes be charted or mapped
from small quadrats. For example, in a square-meter quadrat or frame,
the outline of the crown area of certain species or their basal shoot
systems can often be drawn to scale on a sheet of paper. This can be

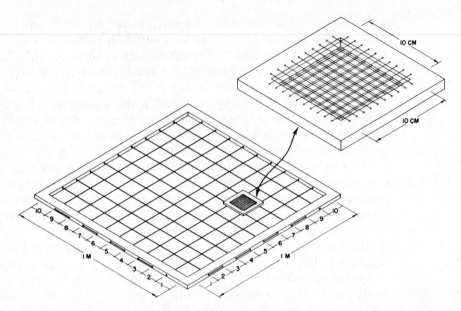

FIGURE 6.4. Square-meter quadrat with 100 dm subsquares for counting local frequency and for charting cover outline of herbaceous plants with solid shoot parts that show almost complete cover within their outlines. A separate decimeter quadrat with 100 subsquares strung from thin nylon thread can be superposed for point frequency measurements of cover. Further explanation in text.

done by subdividing the square meter into 100 square decimeters and by numbering the coordinates of the quadrat frame accordingly, from 1 to 10 (FIG. 6.4). Crown or basal shoot outlines and the area occupied by matted plant forms can thus be transferred quite accurately on a sheet of graph paper.

Where the herbaceous vegetation is taller (for example, in a bunch-grassland) such a 100-square quadrat, when lowered to the ground, may bend down many plants and thus distort or increase their real cover. In that case one can remove a few strings and instead use only twenty-five 2×2 dm squares. Where this still results in the bending and distorting of plants, one can reduce the string-grid to a cross, dividing the frame into four quarters, or one can leave out the string-grid altogether.

Of course, where the plants are so tall that a strong-grid in a square meter frame becomes an obstruction to its positioning on the ground, a 1 m² quadrat may be too small a sample area. In that case one can establish a 2×2 m map-quadrat by repositioning the square meter frame four times. Other combinations of sizes are, of course, possible as well.

It must be emphasized that the quadrat-charting method is primarily useful only for permanent quadrats, because mapping a quadrat is time-consuming. Studies of successional or seasonal changes of a herbaceous plant cover on exactly the same place are ideally done in such quadrats. A square-meter frame can easily be positioned by driving two pegs into the ground at diagonal points and recording the compass direction of the frame. It is wise to survey the pegs for easier relocating by measuring their distance and direction to a nearby landmark, such as a boulder, isolated tree, or cross-road.

Another way of charting plant cover in small quadrats is to use a pantograph. The pantograph system is illustrated in FIGURE 6.5.

Similarly, photographic records can be used, which are merely variations of the quadrat-charting method. WIMBUSH, BARROW, and COSTIN (1967) describe a photographic method of determining plant cover of a bunch-grass vegetation. Photographs were taken at 2 m height above the plant cover at prefixed points along a transect. But, photographs often suffer from unclear background, creating difficulty in subsequent interpretation of plant outlines on prints. This is not so where the tree crown canopy is photographed from below. EVANS and COOMBE (1959) made photographs from the ground upwards to interpret the structure of the crown canopy of forests. For this they used an especially adapted hemispherical lens (called the "fish eye" lens), which is excellent for wide-angle canopy photographs. Such photographs can be taken at prefixed points along a transect. The area-distortion on the prints is easily adjusted and the cover can be calculated from the prints by the point-intercept method by using a transparent dot-grid (SECTION 6.54).

The quadrat-charting method can, of course, also be used to map the position of individual plants whose cover is insignificant. Thus, the method is a universal technique for mapping herbaceous or other small plants on small areas.

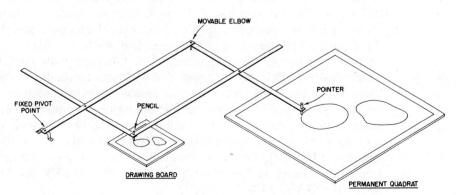

FIGURE 6.5. Pantograph system.

The same method can also be used for trees and other woody plants. But then the quadrat has to be larger, for example, 10×10 m. In that case the quadrat is outlined with string, and a grid of subquadrats can be strung out to convenience. Such larger map-quadrats may have some use in permanent tree plots, but their use is very much more restricted than that of the small chart-quadrats for herbaceous vegetation. The reason is that trees and bushes are long-lived perennial plants on which more detailed periodic measurements are more useful than cover outline and position. The standard detailed measurements on trees include periodic diameter and height measurements.

As a measurement of plant cover, the quadrat-charting method has three major limitations.

1.   The cover of plants is merely shown diagrammatically. It still has to be measured.

2.   The method is limited to plants whose shoot-outline covers nearly 100 percent. It cannot be used for a mixed plant cover, where the shoot systems are heavily intermingled.

3.   The method cannot be used very well to obtain a representative sample of plant species cover over a larger area. It is useful only for a detailed analysis of a small area, i.e., for permanent quadrats.

These three limitations are overcome by the point-intercept method.

6.54   The Point-Intercept Method.   When the cover outline of plants has been drawn to scale from the 100-square quadrat in the field to a sheet of graph paper with 100 squares, the plant cover is easily evaluated by counting filled squares and fractions of squares. This evaluation can, of course, be done directly in the field-quadrat without first preparing a map-diagram.

However, when we count fractions of squares, we are, strictly speaking, still estimating cover, instead of measuring it. A common method to measure cover or area outline on maps is to use a planimeter, which consists of a small wheel with a counter that converts the measure of circumference to area. A much simpler method is to reduce each of the small quadrats to a central point and to count the points that intercept a plant part. The method can be applied by using a transparent dot-grid where each dot represents the area of a small quadrat. This is the most widely used method for measuring small areas on maps, the cover of fungal or algal colonies under a microscope, areas of individual leaves, or areas under a curve. The same principle, namely that of reducing a number of small quadrats each to a point, can be applied directly in the field to the 100-square quadrat (FIG. 6.4). For this reason the method was introduced for vegetation studies as the "point-quadrat" method (LEVY and MADDEN 1933, GOODALL 1952, 1953c).

6.54.1   Application to Herbaceous Cover.   When we lay out the square meter quadrat of FIGURE 6.4 with its 100 dm subsquares on a low herbaceous cover such as a pasture or cut lawn, we could perhaps simply use the cross-points of the string-grid to determine whether or not the vertical projection of a point intercepts with a plant shoot part and to what species each intercepted part belongs.

Apart from the fact that we would then have only 81 cross-points, this method would soon run into difficulties. The main difficulty is that of parallax. By viewing each point formed by the string-grid, two people would probably obtain quite different results. Without a second cross-grid it is possible to aim into various directions under each cross-point. Therefore, it is necessary that a second layer of points is established that matches the first one as shown on the small square-decimeter quadrat on FIGURE 6.4. With such a double grid of cross-points the possibility of aiming at different points under a double-sighting cross is very much reduced.

It would probably be sufficient for most pastures or short-grasslands to reduce each of the 100 dm squares to a single point and to then establish the plant-intercepting points out of 100. As a tool, one could use the small decimeter double-sighting cross by placing it in sequences on each of the 100 subsquares of the square meter frame and by reading off whatever intercepts the central-sighting cross below. To avoid confusion, one can restring the small frame with only one pair of matching crosses in the center or on the side. The position of the double-sighting cross is immaterial. One can also use a knife blade, stick it into the ground and read off the interception, if there is any, at the sharp edge. Such a method, using a bayonet, has been described by POISSONET and POISSONET (1969) for tall-grassland, whereby the bayonet was stuck every 20 cm into the ground along transects. In the decimeter quadrats of the square meter frame this method would not eliminate the possibility of wishful aiming, unless the knife position is prefixed by, for example, the use of a ruler.

The sampling intensity as described by the distance and total number of points is always a question of plant size, variability of species pattern, and objective of the analysis in relation to the time available. The small 100-point frame shown in FIGURE 6.4 was used by one of the authors on species-rich tropical monsoon pastures in Ceylon to measure recovery of the short-grass cover in permanent square meter quadrats following scalping of the grass sod by elephants (MUELLER-DOMBOIS and COORAY 1968).

In the cover analysis, the small 100-point frame was placed on usually six decimeter squares that were each determined randomly from the set of the two coordinates or by stratification from a map diagram. In this way the total cover-sample was 600 points and the percent grass cover in each permanent quadrat was determined for intervals of two months. The small 100-point frame was used only to determine the changing grass cover. Species were not distinguished.

It must be emphasized that the few tools described here have rather limited application, but the point-intercept principle has a very wide application, which extends from microscopic plant life forms (as already mentioned), to trees, and to including the BITTERLICH method for measuring stem cover (SECTION 7.5).

A useful tool for normally sized (20–50 cm tall) herbaceous or dwarf-shrub vegetation, such as meadows or tall-grass, and heath vegetation, including the herbaceous undergrowth layer in many forest stands, is a point-frequency frame such as shown in FIGURE 6.6. The frame shown here, made from wood, is 1 m high and 1 m long. Ten wire pins, or steel rods of the same length as the legs, are slid through holes. These guide holes are bored perpendicular through the two pieces of horizontally fixed lath. The second lath strip eliminates the parallax effect (as does the second cross-grid layer of the small square-decimeter cover frame in FIGURE 6.4). The ten guide holes with their pins are spaced at equal intervals along the linear frame. But the dimensions of the apparatus can be changed to fit the height and spacing of the plants.

The linear frame is mounted with its legs over the strip of herbaceous vegetation to be measured, and the pins are lowered vertically one after the other. Their hits on plant parts are usually recorded by species. Ten placings of the frame result in a record out of 100 sample points. This gives a measure of percent cover for the species that are intercepted by pins. Crown or shoot cover is measured by counting only the first interception or initial touch of each needle with a shoot part; basal area is measured by counting only the hits occurring with stem parts at ground level. The mean height of the plant cover can be determined simultaneously by recording the needle length at each first interception.

For measuring cover, the frame should be held vertically, not obliquely. This requirement becomes even more important if one intends to measure "cover repetition," which was defined by GOODALL (1952) as the number of times a given needle intercepts a plant part when lowered in the same vertical position. If one were to hold the frame obliquely, the number of interceptions would likely be increased, resulting in an overestimate. An estimate of cover repetition may be converted into a measure of yield. Recently, POISSONET (1971) and DAGET and POISSONET (1971) have used the point-frequency method to determine biomass from all interceptions occurring at a point when the wire pin is lowered vertically. Such correlations between number of intercepts per point and biomass must be worked out empirically for each vegetation type, as the relations between volume of standing crop and weight vary from species to species, from place to place, and from time to time.

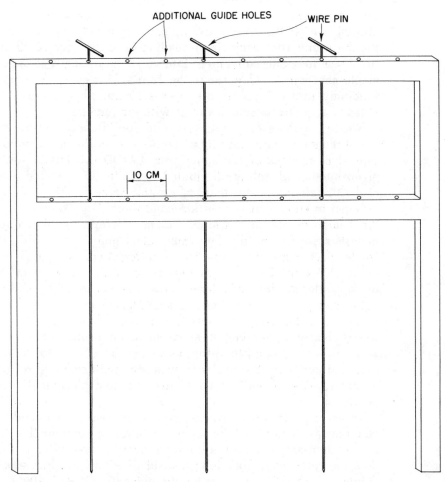

ADDITIONAL GUIDE HOLES   WIRE PIN

10 CM

FIGURE 6.6. Point-frequency frame. Dimensions of frame and number of guide holes or wire pin intervals are adjusted to size and spacing of herbaceous plants. Further explanation in text.

The frame can be held obliquely (at a 45° angle) for easier overviewing, if one is interested merely in the determination of "relative cover," i.e., the percent cover contribution of each species to the total plant cover. In this case, the slight overestimate of absolute cover per species resulting from the probability of increased hits when needles intercept sideways, is of little consequence (GOODALL 1952).

Instead of ten pins, one can also use the same pin for ten readings, but movement of the pin or rod among guide holes slows down the progress.

Where a larger continuous area sample of a herb cover is desirable, one can simply increase the distance between the points as demonstrated in FIGURE 6.6. For example, a single rod may be used by moving the frame after each single reading for a distance of 30 cm, 50 cm, or 100 cm (or other convenient interval). Theoretically, one could use a single sharpened rod without the frame. However, systematically or randomly prefixed points are necessary and this requires the guide holes to avoid unconscious aiming with the rod end.

Narrow spacing of points is useful for intensive, small-area cover measurements. Wide spacing of points serves for more rapid, but less intensive, surveys of the same area. GOODALL (1952) found that a given number of points distributed individually can give a more precise estimate of cover than when the same number of points are grouped in frames under certain conditions. For example, a few scattered large plants, or scattered clumps of plants of the same species, may be recorded in only few frame placements. But within these few frame placements these plants are intercepted by a high number of points. This results in an overestimate of the cover contribution of these scattered clumped plants. This bias can only be eliminated by either increasing the number of frame placements on the same area, or by reducing the grouped points to individual points. The latter alternative is clearly more efficient, since the same accuracy can be obtained with fewer points and therefore less work. This leads to the conclusion that continuous transects of evenly spaced points should result in more accurate measurements of cover than random placements of grouped points.

Transects with evenly spaced points are also recommended, if periodic remeasurements of the same vegetation is the objective. The accuracy in assessing changes in the plant cover is greatly increased, if one uses the same point locations instead of a new random allocation of points at each subsequent remeasurement (GOODALL 1952).

The number of sample points to be recorded should be related to the variability of the cover. The adequacy of cover sampling is governed by the same principles as discussed for density counts in SECTION 5.55. An arbitrary number of 200 points may give satisfactory results in a relatively homogeneous plant cover. Depending on the familiarity with the species, such a record usually takes only about half an hour.

GOODALL (1952) produced data showing the great effect that the diameter of the pin has on the accuracy of the result. For example, with respect to the grass species *Ammophila arenaria,* a pin diameter of 4.75 mm resulted in 71 percent cover, a pin diameter of 1.84 mm in 66.5 percent cover, and a pin diameter reduced to a point with practically no diameter gave only 39 percent cover.

Therefore, the results are only relative and depend on the diameter at the tip of the pin. This is analogous to the frequency determinations in quadrats, where the result depends on the quadrat size. The closest approximation to an absolute measure of cover is achieved with a rod or pin sharpened to a point. Cover measurement with a sharpened rod does not take more time than measurement with a blunt-tipped one. Theoretically, the double-sighting cross method should give the most accurate results. WINKWORTH and GOODALL (1962) described a crosswire sighting tube made of brass, which has been used successfully in Australian tussock grass covers. The tube of 20 cm length and about 5 cm diameter is equipped with rings that hold a fine crosswire at each end. The tube can be held by hand or mounted on a tripod. The crosswires provide for a practically dimensionless point-sample. Of course, such a tool cannot be used to measure "cover repetition" or biomass.

6.54.2   Application to Tree Cover.   The point-intercept method is not restricted to application in herbaceous vegetation. A tree canopy cover can be evaluated by the same principle of counting intercepting points. A simple device that has often been used in forest ecological work is the so-called "moosehorn" crown closure estimator (GARRISON 1949, FIG. 6.7). This is a simple boxlike periscope, which in its bottom part contains a mirror that is fixed at a 45° angle. The top of the periscope is equipped with a glass plate having a grid of 25 dots. The observer views the mirror through a peephole by holding the periscope upright and fixed on a Jacob's staff in front of one eye. The instrument has to be completely levelled, which is done by a 2-way level inside the periscope. The number of dots that are intercepted by a portion of the canopy can then be counted. This is repeated at a certain number of predetermined stations in the stand. Crown canopy photographs taken with the "fisheye" lens are evaluated by the same principle. But the photographs include a wider area per station and can be evaluated more accurately. Such point-intercept analyses are particularly useful if one is interested in seasonal variations of the crown cover. Periodic remeasurements should then be made from the same sampling stations.

   There are various modifications of this method. Recently, MORRISON and YARRANTON (1970) converted a rifle telescope into a point sampler by attaching a right-angle prism that permits reading straight downwards and upwards. The rifle telescope is mounted on a supporting stand consisting of a 3 m long aluminum beam that is held up horizontally by a pair of adjustable legs. The telescope is set up in convenient viewing height at about 1.6 m above the ground. It can be moved horizontally along the beam to any position. Therefore, ten or any other number of positions may be chosen at random or systematically along the beam at each setting of the apparatus. The cross hair in the telescope permits sampling one point at a time. The right-angle prism can be turned around by 180° to read the canopy cover per cent (by species)

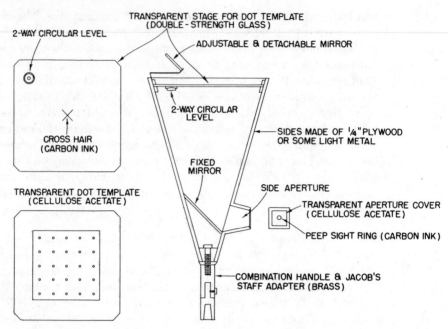

FIGURE 6.7. Sectional view and parts of the "moosehorn" crown closure estimator. (Reproduced with permission from the Journal of Forestry.)

on the same location as the undergrowth vegetation. This appears to be a promising method.

A major disadvantage of the point-intercept method in application to the tree cover is that the height or depth of the crown cover cannot be assessed. Yet, cover layering among trees is ecologically very important. The depth of cover can indirectly be evaluated by measuring light (BUELL and CANTLON 1950, ELLENBERG 1939, EBER 1972). However, light measurements are usually done for different objectives. They are time-consuming and do not offer a proper alternative, because they cannot be expressed in percent cover by species.

This disadvantage does not apply to the line-intercept method, by which the cover of woody plants can be assessed separately for more than one height stratum.

6.55 The Line-Intercept Method. This method for measuring cover was described by CANFIELD (1941). The line-intercept method is based on the principle of reducing the belt-transect, which has two dimensions of length and width, to a line with only one dimension, namely length. A meter tape is laid out on the ground and the crowns that overlap or intercept the line are recorded by species to the nearest 10 cm or whatever accuracy can be recorded conscientiously.

The line-intercept method can only be applied to plants with rather solid, almost 100 percent crown cover or relatively large basal areas. Among herbaceous or low plants these are the same that can be charted from quadrats. But in contrast to the quadrat-charting method, the line-intercept technique is more useful where a cover assessment of a larger area is required. Moreover, the method has a particular advantage for measuring the crown cover of woody plants, shrubs and trees.

Where crowns overlap in layered vegetation, the cover should be measured for each height layer separately. The layers or strata can be defined arbitrarily. In a low-stature forest, convenient strata may be from 0.5 to 2 m, from 2 to 5 m, and from 5 m up. Crown outlines of trees much taller than about 15 m are difficult to assess accurately without a special sighting tool. The upper size limit depends on visibility and the ease of making a vertical projection from the crown outline down to the underlying tape. The accuracy of the method depends largely on the accuracy of the vertical projection. An essential tool in such woody vegetation is a long, thin rod, about 3 m in length. This rod is used to obtain a projection of the crown edge to the tape by holding the rod vertically from the tape to the crown edge.

BORMANN and BUELL (1964) measured trees up to 32 m tall by the line-intercept method. For the vertical projection they used a "cover-sight," described by BUELL and CANTLON (1950). This is the same instrument as the "moosehorn" crown closure estimator described earlier. Except, a single cross-hair was used instead of a grid of dots, and a second cross-hair was mounted in the periscope to eliminate parallax. Furthermore, to facilitate obtaining a true vertical projection of crown perimeters, a plumb bob was hung inside the periscope. LINDSEY (1955) used what he described as a "sighting-level" for locating the vertical projection of the crown outline to the tape. The sighting level is a 5-ft. long stick with a screw mounted into the top end and a carpenter's level mounted at one foot from the lower end of the stick. The carpenter's level is attached in such a way that it can be used to control the stick position in one vertical direction. The stick is held by the observer so that the carpenter's level comes to eye-level position. The top end with the screw is then tilted at a 45° angle towards the observer and the screw is brought in line with the crown perimeter, while the bubble must remain centered. The observer then makes a 90° turn and again brings the three points, crown perimeter, screw and bubble in line by moving to the point along the tape directly beneath the crown outline. The two aiming positions eliminate the need of viewing straight up and down the stick, which would become complicated as a second bubble is required and must be read simultaneously with the first to insure an absolutely vertical position of the stick.

The accumulated length occupied by any one species out of the total meter tape length used for the sample is expressed as the percent cover for that species. The length of tape to be measured depends on the variation in the vegetation segment. However, usually the sample size is limited arbitrarily.

A second, even more important source of error than the relative accuracy of vertical projection is the crown outline itself. As stated before, the method is strictly applicable only to plants with almost 100 percent crown density that at the same time also have a solid or continuous crown outline. If a tree has bushy branches that reach across the line with gaps in between, the gaps should be excluded from measurement for greater precision. But there are many situations where this is difficult to do. Moreover, small within-plant gaps may be ecologically insignificant, and it may be more meaningful to ignore small gaps. DAUBENMIRE (1968) has come out in favor of "rounding out" canopy edges and "filling in" internal gaps on the argument that these gaps may be part of the ecological territory of an individual. Such a decision requires some prior knowledge of the ecological behavior of the species on the part of the investigator. As long as his judgment is consistent and the reasons for his judgment are explained, the method can still be considered objective and reproducible.

The same problem applies to the density of foliage. Where leaves are shed from certain branches, it may be equally valid to ignore such branches as to include them. Here, the judgment of what to include or exclude requires some knowledge of seasonal behavior of the tree species in question. If these two points about crown density and outline are well understood and taken into consideration, the method can be applied to woody vegetations that do not conform to the ideal of solidly covering crown outlines.

The line-intercept method has also been applied to counting individual plants, in conjunction with their measurement of cover (BUELL and CANTLON 1950). However, tree density is more conveniently assessed by other methods (see CHAP. 7).

The Count-Plot Method and Plotless Sampling Techniques

7

7.1 DIFFERENCE BETWEEN COUNT-PLOT AND RELEVÉ METHODS

We describe a plot as any two-dimensional sample area of any size. This includes quadrats, rectangular plots, circular plots and belt-transects (which are merely very long rectangular plots). Belt-transects are often simply called strips or transects.

The count-plot method consists in its simplest form of outlining a sample area in a tree stand and then counting all trees by species in size classes. Thus, a plot is essentially a density-quadrat. Among North American ecologists the count-plot method is well known as the "quadrat method." To the continental European ecologist, however, the term quadrat method would rather imply the releve' method, which is based on a minimal area quadrat as described before. The two concepts are very different, although both can be combined in the analysis of forest communities. Their difference is related to a basic difference in the major vegetation analysis problems that evolved on the two continents.

In continental Europe, the number of indigenous tree species is relatively small and in most European forests the few tree species present are planted. In such plantation stands, the attention was channelled to the undergrowth vegetation, and the undergrowth vegetation was intensively studied as to its response to spatial environmental variations. Therefore, in continental Europe, the undergrowth vegetation in forests presented the greater analytical challenge, and analysis problems of the tree layer were left primarily to the forester.

In contrast, in North America, particularly in Eastern North America, where several of the ecological tree analysis techniques arose, the number of tree species is much greater than in Europe. Here, the quantitative tree analysis techniques were developed from standard timber survey methods, because the natural distribution and diversity of tree species and the stand structure presented the greater analytical challenge. We use "stand structure" here to mean the numerical distribution of differently sized individuals within each tree species of a given stand. Since size in woody plants is related to age, it is possible in many cases to make predictions of stand development from such structural analyses. This form of analysis has always been a preoccupation of North American vegetation ecologists, because of their greater interest in the time changes or dynamics of vegetation over large areas. In contrast, the continental European vegetation ecologist's main interest was always in the small area spatial environmental variations as indicated primarily by herbaceous plants.

7.2   TIMBER SURVEY METHODS

In forest inventory work, systematic sampling is often done by strips or transects which permit continuous sampling within a specified strip-width. The strip-width depends on the size of the trees and their spacing. It must be possible to count the trees conveniently. Therefore, the strip-width will usually vary within limits of 1–5 m to either side of the center line.

The standard records taken are an enumeration by species within diameter classes above a predetermined minimum diameter, starting usually at 1 or 4 inches (2.5 or 10 cm) at breast height (i.e., 1.5 m above the ground). The minimum diameter is arbitrarily determined. Trees

with diameters less than the arbitrary lower limit (saplings and seedlings) are usually enumerated in 1 ft (30 cm) height classes. Where
these smaller trees are densely stocked, they are counted in smaller
subplots. The enumeration by species in relation to strip-width and
strip-length allows calculation of the density (number) of each species
per unit area. The diameter class record provides for subdividing the
density estimate per species by size classes. This information can be
utilized for a structural analysis, which may indicate the trend of development of the tree populations in the community. At the same time
the diameter record permits the conversion to another important
measure, basal area (ba), which is the actual space covered by the tree
stem. This is obtained through the well-known formula, $ba = (\frac{1}{2} d)^2 \times \pi$,
where d stands for diameter.

In North American ecological studies it has become customary to
use tree basal area (stem cover) as an estimate of dominance (CURTIS,
1959). In forestry, however, height is used as an estimate of dominance
and basal area as the basic value for timber volume calculations.

A modification of the strip-width method is the circular plot method,
in which small sample plots are placed at predetermined intervals
along the transect. The intention is to spread the sampling grid across
the segment, wherever the segment is too large for sampling in continuous strips. The sampling intensity will thus be reduced, but the
distribution across the entire stratum is maintained. The size of the
circular plots should be a function of the size and spacing of the trees
to permit accurate enumeration. But the diameter of the circular plot
can be roughly twice the strip-width in the same vegetation, because
the circular plot is usually quartered for easier counting. Such subdividing facilitates keeping track of the tree tally.

In sloping terrain, plot sizes are usually slightly enlarged to allow for
relating the quantitative information to areas on maps, which are horizontal projections. Slope corrections are applied by obtaining the slope
in degrees with an Abney level or other suitable instrument and then by
multiplying the downslope distance of the plot with the secant of the
slope. For example, on a 15° slope, a 10 m long plot length would be
enlarged to $10 \times 1.035 = 10.35$ m. However, where mapping of quantitative data is not the objective, slope corrections should also not be
applied, because such corrections result in an overestimate of the
quantitative parameters on slopes as compared to those on level ground.

The plot boundaries must be accurately located wherever counting
is involved. In contrast, where species quantities are estimated as in
the releve´ method, boundary accuracy is not so critical. To obtain an
accurate right angle in quadratic or rectangular plots, it is useful to use
the Pythagorean principle. For example, from the plot corner, a 4 m

long line may be established in the first direction. Then a 3 m long line is established perpendicular to the first line. The right angle formed by these two lines is then checked by measuring a distance of 5 m between the 4 m and 3 m points of the two plot sides.

Transects or strips, circular plots, rectangular or quadratic plots, all have one important criterion in common. They are two-dimensional area sampling units with specified boundaries that must be laid out in the stand.

7.3   EXAMPLE OF A COUNT-PLOT ANALYSIS

The following example of a tree density, structural and basal area analysis illustrates records typical of the plot or quadrat method.

The quantitative plot analysis relates to a small-area-sample (120 m²) in a tropical rain forest on the Hawaiian Islands.

TABLE 7.1 shows the raw field data, which was recorded by three students in about one hour. All individual trees were measured at their base with a caliper and called out by species in 5 cm diameter classes, using the class-limits or ranges as shown. Basal diameter rather than diameter at breast height was used because many trees were multi-stemmed, branching near the base below breast height. Moreover, basal diameter is the best measure of the true basal area. The problem of where to measure the diameter is, of course, different if one wants to establish the volume of the trees. This was not the objective in this analysis. The record was made in a 6 m wide belt-transect of 20 m length. The requirement for counting individual trees was arbitrarily determined to be a minimum of between 45 to 50 individuals in two general size classes, trees under 2 m tall and trees over 2 m tall. The enumeration was done in 3×5 m=15 m² subplots, one subplot at a time. When the eighth subplot was done, the total enumeration resulted in 48 trees under 2 m tall and 45 over 2 m in stem-height. With this the current sampling objective was accomplished.

The small trees up to 2 m tall, which all had basal diameters of less than 3 cm, were additionally enumerated in five height classes (TABLE 7.2) to analyze the tree-reproduction in more detail.

7.31   Interpretation of Stand Structure.   One objective of the quantitative plot method is to analyze and interpret the trend of numbers of individuals in size classes of the tree species in the stand. However, a trend can only be established when a sufficiently large number of individuals has been recorded; arbitrarily we may say a minimum of 30

TABLE 7.1.   *Example of a Stand-Structure Analysis by the Plot Method. Tropical Rain Forest on Tantalus Mt., Honolulu, Hawaii, at 420 m Elevation. Enumeration in 6×20 m Belt-Transect (120 m²). Raw Data.*

DIAMETER AT BASE CLASS (CM)	RANGE (CM)	NUMBER OF TREES IN SPECIES				
		ACACIA KOA	METROSI- DEROS COLLINA	PSIDIUM GUAJAVA	CITHAR- EXYLUM CAUDATUM	APPROX- IMATE HEIGHTS
1	(0– 2)	0	0	25	23	< 2 m
5	(3– 7)	0	0	19	3⎫	
10	(8–12)	0	1	10	0⎬	2–5 m
15	(13–17)	1	0	4	0⎭	
20	(18–22)	0	0	0	0⎫	
25	(23–27)	1	2	0	0⎪	
30	(28–32)	0	0	1	0⎪	
35	(33–37)	0	0	0	0⎬	5–10 m
40	(38–42)	1	1	0	0⎪	
75	(73–77)	1	0	0	0⎭	
Total		4	4	59	26	93
Total[a]	>3 cm	4	4	34	3	45

[a] Number of trees with over 3 cm basal diameter on 100 m² = 45/1.2 = 37.5. (To obtain an estimate of number of trees per acre, multiply by 40; to obtain an estimate of number of trees per hectare, multiply by 100.)

TABLE 7.2.   *Tree Reproduction <2 m Stem Height in 50 cm Height Classes on Same Area (120 m²) as Stand on TABLE 7.1.*

HEIGHT CLASS	RANGE (CM)	NUMBER OF STEMS IN SPECIES	
		PSIDIUM GUAJAVA[a]	CITHAREXYLUM CAUDATUM
1	< 10	0	0
2	11– 50	7	11
3	51–100	6	4
4	101–150	9	5
5	151–200	3	3
Total		25	23

[a] *Psidium guajava* here had only vegetative reproduction; all from root sprouts.

individuals per species (using FIG. 6.2 as a guide). Therefore, only
the numerical distribution among the size classes of *Psidium guajava*
can be considered as giving a reliable trend for this example (TABLE
7.1). For an adequate numerical trend in *Acacia koa* and *Metrosideros
collina*, the sample area would need to be about 8 to 10 times as large.
Or, 8 to 10 such plots would need to be established and analyzed from
this stand to present a reasonably reliable developmental trend for
these two species.

A brief interpretation of TABLE 7.1 is as follows.

The native *Acacia koa* (Leguminosae) occurs with only four individuals on the 120 m² plot. These four show a wide range of diameters
(from 15 to 75 cm) and therefore can be assumed to be of different ages.
(The tropical rain forest trees do not show annual rings, thus there is
no easy way to determine their ages.) The size-distribution of *Acacia
koa* indicates that the species has maintained itself over a period of
time. Its occurrence is not related merely to one event in time, when
conditions were favorable for reproduction. If that were so, one would
expect the four individuals to be concentrated in one or two size
classes. However, the continued maintenance of *Acacia koa* is questionable from this analysis, because there was no reproduction in the
stand (trees under 2 m).

The same interpretation can be made for the second native tree
species, *Metrosideros collina* (Myrtaceae).

The exotic *Psidium guajava* (Myrtaceae) is present with one tall,
mature individual (in the 5 to 10 m layer), 14 subcanopy trees (up to
5 m tall), 19 saplings (just over 2 m tall) and 25 suckers (here defined
as reproduction, under 2 m tall, TABLE 7.2). The number-trend indicates that this exotic tree species is well established in this rain forest,
and that it is maintaining its position by abundant reproduction. It is
possible that the quantitative importance of *Psidium guajava* may even
increase in the future and that its vigorous reproduction may be a factor
that contributes to the absence of reproduction among the two native
tree species. However, this is merely an indication obtained from this
analysis. Several more plot analyses and, perhaps, experimental research is needed to verify this indication of competitive replacement.

Citharexylum caudatum (Verbenaceae) is represented only by small
trees. Most are under 2 m tall. This species is a recent invader as shown
by its concentration of numbers in the reproduction class, the largest
number of individuals are from 11 to 50 cm tall (TABLE 7.2). Currently,
there were no recently germinated seedlings under 10 cm size. Nevertheless the species seems to establish itself as the second quantitatively
important exotic tree component in this stand.

7.32   Density and Dominance Relations.   A second objective of the plot method is to establish quantitatively the density and dominance relations among the tree species of the stand.

The density relations were already shown on TABLE 7.1. From this it is clear that the two exotic tree species (*Psidium guajava* and *Citharexylum caudatum*) are far more abundant than the two native tree species (*Acacia koa* and *Metrosideros collina*). In TABLE 7.1 it is easy to determine the number of trees by species for any convenient unit of reference area. However, when converting a tree count of a sample area to an acre or hectare, one should be aware that this is merely an estimate. The estimate can be strengthened by increasing the sample size, i.e., the number of plots. How many plots one should use to obtain a reliable estimate per acre or hectare can be determined through the "running mean" (see SECTION 6.42).

According to convention among North American vegetation ecologists, dominance for trees is usually defined as stem cover, and stem cover is the same as basal area. TABLE 7.3 shows the basal-area calculation for the plot-example of TABLE 7.1.

TABLE 7.3 shows that *Acacia koa* is by far the most dominant tree in this rain forest stand. *Psidium guajava* and *Metrosideros collina* are of about equal secondary dominance, and *Citharexylum caudatum* shows only a minor quantitative importance with respect to this parameter. Thus, the density and dominance relations are very different in this stand.

The quantitative plot method can, of course, also include measurements of the undergrowth vegetation. How this is done in standard quantitative field analyses is discussed in SECTION 7.7.

The same stand served for the example of a releve' analysis (TABLE 5.2). A comparison of the two kinds of analyses shows that their information contents are very different.

7.4   PLOTLESS SAMPLING TECHNIQUES

In both the releve' and quantitative plot methods, the basic sampling unit is a two-dimensional reference area. Plotless sampling means sampling without such a prescribed area unit. Plotless methods are available for all three commonly used quantitative parameters:

1. *Frequency.* As we have discussed already, when a frequency frame or sampling quadrat is reduced to a dimensionless point, frequency becomes an absolute measure. The result of such point-sampling is expressed in percent of hits or interceptions. When the number of

TABLE 7.3   Total and Mean Basal Area (cm²) for Each Tree Species on 120 m²
(calculated from TABLE 7.1). Reproduction Ignored.

DIAMETER CLASS (CM)	BASAL AREA (CM²)	NUMBER OF TREES × BASAL AREA[a]			
		ACACIA KOA	METROSIDEROS COLLINA	PSIDIUM GUAJAVA	CITHAREXYLUM CAUDATUM
5	19.6	0	0	312.4	58.8
10	78.5	0	78.5	785.0	0
15	176.7	176.7	0	706.8	0
20	314.2	0	0	0	0
25	490.9	490.9	981.8	0	0
30	706.9	0	0	706.9	0
35	962.1	0	0	0	0
40	1256.7	1256.7	1256.7	0	0
75	4417.9	4417.9	0	0	0
Total		6342.2	2317.0	2571.1	58.8
Number of trees		4	4	34	3
Mean basal area/tree		1585.6	579.3	75.6	19.6
		0.53%	0.19%	0.21%	0.005%

[a] Tree basal area in square meters

$$\text{on } 100 \text{ m}^2 = \frac{\text{overall total}}{1.2} : 10,000 = \frac{11,289.1}{12,000} = 0.94 \text{ m}^2$$

or 0.94 percent total stem cover.

points is high (say at least 100 to 200 points) and the distance between points is closer than the shoot outline of most plants, the point-frequency result becomes a measure of cover. No plot or quadrat is necessary.

A second form of assessing frequency without use of quadrats is to record the presence or absence of plants near points. Frequency near sampling points is often recorded in the distance methods that will be discussed in SECTION 7.6.

2.   *Cover.* As mentioned above, one form of assessing cover without quadrats is through a dense network of frequency points. A second plotless method is the line-intercept method, which—as discussed before—is based on the reduction of a belt-transect to a single line of only one dimension, namely length.

3.   *Density.* The number of individuals of an area or in a stand can be determined by measuring the distance between individuals or between sampling points and individuals. The sampled distances can be converted to two-dimensional units or areas by squaring.

We already discussed two of the important plotless methods, the point-intercept (SECTION 6.54) and the line-intercept methods (SECTION 6.55). The point-intercept method, in its usual form, is mostly applied to herbaceous vegetation. The line-intercept method is perhaps most generally useful for open-grown woody vegetation. In both cases, the result is, of course, applied to a specified area in terms of either absolute or percent cover. Thus, in a wider sense, a plot or relevé is used also in plotless sampling, because the results of the plotless sample must be extrapolated to an area for proper interpretation. This area may also be a map unit.

It remains to discuss two important plotless techniques that evolved from the timber survey methods. One is the determination of stem cover or tree basal area through a modification of the point-sampling technique, the other relates to the determination of stem density through the measurement of distances.

7.5   BITTERLICH'S VARIABLE RADIUS METHOD

BITTERLICH (1948) discovered a remarkably efficient way to measure stem cover in tree stands by applying the point-frequency principle. Since stem cover is the same as tree basal area, and since basal area is one of the basic units for tree volume determination, the method is of great value to forest inventory work (GROSENBAUGH 1952). But also, since stem cover by species is defined as their dominance, and since cover or dominance is one of the most important quantitative parameters in vegetation ecology, the BITTERLICH method has become an important quantitative method, particularly in North American vegetation ecology.

7.51   The Technique.   Trees are counted in a circle from a central sampling point with an angle-gauge. Only trees that are larger in diameter than a specified angle are included in the count. The others are ignored. Therefore, the circular plot around the central sampling point has no fixed radius; instead, the radius varies with the diameter of each tree counted. This also renders the method plotless, because no fixed area-sample is involved.

When trees are counted in this manner with an angle-gauge, their number is proportional to their stem or basal area per unit ground area.

The standard North American angle-gauge is usually made of a 33-inch long stick. Mounted at one end is a 1-inch wide cardboard, plastic, or metal cross-piece and at the other end a similar piece with a notch or peephole. The angle-gauge is held with the peephole or notch at the eye like an Abney level and pointed with the 1-inch wide cross-piece horizontally at each tree surrounding the sampling point. The point aimed at on each tree must be at a fixed height, usually breast height. The same ratio or angle of 1°45′ can be obtained with a 33 cm long gauge and a 1 cm wide cross-piece. When using a round stick, a peep-hole or notch is not even necessary (FIG. 7.1). Only those trees are counted whose diameter exceeds the cross-piece. Therefore, small diam-eter trees are included in the count only if they are close to the observer, while large-diameter trees are included at greater distances away from the observer. With the 1:33 gauge, trees counted will not be further away from the sampling point or observer than 33 times their diameter. Thus, a tree with a 4-inch (10 cm) diameter must be within $4 \times 33 = 132$ inches (3.35 m) of the sampling point, while a 20-inch tree will be in-cluded if it is within $20 \times 33 = 660$ inches (16.8 m).

The selection of the 1:33 ratio for construction of the gauge, or the equivalent sighting angle of 1°45′, was recommended by GROSEN-BAUGH (1952), because the tree count at this angle permits immediate calculation of the basal area in square feet per acre. This is done by multiplying the count by 10. Thus, if 12 trees are counted, the basal area per acre is 120 ft². If, for example, 10 of these are pines and 2 are spruces, then pine occupies a basal area per acre of 100 square feet and spruce 20. Of course, these should be mean values of a num-ber of sampling points to result in a reliable estimate per acre.

BITTERLICH (1948) recommended a gauge ratio of 1.41 cm to 100 cm giving a much narrower sighting angle (50′) and more than twice the tree count per sampling point. At this ratio, the tree count divided by 2 results in the basal area in square meters per hectare. A still simpler ratio for calculation is 2 cm to 100 cm or 1:50, which is equivalent to a sighting angle of 1°10′. The resulting count is directly equal to the basal area in square meters per hectare. The latter angle permits a still

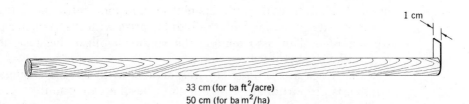

1 cm

33 cm (for ba ft²/acre)
50 cm (for ba m²/ha)

FIGURE 7.1. *BITTERLICH angle-gauge for measuring basal area by count-ing of trees. The gauge is held with the plain end at the eye and pointed hori-zontally with the cross-piece end to each tree surrounding a sampling point. Any tree that appears larger in diameter than the cross-piece is counted, any tree smaller is excluded.*

greater sampling intensity per point than the sighting angle recommended by GROSENBAUGH and used in several North American ecological studies (e.g., SHANKS 1954, RICE and PENFOUND 1955, 1959).

A more sophisticated angle-gauge, developed subsequently by BITTERLICH, is the so-called "Spiegelrelascope." This is a small, compact optical instrument that provides for specified angles by a set of bands that serve as comparison bars. The instrument is equipped with an automatic slope correction that can be switched off, if one is not interested in obtaining basal area data for projecting on maps. The Spiegelrelascope is not so useful under low light intensity, because the visibility through the instrument is then much impaired.

Recently foresters and ecologists have adopted clear-glass prisms as the most popular angle-gauge. When viewing through a prism, tree stems appear displaced to one side. Where the displacement is within the trunkline, the tree is counted; where the displacement is outside, the tree is ignored. A borderline tree is counted as half-tree (DILWORTH and BELL 1972:32). Prisms with angles ground to specifications can be obtained through engineering supply stores.

7.52   The Principle.   To understand how the BITTERLICH method works, we may assume a 10×10 m sample plot stocked with trees. An estimate of stem cover or basal area can be obtained by mapping the stand to scale with the stem areas forming circles. Then, a large number of random points may be superimposed on the map. According to the point-frequency principle, the number of random points that intercept stem areas out of the total number of random points will be proportional to the ratio of stem area to total area.

For example, if 10,000 random points are used and 50 fall into circles, the proportion will be 50 out of 10,000 or 0.005 (0.5 percent). For the 100 m² plot this would result in a stem cover or basal area of 0.005×100=0.5 m². The value of 0.005 also represents the mean number of trees intercepted per sampling point.

Measuring stem cover in this form would be most inefficient, because of the mapping process. For direct field application an impractically large number of sample points would be required to yield an accurate result.

BITTERLICH improved the efficiency of the method by mathematically enlarging the stem area of each tree.

On our map, we may assume a 100 times enlargement of each small circle radius. This would be equivalent to a stem-area or circle-area increase of 100². Note that the ratio of circumference to diameter ($\pi = 3.14$) is maintained in this proportionate enlargement. Probably, then the total map would be covered with these enlarged circles. Moreover, many of the enlarged circles would overlap. Almost each of the 10,000 random points would now intercept a circle or enlarged stem area, and many points would intercept several overlapping circles. The result of this geometric exercise would be that the efficiency of each sampling point in terms of interceptions or hits is increased in proportion to the stem-area enlargement factor, namely by 100^2. The mean number of trees intercepted per sampling point would now be $0.005 \times 10,000 = 50$.

However, the result of interceptions at each point would be fictitious in terms of the real stem area. The overestimate would be the same as the area-enlargement factor, i.e., 10,000 times. Therefore, to reduce the number of interceptions of the sampling points from the fictitious to the real, the number of interceptions needs to be divided by 10,000. This can be expressed as follows:

$$\text{stem area} = \frac{\text{number of interceptions} \times \text{area-enlargement factor}}{\text{total points} \times \text{area-enlargement factor}}$$

Applied to our example,

$$\text{stem area} = \frac{50 \times 10,000}{10,000 \times 10,000} = 0.005$$

This shows that the method has been modified from the standard point-intercept method, but the final answer is the same.

In the field application of BITTERLICH's method, the stem diameters and areas are increased by the angle-gauge. If we put, for example, a 1:50 angle-gauge with its 1 cm wide cross-piece on top of a stick that has a diameter of 1 cm, we have the enlarged radius (R) for that stick. With that radius, formed by the 50 cm piece of the angle-gauge, we can describe a circle area whose diameter is now 100 cm and whose area has been increased by 100^2. The area-enlargement factor is the ratio of the enlarged area ($R^2\pi$) to the actual stem area ($r^2\pi$); i.e., R^2/r^2. In this case it is $50^2/0.5^2 = 100^2$. If the stick was a thin sapling in a field situation, it would just be included in the count, because the sampling point (the observer's position, or exactly the viewing-end of the angle-gauge) just intercepts the fictitiously enlarged stem area. Similarly, if we look over the 1:50 angle-gauge to a tree at some distance, the stem area of that tree is automatically enlarged by 100^2, if the tree is wider than the cross-piece or just covered by it.

The area-enlargement factor of 100^2 obtained with the 1:50 angle-gauge was recommended by BITTERLICH, because the average tree count per sampling point is then equivalent to the stem area or basal area in square meters per hectare. This is so, because the area enlargement factor is the same as the number of square meters contained in a hectare. This is shown by substituting these values in the following basic equation.

$$\text{basal area} = \frac{\text{mean count per sampling point}}{\text{area-enlargement factor}} \times \text{unit reference area}$$

Results per hectare for the 1:50 gauge can be reduced to:

$$\text{basal area in m}^2 = \frac{\text{mean count per sampling point}}{100^2} \times 10,000$$

basal area in m² = mean count per sampling point

GROSENBAUGH (1952) recommended the 1:33 angle-gauge for American foresters, because of their general preference to express basal area results in square feet per acre. For this purpose, the 1:33 ratio is convenient, since the area-enlargement factor ($66^2 = 4356$) is exactly one-tenth the number of square feet in an acre. Therefore, with a 1:33 angle-gauge the basic equation becomes:

$$\text{basal area in ft}^2 = \frac{\text{mean count per sampling point}}{66^2} \times 43,560$$

basal area in ft² = mean count per sampling point × 10

Even though trees are counted in the BITTERLICH method, the basal area estimate does not provide for a density estimate. Neither can one obtain a frequency count from this method, because here the species presence per sampling point is merely a function of diameter size, not of surface area sampled.

Therefore, the method is useful only where a stem cover value alone is satisfactory. This quantity, however, is very rapidly obtained. The method seems particularly useful for tree evaluation in the relevé analysis, when estimate scale values are used for undergrowth plants. Both are rapid survey methods that complement each other (BEN-NINGHOFF and CRAMER 1963)

7.53   Calibration of BITTERLICH Gauge. The simple BITTERLICH gauge as illustrated in FIGURE 7.1 can be used for accurate measurements of basal area per unit ground area, if one knows how to calibrate the instrument. For calibration one selects a nearby tree and views at it over the gauge. The cross-piece or comparison bar must exactly cover or contain the width of the tree. This usually requires change of the observer's position. When the correct distance is obtained, the position is marked on the ground. Then the distance from the position-point to the center of the tree is measured. Secondly, the diameter of the tree is measured. The two values, distance and diameter are then substituted in the calibration equation shown in FIGURE 7.2.

The calibration principle can be used also to measure the diameter of a tree from a distance. This can be useful on steep slopes. In that case, the distance may be obtained by a range finder. How to calculate the tree diameter from a measurement with a BITTERLICH gauge is explained in FIGURE 7.2.

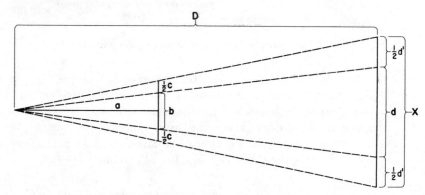

FIGURE 7.2   *Calibration principle of BITTERLICH gauge. Legend: a=length of BITTERLICH gauge; b=width of sighting bar; d=diameter of tree; D= distance from observer to tree. Further explanation in equations below.*

$$a:D = b:d \tag{1}$$

For calibration *(1) is rewritten to read*

$$D = \frac{a}{b} \times d \tag{2}$$

For measuring tree diameter from a distance *(2) is rewritten to read*

$$d = \frac{D}{a} \times b \tag{3}$$

However, a tree may be larger or smaller than b. Let d + d' be X = the diameter of a bigger tree:

$$X:d = (c+b):b \tag{4}$$

$$X = \frac{(c+b)}{b} \times d \tag{5}$$

or

$$X = \frac{(c+b)}{d} \times \frac{D}{a} \times b \tag{6}$$

7.6   THE WISCONSIN DISTANCE METHODS

7.61   Concept of Mean Distance as a Measure of Density.   Related to the timber survey methods are the distance methods for estimating density which were developed by the Wisconsin Plant Ecology Laboratory (WPEL). These were perfected primarily for the tree layer of the plant community (CURTIS 1959).

Plotless techniques were developed which are based on the idea that the number of trees per unit area can be calculated from the average distance between the trees.

If we consider a plantation stand in which the trees are situated at regular intervals of 3 m each, we can quickly determine the number of trees per unit area from these intervals. The spacing of such plantation trees results in a number of 3×3 m quadrats that always form the intervening areas between four corner trees. The quadrats are connected into a contiguous grid of quadrats. We can now imagine a shifting of these quadrats so that a tree becomes the central point in each quadrat. No size-change of the quadrat is involved, and it is clear that each tree occupies an area of 3×3 m $= 9$ m^2. In this plantation stand, the 9 m^2 quadrat is also the mean area per tree. If we now want to establish the number of trees per hectare (ha) of such a plantation stand, we simply divide the mean area into the reference area:

$$\text{number of trees} = \frac{\text{unit reference area}}{\text{mean area}}$$

$$\text{number of trees per ha} = \frac{10{,}000 \text{ m}^2}{9 \text{ m}^2} = 1111$$

Therefore, the problem of determining the number of individuals on an area reduces to finding the mean area of an individual. This mean area can be visualized in a natural (nonregular) stand, not as an even-sided quadrat, but as a quadrangle that is described by four individuals at its end-points.

The important problem in the distance methods is to locate the distance that gives the best estimate of the square root of the mean area per tree. This is done by averaging a number of specifically selected distance-measures in the stand.

In the quantitative plot method this distance is easily established after the number of individuals is known. The mean area per tree is equal to the plot area divided by the number of trees. Applied to the example in TABLE 7.1, the mean area (MA) and mean distance (D) are:

$$MA = \frac{\text{plot area}}{\text{number of trees}} = \frac{120 \text{ m}^2}{45} = 2.67 \text{ m}^2$$

$$D = \sqrt{MA} = \sqrt{2.67} \text{ m}^2 = 1.64 \text{ m}$$

The major advantage of estimating number of individuals through their mean distance rather than through the standard way of counting them in quadrats, plots, or strips is that no plot boundaries are required. This, in many situations, saves considerable time (CURTIS 1959), because tree distances are usually shorter and more easily measured than boundaries.

7.62 Miscellaneous Distance Methods. Different distance methods are proposed in the literature and are currently applied in vegetation and population studies. It seems therefore necessary to introduce at least those methods that have been tried and tested and that are still used in various combinations for methods research.

The choice of pairs of individuals to be measured for their distance is theoretically more complicated in a natural stand than in a regular stand. PIELOU (1959) emphasized that this choice must be truly random. To obtain a truly random choice, all trees in a stand would have to be labelled with a number. Thereafter, one can select individual trees with the aid of a random numbers table. These randomly selected trees may then serve as the trees from which a sample distance is measured to their nearest neighbor. However, the need for prior labelling of all trees would defeat the practicability of the distance method. Therefore, various shortcut methods have been proposed. All of them operate from sampling points, which may be established either randomly or systematically.

One method is to select pairs of individuals nearby randomly selected points. The individuals that are closest together near the point are chosen for sampling the distance between them. This method became known as the "nearest neighbor method" (COTTAM, CURTIS and HALE 1953, COTTAM and CURTIS 1956).

Another method was even simpler. It involved merely to measure the distance from a randomly selected point to the nearest tree. This became known as the "closest individual method" (COTTAM, CURTIS and HALE 1953, COTTAM and CURTIS 1956)

A third method that gained considerable popularity for a while among North American vegetation ecologists, was the so-called "random pairs method" (COTTAM and CURTIS 1949, COTTAM, CURTIS and HALE 1953, COTTAM and CURTIS 1956). Like the nearest neighbor method, the random pairs method involved a distance measure between two individuals instead of a distance measure between a point and an individual. After establishing the sampling point either at random or systematically (at intervals along a transect), one looks for the nearest tree from the sampling point. This tree serves for measuring a distance to a second tree. Facing this first tree, the investigator spreads out his arms to both sides. In this way, two imaginary lines are established; the first line through facing the nearest tree from the sampling point, the second line through the outstretched arms of the observer. The purpose of this is to establish a 180° exclusion angle to exclude any neighboring tree in that sector where the first tree stands for the distance measure. The second tree to which the sample distance is measured is the nearest tree behind the outstretched arms of the observer. The procedure is diagrammed in FIGURE 7.3.

Through empirical testing it was established that these three methods give acceptable mean area and therefore density estimates for random populations, but with certain correction factors. These corrections are for the nearest neighbor method 1.67×the mean distance (D), for the closest individual method 2×D, and for the random pairs method 0.8×D (COTTAM 1955).

In their important research methods paper, COTTAM and CURTIS (1956) tested these three distance methods for their sampling efficiency against a fourth method, the so-called point-centered quarter method. The latter method does not require a correction factor and is as simple in its application as the closest individual method, but four times as sampling-intensive. This also means that it requires less time in the field. The point-centered quarter method was therefore considered the most efficient of the available distance methods. It has since gained wide acceptance.

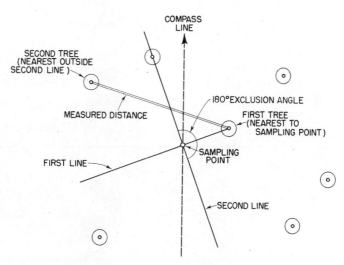

FIGURE 7.3.   Random pairs method.

7.63   The Point-Centered Quarter Method. In the point-centered quarter method four distances instead of one are measured at each sampling point. Four quarters are established at the sampling point through a cross formed by two lines. One line is the compass direction and the second a line running perpendicular to the compass direction through the sampling point. The line-cross can also be randomly established by spinning a cross over each sampling point. The distance to the mid-point of the nearest tree from the sampling point is measured in each quarter (FIG. 7.4).

The four distances of a number of sampling points are averaged and when squared are found to be equal to the mean area occupied by each tree. COTTAM and CURTIS (1956) tested the reliability of this method on several random populations by checking the result with the plot method. They ranked the four quarter (Q) distances of each sampling point by computing the mean of the shortest (Q1), the second shortest (Q2), the third (Q3) and the longest (Q4) distances. The following estimates of the correct mean area per tree (MA) were found to apply to each of the different sets of mean distance.

$$
\begin{aligned}
&\text{Q1 shortest} &&= 0.5 \ \sqrt{\text{MA}} \\
&\text{Q2} &&= 0.8 \ \sqrt{\text{MA}} \\
&\text{Q3} &&= 1.12\sqrt{\text{MA}} \\
&\text{Q4 longest} &&= 1.57\sqrt{\text{MA}} \\
\hline
&\text{Q mean of 4} &&= 1.0 \ \sqrt{\text{MA}}
\end{aligned}
$$

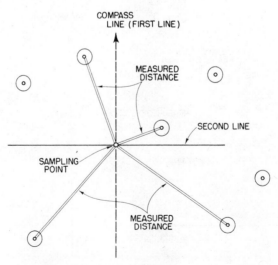

FIGURE 7.4.   Point-centered quarter method.

Therefore, no correction factor is needed when the four quarter distances are averaged; and $MA = D^2$, where D = the mean distance of four point-to-nearest-tree distances taken in each of four quarters. Mathematical proof of the workability of this method has been given by MORISITA (1954).

Of course, the accuracy increases with the number of sampling points, and a minimum of 20 points is recommended (COTTAM and CURTIS 1956).

The method has two limitations (NEWSOME and DIX 1968) for field applications. An individual must be located within each quarter, and an individual must not be measured twice. Therefore, stands with wide spacing of individuals present a problem in using this method. The second limitation applies also to the random pairs method.

The parameters obtained in the distance methods are:

1. Species.
2. Density (from mean distance).
3. Diameter (and therefore basal area and dominance).
4. Frequency (as the occurrence of a species at a sampling point).

The same parameters are also obtained from plots. However, the distance methods have an advantage in that they do not require laying out of plot boundaries. This saves considerable time. It also eliminates to a certain extent the personal error from judging whether boundary individuals are inside or outside the quadrat.

7.64   Example of a Point-Centered Quarter Analysis.   The following example relates to the same tropical rain forest stand that served for the relevé example (SECTION 5.3) and for the quantitative plot example (SECTION 7.3). The point-centered quarter example is shown only for five sampling points to save space (TABLE 7.4). It is recommended to sample at least 20 points per stand. The adequacy of sampling points can, of course, also be determined by plotting the running mean as described in SECTION 6.42.

In the example analysis in TABLE 7.4, trees with basal diameters less than 3 cm were omitted. These included all woody plants under 2 m height. The small trees could, however, be sampled as a second size category from the same sampling points with each four distances. The objective was to determine (from individuals taller than 2 m):

1. the density for each tree species,
2. the dominance of each tree species, and
3. the frequency of each tree species.

A second objective was to convert these absolute values into relative values as an example for deriving the importance value, which will be discussed in SECTION 7.67.

TABLE 7.4 shows the raw data for five sampling points that were arranged in a transect, one point every 5 m. TABLE 7.5 shows the derivation of the mean basal area by species. This value is needed to determine the dominance of the species, which is a combination of number and basal area.

7.65   Limitations of the Distance Methods.   The point-centered quarter method has become well accepted as shown by many vegetation studies (CAPLENOR 1968, HABEK 1968, RISSER and ZEDLER 1968, NEW-SOME and DIX 1968, among others). Apart from its less complicated field application and greater information value per sampling point, the method seems more reliable than the random pairs method. This is based on the observation that the distances of trees to sampling points are more truly random than the distances among trees located through sampling points (COTTAM, CURTIS and HALE 1953, PIELOU 1959).

However, the point-centered quarter method is similarly applicable only to random distributions. Plot studies are more reliable where plant individuals are not randomly distributed (SCHMELZ 1969). Yet plots or quadrats are not fully reliable either. The reason is that a plot may also include either aggregations or underdispersed groupings of individuals in contagiously distributed species combinations. Clumping of individuals or contagious distribution applies to nearly all plant life forms, except trees and annuals. But even among the latter life forms nonrandom distributions are the norm for the individuals of single species in mixed-species stands. Therefore, the method should not be applied to single species in mixed stands. Instead, it should be applied only to broad size classes as shown in the preceding example, where the method was applied to tree individuals of all species taller than 2 m. The density of each species is subsequently established by partitioning the total density estimate.

GREIG-SMITH (1964) has cautioned against applying the point-centered quarter method to herbaceous life forms, such as bunch grass vegetation, because the resulting density values are inaccurate where the distribution of individuals occurs in aggregations. This has been supported by RISSER and ZEDLER (1968) who found in Wisconsin grassland that the point-centered quarter method consistently underestimated the number of individuals in contagiously distributed species. This can be explained by the greater probability of a sampling point to fall between the clumps of individuals than within the clumps in contagious distributions in which the clump diameter is small. By falling

TABLE 7.4. *Quantitative Analysis by Point-Centered Quarter Method. Five Sampling Points, One at Every 5 m Along 110°, Starting at End of Convex, Gently Sloping Ridge Below Pauoa Flats Trail Going Upslope Toward the Trail. Raw Data, March 4, 1972.*

SAMPLING POINT	QUARTER NUMBER	DISTANCE (M)	SPECIES	DIAMETER AT BASE (CM)
1	1	0.7	Psidium guajava	5.5
	2	1.6	Acacia koa	42.5
	3	3.5	Metrosideros collina	17.0
	4	2.0	Metrosideros tremuloides	25.0
2	1	1.1	Psidium guajava	4.0
	2	0.8	Psidium guajava	5.0
	3	1.9	Psidium guajava	5.0
	4	1.8	Psidium guajava	4.0
3	1	1.3	Acacia koa	75.0
	2	0.7	Psidium guajava	3.0
	3	1.5	Metrosideros collina	9.0
	4	2.0	Metrosideros collina	23.0
4	1	3.1	Acacia koa	14.0
	2	1.7	Psidium guajava	6.0
	3	1.1	Psidium guajava	5.0
	4	1.9	Acacia koa	12.0
5	1	2.5	Acacia koa	23.0
	2	2.2	Acacia koa	18.0
	3	1.4	Psidium guajava	5.0
	4	2.8	Metrosideros collina	25.0
		Total 35.6		

Results:

Mean distance (D) = 35.6/20 = 1.78 m

Absolute density = Area/D^2

Where D = mean distance

Number of trees per 100 m^2 = 100/(1.78)^2 = 100/3.17 = 31.5

Absolute dominance = mean ba per tree × number of trees in species

Where ba = basal area

Number of trees in species

SPECIES	NUMBER IN QUARTERS	NUMBER OF TREES IN 100 M^2
Acacia koa	6/20=0.3	0.3 ×31.5= 9.4
Metrosideros collina	4/20=0.2	0.2 ×31.5= 6.3
Metrosideros tremuloides	1/20=0.05	0.05×31.5= 1.6
Psidium guajava	9/20=0.45	0.45×31.5=14.2
		Total 31.5

TABLE 7.5.   Mean Basal Area by Species for the 20 Trees Shown in TABLE 7.4.

ACACIA KOA		METROSIDEROS COLLINA		METROSIDEROS TREMULOIDES		PSIDIUM GUAJAVA	
DIAMETER (CM)	BA (CM²)	DIAMETER (CM)	BA (CM²)	DIAMETER (CM)	BA (CM²)	DIAMETER (CM)	BA (CM²)
42.5	1418	17.0	227	25.0	491	5.5	24
75.0	4418	9.0	64	..	..	4.0	13
14.0	154	23.0	415	..	..	5.0	20
12.0	113	25.0	491	..	..	5.0	20
23.0	415	..	..	..	..	4.0	13
18.0	254	..	..	..	..	3.0	7
..	..	..	..	..	..	6.0	28
..	..	..	..	..	..	5.0	20
..	..	..	..	..	..	5.0	20

Total ba	6772		1197		491		165
Mean ba	1129		299		491		18

Therefore, dominance of   Dominance rank

Acacia koa	$1129 \times 9.4 = 10613$ cm²	1
Metrosideros collina	$299 \times 6.3 = 1884$ cm²	2
Metrosideros tremuloides	$491 \times 1.6 = 786$ cm²	3
Psidium guajava	$18 \times 14.2 = 256$ cm²	4

$$13539 \text{ cm}^2/100\text{m}^2$$

$$\text{Absolute frequency} = \frac{\text{number of points with species}}{\text{total points}} \times 100$$

Acacia koa	$= \frac{4}{5} \times 100 =$	80 percent
Metrosideros collina	$= \frac{3}{5} \times 100 =$	60 percent
Metrosideros tremuloides	$= \frac{1}{5} \times 100 =$	20 percent
Psidium guajava	$= \frac{5}{5} \times 100 =$	100 percent

260 percent

between clumps, the point to plant distances will be longer than average. The longer distances result in an overestimate of the mean area per individual and thus in an underestimate of density.

The opposite, namely overestimation of the number of individuals, is true for regularly distributed individuals. This is shown in FIGURE 7.5. In a regular, quadrangular distribution, such as often found in a planted tree stand, the correct mean area is obtained by squaring the shortest distance between any two trees. This result would be obtained only by sampling point 1 in FIGURE 7.5. Such locating may occur once in a very large number of random point placements or not at all. The most

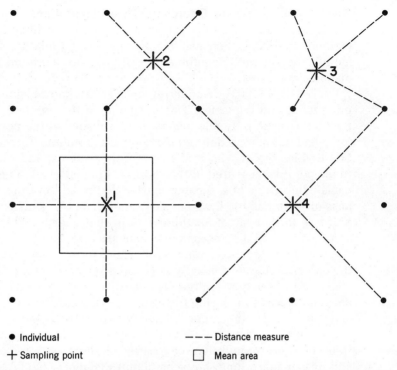

● Individual   — — — Distance measure

+ Sampling point   ☐ Mean area

FIGURE 7.5.   *Application of point-centered quarter method to a regular dis-*
tribution of individuals. Here only sampling point 1 gives the correct estimate
of mean area. Further explanation in text.

common placement would be between trees, such as indicated by points
2 and 3. At these positions the mean distance of four quarters and
therefore the mean area will always be underestimated. This will result
in a considerable overestimate of tree density. Only position 4 would
result in an overestimate of mean distance and thus an underestimate
of density, as is found for contagiously distributed individuals. However, for a sampling point to give this result, not only must the point
fall directly on a tree, but also the quarter dividing lines must pass
through the center of the nearest trees, which would render them in-
valid for inclusion in the sample. This also shows that the boundary
problem, found to be a disadvantage in any plot method, is not entirely
eliminated in the plotless methods. However, it is highly improbable
that position 4 will occur randomly. Instead, tree density can always be
expected to be overestimated by this method when applied to regularly
distributed individuals. This is true also for rectangular and rhombic
regular distributions.

7.66   Modifications to Overcome These Limitations.   Several modifications were suggested to overcome the pattern problem in the distance methods to extend their use to single species population studies. These modifications employ combinations of point-to-plant and plant-to-plant distance measures.

CATANA'S (1963) "wandering quarter method" begins with a sampling point and a quarter. This is similar to the point-centered quarter method. Except, only one quarter is established at the point. This quarter is laid out in a predetermined compass direction. The compass direction divides the quarter into two 45° pie-sections, and the nearest tree to the point is measured in this quarter. Thereafter, this nearest tree becomes the vector of a second quarter that is laid out in the same compass direction as the first one. A second distance is measured from the first tree to its nearest neighbor tree in that quarter. This procedure is continued for 25 distances in one compass direction.

Since the nearest tree may rarely stand in the middle of a quarter on the compass line, but usually is found anywhere within the 90° exclusion angle of the quarter, the distance directions are likely to shift in an irregular zigzag line during the progress. This shifting along the transect is responsible for the name "wandering" quarter method.

If contagious distributions occur in the 25 distances measured along a transect, there should then be a series of short, within-clump distances and one or more long, between-clump distances. CATANA (1963) describes how to detect the two kinds of distances and suggests a correction to obtain a realistic mean distance. However, one problem is the commonly low number of gap- or between-clump distances obtained, which may not give a statistically valid sample for contrasting them to the usually high number of within-clump distances. CATANA therefore suggests sampling four transects arranged to one another in form of a quadrat. The resulting 100 distance-measures should contain a sufficient number of gap-distances for correction if the pattern is contagious.

CATANA tested his method on four artificial populations of 1000 individuals each. In the first truly random population, the wandering quarter method estimated 1025 individuals. In two slightly contagious populations the estimates were 815 and 836 respectively. In a fourth population that tended to be regular, the estimate was 1285 individuals. Thus, the method still underestimates density in contagiously distributed populations and gives strong overestimates, where the distribution tends to be uniform.

Recently, BATCHELER (1971) suggested a further modification. This consists of measuring the distance to the nearest individual from the sampling point. From this individual, the distance is measured to its nearest neighbor, and then a third distance is measured to the next nearest neighbor. Therefore, three distances are measured at a sampling point—one point-to-plant distance and two plant-to-plant distances. No quarter or exclusion angle is used apparently.

BATCHELER points out that the point-to-nearest-plant distance (several times repeated) gives the true mean distance for a random population and that the two additional plant-to-plant distances supply the data for correction of departure in pattern. As correction he suggests dividing the sum of the point distances by either the sum of the nearest neighbor distances or by the sum of the second-nearest neighbor distances and to use this fraction as an exponential function.

According to the point-centered quarter test by COTTAM and CURTIS (1956), the shortest point-to-plant distance gives only 0.5 of the true mean distance in a random population. The true mean distance is obtained only by measuring the point-to-plant distance in a 90° exclusion angle.

BATCHELER's method requires intensive testing, before it can be recommended for general use.

It is apparent that distance methods for the estimation of density of single species are still in the research stage. The methods are not yet reliable for nonrandom populations. They were included in the discussion because further methods-research may soon extend their scope to nonrandom populations.

However, this restriction does not apply to the same extent when all species in a stand are sampled together. Taken together, trees in a stand approach random distribution and then the point-centered quarter method is useful (COTTAM and CURTIS 1956).

Moreover, it is important to realize that two independent sets of data are obtained by the distance methods. The unreliability does not apply to the diameter and frequency measurements, which are independent of the correct mean distance. Therefore, mean basal area per tree can be accurately derived from the diameter measurements (computation as in TABLE 7.3, SECTION 7.32, or TABLE 7.5. SECTION 7.64), but mean basal area per acre or hectare, which is derived through multiplication with density, is dependent on pattern. The point-centered quarter method is widely used in spite of this possible bias in density estimate, because the data is commonly expressed in relative values. This will be further explained in the next section.

LINDSEY, BARTON and MILES (1958) have shown that 0.1 acre (400 m^2) circular plots delimited with a range finder are still more efficient for density evaluation than the point-centered quarter method in stands without view-obscuring undergrowth. The circular plot method also has the advantage that the accuracy of the density count is less affected by departures in pattern from randomness. LINDSEY et al suggest a combination of BITTERLICH's technique for basal area and the circular plot method for density and frequency as the most efficient quantitative method in forest stands. But this suggestion holds only for forest stands in which the stem of each tree is visible near breast-height from a central sampling point.*

For a structural analysis, tree diameters are desirable. These may be measured from the center of the circular plot by using a BITTERLICH gauge as explained before (FIG. 7.2). However, visibility may be limited in stands with dense shrubby undergrowth. In such situations either belt-transects or the point-centered quarter method may be more efficient.

7.67   The Importance Value.   The distance methods yield three quantitative parameters—density, basal area, and frequency. These are, of course, also obtained in the quantitative plot methods.

Any one of the three parameters may be interpreted as an "importance value" (WHITTAKER 1970). This depends on which of the values the investigator considers most important for a particular species, group of species or community. For example, tree seedlings may occur with a high frequency in an undergrowth layer, while in terms of cover, they may be insignificant. However, their high frequency may be of great importance as indicating a new stage of uniformly distributed reproduction. In this case their high frequency may be interpreted as of high "importance."

Yet, it has become common practice, in quantitative descriptive studies that employ the distance measuring techniques, to use the so-called importance value of CURTIS (1959) for the presentation of results. This importance value (I.V.) is defined as the sum of relative density, relative frequency, and relative dominance.

The absolute values for density, dominance, and frequency were defined already in the point-centered quarter example (SECTION 7.64).

The corresponding relative values for the example shown in TABLE 7.4 are shown on the following page.

The importance value may be converted into the so-called "importance percentage" by dividing the importance value by three (RISSER and RICE 1971).

The importance value of a species reaches a maximum of 300 in stands consisting of only one tree species. Two monodominant (single tree species) stands with different numbers of trees per acre and different basal areas will have the same importance value for each species.

---

*To lay out a circular plot, calculate the radius (R) from the area (A) as $R = \sqrt{(A/\pi)}$. Example for 0.1 acre plot $R = \sqrt{(400 \text{ m}^2/3.14)} = 11.3$ m.

1. Relative density $= \dfrac{\text{number of individuals of species}}{\text{total number of individuals}} \times 100$

Acacia koa $\dfrac{9.5}{31.5} \times 100 = 30$ percent[*]

Metrosideros collina $\dfrac{6.3}{31.5} \times 100 = 20$ percent

Metrosideros tremuloides $\dfrac{1.6}{31.5} \times 100 = 5$ percent

Psidium guajava $\dfrac{14.3}{31.5} \times 100 = 45$ percent

$\overline{}$
100 percent

2. Relative dominance $= \dfrac{\text{dominance of a species}}{\text{dominance of all species}} \times 100$

Acacia koa $\dfrac{10{,}613}{13{,}539} \times 100 = 78.4$ percent

Metrosideros collina $\dfrac{1884}{13{,}539} \times 100 = 13.9$ percent

Metrosideros tremuloides $\dfrac{786}{13{,}539} \times 100 = 5.8$ percent

Psidium guajava $\dfrac{256}{13{,}539} \times 100 = 1.9$ percent

3. Relative frequency $= \dfrac{\text{frequency of a species}}{\text{sum frequency of all species}} \times 100$

Acacia koa $\dfrac{80}{260} \times 100 = 30.8$ percent

Metrosideros collina $\dfrac{60}{260} \times 100 = 23.1$ percent

Metrosideros tremuloides $\dfrac{20}{260} \times 100 = 7.7$ percent

Psidium guajava $\dfrac{100}{260} \times 100 = 38.5$ percent

$\overline{}$
100.1 percent

4. Importance value (I.V.) = Relative density + relative dominance + relative frequency

	RELATIVE DENSITY	RELATIVE DOMINANCE	RELATIVE FREQUENCY	I.V.	I.V. Rank
Acacia koa	30.0	78.4	30.8	139.2	1
Metrosideros collina	20.0	13.9	23.1	57.0	3
Metrosideros tremuloides	5.0	5.8	7.7	18.5	4
Psidium guajava	45.0	1.9	38.5	85.4	2

[*] Note, same as number of species occurrences in quarters.

In this case, the importance value does not convey any quantitative difference. Yet, it incorporates quantitative differences as soon as a second tree species appears in the stand. Two stands, each stocked with the same two species, will hardly ever show the same importance values per species. For example, one of the two species may be present with exactly the same number of individuals, the same basal area, and the same frequency, but the second species may show differences in its basal area between the two stands. This renders the importance values of the first species also different for each stand. The disparity between stands increases greatly with each additional species. The summing of the three parameters into one has the effect of increasing the difference between the same species among stands of similar species composition. The importance value therefore underscores the individualistic viewpoint (SECTION 3.13).

The use of relative rather than actual parameters is of limited information value. Densely vegetated and sparsely vegetated habitats can have the same relative densities, relative basal areas, and relative frequencies. Therefore, the importance value gives no idea of species biomass or cover, which are considered of even greater ecological significance in plant distribution than absolute density (FOSBERG 1961, RICE 1967, DAUBENMIRE 1968).

7.7   LITERATURE EXAMPLES OF QUANTITATIVE FIELD ANALYSES IN NORTH AMERICA

In contrast to the example of a semiquantitative relevé analysis of a forest stand given in CHAPTER 5, quantitative field analyses cannot be adequately described by citing only one example. The main reason is that the kind of analysis varies with the objectives—whether the vegetation is to be described for classification, ordination, succession and population-structure, or other purposes.

Measurements of any or all of the three quantitative parameters can certainly be applied to small plots of European relevé size in the same way as they are often applied to the larger sample stands for continuum analysis. But these more accurate measures require more time. It is therefore always necessary to balance the time it takes to establish a quantitative measure against the objectives of the study. If the primary purpose is to describe vegetation through recurring plant assemblages or to portray the spatial variation of a vegetation type, it seems more appropriate to use the time for more relevés with semiquantitative estimates than to present only few relevés with accurate quantitative evaluations. This is based on the observation that vegetation varies from place to place, even if one samples for similarity or constancy in patterns. Moreover, an objective quantitative analysis does not eliminate the fact that selection of a sample area is subjective.

If the objectives are to determine the developmental or succesional trends of the woody plant populations of a forest community, it is necessary to enumerate the different woody plant species in size classes for a structural analysis. In contrast, developmental trends of herbaceous plant species can only be properly evaluated by periodic reassessments in permanent plots. For this, measurements are more appropriate than estimates in most cases. Measurements are also more useful for a close comparison of similar communities.

Five uncomplicated examples of quantitative analyses in forest stands and three in nonforest communities are cited from the literature to bring out the major trends. There are many more variations. In fact, almost any specific problem requires its own modifications of methods. For this reason we suggest that the previously described techniques be used as creative options for specific questions rather than as rigid tools for any situation. The quantitative descriptive methods will be compared to the relevé method in the conclusions (SECTION 7.73).

7.71   Forest Communities.   Here are the five examples of quantitative analyses in forest stands.

7.71.1   Forest Vegetation in Western North America (DAUBENMIRE). In addition to a cover class rating very similar to the BRAUN-BLANQUET scale in value and application (SECTION 5.42), DAUBENMIRE (1968) uses quantitative measures when the objectives of the analysis aim at more than classifying associations. For this he uses plots of 15×25 m in forest vegetation of Washington and Idaho. These 375 m² plots are divided into three strips each of 5×25 m (FIG. 7.6). In these, trees from 1 m height (i.e., from sapling size) upwards are counted by diameter-at-breast-height (*dbh*) classes. For shrubs and herbs, frequency is determined in 20×50 cm (0.1 m²) subplots placed at 1 m intervals along the two sides of the central 5×25 m strip. This results in 50 systematic 0.1 m² frame placings per plot, or in a total sample of 5 m².

The more abundant and uniformly distributed undergrowth plants are objectively evaluated in this way. All plants noted outside the frequency frames are added to the species list. Cover is estimated in each frame placement.

DAUBENMIRE uses a similarly rigorous vegetation segmentation as that applied by KRAJINA (1965, 1969) in western Canada, by GRANDTNER (1966) in Quebec and in European vegetation studies. DAUBENMIRE then places his plots centrally into the tentative vege-

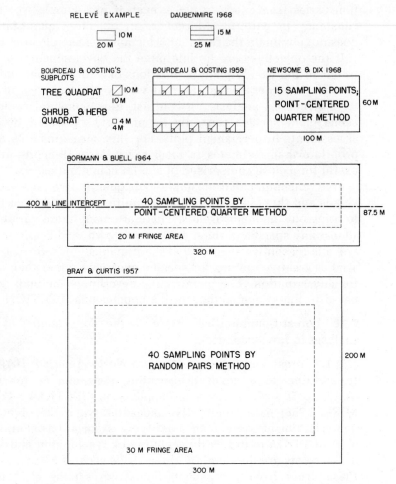

FIGURE 7.6. Comparative sizes of sample stands.

tation segments. Thus, his community studies are in essence relevé analyses in recurring plant communities.

7.71.2.   Eastern Hemlock-Hardwood Forest (BORMANN and BUELL). In most quantitative analyses the emphasis lies on accurate description of the variation throughout broadly defined dominance-communities (i.e., communities defined by dominant species only). Unlike the relevé analysis, in which the limit of homogeneity is defined by the uniformity of the undergrowth vegetation, the sample is spread out over a much larger area.

For example, BORMANN and BUELL (1964) sampled a seven acre (28,000 m²) stand of an old-age hemlock-hardwood forest (*Tsuga cana-*

densis—*Fagus, Fraxinus, Betula, Ulmus, Tilia*) in Vermont as follows: Trees of 10 cm *dbh* or greater were sampled by the point-centered quarter method (SECTION 7.63) at 40 sampling points. The sampling points were located along 12 base lines, each 23 m apart. The sampling points were spaced 20 m apart and none was less than 20 m from a boundary. FIGURE 7.6 shows the probable dimensions of this 28,000 m² sample unit.

Smaller trees (between 2.5 and 10 cm *dbh*) and saplings (between 30 cm height and 2.5 cm *dbh*) were counted at each sampling point in 1×10 m quadrats. Tree seedlings less than 30 cm tall were counted in fifty 0.5×2 m quadrats located 15 m apart along the base lines. Cover was measured by the line-intercept method for two tree layers and one shrub layer. The two tree layers were defined as 3.6 to 12 m and 12 to 32 m tall. These two layers were measured along a 400 m line running across the long dimension of the stand. The shrub cover, probably including all woody plants below 3.6 m height, was measured along 10 m lines at each of the 40 sampling points. Herbaceous plants were assessed in the same 0.5×2 m quadrats as the tree seedlings. But herbs were not counted; instead their cover was estimated in the 50 quadrats. In addition, herb species were listed from the entire stand.

Thus, the sampling layout was systematic to insure a uniform assessment of the 28,000 m² community. In addition to the 160 distance measures, a count sample of smaller trees was made in 400 m², and of seedlings in 50 m². Cover of herbs was estimated for 50 m² and that of woody plants over a length of 400 m.

The objectives of the survey were primarily to describe the stand with modern methods of vegetation measurement, and to determine the successional trend of the stand. The second objective is well accomplished for woody plants by such a structural (i.e., number per size class) analysis.

7.71.3 Live Oak Forest, North Carolina (BOURDEAU and OOSTING). BOURDEAU and OOSTING (1959) studied the live oak (*Quercus virginiana*) forest in North Carolina. This is likewise a broadly defined dominance-community occurring as stabilized vegetation on coastal dunes. Seventeen stands or locations were described by species lists, and five of these subjected to quantitative analysis. These five stands were selected because they were considered large enough. In each stand, an area of 60×100 m (1.5 acres) was outlined and divided into six 10×100 m strips (FIG. 7.6). Of these, two were randomly selected and partitioned into 10 m sections. Along each strip five alternate 10×10 quadrats were then sampled, resulting in a total sample-area of 1000 m². In these, all woody plants above 2.5 cm *dbh* were counted by diameter. Woody plants with less than 2.5 cm *dbh* were not counted, but their crown cover was estimated together with that of the herbs in 4×4 m plots nested in a predetermined corner of each of the ten 100 m² quadrats. Thus, the total area-sample for undergrowth plants was 160 m². Species occurring outside the quadrats were recorded also.

The objectives were to obtain a detailed analysis of the structure of this community as a representative record. No effort was made to distinguish finer patterns in the undergrowth.

7.71.4 Upland Forest of Southern Wisconsin (BRAY and CURTIS). BRAY and CURTIS (1957) sampled the upland forest of southern Wisconsin for a continuum analysis and ordination. This is a mixed hardwood forest containing about 16 prevalent broad-leaved tree species. The whole southwest half of Wisconsin, occupied by this hardwood forest, was considered one community for this purpose. The forest area was mechanically stratified into geographic subsections to provide for a balanced sampling. Within these, 59 stands were selected that were at least 15 acres (6 hectare) in size. (FIG. 7.6). The stands were homogeneous in the tree layer, reasonably undisturbed, and occurred on well-drained soils.

Within each stand the trees, probably meaning all woody plants above 10 cm *dbh,* were measured by the random pairs method at 40 sampling points. Therefore, 40 distances were measured and the 80 trees were recorded by species and diameter. The lines, along which the 40 sampling points were located at predetermined intervals, started always at least 30 m from the edge of a forest. Shrubs and herbs were sampled in twenty 1×1 m quadrats at alternate sampling points for frequency.

7.71.5 Cypress Hills Forest, Alberta and Saskatchewan (NEWSOME and DIX). A description of this forest in Canada was presented by NEWSOME and DIX (1968). In contrast to Eastern North American mixed forests, this Canadian forest is dominated by only three tree species, *Picea glauca, Pinus contorta,* and *Populus tremuloides.* These species often form monodominant stands, but mixtures of Pine–Populus and *Picea–Populus* are common also.

The primary objective was to study the species composition on forest-covered habitats for elucidating patterns of floristic variation by ordination techniques and to explain these in relation to environment.

Six requirements were set for a forest stand to be acceptable for sampling:

  • The tree canopy needed to cover at least 60 percent of the ground.

  • Immature stands were excluded.

- A stand had to extend over at least 0.6 hectare (1.5 acres) (FIG. 7.6).
- Its species composition had to show a certain minimum of homogeneity (subjectively determined).
- Likewise its habitat had to be uniform, for example, stands with microtopographic variations exceeding 5° were excluded.
- Stands obviously disturbed by windfall, fire, or cutting were excluded; less obviously disturbed stands were accepted.

Sample stands were located throughout the Cypress Hills forest so as to include as much variation in species composition and habitat as could be discerned.

Seventy-nine stands were selected in this way. The trees from 9.4 cm (3.7 inches)* diameter at breast height on upwards were sampled by the point-centered quarter method for density (number/area), basal area (dominance), and frequency. The number of point samples was 15. Therefore, 60 trees were measured per stand (The outline and intervals between sampling points are not reported). Fifteen sampling points yielded results that were within 5 percent of 30 sampling points.

Saplings, classified as trees with diameters at breast height from 2.5 to 9.3 cm (1 to 3.6 inches) were sampled at each point in 50.2 m² arms-length quadrats. The quadrat method was used whenever the following two requirements of the point-centered quarter method could not be met:

- That an individual be located within each quarter.
- That an individual must not be measured twice.

Since saplings were sparse in many stands the quadrat method was used in most cases.

Seedlings, shrubs, and herbs were recorded by frequency in 0.5 × 0.5 m frames. These were placed 30 times in each stand; 30 such quadrats gave results, whose mean was within 10 percent of 50 quadrats. Seedlings, in addition, were counted in each quadrat.

7.72 Herbaceous and Low-Shrub Communities. Here are three examples of quantitative analyses in nonforest communities.

7.72.1 Alpine Communities in New Hampshire (BLISS). During his reconnaissance of the vegetation above timberline BLISS (1963) noted continua and discontinua in these alpine communities. On this basis he

---

* This is the lower class-limit of a 4 inch or 10 cm tree, which is a commonly used cutoff point for a broad size class of trees.

delimited the general spatial relations among them. He then located areas of 6×10 m within these segments, which were representative of a specific community type. Therefore, his plots were chosen subjectively. Excluded were rock outcrops and ecotones (transition zones). Within the 6×10 m areas he established 4×8 m plots. In view of the small size of the alpine plants, these 32 m² plots probably satisfied the minimal area requirement. However, for a quantitative analysis, the 8 m side of each plot was subdivided into strips of 1 m width running perpendicular to the slope. Of these, four were selected at random for the sample. In this way, the sample-area was further reduced to 16 m². In each 1×4 strip he counted individuals per species in five 20×50 cm (0.1 m²) frames. These were placed systematically, every 0.5 m, along each of the four 4×1 m strips. Twenty such frame placements were sampled in each plot, resulting in a total area of 2 m²

It is doubtful that this area satisfied the minimal-area requirement. Grasses and shoots of sedges were counted by bunches as individuals. The individuality of heath shrubs could not be ascertained, thus, they were counted by stems. This may have resulted in a mixture of counts of branches and individuals. Cover was estimated by perpendicularly projecting the shoot outline of the plants in each quadrat. Since this was repeated 20 times over a 2 m² area, the average cover percent was probably assessed quite accurately. But it was estimated and not measured. Density and frequency were measured out of 20 quadrats, with frequency a by-product of the density analysis.

We think the sampling applied by BLISS combined some features of the relevé method with that of the more typical North American quantitative methods. The similarities to the relevé method are the fine degree of stratification for recurring plant assemblages, the initial cohesive sample-area, and the small plot size of 32 m² area. These, of course, were necessitated through the time-consuming density analysis.

The purpose of the analysis was the description of plant communities of this alpine region as part of a larger project on plant productivity. The detailed density analysis was definitely justified for the second purpose, the productivity study. But for a mere description of the alpine communities, it probably would have been more efficient to list the plants of the total 32 m² plot area instead of only a 2 m² area within each plot, and to apply a semiquantitative estimate rating. Thereby much time could have been saved, more plots could have been analyzed in the same time, and a more complete and thorough community classification could have been established. This is not meant as a criticism, only as a clarification of methods in relation to the objectives. A quantitative analysis per se is not always better, although it often is favored because it is quantitative.

7.72.2 Herbaceous Wetland Communities, Saskatchewan (WALKER and WEHRHAN). The herbaceous wetlands as defined by WALKER and WEHRHAN (1971) are mixed sedge-grass-forb communities occurring in seasonally wet depressions surrounded by large wheat fields in the Canadian prairies. The depressions range in size from a few square meters to several hectares.

The objective of studying these communities was to analyze their place-to-place floristic variation in relation to edaphic variables.

The sample stands were selected according to the following criteria; (a) absence of any discernable past cultivation, (b) absence of severe grazing or mowing, (c) restricted to nonextreme wetland communities (called marsh-meadow and shallow marsh), (d) low salinity. Thirty-four stands in the vicinity of Saskatoon were selected. In each stand the species were analyzed for frequency in twenty placements of a 0.25 m² frame. The rather small number of frame placements was considered sufficient in view of the great variation between (rather than within) stands. The vegetation data was processed by an ordination technique.

The total sample covered only 5 m² per stand, which is probably smaller than the minimal area of these communities. However, plants found outside the frequency frames were also listed. No sample stand size is reported, except that the communities varied from a few square meters to several hectares in size. It is thus possible that communities with fragmentary species composition (of too small an area) were included as sample stands. The stand outline was probably the total community. No information is given on how the frame placements were arranged. It is possible that they were placed in scattered formation systematically across the stand, concentrated in the center, or randomly assigned throughout the variously sized stands.

7.72.3 Shrub and Grass Communities in Montana (BRANSON, MILLER and McQUEEN). The study by BRANSON, MILLER and McQUEEN (1970) relates to the investigation of dryland community patterns in the semiarid, cool-temperate zone of the mid-western United States. These patterns consist of a mosaic of grass and low-shrub communities of sagebrush and saltbush. In contrast to the interrupted distribution of the wetland communities in Saskatchewan, the dryland communities investigated in Montana occurred in a contiguous pattern.

As in the previously described study, the objective of BRANSON et al. was to investigate the variation of the floristic pattern in relation to edaphic variables in order to find environmental explanations for the local plant distributions.

The authors recognized 14 dominance communities from the start, which were divided into an upland and a lowland group. All communities were sampled by the point-intercept method. The seven upland communities were sampled by a continuous 329 m transect. The lowland communities were more widespread and thus were sampled by systematically spaced 15 m transect sections.

The point-intercept sampling was done with a frame of 10 vertically oriented pins. The pins were spaced 5 cm apart. Therefore, 20 points were used per meter transect distance. This is a very high sampling intensity. The same results could probably have been obtained with 10 or even only 5 points per meter.

The result of the upland community analysis is shown in FIGURE 7.7. Twelve plant species are listed. Their exact distribution along the 329 m (1080 ft) transect is shown, and the species quantities are diagrammed in black by percent cover. The community boundaries were drawn subjectively as the last step in the preparation of FIGURE 7.7. Therefore, the scheme is not an objective classification, but it clearly shows several discontinuities or vegetation boundaries. The reader can also recognize readily the authors' community classification concept, which is based on the spatial dominance (shoot cover)-changes among the species. The first and last Nuttall saltbush communities differ primarily by the high rock component in the first community. Therefore, these are dominance communities rather than communities identified and classified by differential species (see CHAP. 9).

In this study, as in the previous one, no sample stand size is given. Instead, as reference area one may consider any whole dominance-community that was evaluated by a single transect of points.

7.73 Conclusions. Many other variations of quantitative field analyses could be cited, but the main trends and principle differences from the relevé analysis should be quite clear by now.

7.73.1 Forest Communities. Major emphasis in quantitative analyses has so far been put on the tree stratum in forest communities. The tree layer is sampled across a large-sized stand, usually along transects, but it is also often sampled by random points or quadrats. Earlier, 10×10 m quadrats were used for counting the trees by species in diameter classes. More recently, the 10×10 m quadrats have been replaced by the distance-methods, which require less time, although this advantage is not always apparent (LINDSEY, BARTON and MILES 1958). In forests with sparse shrubby undergrowth, belt-transects or circular plots (with range-finder) may be a faster sampling unit for density estimates than points and distances. It should also be remembered that two-dimensional sampling units (plots) may have a greater chance to integrate variations in pattern (departures from randomness) than points and distances. Therefore, plots are more likely to give more accurate density estimates. The smaller woody plants and herbs in

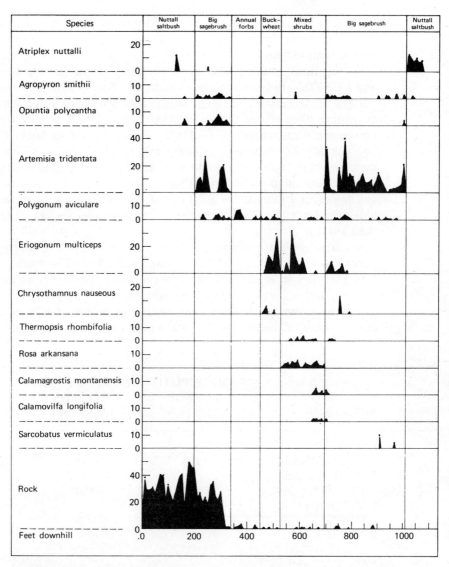

FIGURE 7.7. *Abundance of species and contacts with rocks for upland species. (From BRANSON, MILLER and McQUEEN 1970:393. Reproduced with permission from Ecology.)*

forest communities are usually recorded in quadrats. The quadrats are more or less adapted in size to the respective height strata. Shrubs and small trees are often recorded in 16, 10, or 4 m² quadrats and herbs in 1 and/or 0.1 m² quadrats. For the undergrowth plants, cover is more commonly assessed than density, but, more often, frequency is considered sufficient for descriptive purposes.

In addition to certain differences in technique and objectives, the five cited forest examples differ primarily in sampling intensity and size of sample stand.

Of the three studies employing distance measures for trees, BRAY and CURTIS' study is the least intensive. Forty distance measures involving 80 trees for *dbh* were considered sufficient for a forest cover extending over 60,000 m². In BORMANN and BUELL's study 160 distances were measured in a 28,000 m² forest, and in NEWSOME and DIX's study 60 distances in 6000 m². On the basis of one distance measure, this represents an area of 1500 m² in the first study, 175 m² in the second, and 100 m² in the third.

The lesser intensity in BRAY and CURTIS' study may have been balanced by a greater homogeneity requirement for the tree layer in their survey.

A major departure from the semiquantitative relevé analysis is apparent in the sampling of the lesser vegetation. The herb vegetation was sampled in all studies across the same area outline as the tree stratum, but a much smaller area was actually sampled.

This is shown in the following tabulation, which compares the examples in order of decreasing attention given to herbaceous undergrowth.

REFERENCE	SAMPLE PLOT AREA FOR TREES (m²)	SAMPLE PLOT AREA FOR HERBS (m²)	RATIO: TREE TO HERB PLOT
Relevé (SECTION 5.3)	200	200	1:1
BOURDEAU and OOSTING (1959)	6,000	160	37.5:1
DAUBENMIRE (1968)	375	5	75:1
BORMANN and BUELL (1964)	28,000	50	560:1
NEWSOME and DIX (1968)	6,000	7.5	800:1
BRAY and CURTIS (1957)	60,000	20	3,000:1

The comparison reemphasizes that quantitative analyses of undergrowth or herbaceous layers give information only on a fraction of the sample stand. The less abundant species are never adequately evaluated. In sample stands stratified for undergrowth homogeneity, the minimal area sample should be between approximately 50 to 200 m² in temperate forests (SECTION 5.2). The studies of BOURDEAU and OOSTING (1959) and of BORMANN and BUELL (1964) could be considered adequate from this point of view, but, since the large-sized sample stands were not selected for undergrowth homogeneity (only for tree layer homogeneity), it can be said that not one of the five quantitative forest studies we cited satisfied the minimal area requirement. If plant species outside the quantitative sample quadrats are not also evaluated (or at least listed as present) important information will be lost.

The two continuum studies (NEWSOME and DIX 1968, BRAY and CURTIS 1957) show the least intensive undergrowth evaluation. In BRAY and CURTIS' study this seems to be related to a difference in concept and objectives. While undergrowth species are not ignored, they are regarded as relatively unimportant. CURTIS (1959) considers them merely dependents of the dominants. Although this is true in some situations, many studies have given evidence of their relative independence. For example, NEWSOME and DIX found that many undergrowth species occur outside the forest and most are relatively independent of the species composition of the dominants.

Such a scattered sample of 1 m² in every 3000 m² over sample stands of 60,000 m², as used in the BRAY and CURTIS study, can hardly be expected to yield sufficient data to document the existence of associations among undergrowth plants, or among plants belonging to a vertical cross-section of forest strata.

A rough time-comparison of the quantitative forest analyses with the forest relevé analyses can be made as follows. It took about 45 minutes to complete the 200 m² relevé analysis discussed in SECTION 5.3. The count-plot analysis example (given in SECTION 7.3) took about 1 hour. This covered an area (120 m²) about half the size of the relevé and included a count of about 100 trees—still an insufficient sample. The point-centered quarter analysis example of 5 points with 20 distances (given in SECTION 7.63) took about 30 minutes. In the latter analysis we ignored the small trees under 2 m height, which were included in the count-plot analysis. Thus, the two quantitative methods took about the same time. It is possible that a larger sample would have come out in favor of the distance method.

However, this comparison gives a general indication of the time required for the various analyses described. While a standard relevé analysis may take 1 hour, any quantitative analysis will take more time. Of course, if such details as a rough species list and a determination of tree size classes are made in advance, a quantitative analysis can be done more quickly.

The 15 sampling points (60 trees) in NEWSOME and DIX's study by point-centered quarter method may have taken 1 to 2 hours. The 15 count-quadrats of 50 m² each for the saplings may have added another hour or two. The 30 small (0.25 m²) frequency frame placements also must have added about 2 hours to the analysis. Thus, we estimate that the quantitative field analysis of NEWSOME and DIX took from about 4 to 8 hours per sample plot.

The 40 distance measures of 80 trees at 40 points in BRAY and CURTIS' study must have taken at least 2 hours per stand, because of the wide spacing of each point. Their 20 frequency frame placements must have added another 2 hours field analysis time. Thus, the two studies for continuum analysis must have taken from half-a-day to a day per stand. It is probably not possible to complete more than one such large sample stand per day. In contrast, it is relatively easy to complete four forest releve analyses per day as described in SECTION 5.3.

7.73.2 Herbaceous and Low-Shrub Communities.   In the three quantitative examples of nonforest communities, the same spectrum of parameters (density, cover, and frequency) was assessed as in the forest communities. However, in contrast to the tree stratum, density is the most complicated parameter to assess in communities whose dominant strata are herbaceous. This complication became apparent through the BLISS study of alpine communities, in which only selected species were counted, while the parameters evaluated for all species were cover and frequency. BLISS emphasized that the counting of low-shrubs, bunchgrasses and sedges in his study was for subsequent productivity research. Such a counting effort would not be warranted for a classification or ordination of these alpine communities. For the latter purposes, his cover and frequency evaluation in 20 small quadrats per stand was sufficient.

Species cover in this study was estimated in 20 small (0.1 m²) frames. This poses another question regarding the reliability of a cover estimate: Is it more adequate to (a) evaluate cover accurately in a small part of the plot (2 m² out of 60 m²) or (b) estimate cover with the general BRAUN-BLANQUET scale over the entire plot? Only a comparison of methods would give an answer. For the purposes of a classification or ordination, the second method appears more expedient, because it takes much less time and relates to the whole sample plot.

An evaluation restricted to 20 frequency frame placements per stand was also considered sufficient by WALKER and WEHRHAN for describing the wetland communities in Saskatchewan. If all species present in a "minimal area" were first carefully recorded, and then species quantities were evaluated by such a rapid quantitative method as applied by 20 frequency frame-placements, the difference between the relevé method and this kind of quantitative method would only lie in the accuracy of assessing the more abundant species. It is then still debatable whether 20 frequency placements give a better estimate of species quantity than a cover estimate per species. Many investigators would say that the 20 frame placements give a more objective measure of species quantities, but this judgment should depend on the reproducibility of the results with the given quantitative method.

Depending on the number of species present, the determination of frequency by 20 frame placements would extend the time required for a standard relevé analysis by about 30 to 60 minutes. The question the investigator then faces is, whether he should invest this extra time in evaluating the more abundant species by such a quantitative method, or whether he considers the time more usefully spent in starting another relevé with the BRAUN-BLANQUET scale.

Another problem, discussed in SECTION 6.31, is that frequency in quadrats is not an absolute measure. Without any doubt, a far better quantitative value is cover. The measurement of cover by the point-intercept method as done in the study of grass and shrub communities in Montana by BRANSON et al. is by far the most meaningful quantitative analysis of such nonforest communities for descriptive purposes. In such species-poor communities, as found in this area of Montana, it may even be possible to include in the measure of cover nearly all species that are found in the minimal areas of such communities. But even in species-rich communities, in which again only the more abundant species could be measured adequately by this method, the point-intercept method holds the greatest promise. This is so, because (a) the method gives an absolute measure, (b) the parameter measured is considered ecologically the most significant of the three (SECTION 6.51), (c) cover can be assessed for all plant life forms including the trees (SECTION 6.54.2) and (d) the point-intercept method gives the same measure as is aimed at by the estimate scales (SECTION 5.42).

The sampling intensity of the BRANSON et al. study was very high and consequently the time investment must have been high as well. Depending on species richness of a herbaceous or low-shrub community and the familiarity of the investigator with the species, a sample of 200 points taken with a point-frequency frame (FIG. 6.6, SECTION 6.54) may take between 30 and 60 minutes. In most situations a 200 point sample resulting in a measure of percent cover of the more abundant species seems a better time investment than the placement of 20 frequency frames for the quantitative description of herbaceous and low-shrub communities.

However, both parameters, cover and quadrat-frequency, may complement each other in studies where one aims at more than the description of spatial variation of a plant cover. For example, if one is interested also in the time-variation of a herbaceous cover, one may find species frequency in quadrats to be a less fluctuating parameter than cover in the seasonal behavior of a perennial grassland. The degree of variation of these parameters can in itself be of important information, for example, for the grazing value.

7.73.3   Size of Sample Stands.   Further differences in concept between the relevé method and the quantitative methods are shown by the different sizes of sampling units. It is interesting to note that in FIGURE 7.6 each of the six sample units is subjectively selected, each of them with great care. The sampling within all but the first, however, was done by quantitative methods.

Also, these six sample stands were all chosen for their homogeneity in vegetation cover. They undoubtedly include the minimal area of each community for which they were selected as samples. For this reason they could all be referred to as relevés. However, the four large sample stands are not relevés in the strict sense, because they include a much larger area than the minimal area.

The difference in sample stand size rests on a difference in the homogeneity concept of the authors. Only two degrees of homogeneity have to be recognized to explain this difference. The small-sized sample stands were delimited by the uniformity of the small-sized vegetation—the herbs and small shrubs. The large-sized stands were delimited by the tree vegetation.

A second difference is that the relevé method aims with each sample stand (relevé) at a sample of a near-total species composition of a concrete community. In contrast, the standard quantitative methods aim at sampling the more abundant species. For the reason also, the forest sample stands are large. However, if the minimal area is not used as a guide to size, the guide to size is limited by the author's decision as to how many species he intends to sample adequately. In the count-plot example given in SECTION 7.3 only one tree species (*Psidium guajava*) may be considered adequately sampled with over 30 individuals on 120 m². If the other two important native tree species (*Acacia koa* and *Metrosideros collina*) were to be sampled adequately (with at least 30 individuals per species), the sample stand would have to be enlarged to at least 800 m² (0.2 acre).

However, for a complete stand analysis, both aspects, the near-total species record and an adequate sample of the more abundant species, can and should be combined. In forest stands this can be done as follows: A forest stand sample should always be delimited by the homogeneity or uniformity of the lesser vegetation and the habitat. In mountainous terrain and in level terrain with small-area water table variations, this will result in relatively small sample stands such as shown in FIGURE 7.6 by the relevé example (200 m²) and by DAUBEN-MIRE's sample stand size (375 m²). To obtain an adequate enumeration of the more important tree species, it will then be necessary to add more relevés of the same vegetation type. In level terrain without distinct habitat variations, the relevé should be enlarged for enumeration of the tree layer by watching that the undergrowth-homogeneity is maintained for the larger sized sample area. In the latter case, the original near-total species relevé, which only needs to be a little larger than the minimal area of the community, may form a nested plot within the larger sample stand.

In contrast, quantitative concepts requiring contiguous, large uniform dominance-communities tend to exclude a number of extreme variations in any regional vegetation cover. Many cover variations may be too small for the arbitrarily decided minimum sample area. Therefore, only part of the regional vegetation cover can be described by such an approach. For purposes of classification this would impose a severe disadvantage. For example, BRAY and CURTIS' 15 acre sample stand requirement would be impossible to apply in most mountainous regions.

Yet, continuum analysis is not limited by sample unit size, degree of entitation, or regional characteristics of the vegetation. It can be carried out in narrowly defined communities that are sampled by small plots, if these are distributed over the total geographic range of recurrence of these communities. In this case, the difference between the classification and continuum approach dissolves, because the question is no longer which of the two approaches is more objective or whether continuity or discontinuity is the truer abstraction of the nature of the community. It is merely a question of whether the investigator is interested in the portrayal of the complete vegetation cover of specific regions or whether he is interested in limiting his description to an accurate enumeration of the abundant species only.

Classifying and Ordinating Vegetation Data

Vegetation Structure, Classification Units, and Systems

8

8.1 CONCEPTS OF VEGETATION STRUCTURE

DANSEREAU (1957:147) defines *vegetation structure* as "the organization in space of the individuals that form a stand (and by extension a vegetation type or a plant association)," and he states that "the primary elements of structure are growth form, stratification, and coverage." This definition is still valid today. However, the term vegetation structure is used with different meanings. In the most general sense, the concept of structure is used in all biological research as the complementary concept to function; function relating to physiological processes and structure to the anatomy and morphology of the objects under study. In vegetation ecology one may speak of vegetation structure at least at five levels.

1. Vegetation physiognomy.
2. Biomass structure.
3. Life form structure.
4. Floristic structure.
5. Stand structure.

These five levels of vegetation structure are hierarchically integrated in that the first level includes the second, the second includes the third, and so forth. Thus, the five concepts of vegetation structure merely represent different levels of generalization, with level 1 being the most general and level 5 the most exacting.

FOSBERG (1961) has made a clear distinction between the first two concepts. He defines *vegetation physiognomy* as the external appearance of vegetation. Physiognomy in this sense is the result in part of biomass structure, functional phenomena (such as leaf-fall in forests), and gross compositional characteristics (such as luxuriance or relative xeromorphy). Since physiognomy is the expression of at least three more precisely distinguishable vegetation criteria, of which structure is only a part, physiognomy should not be confused with structure as defined in the second sense.

Biomass structure is a more precise concept of vegetation structure. It relates specifically to the spacing and height of plants forming the matrix of a vegetation cover. The term "vegetation structure," when not otherwise defined, should be used only in this sense of plant biomass structure. An attempt to quantify this concept is shown in SECTION 8.4.

Life form structure relates to the composition of growth forms or life forms of plants in a vegetation. The plant life form concept, which allows grouping individual species with similar gross-morphologies into life form types, is defined in SECTION 8.2. Life form structure or composition is a still more precise concept than biomass structure. Life form structure can be meaningfully quantified (SECTION 8.3).

Floristic structure is usually separated from structure in the traditional sense (i.e., biomass and life form structure) as "floristic composition." It is advisable to retain this separation and not to imply floristic composition when using the word structure without further clarification. In the case of life form structure one can also speak of life form composition. The word composition, when not further qualified, usually implies floristic composition at the species level.

KERSHAW (1964:5) distinguishes three components of vegetation structure: (a) vertical structure (i.e., stratification into layers); (b) horizontal structure (i.e., spatial distribution of species populations and individuals), and (c) quantitative structure (i.e., abundance of each species in the community). The same three divisions of vegetation structure have been accepted by SHIMWELL (1972). SHIMWELL extends the term to include "structure in time," referring to succession and climax. However, we are here concerned only with spatial structure.

We consider vertical structure part of biomass structure as defined above. The other two components of structure as defined by KERSHAW are parts of "floristic structure" since they are seen at the level of species rather than life form. KÜCHLER (1967, 1972) likes to reserve the term structure in the sense of life form structure, but, as shown by KERSHAW's definition and that of other ecologists (TÜXEN 1970, WHITTAKER 1970), the term is understood today in the broad sense of vegetation form or morphology as opposed to function in the physiological sense. In this broad sense of structure, species patterns (in the sense of random, regular, or contagious distributions), species quantities, species diversity (PAINE 1971), and mapped patterns or mosaics of species associations, are all aspects of structure. We may add that a properly recorded relevé conveys structure at least at two levels, (a) by the recording of species in vertical strata or layers, and (b) by showing the abundance of each species in each stratum. A relevé record also shows the number of species on a given area (i.e., species diversity, which is considered an aspect of community structure). One can add the life form or growth form of each species and thus show a fourth aspect of structure on a relevé record sheet.

We already discussed *stand structure* in the preceding chapter, SECTION 7.31. In that section we were concerned with the distribution of numbers of individuals in different size-classes of certain tree species in a forest stand. When individual species are analyzed in this way, one can speak of a "population structure" analysis. When the population structure curves of several species of the same stand are compared with one another, one can speak of a "stand structure" or "community structure" analysis. This type of structural analysis is only a more detailed quantitative approach to the recording of species in different vertical strata in a relevé with an estimate of species quantity in each stratum (which was discussed as an aspect of floristic structure).

Therefore, to avoid confusion it seems necessary to subdivide vegetation structure into at least the five current understandings listed under the first paragraph.

Vegetation can be classified by structure without reference to species names. This has proved to be particularly helpful in floristically un-explored areas, and in areas where vegetation cannot be classified easily by dominant species, for example, in the humid tropics (WEBB 1968, WEBB et al. 1970). Moreover, structural descriptions are highly complementary to floristic descriptions of vegetation. A species list of a community, even when the quantities of each species are shown, does not in itself convey the morphological and anatomical features of that vegetation segment. The latter features can, however, be portrayed in a structural classification. Such classification systems are referred to as "structural systems." Structural systems can be contrasted with classification systems based on species composition. The latter are re-ferred to as floristic classification schemes. The primary attributes in a structural classification scheme are not species, but plant life forms, vertical stratification of plant biomass and its horizontal coverage within these strata, and phenological or seasonal aspects. These vegeta-tion attributes will be discussed next.

8.2 PLANT LIFE FORMS

Plants are classified taxonomically into families, genera, species, vari-eties, etc. This, however, is not the only way to classify plants. Species and individuals can be grouped into life form or growth form classes on the basis of their similarities in structure and function. A plant life form is usually understood to be a growth form which displays an obvious relationship to important environmental factors. For example, a deciduous tree is a plant life form that responds to an unfavorable season by shedding its leaves.

The life form of a plant species is usually a constant characteristic. But the same species may assume a different life form when growing under very different environmental conditions. For example, certain high altitude tree species assume a creeping growth habit, called "krummholz," near their upper limit of distribution, while they grow as perfectly normal trees below. A species goes through differing life form stages in its life cycle from seed to maturity. Here, we are con-cerned primarily with the life form at maturity. Plants of the same life form growing together are likely to compete directly for the same space or niche. Their similarity in structure and form indicates a similarity in adaptation to the utilization of the environmental resources offered in a given space. The most extreme example of life form similarity is shown among individuals of the same species. Wherever they grow close together, they are also the strongest competitors, because they are adapted to use the environmental resources in the same manner.

Of course, species of very unrelated families may also be of the same life form. For example similar stem-succulents evolved in the families Cactaceae, Euphorbiaceae, Asclepidiaceae and Liliaceae.

The composition of life forms, in addition to the composition of species, in a plant community is of special interest, for it may give information on the response of a community to particular environmental factors, on the utilization of space (if combined with cover values), and on the probable competitive relations within a community.

Information on plant life forms is of value also for the reader who cannot visualize from the name how a species looks. Moreover, the life forms are not yet known for all species. This is particularly true for the functional life form characteristics, such as deciduousness or evergreenness in woody plant species.

Since VON HUMBOLDT (1806), many differing classifications of plant life forms, "growth forms," or "basic forms" have been developed. A literature review has recently been presented by LACZA and FEKETE 1969 (see also FEKETE and LACZA 1970). The most detailed extension of VON HUMBOLDT's scheme was presented by GINZBERGER and STADLMANN (1939). However, their primary objective was to give a plant physiognomical tool for the description of different landscapes and not so much for the investigation of the vegetation. In vegetation research, RAUNKIAER's (1934, 1937) life form system and its further extension by BRAUN-BLANQUET (1928) has found widest application. RAUNKIAER's system is ecologically oriented and based primarily on the position of the buds or organs from which new shoots or foliage develop after an unfavorable season. In the temperate zone this unfavorable season is the winter. Yet, unfavorable seasons occur also in areas with drought seasons in the subtropics and tropics. Plant behavior during an unfavorable season in warm climates is very similar to plant behavior during winter in the temperate climates. Therefore, RAUNKIAER's system can be applied also to areas outside the temperate zone. However, plant behavior during the growing season was largely disregarded by RAUNKIAER.

As with RÜBEL (1922), however, we consider it necessary to include plant behavior characteristics of the favorable season. The structure and seasonality of the crown, foliage, and shoot systems are emphasized in our system (ELLENBERG and MUELLER-DOMBOIS 1967a), APPENDIX A.

In contrast to BRAUN-BLANQUET (1928), whose system is otherwise reflected in ours, we also follow RÜBEL by beginning with the dominant and large-sized life forms, while mosses, lichens, and algae are treated at the end in our scheme. The life form system for thallophytes, parasites, and heterotrophic plants has not been carried through in the same detail as for the more important primary producers. We are presenting the life form system in form of a key that is applicable to the range of plant life forms occurring from the tundra to the desert and humid tropics. However, its workability for all situations awaits further testing. We welcome any comments that may be used to improve the key (APPENDIX A).

The key separates the five basic life forms of RAUNKIAER, the phanerophytes, chamaephytes, hemicryptophytes, geophytes, and therophytes. But then it is expanded to include a total of 23 major life forms. The first level of separation of these 23 major life forms is functional, namely into autotrophic, semiautotrophic, or heterotrophic plants. Within each of the three trophic groups, the second level of separation is on the basis of anatomical structure, into vascular (kormophytes) and nonvascular (thallophytes) plants. Within these groups, the third level of separation is on the basis of the support-structures of plants.

For example, among the autotrophic vascular plants three groups are distinguished, structurally self-supporting plants to which the five basic life forms of RAUNKIAER belong, plants that support themselves structurally on others such as lianas and epiphytes, and plants whose photosynthetic organs are structurally supported by water in aquatic habitats.

On a fourth level one can distinguish perennial and annual plants, which exist among both groups, the kormophytes and thallophytes.

The range of adaptation in plants from perennial to annual existence has been well recognized by RAUNKIAER in his five basic life forms. This range of adaptation relates to the mode of shoot-withdrawal in the unfavorable season. It extends in perennial plants from (a) no withdrawal in trees, shrubs, and herbs in the humid tropics to (b) foliage-withdrawal in deciduous woody plants and shoot-reducing shrubs and herbs (chamaephytes and hemicryptophytes) to (c) total shoot reduction in geophytes. The most complete form of shoot reduction is represented by the annuals or therophytes, which survive only in the form of seeds.

The key does not only recognize further variations in the modes of shoot reduction, but also emphasizes the basic separation into scapose (single stemmed), caespitose (multistemmed or bunched), and reptant (creeping or matted) support-systems that are found in all five of RAUNKIAER's life forms. These different support-systems are important adaptations in the competition and coexistence of the plant life forms in communities. A separation into caespitose and matted forms is also indicated among the mosses and lichens.

Since the structural differentiations in nature are greatest within the phanerophytes and chamaephytes, these groups are separated in the key into more subforms than the others. For example, the phanerophytes (plants > 1 m tall) are separated into four major height classes (nano-, micro-, meso- and megaphanerophytes) and into five main stem or trunk forms (normal woody, tuft trees, bottle trees, succulent, and herbaceous-stem trees). Within these groups rooting and bark characteristics and the organs that are active during the growing seasons are further differentiated, for example, by leaf size, shape and texture, and crown form. An open-ended numerical system of all recognizable structural and functional plant properties in the key allows for a detailed life form classification in the field that can be extended where necessary and adapted to computer processing.

8.3 LIFE FORM SPECTRA

When the life forms of all species are listed for a community, one can show the result in form of a spectrum. The simplest form of a life form spectrum is to classify all species on a list into the five basic RAUNKIAERIAN life forms, to sum the number of species by life form class, and to express the results in percent. Life form spectra of this sort were used by RAUNKIAER to compare different floristic regions. The next step was the application of a more refined system of life form types for the comparison of communities (BRAUN-BLAN-QUET 1928). A still more detailed application of life form spectra is to incorporate species quantities as done by TÜXEN and ELLENBERG (1937), STERN and BUELL (1951) and others. One can use cover-abundance ratings or any measured quantitative value (density, cover, frequency).

FIGURE 8.1 shows life form spectra of forest communities near Hamburg. This example does not yet incorporate all variations of life forms shown in the key. It is restricted largely to the forms recognized by RAUNKIAER. However, in contrast to the more common way of representation, the kormophytes were here separated from the thallophytes as first suggested by GAMS (1918). In addition, the proportion of evergreen species within the different life form types, and those that carry on photosynthesis only in the spring, are shown by special markings on the histograms. This brings out more clearly the ecological differences. The beech forest on moraine is characterized particularly by spring- and summergreen rhizome-geophytes, while evergreen plants and thallophytes are little represented. The latter two life form types are much more abundant in the less shaded oak-birch forest. They usually increase under the same tree cover and climatic conditions with increasing soil acidity and decreasing nitrogen supply and are practically the only ground vegetation in the artificial spruce forest with its seasonally nonvarying light relations. Also, certain characteris-

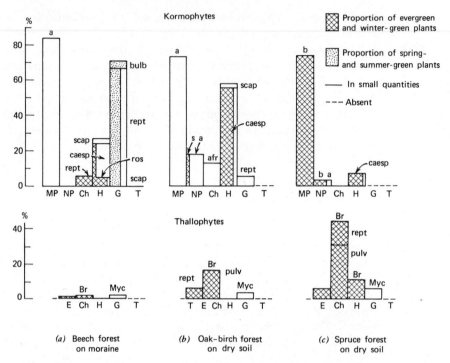

FIGURE 8.1   *Life from spectra of forest stands on nutritionally rich (a) and nutritionally poor soils (b and c). Legend: MP=Meso- and Microphanerophytes; NP=Nanophanerophytes; Ch=Chamaephytes; H=Hemicrytophytes; G=Geophytes; T=Therophytes; E=Epiphytes (here only among Thallophytes); a=aestivo, i.e., summer-green or cold-deciduous; b=belonido, i.e., needle-leaved; fr=frutescent, i.e., woodiness completed into branch-tips; s= sclerophyllous, i.e., hard-leaved; rept=reptant, i.e., creeping; caesp=caespitose, i.e., branched from near the base; scap=scapose, i.e., single-stemmed; ros=with rosette; Br=Bryophytes; Myc=Fungi.*

tic differences with regard to thallophytic vegetation are shown between hardwood and softwood stands. Mat-forming mosses (*Bryochamaephyta reptantia*) are more abundant than cushion-forming mosses (*Bryochamaephyta pulvinata*) in the oak-birch forest. Mat-forming mosses grow faster and can cope better with the annual litter fall of the deciduous forest. Pulvinate, or cushion-forming mosses, are more abundant in the conifer-forest. But some mosses with a flat-appressed growth habit are found here too, such as *Plagiothecium curvifolium* and some liverworts. These are completely absent from the hardwood forest.

In the life form spectra, the individual groups are designated by abbreviations of their scientific names, which are given in FIGURE 8.1 and in APPENDIX A. The abbreviations or life form symbols are commonly used also in relevé tables. This enables the reader unfamiliar

TABLE 8.1   Tabulation of BRAUN-BLANQUET Cover-Abundance Values Converted to Mean Cover Degree.

MAGNITUDE AFTER BRAUN-BLANQUET	RANGE OF COVER DEGREE VALUES (PERCENT)	MEAN COVER DEGREE (PERCENT)
5	75–100	87.5
4	50– 75	62.5
3	25– 50	37.5
2	5– 25	15.0
1	} < 5	2.5[a]
+		0.1[a]
r	Value ignored	

[a] Arbitrarily determined.

with the species to obtain an idea of the structure of the communities and of the ecological role of the species.

The percent cover values in FIGURE 8.1 were derived from BRAUN-BLANQUET scale values. When using the BRAUN-BLANQUET scale it is necessary to convert the rating symbols into cover percentages, because the scale values cannot be added. When converted to percent cover values, the species quantities in the same life form class of a relevé are added for presentation in a life form diagram. The mean values given in TABLE 8.1 were used for FIGURE 8.1 (after TÜXEN and ELLENBERG 1937).

This conversion, however approximate, is usually sufficient for obtaining a good representation for life form spectra.

8.4   BIOMASS STRATIFICATION AND PROFILE DIAGRAMS

Plant-height is used as a criterion in the life form classification. Therefore, the life form spectra also give a certain idea of stratification or layering in communities. However, biomass stratification can also be indicated diagrammatically (after HULT 1881), if both height and cover of each layer have been recorded during field analysis. FIGURE 8.2 shows layer-diagrams of a *Fagus* (beech) forest and a *Quercus-Carpinus* (oak-hornbeam) mixedwood forest. The cover-abundance values of the BRAUN-BLANQUET scale and their corresponding cover percentages were used instead of the earlier abundance classes of HULT.

Instead of using abstract diagrams, it is possible to represent layering of the species also in less abstract or semischematic illustrations as

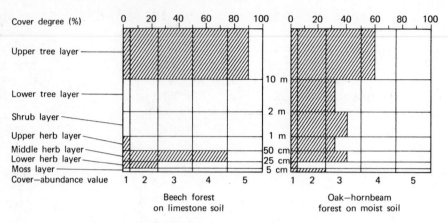

FIGURE 8.2.   Layer diagrams of beech and oak–hornbeam forests.

profile diagrams. This can be done with differing degrees of exactness that must be in balance with the size of the community, its variability and the objectives of the study. Moreover, profile diagrams can be used to illustrate details in vertical spacing of species, which are not representable in layer diagrams. The profile-diagram method has been used effectively by BEARD (1946) for portraying the structural characteristics of tropical rain forests in South America.

FIGURE 8.3 shows a profile diagram of a closed forest community and FIGURE 8.4 a profile diagram of a more or less open beach community.

Since spacing is a three-dimensional property, it is important to choose a transect-width for the profile diagram that conveys the correct plant spacing of the community. The width should usually not exceed a few meters.

For a detailed analysis, a transect map can be established by drawing to scale the position of the more dominant individuals and by measuring their heights. The transect width depends on the size and spacing of the dominants. A three meter wide strip to either side of a compass line can be used in many normally stocked woody plant communities. For an exact representation of crown cover, the line-intercept method is a useful tool in preparing profile diagrams.

Most plant communities not only show stratification above the soil but also within. Unfortunately, little is known about root stratification since it is not readily observed. The classification of species into rooting-depth groups as done by ELLENBERG (1950, 1952) can be used as a guide.

It would be very desirable to investigate as many as possible of the more important plant communities by profile diagrams showing the layering above and below ground. Often a knowledge of root-strati-

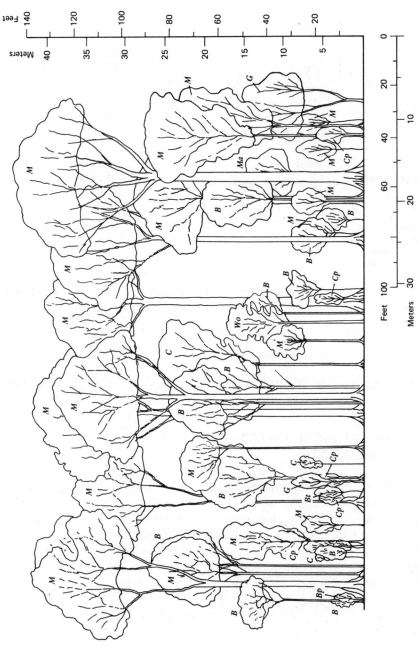

FIGURE 8.3. Profile diagram of a closed evergreen forest in Trinidad, British West Indies. Mora excelsa community. The diagram represents a strip of forest 61 m long and 7.6 m wide. Legend: B=Clathrotropis brachypetala; Bp=Swartzia pinnata; Bt=Rudgea freemani; C=Carapa guianensis; Cp=Brownea latifolia; G=Eschweilera subglandulosa; M=Mora excelsa; MA=Sterculia caribea; Wo=Terminalia amazonia. (From BEARD 1946. Reproduced with permission from Ecology.)

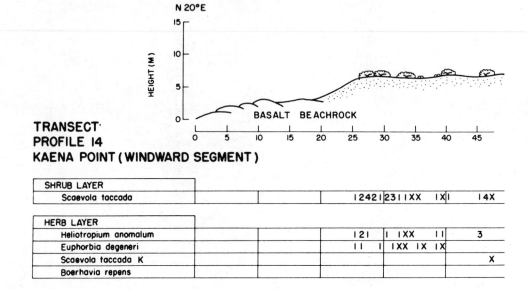

SHRUB LAYER				
Scaevola taccada			I 2 4 2 I	2 3 I I XX I X I I 4 X

HERB LAYER				
Heliotropium anomalum			I 2 I	I I XX I I	3
Euphorbia degeneri			I I	I I XX I X I X	
Scaevola taccada K					X
Boerhavia repens					

TRANSECT·
PROFILE 14
KAENA POINT (WINDWARD SEGMENT)

FIGURE 8.4   Profile diagram of an open coastline community on Oahu, Hawaiian Islands. Width of the belt-transect was 2 m. The tabulated species values are BRAUN-BLANQUET scale values, × here stands for + (see page 60); K = seedlings. (From RICHMOND and MUELLER-DOMBOIS 1972.)

fication provides the key to an understanding of the ecological relationships among the species and communities studied.

8.5   PHENOLOGICAL ASPECTS IN PLANT COMMUNITIES

Nearly all plant commmunities of the temperate, subtropical, and tropical dry zones have more or less pronounced seasonal aspects. Phenological community changes were already well described in earlier classical studies, particularly by VON KERNER (1863).

The phenology of certain species can be assessed quantitatively by periodic observations in permanent quadrats and by marking individuals, or even individual branches and flowers on trees.

The phenology of all species of a community can be represented diagrammatically as done by GAMS (1918) or ELLENBERG (1939), among others. Recently, DIERSCHKE (1970) reviewed and improved the diagrammatic method of representing community phenology. FIGURE 8.5 shows a phenological diagram of a herbaceous community which, as a community-type, occurs along forest borders under shrubs

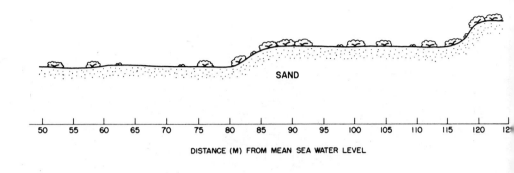

DISTANCE (M) FROM MEAN SEA WATER LEVEL

| X32 | I I | | X X | I | 25555 | 554 | I I 2 I | I X I I | X | I I | I | I 2 3 | 4 4 |

X X		2	I I		XX I		X	2	I I
	X		X2 I		X				I
X	X	X	X	X	XXX				XX X	
				X			X		X

in northwest Germany. The diagram refers to the observation of seasonal changes in a permanent 5 m² quadrat in such a community. The length of the horizontal strip next to each plant name refers to the length of time, between April and October, during which the shoot parts are observed above the ground. The width or height of each strip, which varies through the year, refers to the quantity in terms of BRAUN-BLANQUET values from r to 2. All species are therefore present in small quantities only. The vertical lines inside each strip refer to the vegetative phenological phases which are defined as follows:

0– 2	First leaf unfolding	
3	Two to three leaves unfolded	Shoot system
4– 5	Nearly all leaves unfolded	developing
6	Plant fully developed	
7	Stem or first leaf turning yellow	
8	< 50 percent of shoot system yellow	Shoot system
9–10	> 50 percent of shoot system yellow or plant dead	dying

When the symbols are understood, the diagram shows with one view the integration of the vegetative phases among the species in the community.

The generative development is shown only for the flowering phases on the diagram, by time and quantity. In addition the color of the flowers is indicated.

It is often useful to take brief notes about the phenological condition of the species during a relevé analysis. This can be done by add-

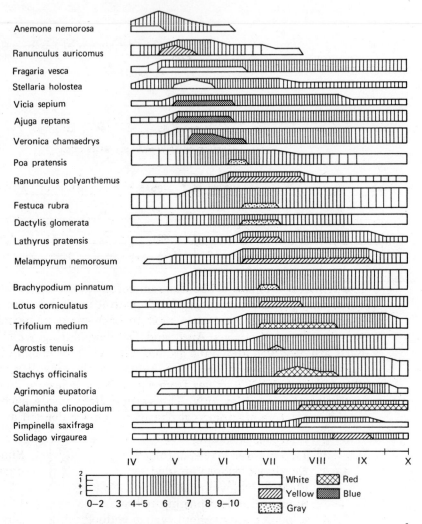

FIGURE 8.5.   *Phenological diagram of selected species in a permanent relevé of the Trifolium–Melampyrum nemorosum community near Göttingen, Germany. Explanation in text. (From DIERSCHKE 1972.)*

ing the following symbols to the plant names listed:

g	germinant	sp	sprouting (in case of bi- and
s	seedling		perennials)
st	remaining sterile	w	withering or dying
b	budding	f	foliage shed
fl	flowering	d	dead, or aerial shoots dried up
fr	fruiting	sd	only present as seed

For more exact phenological observations it is advisable to use a numerical scale as, for example, explained above and designed by ELLENBERG (1954) for microclimatological field investigations.

8.6   CLASSIFICATION: LIMITS AND POSSIBILITIES

8.61   Plant Distribution and Classification. All who have analyzed a larger number of vegetation releves will have made the following three observations:

    1.   That similar species combinations recur under similar habitat conditions, even in geographically quite separated locations.

    2.   That no two releve analyses are exactly alike and that even stands in close proximity on seemingly equivalent habitats show certain deviations from one another.

    3.   That the species assemblage changes more or less continuously if one samples a geographically widespread, major community throughout its range of distribution.

Therefore, recurring species combinations are obviously correlated with their environment. However, these species combinations are variable, and their variability is unlimited in principle, since the plant community lacks organismic character.

These three generalizations, which are substantiated over and over again during extended field work, define the possibilities and limits of a systematic classification of plant communities. It is undoubtedly possible to order the numerous and variable plant communities into a comprehensive system. However, such an order can be achieved only if one abstracts particular properties from individual communities. Moreover, such an orderly arrangement is applicable only to a limited geographic area; that is, it is only of regional significance and cannot be extrapolated to another area without further study.

The last point brings out a major difference to the taxonomic classification of plants and animals. The latter are abstractions as well, but they refer to populations of individuals with genetic relationships. In contrast, the relationships of plant communities in a vegetation class consist of certain similarities in plant combinations and their causes of development. A vegetation class or community type is the result only of abstracting certain similarities of a number of concrete communities.

As in the classification of species, rocks, land forms, or other study objects, the classification of plant communities depends on the questions asked and on general agreement. Classification is necessary for scientific communication, but it can never become absolutely objective or accomplished without the element of personal judgment.

8.62 Criteria and Systems for Classifying Vegetation. Plant communities can be classified by the following criteria depending upon which of the properties are emphasized:

I. Properties of the vegetation itself
   A. Physiognomic and structural criteria
      1. Certain life or growth forms
         a) Dominant life forms
         b) Combination of life forms
      2. Vertical stratification (layering) and "organizational development" (i.e., the complexity in structure as produced by arrangements of different life forms)
      3. Periodicity (i.e., leaf-fall, etc.)
   B. Floristic criteria
      1. A single plant species (in special cases also 2–3 species)
         a) The dominant species (in terms of height or cover or a combination of both)
         b) The most frequent species (or most numerous species)
      2. Certain groups of plant species
         a) Statistically derived plant groups
            (1) Constant (always present) species
            (2) Differential (or separating) species
            (3) Character (indicator or diagnostic) species
         b) Plant groups derived *without* using vegetation statistics
            (1) Species of the same ecological significance
            (2) Species of the same geographical distribution
            (3) Species of the same dynamic significance
   C. Numerical relation criteria (community coefficients)
      1. Between different species
      2. Between different communities
II. Properties outside the vegetation
   A. The presumed final stage in vegetation development (climax)
      1. Defined by life form combinations
      2. Defined by floristic criteria
   B. The habitat or environment
      1. Certain site factors
         a) Climate
        b) Water relations
        c) Soil
        d) Anthropogenic influences (management practices)
      2. Combination of site factors
   C. Geographical location of communities

III.   Properties combining vegetation and environment
   A. By independent analysis of vegetation (in sense of I) and independent analysis of environmental components and subsequent correlation (for example, through matching of map-units).
   B. By combined analysis of vegetation and environment and emphasis on interdependencies in the functional sense.

Interest was focused initially on single or only a few criteria for classifying, as in the taxonomic classification of plant and animal species. The best known systems were the following:

1.   The physiognomic classifications of GRISEBACH (1872) and DRUDE (1902).

2.   The environmentally oriented classifications of WARMING (1909), GRAEBNER (e.g., 1925) and SUKACHEV (1932).

3.   The physiognomic-ecological classification of SCHIMPER (1898), DIELS and MATTICK (1908), BROCKMAN-JEROSCH and RÜBEL (1912), DU RIETZ (1921) and RÜBEL (1933). (DU RIETZ also used dominant species.)

4.   The areal-geographic-floristic classification of SCHMID (1963) (based mainly on the geographical distribution of species).

5.   The dynamic-floristic classification of CLEMENTS (1916, 1928), TANSLEY and other American and British ecologists (mainly based on the final stages of vegetation development).

6.   The floristic-structural classification of CAJANDER (1909) and BRAUN-BLANQUET (1928). (The latter uses besides statistically derived species groups only the property of organizational development of communities.)

Each of these classifying systems has certain advantages and disadvantages. Their common disadvantage is a certain inflexibility in adhering to a predefined consistency. Consequently, they cannot do justice to the natural variability of vegetation. Even the most successful system among those mentioned, the BRAUN-BLANQUET system is no exception in this regard. It is also to be considered an "artificial" system.

Systematic taxonomy after LINNÉ (who already made attempts in this direction) progressed from an "artificial" to a "natural" classification by stressing not only flower characteristics, but also others, where desirable. A similar development is indicated for vegetation science; that is, the derivation of "natural" classifications that take account of many criteria by permitting as much flexibility as is adequate for the variation in purpose and study object.

It is suggestive for broad vegetation units to emphasize physiognomic-ecological or structural criteria, without which worldwide vegetation comparisons could not be made. Floristic criteria gain significance on a narrower geographic scale. Depending on the *purpose* of the investigation, one can also use all other criteria mentioned above.

However, one basic principle should never be violated and that is, to derive the conclusions only from facts observed in the field and not from preconceived notions.

The viewpoints underlying the six above mentioned systems have not lost any significance. They are the important underpinnings of present-day vegetation ecology. We will discuss all of these in the following pages.

The purely physiognomic system has its continuation in the much tighter, structurally defined vegetation system for general purposes by FOSBERG (1961) (SECTION 8.73). The environmentally oriented classification is further explained by the example of KRAJINA's biogeoclimatic zonation scheme (section 8.81), and specific procedures for vegetation-environment-correlation studies form the body of CHAPTER 11. The physiognomic-ecological classifications of SCHIMPER, RÜBEL and others has its continuation in the recently established UNESCO system for classifying world vegetation (SECTION 8.74). The areal-geographic-floristic orientation suggested the need to recognize the importance of history and evolution in vegetation ecology (CHAPTER 12). The dynamic-floristic classification of CLEMENTS, TANSLEY and other Anglo-American ecologists, though not any more applied in the earlier simplistic sense of CLEMENTS, has given rise to the dynamic viewpoint and research in vegetation ecology, which pervades many areas. Its central concepts are discussed in CHAPTER 13. The floristic-structural classification of BRAUN-BLANQUET has influenced or stimulated many of the specific techniques of processing vegetation data. The basic method is shown in CHAPTER 9 and mathematical approaches and variations of the floristic ordering theme are discussed in CHAPTER 10.

We will now select from the list of classifying criteria those that are currently more widely used in vegetation ecology. We begin with structural type concepts and systems which allow us to make worldwide comparisons in the vegetation. Then we will discuss combined vegetation-environmental or ecosystem schemes that are likewise of worldwide application. Finally, we will introduce floristic type concepts and systems, which are intended to set the framework for detailed regional ecological studies. This final part leads to CHAPTERS 9 and 10, which treat the mechanics of floristic and numerical ordering and classifying of vegetation data.

8.7   STRUCTURAL VEGETATION UNITS AND SYSTEMS

8.71   The Formation Concept. Nearly all earlier attempts in classifying vegetation were based on physiognomic criteria that were more or less closely associated with features of the environment. Plant communities that are dominated by one particular life form, and which recur on similar habitats, are called formations (in the physiognomic-ecological sense). Examples are the tropical rain forest, the mangrove swamp, the cacti desert, the grass steppe, the high moor, and the dwarf-scrub heath. Recognition of such types serves for initial orientation in setting subsequent studies into the proper perspective.

The European tendency has been to define the term formation physiognomically; that is, through properties of the vegetation itself. Environmental attributes were added for closer description only. The American tendency has been to define the same concept geographically and climatically; that is, through properties outside the current vegetation cover. In the latter sense, the physiognomy of vegetation in certain areas of a macroclimatic or geographic region was used only as a general indicator for the entire region. This has led to quite a different understanding of the same term. According to CLEMENTS (1928) a formation is the general plant cover of an area which may include several physiognomic variations. These variations are inferred to belong to the prevailing, climatically controlled, physiognomic type. For example, the prevailing physiognomic type may be grassland, though the area may show stands of scrub and open forest. These would still be part of the grassland formation if occurring in the same so-called grassland climate. The same idea in the European understanding is not a formation, but a vegetation region. A vegetation region usually contains a mosaic of actual vegetation types.

CLEMENTS did recognize this zonal or regional vegetation mosaic, but he added to the confusion of the term formation, by interpreting this vegetation mosaic as consisting of different developmental stages of the same formation. CLEMENTS converted the spatial side-by-side variation of vegetation by mentally grouping the different vegetation types or stages into a successional series, i.e., a time sequence. He believed that the regional side-by-side variations in time all lead to the same climax formation. This has led to some erroneous assumptions. Such a system that links all vegetation units to the final stage in succession provides for a gigantic outline. However, it must resort to tenuous assumptions in many places, even in a country such as North America where much original vegetation was present. The system is inclined to force certain communities into preconceived positions.

Such a system would be accompanied by many uncertainties in regions where almost all the vegetation is anthropogenically influenced. These uncertainties may be sufficient to make the system of little scientific value.

In contrast to CLEMENTS' interpretation, the term formation as originally intended refers to a real mappable vegetation unit, that is easily recognized by a characteristic dominant life form or life form combination. This meaning of the term formation is used by the two recent formation systems of FOSBERG (1961) and UNESCO (see ELLENBERG and MUELLER-DOMBOIS 1967a) discussed in SECTIONS 8.73 and 8.74. The regional or zonal vegetation mosaic "formation" of CLEMENTS is more appropriately termed "biogeoclimatic zone," as defined by KRAJINA (1965). (See SECTION 8.81.)

8.72   The Structural Classification Schemes of DANSEREAU and of KÜCHLER.   A well-known structural scheme is that of DANSEREAU (1957). His scheme employs six categories: (a) plant life form, (b) plant size, (c) coverage, (d) function (in the sense of deciduousness or evergreenness), (e) leaf shape and size, and (f) leaf texture. Each of these six categories contains a number of criteria that can be used to characterize a vegetation segment in the field. For example, his plant life form category includes six general life form groups: trees, shrubs, herbs, bryoids, epiphytes, and lianas; his size category includes three height classes: tall, medium, and low, which are defined quantitatively for certain life forms (for example, low trees range from 8 to 10 m in height); his coverage category includes four criteria: barren or very sparse, discontinuous, in tufts or groups, and continuous. Each criterion is designated by a letter symbol. The letter symbols can be combined to describe and differentiate formations as units in the field, on air photos, or on a map. In addition, the map units can be further interpreted by schematic profile diagrams. These profile diagrams are established from a system of diagrammatic symbols, whereby each symbol denotes a structural criterion. DANSEREAU's profile-diagram method—though logical—has not been applied widely in spite of its worldwide scope. The reason appears to be simply that the diagrammatic symbols render the profile diagrams too abstract, while less schematic profiles, such as the one of BEARD (FIG. 8.3), are easier to interpret.

Another well-known structural system is that of KÜCHLER (1967), which provides for a hierarchical approach. It begins with a separation into two broad vegetation categories; (a) basically woody vegetation, and (b) basically herbaceous vegetation. Within the first category, KÜCHLER distinguishes seven woody vegetation types (B = broadleaf evergreen, D = broadleaf deciduous, E = needleleaf evergreen, N =

needleleaf deciduous, A = aphyllous, S = semideciduous $(B + D)$, and M = mixed $(D + E)$). In the second category he distinguishes three herbaceous vegetation types $(G$ = graminoids, H = forbs, and L = lichens and mosses). These ten basic physiognomic categories can be further differentiated by whether or not they show a dominance of specialized life forms. The specialized life forms given in the system are five: C = climbers, K = stem-succulents, T = tuft plants, V = bamboos, and X = epiphytes. A third major distinction in KÜCHLER's system is based on prevailing leaf characteristics in the vegetation segment $(h$ = hard (sclerophyll), w = soft, k = succulent, l = large $(>400 \text{ cm}^2)$, and s = small $(< 4 \text{ cm}^2))$. Further structural separations are made on height (stratification) and coverage of the vegetation. For height KÜCHLER gives eight classes $(1 = < 0.1 \text{ m}; 2 = 0.1 - 0.5 \text{ m}; 3 = 0.5 = 2 \text{ m}; 4 = 2 - 5 \text{ m}; 5 = 5 - 10 \text{ m}; 6 = 10 - 20 \text{ m}; 7 = 20 - 35 \text{ m}; 8 = > 35\text{m})$, and for coverage six $(c$ = continuous $(> 75 \text{ percent}); i$ = interrupted $(50 - 75 \text{ percent}); p$ = parklike or in patches $(25 - 50 \text{ percent}); r$ = rare $(6 - 25 \text{ percent}); b$ = barely present or sporadic $(1 - 5 \text{ percent})$; and a = almost absent or extremely scarce $(< 1 \text{ percent}))$.

With this set of categories and criteria, any vegetation segment may be characterized structurally by a formula composed of the letter and number symbols given. KÜCHLER gives various concrete examples and claims that the system can be applied to all map scales.

8.73   FOSBERG's Structural Formation System.   FOSBERG presented a first (1961) and more recently a second (in 1967) approximation of a general structural classification of vegetation, which was adopted as a guide to mapping vegetation for the International Biological Program (IBP). One of the main features of FOSBERG's system is that it is based—like the schemes of DANSEREAU and KÜCHLER—strictly on existing vegetation and purposely avoids incorporation of environmental criteria. This has the advantage that the vegetation units established in this manner can be objectively correlated to independently established environmental patterns, because the vegetation boundaries are not in part delimited by environmental features. Where vegetation units are delimited in part by environmental features, correlation of such a vegetation map to environmental maps of the same area becomes problematic as this may result in circular reasoning.

The objective of FOSBERG's scheme is to subdivide the vegetation cover of the earth into units that are meaningful for a large variety of purposes by criteria that are applicable on a worldwide basis. These criteria cannot be floristic, because the distribution of plant species is geographically restricted (as discussed in SECTION 8.61). Therefore, they must be primarily structural.

FOSBERG makes a distinction between physiognomy and structure.

Physiognomy refers to the external appearance of vegetation and to its gross compositional features, implying such broad units as forests, grasslands, savannas, and deserts among others. Structure relates more specifically to the arrangement in space of the plant biomass. In addition, FOSBERG uses function in the sense of seasonal leaf shedding versus retention of leaves and specific aspects of growth or life form as important criteria for classifying the vegetation cover.

The vegetation is classified by use of a key. It begins with a breakdown into three alternatives—closed, open, or sparse vegetation. Thus, first consideration is given to spacing or cover of the plant biomass. Closed is defined as crowns or shoots interlocking, open as not touching, and sparse as separated by more than the plant's crown or shoot diameters on the average. Sparse vegetation is equated with the term desert, which is further defined as vegetations where plants are so scattered that the substratum dominates the landscape.

This first separation results in the first rank of vegetation units, which are called the primary structural groups, namely, closed, open, and sparse. Within each of these, the second rank of vegetation units are separated and called formation classes.

For example, in the closed primary structure group, individual formation classes are distinguished as forest, tall savanna, low savanna (tall and low referring to height of grass layer), scrub, dwarf scrub, tall grass, short grass, broad leaved herb vegetation, etc. Therefore, in the formation class breakdown, primary consideration is given to differences in the heights of vegetation layers and their continuity or discontinuity. But at least one of the layers in a vegetation unit must be continuous, or closed, to distinguish all of these formation classes from those in the open primary structural group.

Thirty-one formation classes are distinguishable in the first key. The individual formation classes are then further subdivided in separate keys. The first subdivision within each formation class key is by function, implying whether the foliage is evergreen or whether there are leafless periods for the dominant layer. This functional separation distinguishes the third rank, called formation group. A further separation within the formation groups leads to the actual map units, referred to as formations.

These are distinguished on the basis of dominant life form with emphasis on leaf texture (sclerophyllous, orthophyllous = ordinary leaf texture as opposed to sclerophyllous); leaf size (megaphyllous = at least 50 cm long and at least 5 cm wide, mesophyllous = leaves of ordinary size, and microphyllous = for trees 2.5 cm greatest dimension and for shrubs 1 cm or less); leaf shape (narrow versus broad); thorniness; and growth form (gnarled versus straight, succulence, graminoid, etc.).

Occasionally, the formations, which represent the fourth rank, are subdivided into subformations—the fifth and ultimate division. For example, the formation "gnarled evergreen forest" is subdivided into two subformations, "gnarled evergreen mossy forest" and "gnarled evergreen sclerophyll forest."

One or more examples of vegetation that fit the structural definition are supplied for each formation and subformation. A glossary defines all technical terms.

FIGURE 8.6 shows a summary of FOSBERG's system. The three primary structural groups (closed, open, and sparse vegetation) are shown on the left with their formation classes. Each of the primary structural groups also contains two aquatic formation classes—submerged and floating meadows. Further subdivisions and definitions are shown by symbols on the right side of each formation class name. The main life form types are listed above each column. The black squares indicate which of the vegetation strata are closed in the 17 different closed formation classes. The parallel criterion characterizing open and sparse vegetation is also indicated in a like manner for the important structure-giving layer in the two other primary structural groups. Additional symbols give further information about the remaining layers in the formation classes.

The classification system is necessarily artificial, because, for example, the primary criterion of spacing may separate some environmentally or floristically closely similar vegetations into different primary structural groups. Yet, it serves as a practical tool for mapping and organizing vegetation data for general purposes. Floristic associations can be studied within and across the structural frame given by the units. The structural vegetation patterns can be compared to climate, soil, history, and other environmental maps from which one can derive the major regional or zonal ecosystems.

8.74   UNESCO's Structural-Ecological Formation System. This classification system was recently published (ELLENBERG and MUELLER-DOMBOIS 1967a, UNESCO Standing Committee on Classification and Mapping 1969, 1973). It was also republished in an abbreviated version by REICHLE (1970). The system is included as APPENDIX B. Its purpose is to serve as a basis for mapping world vegetation at a scale of 1:1 million in terms of vegetation units that indicate parallel environments or habitats in different parts of the globe. Existing classifications were reviewed and these have influenced the thinking of the committee (notably RÜBEL's system). But none of the existing systems are found entirely suitable for the intended purpose. As in FOSBERG's system, structure forms the main separating criterion. However, terms referring to climate, soil, and landforms were included in the vegetation names and definitions, wherever they aided in the identification of the units. The reason for this is that significant ecological differences in habitat

FIGURE 8.6 — Summary of FOSBERG'S formation system

Legend: ■ Closed · O Open · S Sparse · x Absent to closed · ao Absent to open · s Absent to sparse · (blank) Absent

			Floating aquatic	Submerged aquatic	Bryoid	Broad leaved herbs	Short grass	Tall grass	Dwarf shrub	Shrub	Tree
Closed vegetation											
1	A	Forest			x	x	x	x	x	x	■
	B	Scrub			x	x	x	x	x	■	
	C	Dwarf scrub			x	x	x		■		
	D	Open forest with closed lower layers			←——— Closed ———→						O
	E	Closed scrub with scattered trees			x	x	x	x	x	■	S
	F	Dwarf scrub with scattered trees			x	x	x		■		S
	G	Open scrub with closed ground cover			←——— Closed ———→					O	
	H	Open dwarf scrub with closed ground cover			←— Closed —→				O		
	I	Tall savanna			←——— Closed ———→				s	s	S
	J	Low savanna			←— Closed —→			s	s	s	S
	K	Shrub savanna			←——— Closed ———→				Sparse		
	L	Tall grass			x	x	x	■			
	M	Short grass			x	x	■				
	N	Broad leaved herb vegetation			x	■	ao	ao			
	O	Closed bryoid vegetation			■	s	s				
	P	Submerged meadows	ao	■							
	Q	Floating meadows	■	x							
Open vegetation											
2	A	Steppe forest			ao	ao	ao	ao	ao	ao	O
	B	Steppe scrub			ao	ao	ao	ao	ao	O	
	C	Dwarf steppe scrub			ao	ao	ao		O		
	D	Steppe savanna			ao	ao	←— O —→		ao	ao	S
	E	Shrub steppe savanna			ao	ao	ao	ao	s	S	
	F	Dwarf shrub steppe savanna			ao	ao	O		S		
	G	Steppe			←——— O ———→						
	H	Bryoid steppe			O						
	I	Open submerged meadows	s	O							
	J	Open floating meadows	O	s							
Sparse vegetation											
3	A	Desert forest			s	s	s	s	s	s	S
	B	Desert scrub			s	s	s	s	←— S —→		
	C	Desert herb vegetation			←——— S ———→						
	D	Sparse submerged meadows		S							

FIGURE 8.6. *Summary of FOSBERG'S formation system. (From PETERKEN 1967).*

are not always reflected by easily definable structural or physiognomic vegetation responses. For example, tropical lowland rain forests differ ecologically from tropical montane rain forests. Yet, their structural differences are apparent only in certain regions and not on a world-wide scale.

In spite of environmental names, the units defined here are real vegetation units, because they can only be mapped if the vegetation in an area exists in the form defined here.

The vegetation units are listed in hierarchical order under each of the following seven *formation classes.*

1.   Closed forests.
2.   Woodlands or open forests.
3.   Scrub or shrubland.
4.   Dwarf-scrub and related units.
5.   Terrestrial herbaceous communities.
6.   Deserts and other sparsely vegetated areas.
7.   Aquatic plant formations.

Thus, spacing and height of dominant growth form are treated as parallel criteria in distinguishing formation classes. Each woody formation class is subdivided into *formation subclasses* on the basis of whether the vegetation is mainly evergreen or deciduous. These are then further separated into *formation groups* by the macroclimate in which they occur. For example, distiguished among closed forests that are mainly evergreen are tropical ombrophilous (or rain) forests, tropical and subtropical seasonal forests, tropical and subtropical semideciduous forests, temperate rain forests, etc. The next lower subdivision is the *formation.* Formations in tropical rain forests are tropical lowland rain forest, submontane and montane rain forests, tropical cloud forests, tropical subalpine rain forests (usually transitory to woodlands), tropical alluvial forests, tropical swamp rain forests, and tropical bog forests. The next lower level represents the *subformation,* which, together with the formation, is considered the main map unit. For example, the tropical cloud forest is subdivided into a broad-leaved subformation (the most common form) and a needle-leaved or microphyllous subformation. Definitions of broad-leaved and microphyllous follow those given in the plant life form key (APPENDIX A).

The classification gives an outline of all better-known formations of the earth. The system is flexible and allows inclusion of additional units if this should become necessary. It provides a framework that permits accommodating an unlimited number of floristically quite different units (that occur in various localities scattered over the earth's surface) into physiognomically and ecologically equivalent abstract categories.

Both the UNESCO classification and FOSBERG's scheme can be applied to categorize vegetation in the field and on maps in comparative terms within each scheme and also between them.

FOSBERG's scheme provides a ready field tool for mapping. It allows establishing pure vegetation units for correlations with environmental units mapped independently at the same scale. Because of its strictly structural orientation it may group ecologically quite different vegetations into the same unit. For example, tropical lowland and montane rain forests may form one vegetation unit. However, the ecological difference would become apparent by comparing the vegetation units to environmental maps of such an area, and there would be no danger of circular reasoning.

The structural-ecological scheme of UNESCO gives some environmental-geographic information at the start and therefore conveys an immediate orientation that appears useful for a worldwide inventory. It provides for an outline of available types and a general overview that can serve for immediate statistical purposes. For example, endangered vegetation in different parts of the world may be singled out for conservation. Specific mapping criteria may have to be worked out regionally within this framework. These can then be conveniently based on a combination of regionally significant structural and floristic criteria.

Both systems are artificial. For example, an open forest or woodland may differ from a closed forest only because of some disturbance. However, the primary objective of both schemes is identification of given vegetations. An arrangement according to ecological, sociological, or historical relationships would handicap the diagnostic value of such a classification. Moreover, it would hardly ever be completed, since ecological, sociological, and historical relationships are the objects of continuing research and readjustment.

8.75 The Synusia: Transition to Floristic and Functional Unit Interpretation. Each plant community consists of several groups of life forms (see SECTION 8.3). These have a certain individuality of their own in relation to the rest of the community. Therefore, they were considered as the actual basic units of vegetation by some investigators, particularly by GAMS (1918).

A group of plants of the same life form type occurring together in the same habitat, is called a synusia or "union." A synusia may be composed of quite unrelated species, which may in part replace each other in different habitats.

Simple plant communities consist of only one synusia, for example, certain crustose lichen communities on rocks or bark. However, most habitats are occupied by several synusiae, which may grow above each other in layers, beside each other, or in mixture. For example, the Northwest German *Calluna*-heath community consists of an evergreen dwarf-shrub *layer* (s Ch *frut*, synusia 1); in its shade grows a moss-carpet synusia (*Br Ch rept*, synusia 2), which occurs in very similar composition also on acid forest humus. On more open ground, caused by interruptions in the heath layer, occur fruticose lichen unions (*Li Ch*, synusia 3), which have even greater individuality. Epiphytic lichens (*Th E*, synusia 4) are often found on older branches of the heath plants. Occasionally, hemicryptophyte-unions (synusia 5) penetrate into the *Calluna* communities, whose main distribution center is otherwise in grassland communities. This fifth synusia-combination is found in the more continental climate of East Germany and Poland and then only in semishade conditions beneath open tree crown canopies. These are formed usually by pines, which represent a sixth synusia. Broad-leaf mixedwood communities consist often of more than 10 synusiae.

The advantages of the synusia concept are quite obvious: synusiae are easily recognized, even without knowledge of the species names. Descriptions of their combinations portray a clear picture of the communities and provide a certain idea of the habitat conditions. Synusial combination can be traced even across the limits of different floristic regions and permit recognition of ecological relationships. Therefore, they are useful for worldwide comparisons—as are the formations.

However, if they were used as basic units for classifying vegetation, one would arbitrarily separate the topographical and ecological unity of all those communities that consist of several synusiae, such as forest stands or heath communities. Synusiae should be treated as structurally definable subunits (i.e., layer-communities or life form types) within a plant community. The nature and degree of their interdependency is sometimes easily noted, but more often could become a useful subject of research. One cannot ignore species composition in any detailed study of vegetation. Therefore, most investigators consider floristically defined vegetation units as more useful than structurally defined ones. An assessment of synusial structure and floristic composition, however, would combine the advantages of the two approaches. For example, similar forest formations (such as evergreen montane tropical rain forests) in different floristic regions could be compared by the synusiae they contain. The species with their quantities could then be listed within the synusiae, and the species diversity could be established for each synusia. In this way, comparisons can be made on a sound ecological basis.

The synusial concept has a further advantage in that it leads to the functional concepts of "ecological species group" (SECTION 11.1) and "general niche" (SECTION 12.73). As is well known, the members of the same species are strong competitors for the same basic set of resources available in a habitat, because they are morphologically, anatomically, and physiologically similarly adapted. On a more general level, the members of a synusia (species individuals of different taxa growing together, but belonging to the same life form type) may be suspected to compete for a generally similar set of resources offered in a given habitat. The relation of member species of a synusia can be suspected to be of greater competition than that among species of the same community belonging to different synusiae. However, this judgment from observable structural relationships among the member species of a community must be viewed as a working hypothesis. Direct evidence of such interrelationships can only be established by exact measurement and experimentation.

8.8   COMBINED VEGETATION-ENVIRONMENT OR ECOSYSTEM SCHEMES

8.81   KRAJINA's Biogeoclimatic Zonation Scheme.   The concept of biogeoclimatic zones is derived from the concept of vegetation zones. Vegetation zones refer to the vegetation cover found in a specified geographic region or zone which has a uniform macroclimate. And, the vegetation cover of a macroclimatic region or zone usually shows a number of different plant communities as a vegetation mosaic. Therefore, the zonal concept is entirely different from the formation concept, which refers to a specific, structurally or physiognomic-ecologically defined community.

The term biogeoclimatic zone infers a zonal recognition not only of the vegetation and climate, but also of the animals, the soil, and the geological substrate. KRAJINA (1965) defines a biogeoclimatic zone as a geographic area that is predominantly controlled by the same macroclimate and characterized by the same soils and the same zonal (climatic climax) vegetation. In essence, therefore, a biogeoclimatic zone can be considered a geographically rather large ecosystem of zonal or regional extent, which contains a number of smaller ecosystems. Therefore, KRAJINA's system is not a vegetation scheme but an ecosystem scheme.

The smaller ecosystems are the biogeocoenoses in the sense of SUKACHEV (1945). The geographic outline of a biogeocoenosis is indicated by the phytocoenosis, which is a rather narrowly defined, mappable plant community. Narrowly defined means that the vegetation must be homogeneous in all its layers, not only in the dominant layer. In a forest, for example, not only the tree species composition, but also the undergrowth must be homogeneous. Therefore, sample stand selection is the same as in the releve´ method (CHAP. 5).

The geographic outline of a biogeoclimatic zone is defined climatically, vegetationally, and pedologically. The zones are recognized in the field by the same zonal vegetation, the same zonal soils, and the same (zonal) macroclimate. The zonal vegetation and soils occur only on such mesic habitats where the soils have matured for several thousands of years and the vegetation is basically undisturbed for several hundred years. The mesic habitats in a region, strongly affected by glaciation, are formed from deep glacial or alluvial soils with silt and fine sand particles predominating. They are not subject to excessive drainage and they are not affected by prolonged duration of water derived from seepage. Their soil moisture content is entirely dependent on water from precipitation (mainly from rain and snow).

Other habitats are either drier or wetter than mesic ones. In both cases such habitats are edaphically different. They are either more easily drained, because their soils are shallow or formed from coarse material, or they are supplied by surplus water transported laterally from some other habitat. Such ecosystems, edaphically different from the mesic ones, are still corollary to them in the biogeoclimatic sense through the macroclimate. Therefore, they cannot be excluded from the characterization of the zones. Their vegetation may be so different, that floristically they may belong to some other alliances and orders (sensu BRAUN-BLANQUET) than the vegetation growing on mesic sites.

Every biogeocoenosis of a certain biogeoclimatic zone has its characteristic development, represented by successive phytocoenoses. Some of the biogeocoenoses may develop similarly in several zones provided that some compensatory environmental factors function there.

Because zonal soils may develop from a deep and loose parent material only if they mature at least for several thousand years, and because the zonal vegetation is properly developed when usually undisturbed for several hundred years, not every mesic site is represented by the zonal soil and zonal vegetation. Therefore, relatively young alluvial terraces may be represented by much greater and richer soils and younger and usually richer vegetational units or biogeocoenoses.

Some biogeoclimatic zones may have some similarities in their zonal vegetational and soil characteristics. However, they might be distinguished as different zones by their different "corollary" ecosystems, that is, by wetter or drier ecosystems that occur in the same zones with the mesic ones. Therefore, biogeoclimatic zones are defined not only by their mesic and mature ecosystems but also by their other ("corollary") ecosystematic units, which occur in the same macroclimate.

A controlling role is played by macroclimate, while both vegetation and soil are considered dependent upon climate. KRAJINA characterizes each zone by nineteen climatic parameters. As a major tool he uses the bioclimatic zonation scheme of KÖPPEN (1936), which results in mappable units based mainly on a combination of the annual distribution of mean monthly temperature and precipitation. KRAJINA cites several extreme climatic parameters, such as absolute maximum and minimum temperatures, wettest and driest month precipitation, annual snowfall, cloud occurrence, etc., which help to characterize and separate the biogeoclimatic zones. In addition to the nineteen climatic parameters, each zone is described by elevation, latitude, its zonal soil, and its zonal soil forming process. The zonal soil corresponds more or less to the great soil group concept used in soil classification.

In the actual mapping of the biogeoclimatic zones, topography is the major criterion as it integrates all the parameters of climate (temperature, rainfall, snow duration, wind exposure, cloud occurrence) and even vegetation and soil.

The biogeoclimatic zonation system is built up from the detailed study of individual biogeocoenoses. The number of biogeocoenosis types varies from about eight to twenty for a zone and each biogeocoenosis type is based on the detailed ecological study of at least ten individual concrete biogeocoenoses in the field. The biogeocoenosis units are considered integrated by a common macroenvironment into a biogeoclimatic zone. KRAJINA therefore emphasizes the biogeoclimatic zone as an important upper level of integration above that of biogeocoenoses. The system is quite different in approach from the previously described structural classification schemes, which start from the broad aspects and proceed to the more detailed. Moreover, in those schemes mapping is based on actual vegetation boundaries. In the biogeoclimatic zonation scheme the boundaries are based on soil, climate, and vegetation, using topography as the integrator. KRAJINA's system requires considerable knowledge of the entire vegetation ecology of an area before zonal boundaries can be assigned in detail. In fact, KRAJINA's system is the outcome of many years of intensive study of the vegetation ecology in British Columbia (KRAJINA 1960, 1965, 1969). The system has also been interpreted for the general Biology Textbook Series (yellow version) by PHILLIPS (1964).

8.82   A Classification of World Ecosystems.   Recently, ELLENBERG (1973) presented a scheme for classifying the world into a hierarchy of ecosystems from a functional viewpoint. The largest and all-encompassing ecosystem is the "biosphere", i.e., the outer skin of our planet (soil, water, and atmosphere) as far as it is the life medium of organisms. It includes the oceans to their maximal depths. The biosphere is subdivided into two main groups according to type of energy source: (a) *natural or predominantly natural ecosystems*, i.e., those whose functions depend directly on the sun as energy source; and (b) *urban-industrial ecosystems*, whose functions depend on reconstituted energy (fossil fuel and, recently, atomic energy).

Six main separating criteria are used at different levels in the hierarchy.

1. Prevailing life-medium (air, water, soil, buildings).

2. Biomass and productivity of the primary producers.

3. Factors limiting the activity of primary producers, consumers, and decomposers.

4. Regulating mechanisms of matter or nutrient gain or loss.

5. Relative role of secondary producers (i.e., of the herbivores, carnivores, parasites, and other mineralizers).

6. The role of man in the ecosystem (i.e., his role in the origin, development, energy flow, and mineral cycling of the ecosystem, particularly his function in supplementing energy sources).

A hierarchical order is obtained by defining successively smaller ecosystems within larger ecosystems. Starting with the biosphere, the next lower level is referred to as *mega-ecosystems*. Five mega-ecosystems are recognized by the life-media (criterion 1) that they represent (capital letters as used in ELLENBERG's key):

M	Marine ecosystems (saline water)	⎫
L	Limnic ecosystems (fresh water)	⎬
S	Semiterrestrial ecosystems	Predominantly
	(wet-soil and air)	natural
T	Terrestrial ecosystems	
	(aerated soil and air)	⎭
U	Urban-industrial ecosystems	
	(the creations of man)	Artificial

Macro-ecosystems represent the next lower level within each mega-ecosystem. The macro-ecosystems are still very broad or inclusive units that are separated mainly by the criteria 2 to 4 (for example, forests).

Meso-ecosystems are considered the basic units of this scheme. They are the "ecosystems" in the most commonly understood sense. A meso-ecosystem is considered a relatively uniform or homogeneous system with respect to the abiotic conditions as well as the life forms of the prevailing primary and secondary producers (for example, a cold-deciduous broadleaf forest with its animal life).

Micro-ecosystems are subdivisions of meso-ecosystems, which depart with respect to a certain component (a lowland, montane, or subalpine cold-deciduous broadleaf forest with its animal life).

Nano-ecosystems are considered to be small ecosystems that are spatially contained within larger ecosystems and that exhibit a certain individuality of their own (a wet depression in a montane deciduous broadleaf forest).

Within almost all ecosystems one can recognize strata or other *partial systems,* which can be analyzed individually. At least three partial systems can be recognized generally:

• *Topo-partial system*—a layer or other topographically stratified segment within an ecosystem (the topsoil in a forest)

• *Substrate-partial system*—a small islandlike community within an ecosystem (a moss-covered log in a forest)

• *Pheno-partial system*—a partial system that appears only during a certain time of the year (an algal bloom at the surface of a lake)

The classification scheme includes a special scale for defining the kind and degree of human influences for each ecosystem to be classified. Four kinds of human interferences are recognized:

1. *Harvesting* of organic materials and minerals, which are significant for the metabolism of an ecosystem.

2. *Adding* of mineral or organic materials or organisms.

3. *Toxification*—adding of substances which are abnormal for the metabolism of the ecosystem and which are detrimental to important organisms or organism groups.

4. *Changing of the species composition*—by suppressing existing species or by introduction of alien species into the ecosystem.

The degree for each of the types of human interferences is expressed by a scale of increasing severity from 1 (no harvesting) to 9 (destructive harvesting).

For worldwide comparisons of ecosystems the scheme also includes a biogeographic separation into nine regions, such as tropo-American, tropo-African, tropo-Asian, Australian, etc. Each of these biogeographic regions can be further subdivided into biogeographic subregions or provinces.

All criteria are identified in the scheme by letter symbols and a decimal system. These provide for classifying any ecosystem by a short formula on a worldwide basis.

An overview, in form of a key, shows the four predominantly natural mega-ecosystem types (M, L, S and T) subdivided to meso-ecosystems and in some examples to the level of nano-ecosystem and partial system (where well known). The scheme can be completed with derivation of further knowledge.

The key makes a major division between aquatic (M + L) and land ecosystem (S + T) on the basis of structure. The vertical extent of predominantly natural land ecosystems (in contrast to aquatic ecosystems) is not determined by their life medium (soil and air) and the availability of light, but by the height growth of the dominant vascular plants. It follows that the terrestrial ecosystems are defined primarily by vegetation structural criteria, and their classification is based on the UNESCO structural-ecological formation system (in APPENDIX B). Therefore, meso- and micro-ecosystems are divisions somewhat parallel to formation and subformation types, but they are described in functional terms (criteria 2 to 6 as far as these are known). It may also be noted that the second structural unit-concept of synusia has given rise to the functional concept of partial system as used in this ecosystem scheme.

While the scheme is based entirely on structural-functional criteria, it is clear that any exact investigation of ecosystems cannot ignore the species composition that forms the living matrix of the system. On the contrary, for any detailed investigation of ecosystems it is desirable to derive as complete as possible a species list of the participating plants and animals. Moreover, abundance determinations should be made for at least those species of plants and animals that are significant for the productivity and maintenance of the ecosystem. These lists are then usefully ordered or classified according to animal- and plant-sociological viewpoints.

8.9   FLORISTIC VEGETATION UNITS AND SYSTEMS

8.91   Species Dominance Community-Type Concepts: The Sociation and Consociation.   Single, easily noticed plant species provide the simplest floristic tool for attaining a certain order in the great variability of plant communities. These have always been used (even by untrained persons) in differentiating forest stands (beech forest, pine forest, etc.). Such a simple classification can also be very satisfactory for scientific purposes, if the area is floristically poor. In Scandinavian countries the most abundant or the most dominant species are used for distinguishing the so-called sociations.

DU RIETZ (1921) considered the *sociation* the basic unit of vegetation classification and defined it as a recurring plant community of essentially homogenous species composition with at least certain dominant species in *each* layer.* For example, the East German pine-heath communities form a *Pinus sylvestris-Calluna vulgaris-Cladonia* sociation, certain beech forests a *Fagus sylvatica-Allium ursinum* sociation, etc.

DU RIETZ speaks of a *consociation* if only the *upper stratum* of a several-layered community is dominated by one species. As a type concept, a consociation can also be understood as a class composed of individual concrete sociations, whose upper strata are dominated by the same species, while the lower strata may be dominated by different species in each vegetation sample. The term consociation was used also by CLEMENTS, TANSLEY and RÜBEL in a very similar way. Consociations are more common than sociations particularly in species-rich areas. An example is the oak forests of England, which, according to TANSLEY, represent a consociation with very variable undergrowth. Few oak forests have the same dominants in the herb layer; one example is the *Vaccinium*-oak forests on acid soils. A typical consociation is shown in the profile diagram of FIGURE 8.3.

PETERSEN (1927) tried to apply the consociation concept to the classification of meadow communities in Central Europe. He distinguished meadow types by the dominance of certain grass species, one dominant species characterizing a meadow type. At the same time, these dominants were to serve as indicators of the feed value and habitat of such types. However, because of the great number of species in Central and South Europe, there are rarely meadows with only one dominant species. Therefore, it would be necessary to consider most communities as mixed types or they would not fit into PETERSEN's system at all.

This difficulty with regard to the sociation and consociation concepts exists in all regions with large numbers of species, where many species compete for the same habitat. A good example is the tropical rain forest in continental lowland areas. Therefore, sociations and consociations have no universal applicability as units in vegetation classification.

---

* DU RIETZ (1921:307) speaks of "constant" species, i.e., those that recur regularly as dominants. However, this requirement is redundant, because a community without a dominant species cannot be considered a member of a sociation according to definition.

However, even in such communities where single plant species have become dominant it is often not satisfactory to classify them as belonging to a certain consociation type. It was found that the same species may become dominant under different habitat conditions, whereby the associated flora may differ considerably in response to the differences in environment. For example, the tall reed grass *Phragmites communis* may grow in pure stands at the margin of larger lakes with occasional admixture of *Scirpus lacustris* or other tall semiaquatic plants. *Phragmites* is found to also form vigorous stands at river margins in the tidal ranges, in habitats with considerable daily and annual fluctuations in water level. The associated plants named above cannot grow under these conditions. Instead, a more or less rich geophyte-flora is found growing there in the spring, specially the yellow-flowering *Ranunculus ficaria* and *Caltha palustris*. It is obvious that the two *Phragmites* consociations can be considered one unit only for very superficial reasons. The *Calluna* sociations may be mentioned as another example. They occur in the Northwest German lowland on dry as well as wet soils and they recur also in alpine zones in very different environmental conditions.

Thus, community types defined by a single dominant species (consociations) may lump together very different habitats. Moreover, the single dominant species concept cannot be applied in many regions. It is better to use a more flexible concept of floristic dominance types, where community types can be recognized by one or more dominant species in the prevailing synusia. This in fact, is the most widely used community-type concept in North American vegetation studies (WHITTAKER 1962). Because more than one dominant species are often used to designate these dominance community types, they have been called "associations" by CLEMENTS. These so-called "associations" are usually very large and heterogeneous in habitat conditions and they differ entirely from the European association concept, which is discussed in the next section.

8.92   The Association Concept. It is quite possible to differentiate several vegetation units in the above examples, if one considers the associated, as well as the dominant species. Units that are floristically defined in this manner are called associations. In contrast to a sociation, an association does not have to show a single dominant species in each layer. Instead more than one species per layer may be used to define an association.

Following a resolution of the International Botanical Congress in Brussels in 1910, it was agreed to apply the term association only to communities "of definite floristic composition, uniform physiognomy and when occurring in uniform habitat conditions." In the continental European understanding, an association refers to a relatively small vegetation unit, a unit below the level of consociation. The 1910 International definition of the term association was rather strictly interpreted in continental Europe. However, an exact fulfillment of the three requirements (definite flora, uniform habitat, and physiognomy) is not always possible.

Particularly the requirement referring to a "uniform" habitat is difficult to fulfill. A uniform habitat may be found in several field situations; but the vegetation samples to be grouped into an association-type can never have identical habitats, because no two places on the earth's surface have exactly the same combinations of site factors. Likewise, the criterion of definite flora needs closer definition. In classifying, it is impossible, even though ideal, to consider all species to be of equal significance. Because of the great variability of communities, one would have to distinguish as many "units" as there are plant communities. Even two closely similar vegetation samples will not have identical species lists. Yet, closely similar vegetation samples will have a certain proportion of species in common. Therefore, it is possible only to emphasize certain groups of species, namely those that recur commonly in different locations of a region. Only those communities are put into a type that show the same groups of species. Such groups can be distinguished either by comparing a large number of vegetation samples (i.e., by tabular comparison) or they may be distinguished in other ways. We will discuss the table-technique first (CHAPTER 9), because it has received much emphasis and recognition at least in European phytosociology under the influence of BRAUN-BLANQUET. An association-type therefore can be defined as a unit of vegetation derived from a number of vegetation samples or relevés that have a certain number of their total species in common. An individual association member, i.e., a concrete community, can be recognized in the field by the presence of certain species of a diagnostic group.

BRAUN-BLANQUET has narrowed the concept of association by adopting it as the basic unit of his classifying system. It may be well to point out that the association does not represent the smallest recognizable vegetation unit. An association is recognized by its species composition, in particular by own or preferably own "character species" (i.e., species unique to the association-type), but also by several "differentiating species" as explained in CHAPTER 9. The number of these species is not specified.

Unfortunately, the Brussels definition does not really specify the criteria that were meant to be applied in the distinction of an association. As a result, two entirely different association concepts evolved in continental Europe and North America. The only criterion common to both these different interpretations is that an association name is made up of a combination of species names. In North America, CLEMENTS (1928) interpreted the term association very broadly to refer to the first subdivision of a formation. This broad association concept is still widely used in the United States. Since CLEMENTS' "formation" was actually the general plant cover in a given macroclimatic region (i.e., a vegetation mosaic), his association concept was more or less a climatic subregion of which a selected vegetation cover was used as an indicator. For the whole of North America CLEMENTS recognized three so-called climaxes—a grassland, scrub and forest climax. Each climax was subdivided into a few "formations" (regions) and each "formation" was subdivided into two or more "associations." For example, in the forest climax, the Pacific coastal forest (region) was called the *Thuja–Tsuga* formation. This formation was subdivided into two associations, the *Thuja–Tsuga* association and the *Larix–Pinus* association. CLEMENTS defined an association floristically by joining the names of two regionally dominant species and then implied that an association was a grouping of two or more consociations. Thus, CLEMENTS' association concept was even more inclusive than the consociation concept, which defines community types by single dominant species.

8.93   BRAUN-BLANQUET's Floristic Association System. The same technique used in deriving associations is used also for other vegetation units, which are above or below the rank of association in BRAUN-BLANQUET's (1928, 1932, 1965) system.

In brief, the system consists of preparing species lists in relevés and then processing these lists in synthesis tables. In these tables, the species common to several relevés are identified and emphasized. The species unique to each relevé are not ignored, but they are not given the same value as the species that recur together in a number of relevés. These common species groups are the key to the identification of vegetation units. The technique will be explained in detail in CHAPTER 9.

The association, as previously defined in the continental European sense, is considered the basic unit in BRAUN-BLANQUET's system. Therefore, his system can be called a floristic association system. Other vegetation units are recognized by the same tabulation technique, but as units above or below the rank of association. In this way all units are interconnected in form of a hierarchy, but each unit is identified by certain common groups of species.

The different ranks are usually designated by a particular ending added to the root of the scientific genus name of an especially char-

RANK	ENDING	EXAMPLE
class	-etea	*Molinio-Arrhenatheretea*
order	-etalia	*Arrhenatheretalia*
alliance	-ion	*Arrhenatherion*
association	-etum	*Arrhenatheretum*
subassociation	-etosum	*Arrhenatheretum brizetosum*
variant	no ending	Salvia variant of the *Arrhenatheretum brizetosum*
facies	-osum	*Arrhenatheretum brizetosum bromosum erecti*

acteristic species.* The summary in the table appears sufficient as an outline.

The lowest unit in this system, the facies, is not characterized by exclusive species (i.e., "character species"), but merely by the dominance of a certain (or several) species. Therefore, it corresponds in some respects to the consociation or sociation. However, it is viewed here in relation to the other ranked units, whose geographic coverage is progressively larger.

In contrast to the association, subassociations and variants have no character species of their own. However, they differ by groups of differentiating species; that is, by plants that form certain subunits within an association, but which occur also in other associations.

Recently, the tendency has developed to distinguish associations merely by differentiating species. This implies dispensing with the requirement of character species for an association. This development results from the experience that there are only few character species in the strict sense. However, the alliances retain their own character species, while orders and classes usually show numerous character species.

The segregation of different vegetation units by differential species is based on tabular comparison of vegetation releve's. Therefore, it is based on a purely inductive method. However, ranking of the units into the previously discussed system, that is, in particular the solving of the question as to which of the units can be considered associations, depends on the personal judgment of the investigator. Both work phases are therefore strictly separated in the following discussion of the table technique. More recent methods of vegetation classification will be discussed in CHAPTERS 10 and 11.

---

* The specific epithet is added only in case there could be misunderstandings. In that case it is used in the genetive (e.g., *Pinetum mughi*). If two genus names are used, the first is usually supplied with the ending -o (e.g., *Querco-Betuletum*). The ending -etum was originally used in Latin for naming plant communities in which one species was dominant. However, in contrast to the sociation and consociation, the association (*sensu* BRAUN-BLANQUET) is not characterized by this criterion.

Classifying Vegetation by Tabular Comparison

9 In classifications based on floristic criteria, i.e., species composition and their quantitative variations, the problem of separating vegetation into units can be studied and resolved after the species lists of all sample stands are transferred into a single table. Such a table, showing the floristic information of all relevés under comparison, is conveniently referred to as a synthesis table. A synthesis table, in addition to offering an aid in classification, often reveals information that was not realized during the field work. Moreover, a synthesis table serves as a means of documenting the floristic information of a vegetation study.

Tabulation of species lists of a number of individual vegetation samples can be done in several ways. Our objective will be to arrive at a synthesis table that shows the relevé data in a well-organized form so that important trends of species distribution between the sample stands are immediately recognized. We want to isolate groups of species that show similar distributions among the relevés under comparison. Also, we want to place those relevés side by side in the table that have similar species compositions.

This is conveniently done in several work phases. These will be briefly indicated here. Thereafter, these phases will be discussed in detail in the following sections.

First, the "constant" species are determined. These are species that are present in a high percentage of the relevés. Arbitrary limits can be set. For example, constant species may be defined as those present in more than 60 percent of the relevés.

Second, the "differential" (or differentiating) species are determined. These are found among the species with restricted distributions or amplitudes in the relevés under comparison. But excluded from these are the species of low constancy, i.e., the rare ones. Again, arbitrary limits can be set. For example, low constancy may be defined as less than 10 percent presence in the relevés under comparison. Among the species with an intermediate range of constancy (10–60 percent), we try to locate those that occur together in several relevés. Such a group is called a differential group of species, because it differentiates certain relevés from others.

The aim of the tabulation technique is to come up with a "differentiated table." A differentiated table shows the differential species and differential species groups sorted into blocks and separated from the other species, which are listed in order from high to low constancy. The other species are the constant and rare ones and those of the intermediate constancy range that showed a low or no degree of association with the differential species.

A differentiated table can then be compared to similarly derived differential tables for vegetation of other areas or regions. This may allow the recognition of characteristic tendencies of species distributions and the distinction of "character species" within the same broad floristic province. A character species is one that shows a distinct maximum concentration (quantitatively and by presence) in a well-definable vegetation type. We will say more about the character species concept later (SECTION 9.6).

The three phases of the tabulation technique for arriving at constant, differential, and character species can be outlined in six steps:

1. Construction of a "raw table."
2. Rewriting the same table into a "constancy table."
3. Extracting species of intermediate constancy and from these, determining differential species with the aid of "partial tables."

4. Rearranging the relevés (table columns) and species (table rows) in relation to the presence or absence of groups of differentia species into a "differentiated table."

5. Determining character species through "summary tables."

6. Rearranging the differentiated table into a "characterized table."

9.1   CONSTRUCTION OF A RAW TABLE

Comparing data, by merely placing note sheets side by side, becomes difficult where more than five vegetation samples of similar communities are available. It is therefore practical to assemble all relevés into a table. This is conveniently done on graph paper in such a way that each evaluation figure—regardless of whether this involves one or several per species—is written in only one square (a space of approximately 0.5 × 0.5 cm).

Before transferring the data into a table, the individual relevés may be sorted according to any viewpoint. For example, they may be sorted in relation to the increasing intensity of a certain site factor, the increasing altitude, the dominance, the presence or absence of certain species, the species number per relevé, or simply in sequence of the analysis dates. Certain species can be emphasized right away, for example, presumed character or differential species or species whose ecological importance was already determined in connection with other observations. In forest communities it is advisable to keep the strata separate. In grassland communities one should do the same for the more abundant or important life forms or taxonomic groups, such as the grasses, legumes, and other herbs. However, the work usually progresses fastest if the species are transferred in the originally recorded sequence. This results in a so-called "raw table" (see TABLES 9.1 and 9.2).

The different work phases are explained from data accumulated by students during a field course. The vegetation samples refer to *Arrhenatherum elatius* meadow communities, which were evaluated according to the percent biomass estimating method of KLAPP (see SECTION 5.43). The communities occurred in the broad Danube lowland downstream of Ulm, in an area 5 × 5 km. Thus, they are rather close together. The meadows here are generally poorly fertilized and receive little attention. Therefore, they deviate to a certain extent from the more typical, i.e., better fertilized, *Arrhenatherum elatius* communities. Before preparing the raw table, the individual samples were sorted according to the increasing quantity of *Arrhenatherum elatius*, the most obvious tall-grass.

TABLE 9.1 Start of a "Raw Table." The First Relevé of a Meadow Community, Evaluated by the Method of KLAPP, Is Transferred. Enough Space Was Allotted for Transferring Further Relevé Lists.

Relevé number	1	2	3	4	5	6	7	8	9	10	etc.
total no. of species	31										
Grasses:											
Arrhenatherum elatius	$+^0$										
Dactylis glomerata	5										
Helictotrichon pubescens	1^0										
Bromus erectus	50										
Festuca ovina	2										
Poa pratensis	4										
Briza media	1										
Koeleria pyramidata	3										
Festuca rubra	15										
Sedges:											
Carex flacca	2										
Legumes:											
Trifolium pratense	+										
Trifolium repens	+										
Medicago lupulina	1										
Others:											
Achillea millefolium	6										
Daucus carota	1										
Campanula rotundifol.	1										
Plantago lanceolata	1										
Heracleum sphondyl.	+										
Galium molugo	3										
Chrysanthemum leucanth.	1										
Scabiosa columbaria	+										
Linum catharticum	+										
Rumex acetosa	+										
Ranunculus acer	+										
Thymus serpyllum	1										
Cerastium caespitosum	+										
Centaurea jacea	$+^0$										
Taraxacum officinale	$+^0$										
Campanula glomerata	+										
Veronica chamaedrys	+										
Plantago media	+										

In the raw table, one vertical column is allotted for each relevé or sample stand. The relevé number or an abbreviated relevé name is written at the top of each column. The number of species found in the relevé is entered beneath. This provides for a check in subsequent rearranging of the relevé sequence in other tables. At this stage it is practical to omit the detailed location and habitat descriptions at the head of the table. These can be entered into the final characterized table. TABLE 9.1 shows, better than words, how the first relevé is transferred into the initial raw table. The second relevé record (TABLE 9.2, relevé 2) is transferred into the next column. Species names, not yet recorded in the first transfer, are added. The other relevés are transferred into the following columns in the same manner.

The number of species added with each new relevé transfer will decrease consistently, if the sample stands are relatively similar. At approximately the tenth relevé, the number of new species to be added was rather small. A relatively low degree of similarity is indicated if the number of species to be added with each new transfer remains high. Such communities should not be combined into a single vegetation unit. Occasionally, there are some relevés that fall out of line by showing major floristic inconsistencies. This applies, for example, to relevé 19 (TABLE 9.2), which contains two species with relatively high biomass values (*Glyceria fluitans and Phalaris arundinacea*) that are unique only to this relevé. Such a sample stand can be crossed out and later regrouped with others, with which it shows greater similarity. At this stage, however, we may leave relevé 19 in the table.

9.2 CALCULATING DEGREE OF CONSTANCY

On first view, the finished raw table is difficult to comprehend. Frequently and rarely occurring species are in irregular sequence, because the species were written down in order of their chance appearance on the stand analysis sheets. Therefore, the next step involves sorting of the species according to their degree of constancy. The term "absolute" constancy refers to the number of relevés in which a given species occurs (e.g., in TABLE 9.2: *Arrhenatherum* 25, *Brachypodium* 1). For comparing different tables it is convenient to convert absolute constancy into relative or percent constancy. For example, in TABLE 9.2 the constancy of *Arrhenatherum* is 100 percent, while that of *Brachypodium* is 4 percent.

After determining the constancy for all species (last column, TABLE 9.2), the species can be transcribed in a new order from high to low constancy in a "constancy" table (TABLE 9.3). A constancy table greatly facilitates comparison of the individual relevés. With it, one can more easily distinguish the similar from the less similar relevés.

TABLE 9.2 Twenty-five Relevés of Meadow Communities from the Danube Valley South of Ulm. They Are Arranged According to Increasing Abundance of Arrhenatherum elatius. The Species Were Recorded in Four Groups in the Field: Grasses, Sedges, Legumes, and Other Species. This Grouping Served as a Convenient, Tentative Order.

	1	2	3	4	5	6	7	8	9	10	11	12	13	14	15	16	17	18	19	20	21	22	23	24	25	
Relevé number	1	2	3	4	5	6	7	8	9	10	11	12	13	14	15	16	17	18	19	20	21	22	23	24	25	
total no. of species	31	25	32	28	32	32	36	34	32	31	28	34	37	33	34	37	35	29	29	30	35	27	29	30	28	
Grasses:																										
Arrhenatherum elatius	+°	1	2°	2	4	4	4	4	5	8	9	10	10	12	15	15	22	22	24	25	25	26	30	35	25	25
Dactylis glomerata	5	5	15	5	12	12	4	10	2	6	12	32	15	10	6	15	15	18	1	5	18	8	8	8	18	25
Helictotrichon pubesc.	1°	1		20	8	3		4	+		1	4			13	4	4		28	+	2		1		16	
Bromus erectus	50		35	74					47	21				37									10		7	
Festuca ovina	2	1							1	2															4	
Poa pratensis	4	74	10	5	4	2	3	4	10	8	6	25	2	5	15	10	5	10	1	6	1	9	16	20	10	25
Briza media	1	1							2	1						2									5	
Koeleria pyramidata	3			2					3	3															4	
Festuca rubra	15			2	+		+	3	4	2	2	1		6		2		+	2			+	2		15	
" pratensis		5	3		20	3	2	8	5	2	10	10	2	2	6	5	28	12	10	2	15	15	2	2	3	23
Trisetum flavescens		2	5		8			3				6		4	10	5	5	5	8	8	16	4	2			15
Alopecurus pratensis				2	8			6	4						2			15	10				1		9	
Holcus lanatus				1	1	1	2		2		1	2		2	+		2	+					15		12	
Deschampsia caespitosa						11	2		28				1			10	2	5							7	
Poa trivialis						2																			1	
Phleum pratense									+																1	
Festuca arundinacea													1												1	
Lolium perenne																2									1	
Glyceria fluitans																		20							1	
Phalaris arundinacea																		28							1	
Phragmites communis																					+				1	
Brachypodium pinnatum																								5	1	
Sedges:																										
Carex flacca	2		3										2	1	3	2									6	
" acutiformis						10			4									2		1				2	5	
" hirta									6							2									2	
" panicea																2									1	
" gracilis																		4						1	2	
Legumes:																										
Trifolium pratense	+	+	1	+	4	1	1	+				+		+	1		+	+°	+	+	+	2	2		18	
" repens	+				2	+		+	1			+	+	+	+		+		+	+		1		+	14	
Medicago lupulina	1		+	+	2	+			+	+	+	+	+	+	+	+	+					1	1		17	
Vicia sepium			+	+	1	+		+							+		+								7	
Lotus corniculatus			+				+						+										+		4	
Lathyrus pratensis					1	+	+	+		+	+	+						1	+	+					10	
Vicia cracca									+				1	+	+								+		5	
Others:																										
Achillea millefolium	6	+	3	1	2		1		2	8		1	4	16	2	5	+	12		6	+	4	+	3	+	21
Daucus carota	1	1	1	+	+	1	1	5	1	1		+	2		+	+	+	1		1	+	2				20
Campanula rotundif.	1	+	1			+	+	1		1		1	1	+								1	+	+		13
Plantago lanceolata	1	1	1	2	1	4	4	1	1	+	1	4	2	1	2	+	1	1			2	8	+	+		25
Heracleum sphondyl.	+	1°		3	+	4										26			1		+	1	+	+	+	14
Galium mollugo	3	2	7	1	3	24	6	4	2	12	5	6	10	14	3	5	3	6	2	5	1	3	12	3	2	25
Chrysanthemum leucant.	1	+	3	+	+	4	2	1	2	5	+	3	1	2	1	1		+	3	+	1	1	6			23
Scabiosa columbaria	+	2	1						+						+										5	
Linum catharticum	+	+	+						+							+									5	
Rumex acetosa	+			+	+	2	1	3		2	1	1	2	+	1	+		1	1	+	1	+		1		19
Ranunculus acer	+	+	+		1	2	2	+		+	3	1	+	2	1	+	+	+		+	1	+	2	+	+	+	23

Relevé number	1	2	3	4	5	6	7	8	9	10	11	12	13	14	15	16	17	18	19	20	21	22	23	24	25	
Thymus serpyllum	1	+	+	+					2																	5
Cerastium caespitosum	+	+		+	+			+°	+		+		+	+	+					+	+	+				13
Centaurea jacea	+°	1	6			2	3					1		4	+	2	2			2	+			+	2	14
Taraxacum officinale	+°		+°	+	3	1	3	1		+		+	2	+	+		+	1		+			+	4	+	19
Campanula glomerata	+	1	1					1			1			1			+							+		7
Veronica chamaedrys	+	+°	+	+	1	1	+	+	1	+	1	+		+	1	+		+	1	1	+	2	+	1		22
Plantago media	+	+	+	+	+		1	+			1	+		+	+		+			+				+	+	14
Silene inflata	2					+									+											3
Leontodon hispidus	1	+				4										1										4
Crepis biennis	+	+°		2	1		8			+	1		+			+	1		1	+	1	6	+			15
Myosotis arvensis	+					+																				2
Ajuga reptans	+			+	+	1	3	1		1	1	+	+			+	+		+	1	+				+	16
Salvia pratensis		4						2	5			1												4		5
Knautia arvensis		1°											+			+										3
Viola hirta		+						2	3				+			+								+		6
Bellis perennis		+		+	+		+	+	1		1	+			+	+	+									11
Dianthus superbus		+	1										+			+										4
Pimpinella saxifraga			+						+																	2
Galium boreale			+			1		1				1														4
Cirsium oleraceum					+	12	20	20			3		+	18		2		+	+	2	+		1		3	14
Tragopogon pratensis		2	1	+								+	1		+	1										7
Glechoma hederacea		+		+	+			+	+											+						6
Anthriscus silvestris		2						+				+												+	+	5
Filipendula ulmaria			3		2							+	+													4
Geum rivale			2	1	5					3	+	+				1	+	+	1	+	1	+			+	14
Melandrium diurnum			2	1							+	+	1			1	+		+	+	+			4		11
Angelica silvestris			1	2								+	1										1		+	6
Lysimachia nummul.				+		1				+	+					+			+	+					+	8
Prunella vulgaris					+	2	+	+				+			+	+	+			1	+	1	+			12
Pimpinella magna					+								1			+			1							4
Polygonum bistorta					1						+°						1									4
Lychnis flos cuculi						+				+				1	+		+		+						+	7
Senecio jacobaea							+		+	+			+			+			+							6
Potentilla reptans									1			+							1							3
Cardamine pratensis											+						+									2
Myosotis palustris											+											+				2
Geranium pratense											+															1
Pastinaca sativa													1	+				3								3
Galium uliginosum											+															1
Sanguisorba officinalis											+															1
Galium verum													1											+		2
Silaus pratensis																1	+		+							3
Ranunculus repens																+										1
Euphrasia odontites																		+					+			2
Lamium album																				+						1
Rumex crispus																				+					+	2
Polygonum convolvulus																				+						1
Chenopodium album																				+						1
Alchemilla vulgaris																								+		1

9.3 RECOGNIZING DIFFERENTIAL SPECIES

For grouping the relevé series into vegetation classes, neither the species with a high constancy nor those with a low constancy are useful. The species with a high constancy are more or less characteristic

TABLE 9.3 Section from a "Constancy Table." This Table Section Is Extracted from TABLE 9.2 and Shows the Species in Order of Absolute Constancy. Species Present in More Than 15 Relevés and in Less Than 3 Are Not Shown. Species That Seem to Indicate One Subunit Are Underlined by a Straight Line, Those That Seem to Indicate Another Subunit Are Underlined by a Wavy Line. To Save Time, Species Names Are Abbreviated.

Relevé number	1	2	3	4	5	6	7	8	9	10	11	12	13	14	15	16	17	18	19	20	21	22	23	24	25	
Fest. rubra	15			2		+		+	3	4	2	2	1		6		2		+	2			+	2		15
Triset. flav.		2	5		8			3				6		4	10	5	5	8	8	16	4	2				15
Crepis b.			+	+°	2	1		8		+	1		+			+	1		1	+	1	6	+			15
Trif. rep.	+			2	+		+	1				+	+	+	+		+		+	+		1		+		14
Heracl. sphond.	+		10		3	+	4						26			+	1	+		1		+	1	+	+	14
Cent. jac.	+°	1	6			2	3					1		4	+	2	2			2	+		+	2		14
Plant. med.	+	+	+	+	+			1	+			1	+	+	+								+	+		14
Cirs. oler.						+	12	20	20			3		+	18		2		+	+	2	+	1		3	14
Geum riv.						2	1	5				3	+	+		1	+	+	1	+	1	+			+	14
Camp. rot.	1		+	1			+	+	1		1		1	1	+							1	+	+		13
Cer. caesp.	+	+		+	+		+°	+		+		+	+	+			+	+	+							13
Holc. lan.					1	1	1	2		2		1	2			2		+		2		+		15		12
Prun. vulg.						+	2	+	+				+		+	+	+			1	+	1	+			12
Bellis per.			+		+	+	+	+	1			1	+		+	+	+									11
Melandr. d.					2	1					+	+	1		1	+				+	+				4	11
Lathyr. pr.					1	+	+	+			+	+	1						1	+	+					10
Alop. pr.					2	8		10			6		4				2			15	10			1		9
Lysim. numm.						+		1			+	+					+			+	+				+	8
Brom. er.	50		35	74				47	21					37								10				7
Desch. caesp.							11	2			28			1			10	2	5							7
Vic. sep.			+	+	1		+		+					+			+									7
Camp. glom.	+		1	1				1			1			+									+			7
Tragop. pr.				2	1	+							+	1		+	1									7
Lychn. fl. c.							+			+					1	+	+		+						+	7
Car. flacca	2			3									2	1	3	2										6
Viola hirta			+						2	3				+			+					+				6
Glech. hed.					+		+	+			+	+								+						6
Angel. silv.						1	2						+	1								1			+	6
Senec. jac.									+		+	+		+			+			+						6
Briza m.	1	1							2	1				2												5
Car. acut.							10				4							2		1				2		5
Vic. cracca											+		1	+	+								+			5
Scab. col.	+	2		1							+			+												5
Linum cath.	+		+	+					+					1	+											5
Thym. serp.	1	+	+	+							2															5
Salvia pr.			4						2	5				1									4			5
Anthr. silv.				2					2			+		+									+		+	5
Fest. ov.	2	1								1	2															4
Koel. pyr.	3			2					3	3																4
Lot. corn.			+				+							+									+			4
Leont. hisp.		1	+			4								1												4
Dianth. sup.			+	1										+				+								4
Gal. bor.				+			1		1					1												4
Fil. ulm.					3	2							+	+												4
Pimp. magn.						+							1					+		1						4
Pol. bist.						1							+°					1						4	4
Sil. infl.		2			+									+												3
Knaut. arv.		1°												+		+										3
Pot. rept.									1					+				1								3
Past. sat.													1	+				3								3
Sil. prat.														1	+			+								3

for the entire releve' series under comparison. The species with a low constancy may be considered as more or less accidental occurrences.

Instead, among the species present with an intermediate range of constancy may be some that are useful for distinguishing groups of releve's as subdivisions of the series. Therefore, we can now work with the section of the constancy table (TABLE 9.3) which only shows the species present with an intermediate constancy. We define intermediate constancy as the range between 10 to 60 percent, which applies to those species present between 3 and 15 times in the 25 releve's. Other limits can be set depending on the characteristics of the vegetation and the judgment of the investigator.

What we are now looking for in this section of the constancy table are species that occur together in several releve's. We want to distinguish such mutually associated species from others that seem to avoid the association of these first species by being present in other releve's. If we find two or more such groups of species, we have determined differential species groups.

We cannot expect any two species to have exactly the same distribution, but we can expect two or more species to have similar distributions or amplitudes. In some situations we may have to be satisfied with single differentiating species rather than groups of differentiating species. The isolation of differentiating species therefore requires judgment. This can be translated into arbitrary limits. Species useful for grouping of releve's into vegetation classes or community types should be present in at least 50 percent of the releve's intended for grouping. But they should be absent or only sparingly present in others; that is, with 10 percent constancy or less.

The association tendencies of certain species may have already been noted in the field. Moreover, if the releves were selected for recurring combinations of species, these become the most likely candidates for differential species groups. But the tabular comparison usually brings out additional ones.

Species that appear to be useful as differential species (from carefully scanning the constancy table with the above criteria in mind) can be so designated by underlining (suitably in color). Such species have been underlined in TABLE 9.3.

The underlined species are scattered throughout the table body. It is therefore useful to rearrange the table once more. Experience has shown that this is done best in three steps:

First, a "partial table" is prepared. This involves extracting only those species that appear possible as differential (i.e., underlined) species into a new table and leaving out the others. The differential species with similar distributions across the releve series are moved together, below each other, in the new partial table (TABLE 9.4). This results in a new order of species into groups with similar distributions

TABLE 9.4—Example of a "Partial Table." Extracted from TABLE 9.3 by Rearranging the Species Sequence; i.e., Those Underlined by Straight Lines (Group I) and Those Underlined by Wavy Lines (Group II) Have Been Moved Together. Relevés with Common Differential Species Can Be Emphasized by Enframing Their Associated Species Groups. The Table Can Now Be Arranged by Moving Those Relevés Next to Each Other That Are Tied Together by the Same Group of Associated Species. In Addition, the Relevés Can Be Further Ordinated within the Groups by Arranging Them in Sequence of Their Numbers of Differential Species.

Relevé number	1	2	3	4	5	6	7	8	9	10	11	12	13	14	15	16	17	18	19	20	21	22	23	24	25
I. Brom. er.	50		35	74				47	21						37									10	
Camp. glom.	+		1	1				1				1			+									+	
Viola hirta				+				2	3					+		+								+	
Briza m.	1	1						2	1					2											
Scab. col.	+	2		1					+					+											
Linum cath.	+		+	+				+						+											
Thym. serp.	1	+	+	+					2																
Salvia pr.			4					2	5			1												4	
Koel. pyr.	3			2				3	3																
Fest. ov.	2	1						1	2																
II. Cirs. oler.					+	12	20	20			3		+	18			2		+	+	2	+	1		3
Geum riv.						2	1	5			3	+	+		1	+	+		1	+	1	+			+
Holc. lan.					1	1	1	2			2		1	2			2	+		2		+			15
Melandr. d.						2	1					+	+	1		1	+		+	+	+				4
Alop. pr.					2	8		10			6		4				2			15	10				1
Lysim. numm.						+		1			+		+				+			+	+				+
Desch. caesp.							11	2			28			1			10		2	5					
Lychn. fl. c.								+			+				1	+	+		+	+					+
Glech. hed.					+		+	+			+		+									+			
Angel. silv.						1	2						+	1									1		+
Car. acut.							10				4								2	1					2
Fil. ulm.					3		2						+	+											
Pimp. magn.							+							1						1					
Pol. bist.							1					+°				+		1							4
Past. sat													1	+				3							
Species numbers group I	8	4	6	6				8	8			1		4	3		1							4	
" II					4	8	10	9			9	5	9	7		3	8	3	8	3	10	5	3		10
New running no.[1]	1	8	5	4	14	17	23	20	2	3	22	11	21	16	6	9	18	10	19	12	24	15	13	7	25

[1] Relevé 1 remains at the first place, relevé 2 moves to the 8th place, relevé 3 moves to the 5th place, etc.

or amplitudes. The extracted species may be called "distribution groups," "social groups," or groups of associated species.

The extraction is much facilitated by placing the same size of graph paper on the constancy table at the line immediately beneath the species to be transferred. If the partial table becomes too long, it can be folded at every tenth line. TABLE 9.4 is the completed partial table.

Our example shows that at least three vegetation units or types can be recognized among the meadow communities: one, in which *Bromus erectus, Salvia pratensis,* and others are found together, a second one in which *Cirsium oleraceum, Alopecurus,* and others are found together. A third vegetation unit or relevé group can be recognized in which both species groups are represented, each with only few species (relevés 12, 16, 18) or in which only one species group is sparingly represented (relevés 20, 23). In the latter case the number of differential species is considered too few (with only three each) to include these relevés into the *Cirsium–Alopecurus* type.

The vegetation units can be further clarified by rearranging the relevé order in an "ordinated partial table." Its purpose is to show the relevés in order of their diagnostic-floristic similarity.

The idea of diagnostic-floristic similarity applies to the number of differential species per group that are present in a relevé. It thus differs from the idea of total floristic similarity, as evaluated for standard floristic similarity relationships (see SECTION 10.1). For example, relevés in which the *Bromus erectus* group is represented with the greatest number of species are written at the beginning of the sequence and those having most representatives of the *Cirsium oleraceum* group are written at the end. The new relevé sequence is established from TABLE 9.4. The number of species in each distribution group is summed at the bottom of TABLE 9.4 for each relevé. The new relevé sequence is then obtained by ordinating the plots according to their number of differential species within each species group.

Where the number of species per group is the same in two or more (e.g. in relevés 1, 9, and 10, each with eight species), their sequence is arranged in order of the greater quantitative sum of the differentiating species in the group. The new relevé sequence is entered at the bottom of TABLE 9.4.

The table work described so far is easily done by one person. However, preparation of the ordinated partial table (TABLE 9.6) and of the "differentiated table" (TABLE 9.7), which will be discussed below, is much facilitated by having a second person dictating the data.

It is useful to prepare a sheet of graph paper for the ordinated partial table by entering the new relevé sequence (from the bottom of TABLE 9.4) at the head of the next table (TABLE 9.6). In the first row, a series of new "running numbers" (from 1 to 25) is recorded; in the

TABLE 9.5 Dictation-Strips.

A (for the first partial table, see TABLE 9.4)	1	8	5	4	14	17	23	20	2	3	22	11	21	16	6	9	18	10	19	12	24	15	13	7	25
B (for the ordinated partial table, see TABLE 9.6)	1	2	3	4	5	6	7	8	9	10	11	12	13	14	15	16	17	18	19	20	21	22	23	24	25

TABLE 9.6 Ordinated Partial Table (Compare to TABLE 9.4) Obtained by Rearranging the Releve' Order According to the New Sequence Shown at the Bottom of TABLE 9.4. The Sequence of Species Remained the Same as in TABLE 9.4. After Rearrangement, Species Can Be Noticed in Groups I and II That Have More Restricted Amplitudes (Underlined or Underlined with Dots, Respectively) Than Others.

New running numb.	1	2	3	4	5	6	7	8	9	10	11	12	13	14	15	16	17	18	19	20	21	22	23	24	25
Relevé number	1	9	10	4	3	15	24	2	16	18	12	20	23	5	22	14	6	17	19	8	13	11	7	21	25
I. Brom. er.	50	47	21	74	35	37	10																		
Camp. glom.	+	1		1	1		+					1													
Viola hirta		2	3		+	+	+		+																
Briza m.	1	2	1				1	2																	
Scab. col.	+		+	1		+		2																	
Linum cath.	+	+		+	+				1																
Thym. serp.	1	2		+	+			+																	
Salvia pr.		2	6		4	1	4																		
Koel. pyr.	3	3	3	2																					
Fest. ov.	2	1	2				1																		
II. Cirs. oler.[1]								+		1	+	+	14	12	2	+	20	+	3	20	2	3			
Geum rivale								1	+	+	+		+		2	+	1	5	+	3	1	+	+		
Holc. lan.											+	1		2	1	2	+	2	1	2	1	2	15		
Melandr. d.								1	+				+	1	2	+	+				1	+	4		
Alop. pr.									2				2	10		8			10	4	6		15	1	
Lysim. numm.								+	+						+	+		1		+		+	+		
Dasch. caesp.														1		10	2	2		20	11	5			
Lych. fl. c.								1							+	+	+			+		+	+		
Glech. hed.									+			+	+				+		+	+					
Angel. silv.											1			1	1				+		2		+		
Car. acut.																2				4	10	1	2		
Fil. ulm.															+	3			2	+					
Pimp. magn.																+				1		+	1		
Pol. bist.											+o							·	1			1	4		
Past. sat.									3							+				1					
Species numbers group Ia	5	5	5	4	3	3	2	3																	
" Ib	3	4	2	2	3	1	2	1	2	1	1														
" IIb									3	3	4	3	1	4	5	3	5	5	5	6	6	6	4	6	6
" IIa											1		2			4	3	3	3	3	3	6	4	4	
Changed runn. no.[2]	3	4	2	1	6	5	7	8	9	10	11	12	15	13	14	24	19	18	16	21	17	20	25	22	23

[1] When occurring with very low biomass (+), *Cirsium oleraceum* is not included in group IIa because its occurrence as a single individual is not considered sufficient for establishing a trend.

[2] Criteria for fixing the definite order of releves are not only the relations of species numbers in groups Ia, Ib, IIb and IIa, but also the biomass of important species, i.e., *Bromus erectus* in group Ia as well as *Cirsium oleraceum* in group IIa.

second row, the original relevé numbers are entered in their new sequence. Within each group, the species order can be further improved by rewriting it in sequence of constancy from high to low. A second person dictates the values for each species from TABLE 9.4 by using a dictation strip (of the same graph paper) that shows the new sequence of original relevé numbers (TABLE 9.5A). The recording person uses a strip that contains the relevé numbers in simple running order from 1 to 25 (TABLE 9.5B), which he moves downward from species row to species row. The person dictating always calls out the new running number first and then the corresponding quantitative value for each species from the constancy table.

The result of the transfer is the ordinated partial table (TABLE 9.6). From this it can be seen that within each, the *Bromus-* and the *Cirsium-*group, there are a few species of wider and a few of narrower distribution. The latter are underlined in TABLE 9.6.

Next, an improved species sequence can be obtained by relisting the underlined species and the nonunderlined ones in separate subgroups. This is best done in another partial table (not shown here). Furthermore, within each amplitudinal subgroup, the species order can be improved by writing the grouped species in sequence of their abundance, from high to low.

Upon completing the final partial table, the entire raw table is converted into a "differentiated" table (TABLE 9.7). In this table, the differential and all other species are written down in order of their constancy, and the relevés are grouped into three units. A dictation-strip is used also for this transfer.

Even in this stage the table is not perfectly ordinated. For example, the relevé sequence could be further improved by moving relevé 14 (running No. 24) more to the left on the table. Relevé 19 (running No. 16) was eliminated, because it shows a considerable gap in the group of "other species" and because it contains more than five species not found in any other plot. Apparently, this community has little similarity to the rest of the *Arrhenatherum* communities in spite of abundant *Arrhenatherum*. This is probably due to the fact that it occurs near a small tributary of the Danube where the site is periodically flooded and subjected to some sedimentation. However, such eliminations should be done with great caution to avoid arbitrary elimination of all "unfitting" material. Sample stand elimination from a table, however, does not mean disregarding the variation displayed by such a stand. It means merely that the stand is considered to represent a category by itself and may be grouped with other more similar stands, once the survey is expanded to include a more complete assessment of the *Arrhenatherum* grass cover communities of that geographic area.

TABLE 9.7 Example of a "Differentiated Table." The Relevé Sequence and the Order of the Tentative Differential Species (Groups Ia, IIa, Ib, and IIb Resulted from Table 9.6.)

Unit	A								B						C										
Preliminary name	Bromus-Arrhenatherum community								Geum-Arrhenath. commun.						Cirsium-Arrhenatherum community										
Changed running number	1	2	3	4	5	6	7	8	9	10	11	12	13	14	15	16	17	18	19	20	21	22	23	24	25
Relevé number	4	10	1	9	15	3	24	2	16	18	12	20	5	22	23	19	13	17	6	11	8	21	25	14	7
Relevé size (m²)	10	10	10	10	16	25	16	10	16	25	16	10	25	16	16	10	10	25	10	25	16	10	16	16	10
Openings (in % of area)	50	55	45	40	40	50	60	60	60	45	35	65	20	30	35	25	45	50	45	40	35	45	45	55	60
Hay yield (kg·100/ha)	P	<P	<P	P	P	<P	P	P	P	<P	P	P	10	38	18	20	25	20	25	12	30	30	42	45	8	28
Grasses (total biomass %)	90	55	83	85	89	70	80	90	70	75	80	70	71	85	56	89	45	85	38	75	40	89	82	33	30
Sedges "	3		2		1				5						6		4		10			1	3	2	10
Legumes "	+	+	1	1	+	1	2	+	1	+	1	9	1	+	+	+	+	2	+	+	+		1	1	
Other species "	7	45	14	14	10	29	18	10	24	25	20	24	20	14	40	5	55	10	60	15	60	10	15	64	58
Total number of species	28	29	31	32	34	32	31	25	27	29	34	40	32	22	29	20	31	35	32	28	34	35	28	33	36
Ia Bromus erectus	74	21	50	47	37	85	10																		
Scabiosa columbaria	1	+	+		+		2																		
Thymus serpyllum	+	2	1			+		+																	
Salvia pratensis	5		2	1	4	4																			
Koeleria pyramidata	2	3	3	3																					
Festuca ovina		2	2	1			1																		
Ib Campanula glomerata	1		+	1		1	+		+		1														
Viola hirta	3		2	+	+	+				+															
Briza media		1	1	2								1	2												
Linum catharticum	+		+	+		+			+																
IIb Geum rivale									1	+	+	+		+	+	+	+	2	3	5	1	+		1	
Holcus lanatus													1		+	+	1	2	1	2	2	2	15	2	1
Melandrium diurnum									1		+			+	+	+	+	2			+	4	1	1	
Alopecurus pratensis									2			2	10				4	8	6	10	15	1			
Lysimachia numm.									+	+							+	+	+	1	+	+			
Lychnis flos cuculi									1						+			+	+	+	+				
Glechoma hederacea										+		+	+					+	+						+
IIa Cirsium oleraceum										+	+	+		1	+	+	2	12	3	20	2	3	18	20	
Deschampsia caespitosa															2		10	28	2	5		1	11		
Angelica silvestris														1		+		1				+	1	2	
Carex acutiformis															2			4		1	2		10		
Filipendula ulmaria																+		3		2		+			
Pimpinella magna																1	+			1		+			
Polygonum bistorta													+°			1				4			1		
Arrhenatherum elatius	2	8	+°	5	15	20	30	1	15	22	10	24	4	25	26	32	10	15	4	9	9	25	35	12	4
Dactylis glomerata	5	6	5	2	6	15	8	5	15	18	32	5	12	8	P	1	15	15	12	12	10	18	18	10	4
Galium mollugo	1	12	3	2	3	7	3	2	5	6	6	5	3	3	12	2	10	3	24	5	4	1	2	14	6
Poa pratensis	5	8	4	10	15	10	20	74	10	10	25	6	4	9	16	1	2	5	2	6	4	1	10	5	6
Plantago lanceolata	1	1	1	1	1	1	+	1	2	1	1	2	2	P	+	4	+	1	+	4	1	+	2	4	
Festuca pratensis		2		5	6	3	2	5	5	12	10	2	20	15	2	10	2	2P	3	10	P	15	3	2	2
Chrysanthemum leucanthem.	+	5	1	2	1	3	6	+	1		3	3	+	1	1	+	1	1	4	+	1	+		2	2
Ranunculus acer	3	+	+	+	+	+	+	+	+	1	+	1	2	+	+	2	+	2		+	+	+	+	1	2
Veronica chamaedrys	+	1	+	+	+	+	+			1	1	+		2	+	+	1	+	1	+	1	1			
Achillea millefolium	18	6	2	2	3	3	+		5	12	1	6	2	4			4	+			+		16	1	
Daucus carota	+	1	1	1	+	1		1	+	1	+	1	+	2		2	+	1		5	+			1	
Rumex acetosa	+		+		+				1		1	1	+	1	+	1	1	+	2	2	3	+	1	2	1

TABLE 9.7 (Continued) The Remaining Species Have Been Transferred by Dictation from TABLE 9.2 According to the Changed Relevé Sequence. In Addition They Were Ordered According to Constancy. The Vegetation Units Are Separated by Vertical Lines. The Table Also Shows the Complete Introductory Information.

	A								B								C								
Running number	1	2	3	4	5	6	7	8	9	10	11	12	13	14	15	16	17	18	19	20	21	22	23	24	25
Taraxacum officinale	+	+	+°		+°	+			1	+	+	3	+	4		2	+	1		1			+	+	3
Trifolium pratense	+		+		+	1	2	+	1	+		+	4	+	2	+°	+		1		+	+			1
Medicago lupulina	+	+	1	+	+	+			+	+	+		2	1	1		+	+	+	+					+
Helictotrichon pubescens			+	1°	4			1	1	1	13	4	1	20	20	2		4	4	8			+		3
Ajuga reptans				1				+		+	1	+	+	+			+	+	+	1	3	1	+	+	1
Trisetum flavescens			3	4	5	2	2	10	5		8	8	16	4	5	6	5			8					
Crepis biennis				+°	+	+			1	1	1	2	1	6			+	1	+	8	+		+		
Trifolium repens			+	1	+				+	+		+	2		1		+		+		+	+	+	+	
Heracleum sphondylium			+			1°	+		+	+		1	3	+	1		26	1	+				+		4
Centaurea jacea			+°		+	6	+	1	2		1	2			+			2	2			+		4	3
Plantago media	+		+		+	+	+	+	+	+	1		+		+	+		+			+				
Campanula rotundifolia	1	1	1	+	1	+	+		+		1				1				+		+	+	1		
Festuca rubra	2	4	15	3	6		2				2	2	+		+	1	1	2		2	+				
Cerastium caespitosum	+	+	+	+	+		+			+	+		+	+		+	+		+°	+		+			
Prunella vulgaris		+		+		+			+	+			+	1		+	+			2	1				+
Bellis perennis		1		+		+			+	+	1		+			+	+	+	+						
Lathyrus praetensis			+						+	1		+			+		1	+	+	+					+
Vicia sepium	+	+			+	+			+			1													+
Tragopogon pratensis			+					1		+	2			1	1										+
Carex flacca	3		2		1			3								2					2				
Senecio jacobaea			+	+					+	+							+		+						
Anthriscus silvestris						+					2		+		+						+				
Vicia cracca			+		+	+	+													1					
Lotus corniculatus					+	+													+		+				
Leontodon hispidus					+	1	1																		4
Dianthus superbus	1			+		+		+		+															
Galium boreale	+		1													1		1							
Silene inflata				+		2									+										
Knautia arvensis				+	1°		+																		
Potentilla reptans		1					1																+		
Pastinaca sativa							3					1										+		
Silaus pratensis						1		+					+											
Carex hirta													2	6											
" gracilis										4						1								
Myosotis arvensis					+																				+
Pimpinella saxifraga	+		+																					
Cardamine pratensis					+								+										
Myosotis palustris									+								+					
Galium verum				1	+																	
Euphrasia odontites						+						+									
Rumex crispus													+					+				
Poa trivialis																					2
Phleum pratense										+										1	
Festuca arundinacea									2											
Lolium perenne																				
~~Glyceria fluitans~~													20							
~~Phalaris arundinacea~~													20							
Phragmites communis																	+			
Brachypodium pinnatum						5														
Carex panicea					2															
Geranium pratense								+											
Galium uliginosum																			+
Sanguisorba officinale																			+
Ranunculus repens																	+			
~~Lamium album~~													+						
~~Polygonum convolvulus~~													+						
~~Chenopodium album~~													+						
Alchemilla vulgaris					+														

Repeated tabular rearrangement aids greatly in clarifying the similarity relations among individual communities. Therefore, this aspect should not be neglected. However, it increases the danger of errors during transfers. To avoid errors it is usually sufficient to check the total species number of each releve´ with that shown in the first table. The use of dictation-strips may permit an occasional side-wise distortion during data transfer, but a deviation up or down from the species row is practically impossible.

The technical part of the table work was discussed in such detail because a rational technique can save much time. Moreover, we wanted to show that the so-called "vegetation statistical" or tabulation technique has little to do with statistics in the usual sense. Rather it is a method of orderly comparison, in which more emphasis is given to qualitative criteria, such as the presence or absence of a species, than to quantitative criteria.

It may be pointed out that the original field data has not been changed or converted in any form. Only the releve´ and species orders have been reshuffled several times to bring out the most important qualitative differences between all plots.

The arrangement of the associated species groups is strictly an objective procedure that can be followed through and understood and repeated with the same data by any person not acquainted with the area and vegetation. The assignment of class limits, however, is based on judgment. For example, another person may want to set somewhat different criteria, and the assignment of a marginal plot to either one of two categories is open to argument. An objective assignment of class limits is very difficult to accomplish because of the ever present uneven variability among releves´. Moreover, any attempt to strive for an even amount of variability within each class—as would be necessary in an objective assignment of class limits—is usually of doubtful ecological significance.

We may summarize the tabulation procedure from the assembled field releve´ sheets to the final differentiated table in seven steps:

1.   Enter releve´ data into one table, i.e., the *raw table*. Species values are recorded in rows, releves´ in columns. Order of entry inconsequential.

2.   Count the number of times each species is present in the entire releve´ series and calculate the *constancy value* for each species. Record this value at the end of each species row.

3.   Enter all species into a new table in order of high to low constancy (*first change of species order*). This is the constancy table. You can save time by entering only the species whose constancy value lies

in an intermediate range. This range may be set arbitrarily as between 60 to 10 percent, 70 to 25 percent, etc. This depends on the vegetation and your judgment.

4. Search among these species of intermediate constancy for mutually associated ones that occur together in more than one relevé. These may serve as *differentiating species.* Also search for those species that are mutually exclusive to the first group. They may serve as another group of differentiating species. Underline each species that falls into the same category.

5. Extract these underlined species into a new *partial table* by moving the species of each group together (*second change of species order*).

6. *Change relevé order* by moving the relevés together in which the mutually associated species occur. This results in *an ordinated partial table.* At this stage further refinements in relevé and species order can be introduced, such as ordering the relevés by number of differential species per group and ordering the differential species from broad to narrow amplitude and from high to low constancy within each differential species group.

7. Final step. Record all species into a *completed differentiated table* by first writing down the differentiating species as found in steps 1 to 6. These are followed by the other species in order of constancy from high to low. The latter values were obtained in step 2.

9.4   UNRANKED ABSTRACT PLANT COMMUNITIES

One important result of the tabulation process, terminating with the differentiated table (TABLE 9.7), is a combining of the relevés into several "vegetation units." But their systematic rank or hierarchical status in relation to already established community types has not yet been determined. Any other vegetation formation could probably be subdivided into smaller units in the same way. Such vegetation units that are not yet placed into existing type concepts or categories and that are the result of grouping relevés or sample stands on the basis of floristic similarity, are designated simply as abstract plant communities or community types. They may also be called noda in the sense of POORE (1955). The first vegetation unit (*A*) on TABLE 9.7 is identified by the presence of several members of the *Bromus erectus* group (I*a*). The second unit (*B*) is recognized by the presence of several members of the *Geum rivale* group (II*b*) and the simultaneous absence of several members of the *Cirsium oleraceum* group. The third unit (*C*) is identified by the presence of several members of the *Cirsium oleraceum* group (IIa) and the *Geum rivale* group (IIb).

These vegetation units were established merely from floristic criteria through tabular comparison of species compositions. However, by inspecting their habitats more closely, it is found that they can be characterized also by ecological and economic criteria. In TABLE 9.7 they form an ecological series of increasing soil moisture. The productive capacity increases in the same direction upon uniform application of fertilizers (compare to productivity estimates recorded at the top of TABLE 9.7). However, the second vegetation unit is the most valuable with regard to feed value.

Such vegetation types are useful as work-units for the ecologist and the manager, in spite of the fact that no systematic unit-rank (as in SECTION 8.93) has been assigned as yet. Their distribution can be mapped even without awaiting their classification into a systematic category. For easier communication they can be given short names, such as:

A. *Bromus–Arrhenatherum* meadow type
B. *Geum–Arrhenatherum* meadow type
C. *Cirsium–Arrhenatherum* meadow type

Common names may be preferred for certain purposes. Such names are simply symbols for certain combinations of species groups. This applies to all names derived for floristically defined units. Here, they do not necessarily imply that the species used in the names are dominants or that they are present in each relevé. For example, relevé 2 in TABLE 9.7 (running No. 8) can still be considered a *Bromus–Arrhenatherum* meadow in spite of the current absence of *Bromus erectus*. This may appear strange to the uninitiated person and indeed can lead to confusion. But the reason for including relevé 2 still in the *Bromus–Arrhenatherum* community type is that this relevé contains three members of the differential species group named after the leading species, *Bromus erectus*. Moreover, relevé 2 lacks any differentiating species of the other two community types. It is also likely that a larger-sized releve superposed on the present releve size of 10×10 m would yield specimens of *Bromus erectus*. In that case the membership of relevé 2 to the first community type would be, of course, more obvious. Certainly, relevé 2 is a border type relevé and its placement into the *Bromus–Arrhenatherum* community type is a matter of personal judgment and therefore open to argument. It may, like relevé 19, be considered a category of its own that could be crossed out on the table.

The systematic rank of the vegetation units derived by tabular comparison cannot be determined from TABLE 9.7 without further knowledge of the regional vegetation. We are therefore not in the position as yet to say whether these units are variants, subassociations or associations in the accepted understanding of BRAUN-BLANQUET's floristic vegetation type concepts. We can only give a class-designation according to the distributional amplitudes of the differential species groups. Units *B* and *C* are unified by the *Geum rivale* group and separated by the *Cirsium oleraceum* group. Therefore, *B* and *C* could be considered as one community type including all *Geum–Arrhenatherum* meadows. The combined unit could be given equal rank to the *Bromus–Arrhenatherum* community type (*A*).

When considerable data on plant communities and plant community types has been accumulated in a well-investigated larger region or subcontinent, it is of considerable interest to relate individual community types to a more encompassing scheme of vegetation type concepts. We may therefore ask how our table-derived community types can be fitted individually and as a group (i.e., the *Arrenatherum* grass meadow unit as a whole) into the scheme of BRAUN-BLANQUET's categories that are so widely applied throughout Europe. This will be discussed in the next section.

9.5   RANKING OF ABSTRACT PLANT COMMUNITIES

The assigning of vegetation units to any of the floristic categories of BRAUN-BLANQUET's hierarchical system (compare SECTION 8.93) is much more a matter of personal judgment than is the derivation of unranked vegetation units. However, it is not entirely arbitrary either when based on a large number of relevés, on detailed tabular comparisons, and on a thorough knowledge of the vegetation of a larger geographic area. The philosophical basis is exactly the same as in any more widely applied soil classification scheme. The assignment of individual entities and group-entities is based on general agreement with the definitions of the categories, on knowledge and familiarity with the entities, and on common sense.

Comparative evaluation of a large mass of data requires a special technique, because tabular representation of each individual relevé would soon become rather unwieldy, when the total number of relevés exceeds 100 to 200. The units distinguished in the detailed tables are summarized into one column for the preparation of a "summary table." The summary can be done in different ways:

   1.   Simply by stating the percent constancy of each species.

2. By showing percent constancy together with the mean magnitude value or mean plant biomass estimate, as the case may be.

3. By calculating the "cover degree value" (i.e., mean percent cover) for each species.

It should be emphasized that whenever a mean quantitative value is calculated for a species in a group of relevés (as in point 3 above), a second value indicating how representative the mean is, should also be included. Otherwise, there is no way of telling anything about the variation within the unit. This second value is quite conveniently given by constancy.

Here we will discuss only the first two ways, since they require the least amount of work and are usually quite satisfactory for summary purposes.

Constancy can be expressed either in percent or more simply still in five classes that are designated by Roman numerals as follows:

  I.   Up to 20 percent constancy (rare)
 II.   20.1–40 percent low constancy
III.   40.1–60 percent intermediate constancy
 IV.   60.1–80 percent moderately high constancy
  V.   80.1–100 percent high constancy

The magnitude values in the BRAUN-BLANQUET rating scale (SECTION 5.42) cannot be averaged without reconverting them to cover percentages. The cover percent ranges corresponding to each magnitude value in the rating scale are not equal. Therefore, it is practical to either state the range of magnitude values for each species, or to show the most commonly recorded magnitude value (mode) or both as summation values for relevés forming a vegetation type. The estimated plant biomass percentages, however, recorded in TABLE 9.7, can be averaged directly. This can be done by omitting the relevés in which the species were not recorded (see TABLE 9.8), because the presence or absence of a species is already evaluated by the constancy percent of each species.

For explaining the summary procedure, we will use the data of TABLE 9.7 by rearranging it into a summary table (TABLE 9.9). Unfortunately most vegetation data is published in form of summary tables only, because of publication costs. Summary tables are, of course, less suitable for critical examination of the data, because the variation among relevés is not readily evident.

TABLE 9.9 shows a further step in treating the material. It represents already a so-called "characterized table." All those species have been

TABLE 9.8. Beginning of a "Summary Table." Calculation of Constancy and Mean Biomass Value for Each Plant Species. The Two Blocked-Out Columns Are Transferred into the Summary Table (TABLE 9.9).

Running no. of TABLE 9.7	1	2	3	4	5	6	7	8	Constancy		Biomass(%)	
Relevé number	4	10	1	9	15	3	24	2	absolute	in %	sum	mean
I$_a$. Bromus erectus group												
Bromus erectus	74	21	50	47	37	35	10		7	88	274	39
Scabiosa columbaria	1	+	+		+			2	5	63	3	0.6
Thymus serpyllum	+	2	1			+		+	5	63	3	0.6
Salvia pratensis			5		2	1	4	4	5	63	16	3.2
Koeleria pyramidata	2	3	3	3		·			4	50	11	2.8
Festuca ovina		2	2	1			1		4	50	6	1.5

combined into groups that are presumed to be useful as character species for certain associations, alliances, orders, or classes in Southwest Germany. It should be emphasized once more that this arrangement into groups of character species has not resulted from tabular comparison of the data analyzed in TABLE 9.7, but that it is the result of the comparison with other communities. Therefore, the individual vegetation samples in TABLE 9.7 do not convey the significance of the new arrangement in species sequence as shown in TABLE 9.9. The ranked order of character species in TABLE 9.9 can be viewed merely as a working hypothesis. It was therefore applied only to a limited geographic area. A final ranking into specific vegetation units can only be made after the entire spectrum of European plant communities, or at least those of Central Europe, is known without any gaps.

A good possibility for comparison is offered in TABLE 9.10, in which we purposely did not show plant communities from South Germany, but plant communities from Northwest Germany that were described by TÜXEN (1937). Each community number on TABLE 9.10 refers to an abstract community. The comparison serves the simultaneous purpose of asking whether or not the distribution of species that resulted from a study of a limited area in Southwest Germany is similar to another more thoroughly investigated area of Central Europe.

The answer is yes. In comparing TABLE 9.10 with TABLE 9.9 one can recognize clear parallels in the distribution of most species. However, there are certain differences in abundance. For example, Geum rivale and Melandrium diurnum are less common in Northwest than Southwest Germany. Also, Bromus erectus and Salvia pratensis

TABLE 9.9 Example of a "Characterized Summary Table." Derived Through Calculation of Percent Constancy and Mean Plant Biomass Values of the Data in TABLE 9.7 as Explained in TABLE 9.8. The Group of Other Species Shown in TABLE 9.7 (Including Some Species of Groups Ib, IIa, and IIb in TABLE 9.7) Was Classified into Four Species Groups (d,e,f,g), According to Their Systematic Rank as Differentiating Species for Other Vegetation Types. This Separation Is Only of Local Validity. Species with a Low Constancy Were Omitted from Group g.

Unit	A		B		C	
Number of relevés	8		6		10	
Mean yield						
Mean number of species						
Constancy (const), mean Biomass	const.	m.B.	const.	m.B.	const.	m.B.
a. Differential species of the Bromus-Arrhenatherum unit (A)						
Bromus erectus	88	39				
Scabiosa columbaria	63	0.6				
Thymus serpyllum	63	0.6				
Salvia pratensis	63	3.2				
Koeleria pyramidata	50	2.8				
Festuca ovina	50	1.5				
b. Differential species of the Geum-Arrhenath. unit (B + C)						
Geum rivale			83	+	80	1.5
Holcus lanatus			17	1	100	2.8
Melandrium diurnum			50	0.3	70	4.1
Alopecurus pratensis			50	4.7	60	7.3
Lysimachia nummularia			33	+	60	+
c. Differential species of the Cirsium-Arrhenath. unit (C)						
Cirsium oleraceum			50	+	100	8.1
Deschampsia caespitosa					60	9.5
Angelica silvestris					60	0.8
d. Character species (?) of the Arrhenatheretum elatioris grassland communities						
Arrhenatherum elatius	100	7.9	100	16.7	100	14.4
Galium mollugo	100	4.1	100	4.7	100	8.1
Trisetum flavescens	63	3.2	83	9.4	40	5.8
Crepis biennis	38	+	83	4.2	70	2.1
Heracleum sphondylium	38	0.3	83	0.8	60	5.3
Tragopogon pratensis	13	+	50	1	30	0.7
Anthriscus silvestris	13	+	17	2	30	+
Pastinaca sativa			17	3	20	0.5
Geranium pratense			17	+		
e. Character species (?) of the Arrhenath. and Lolium comm.						
Chrysanthemum leucanthem.	100	2.3	83	1.6	90	1.3
Veronica chamaedrys	88	+	83	0.6	90	0.7
Daucus carota	88	0.7	100	0.3	70	1.6
Trifolium repens	38	0.3	67	0.5	70	+
Bellis perennis	38	0.3	67	0.3	40	+
Lolium perenne			17	2		

Unit	A		B		C	
f. Character species of grass-land class						
Dactylis glomerata	100	6.5	100	6.5	100	12.2
Poa pratensis	100	18.3	100	10.7	100	5.7
Festuca pratensis	75	3.8	100	10.7	100	7.5
Ranunculus acer	88	0.4	100	0.8	90	0.7
Rumex acetosa	38	+	83	0.8	100	4.2
Trifolium pratense	75	0.5	83	1	60	0.7
Helictotrichon pubescens	63	1.4	100	14.3	50	3.8
Centaurea jacea	63	1.4	50	4.7	60	4.8
Festuca rubra	75	5.3	50	4.3	50	1
Cerastium caespitosum	63	+	33	+	50	+
Lathyrus pratensis	13	+	50	0.3	60	+
Vicia cracca	38	+	17	+	10	1
Filipendula ulmaria					40	1.3
Pimpinella magna					40	0.5
Polygonum bistorta			17	+⁰	20	2.5
Silaus pratensis			33	0.5	10	+
g. Species not strongly corre-lated with grassland						
Plantago lanceolata	100	0.9	100	6.5	100	2.4
Achillea millefolium	100	3.1	100	5	70	3
Taraxacum officinale	75	+	83	0.8	80	1.4
Medicago lupulina	75	+	83	0.6	60	+
Ajuga reptans	25	0.5	83	+	90	0.7
Plantago media	75	+	67	0.3	40	0.3
Campanula rotundifolia	88	0.6	33	0.5	40	0.5
Prunella vulgaris	38	+	50	+	60	0.7
Vicia sepium	50	+	33	0.5	10	+
Campanula glomerata	63	0.6	33	0.5		
Carex flacca	38	2	17	3	20	2
Viola hirta	63	1	17	+		
Senecio jacobaea	25	+	33	+	20	+
Briza media	50	1.3	17	2		
Linum catharticum	50	+	17	+		
Lotus corniculatus	25	+			20	+
Leontodon hispidus	25	0.5	17	1	10	4
Dianthus superbus	25	0.5	33	+		
Galium boreale	25	0.5			20	1
Silene inflata	25	1			10	+
Knautia arvensis	25	0.5	17	+		
Pimpinella saxifraga	25	+				
Galium verum	25	0.5				
Carex acutiformis					40	4.3

TABLE 9.10. Summary of the Distribution of Some Species Among Community-Types of Northwest Germany. Summary of Abbreviated Extracts from Tables by TÜXEN (1937). The Figures are Percent Constancy Values. The Grouping of Species Corresponds to TABLE 9.9. Only the Three Most Constant Species Were Chosen from Groups f and g.

Plant formations (in North West Germany)	Dry grass communities			Fertilized pastures			Fresh fertilized pastures						Managed moist meadows							
Community type, see footnote	1	2	3	4	5	6	7	8	9	10	11	12	13	14	15	16	17	18	19	20
Page in TÜXEN 1937	56	70	69	103	101	102	109	106	108	105	104	83	92	80	91	81	90	95	89	100
a. Thymus serpyllum	68	87	78				38	33			27				57					
Festuca ovina	90	85	56					36												
Scabiosa columbaria		78	33																	
Koeleria pyramidata		39	22																	
Bromus erectus		17																		
Salvia pratensis	6																			
b. Holcus lanatus				100	72	82	50	18	86	90	91	86	100	20	100	94	100	100		43
Alopecurus pratensis				75	50	46	87	50	14	90	45	43	40	20			38	23	44	64
Lysimachia nummularia									27	40	36	24	80	40	43	62			22	50
Geum rivale												27	v		43		v			
Melandrium diurnum							25							20						
c. Deschampsia caespitosa				37	33	27	25		9		82	29	100	20	7	6		16	67	21
Angelica silvestris												27	67	20	80	71	88		78	7
Cirsium oleraceum												36	39	20	10	14	44		56	
d. Arrhenatherum elatius					22		18		9	82	85	64	17	20	20					
Crepis biennis					25	27			9	68	75	45	11	20	10		12		11	
Tragopogon pratensis						9			18	86	60	18	4	20						
Pastinaca sativa						18				59	60	45								
Anthriscus silvestris					17	9			27	82	90	18	36	20	80		6			
Heracleum sphondylium					8	27	25		83	64	86	95	36	57		60			8	
Trisetum flavescens						9		75	83	73	91	85	55	25			14	6	8	
Galium mollugo	24	22	25			36	50	33	45	50	55	73	v	60				71		
e. Chrysanthemum leucanth.				82	50	91	87	83	82	100	80	100	32	60		57	12	31	v	
Veronica chamaedrys				62	v	46	87	100	100	86	65	64	45	80	40	57	38		33	
Daucus carota	50	89		37	8	36			9	64	60	91	25	20	10	14	6			7
Bellis perennis	28			62	75	82		67	73	91	90	73	93	40	20		78	100	67	29
Trifolium repens			22	75	100	100	50	17	9	18	15	18	14	20		19	92	v		57
Lolium perenne				37	100	36				64	75		36					46		21
f. Festuca pratensis				100	100	18	13	17	18	23	45	82	64	100		43	50	69	56	50
Poa pratensis	20	28	45	75	83	55	38	67	36	91	80	73	79	60	40	43	62	62		62
Dactylis glomerata				45	50	25	46	87	67	64	100	95	91	32	20	90	71	12	22	7
g. Plantago lanceolata	48	65	45	100	75	100	75	100	100	95	85	100	86	80		71	62	85	56	
Achillea millefolium	70	52	45	100	83	73	63	83	82	100	95	91	61	40		85		62		
Taraxacum officinale		20	67	62	83	100	50	100	82	95	95	73	79	80	60	29	38	69	78	64

	other moist and wet meadows							Sedge commun.	on sand		on loam		on wet soils						Cut-Over areas		Nitrate rich sites		other					Crop fields		
21	22	23	24	25	26	27	28	29	30	31	32	33	34	35	36	37	38	39	40	41	42	43	44	45	46	47	48	49	50	51
97	79	93	94	87	84	86	75	53	132	131	157	166	161	163	158	149	148	147	34	36	38	26	21	22	24	23	26	31	30	17
								46	75																					
v		33	73	84	29	86	30												63	50			35							
33	17	62	27				v																							
50		67	v		29			33																						
														33																
																						33								
17	42	v	55	46	14	57		33			41	70	81	92	63	53	46	46			67									
v	33	12	18	8	71	36	30							66	23	32	54	39		v	67	v								
	42				29									25	46	v	v				33	v								
	8																													
			9																											
																			v	33	v	50								
																					50	23								
	50			23			v																							
					14	7																								
							23																							
				15	14	7																								
17			30	18	8																									
50			33	9	8	14	7	30																47	90	55	24		44	
																			13		17					80	6			
17	25	50	46	23	14																									
50		50	22	46	29															v	29		v							
	50											72	55	52	42	46	21		43	20	41									
	v	36	61	43		40														59	59	45	v		28	44	v			
		38							42											27	59						43	22	55	
v		38								12								28	v		64	89	63	80	35	v	22	67		

201

are less common in Northwest Germany, and they never occur here together with *Arrhenatherum* or any other species of group *d* (TABLE 9.10) except with *Galium mollugo*. Nevertheless, *Bromus* and *Salvia* can also be used in this region as indicators of dry meadows, and likewise the *Cirsium oleraceum* group *c* (TABLE 9.10) shows an affinity to moist habitats in Northwest Germany. The major distribution center of both groups lies, in the North as well as South, outside the *Arrhenatherum* meadows. Thus, they represent only differential species and not character species within the *Arrhenatherum* community type.

TABLE 9.10 now permits a decision as to which species can be considered a character species for certain associations, alliances, orders, or classes. Among character species, BRAUN-BLANQUET distinguishes

---

Explanation of community-type numbers (for more complete information see TÜXEN 1937): 1. Dry-grass community on acid sand. 2. Semi-dry-grass community on limestone soil. 3. Communities on barren limestone slopes. 4. Moist *Lolium perenne* pasture. 5. Fresh *Lolium perenne* pasture. 6. Moderately dry *Lolium perenne* pasture. 7. Acid *Trisetum flavescens* meadow (mountain meadow). 8. Moderately acid *Trisetum flavescens* meadow. 9. *Trisetum flavescens* meadow on limestone soil. 10. Moderately dry *Trisetum flavescens* meadow. 11. Fresh *Trisetum flavescens* meadow. 12. *Molinia coerulea–Trisetum flavescens* meadow. 13. Moderately moist *Cirsium oleraceum* meadow. 14. *Molinia coerulea* meadow on limestone soil. 15. *Petasites officinalis–Cirsium oleraceum* meadow. 16. Slightly acid *Molinia coerulea* meadow (with *Parnassia*). 17. *Cirsium oleraceum* meadow with sedges. 18. Moderately moist *Juncus filiformis* meadow (an acid, fertilized meadow-type); 19. Typical *Cirsium oleraceum* meadow. 20. Moderately moist *Ranunculus repens* grass community (occasionally flooded). 21. Frequently flooded *Ranunculus repens* grass community. 22. *Filipendula ulmaria* creek-shore community. 23. *Juncus filiformis* meadow. 24. Wet *Juncus filiformis* meadow. 25. Acid *Nardus stricta–Molinia coerulea* meadow. 26. Acid *Molinia coerulea* meadow. 27. Wet, acid *Molinia coerulea* meadow. 28. Short-sedge meadow on limestone soil. 29. Tall-sedge meadow. 30. *Quercus robur–Betula pendula* forest rich with *Viola reichenbachiana*. 31. *Nardus stricta* mountain meadow. 32. Typical *Quercus robur–Fagus sylvatica* forest. 33. Moist *Fagus sylvatica* mixed-wood forest. 34. Moist *Quercus–Fagus* forest. 35. Wet *Quercus–Fagus* forest. 36. *Fraxinus excelsior* forest with sedges. 37. *Fraxinus excelsior* forest. 38. *Alnus glutinosa* swamp forest with moving ground water. 39. *Alnus glutinosa* swamp forest. 40. Cut-over communities on limestone soils. 41. The same on acid soils. 42. The same on half bog soils. 43. Alluvial bottomland communities. 44. Nitrate-rich wayside communities. 45. Nitrate-poor wayside communities. 46. Foot-path communities on moist soils. 47. The same on fresh loamy soils. 48. Flood plain and lakebed communities. 49. Hoed-vegetable fields on sand. 50. Cereal crop fields on moist sand. 51. The same on limestone soil.

the following three degrees of exclusiveness or "fidelity" in relation to a given vegetation unit:

1. Absolutely restricted (fidel), meaning that the species is exclusively or almost exclusively found only in a single association.

2. Strongly associated, meaning that the species is represented also in other associations, however, much more sparsely.

3. Favorably associated, meaning that the species is represented more or less commonly in several associations, but that it is optimally developed or abundant only in one particular association.

Arbitrary limits can easily be established for each class of character species.

If TABLE 9.10 is inspected with these criteria in mind, it becomes evident that only very few species can be considered exclusive (or fidel) in Northwest Germany. Among meadow plants, only *Bromus erectus* and *Salvia pratensis* are absolutely restricted to the *Mesobrometum erecti* association (TÜXEN 1937:70) in the subatlantic northwest of Germany. (However, this association also has other species that are not listed in TABLE 9.10; among them *Cirsium acaule*, *Brachypodium pinnatum* and *Gentiana ciliata*.) All other species of TABLE 9.10 can be considered merely as strongly or favorably associated character species for certain plant associations.

Strongly associated character species of the *Arrhenatherum* meadows are in Northwest Germany only: *Arrhenatherum elatius*, *Crepis biennis*, *Tragopogon pratensis*, and *Pastinaca sativa* of group *d*, TABLE 9.10. These species occur with a high degree of constancy in three of the vegetation units differentiated by TÜXEN, namely, in the two subassociations of the *Arrhenatheretum elatioris* association (communities Nos. 10 and 11 in TABLE 9.10) and in the "NW German variant" of the *Molinietum caricetosum tomentosae* association (community No. 12). The latter is very similar to the *Arrhenatheretum* and would probably be better combined with this association than with the moist meadows.

It may be noted that particularly this association (No. 12, TABLE 9.10) shows the greatest similarity to our unit *C* in TABLE 9.9. This similarity is not accidental, but reflects the similarity of the habitats. The relevés of TÜXEN were established in the hill country of South Hannover and its northern border zone. The climate here is relatively continental and the soils are calcareous. Both conditions apply also to the Danube lowland south of Ulm. Another common point is the fact that both *Cirsium* meadow types in the north and south have originated from *Molinietum* communities (*Molinia coerulea* litter meadows) through fertilization and twice annual cutting.

In comparison to the first four species of group *d* (TABLE 9.10), the other representatives of the same group show a much wider distribution. *Anthriscus silvestris* is found in many localities that are rich in nitrogen (communities 42, 43, 44, TABLE 9.10). The species also invades well-fertilized, moist meadows and attains for example, a constancy of 80 percent in a subassociation of the *Cirsium oleraceum–Angelica silvestris* association (running No. 15, TABLE 9.10) that occurs near streams. Therefore, this species can only be considered a favorably associated species of the *Arrhenatherum* meadows.

Heracleum, *Trisetum*, and *Galium mollugo* occur at higher elevations in mountainous areas than *Arrhenatherum*, *Crepis*, *Tragopogon*, and *Pastinaca*. Therefore, the former are of local significance for the *Trisetum meadows*, (running Nos. 7–9, TABLE 9.10). If these units are considered a different association from the *Arrhenatheretum* association (Nos. 10 and 11), *Heracleum* and *Trisetum* can be used as strongly associated character species of an alliance (*Arrhenatherion*) (running Nos. 7–12, TABLE 9.10) of the *Arrhenatherum* and *Trisetum meadows*. *Gallium mollugo* can be considered only a favorably associated character species of this alliance, because it is found also in several other associations.

A much wider amplitude is exhibited by the species group *e* (TABLE 9.10). *Chrysanthemum*, *Veronica chamaedrys*, *Daucus*, and *Bellis* are found in nearly all meadows and pastures, which are fertilized and are neither dry nor moist. They attain their greatest constancy in the community types of fresh, regularly fertilized meadows and pastures (running Nos. 4–12, TABLE 9.10). Therefore, they can be considered as strongly or favorably associated character species of an order (*Arrhenatheretalia*), which includes the alliance of the regularly fertilized, fresh meadows (running Nos. 7–12) and the alliance of the regularly fertilized pastures (running Nos. 4–6). *Trifolium repens* and *Lolium perenne* could be regarded as favorably associated species of the last named alliance, because they show, according to constancy, a definite optimum in the regularly fertilized pastures, in spite of their much wider distribution.

The amplitude of group *f*, which only shows three species on TABLE 9.10, is even wider. A similar distribution is shown also by *Holcus lanatus* and *Alopecurus pratensis* in group *b*. These five species can be considered as character species of a class of agriculturally managed or cultivated grasslands (*Molinio–Arrhenatheretea,* running Nos. 4–20). Among the species shown in TABLE 9.10, *Festuca pratensis* and *Alopecurus pratensis* are exclusive to this class, the others, however, are only favorably associated.

Group g in TABLE 9.10 gives three examples of species, which are present in numerous communities. Therefore, they are considered only as vaguely associated with grass cover communities. However, these species are not ubiquists in the proper sense, because they are rarely present, for example, in forests and on most wetland soils. Such species are usually classified as "companions" in phytosociological tables, which form a final and very heterogeneous group of species.

We discussed TABLE 9.10 in detail to give an idea of the degree of certainty with which one can recognize character species and abstract communities of a certain categorical rank. The same method of tabular comparison which was explained from several releves of meadow communities, is used for ranking the initially unranked abstract communities into a certain hierarchical order of vegetation units.

9.6 DETERMINATION OF PLANT ASSOCIATIONS AND CHARACTER SPECIES

Hierarchical grouping of vegetation units remains relative because it depends on the delimitation of the basic units, the associations. For example, if the *Arrhenatherum*-meadows are considered to form an alliance rather than an association—whereby the alliance would include several local associations—then the current alliance including the *Trisetum*- and *Arrhenatherum-meadows* would become an order. If, on the other hand, both the *Trisetum*- and *Arrhenatherum*-meadows are combined into one association—perhaps for the reason that the *Trisetum*-meadows have hardly any character species of their own in the lower montane vegetation belt—then the higher hierarchical units would also move down one step. Both treatments would be just as defendable as the one used in the previous section. These treatments are purely a matter of personal judgment.

It was customary in the first evalution of the vegetation of a given region to only distinguish a few, rather broad associations. These are identifiable by many character species, which provided a rapid, but rather coarse outline of the plant cover of such a region.

However, with increasing knowledge arose the desirability to subdivide the original association for more adequate description of the many variations in habitat and geographical locations. The unit-sequence: association-subassociation-variant-facies soon became insufficient for encompassing the multitude of variations into one system. The most commonly used evasion of this problem was to raise the sequence of units: associations became alliances, these in turn became orders and finally classes. This gave room for new associations and subunits. This trend can easily be noticed when comparing the works of BRAUN-BLANQUET, KLIKA, TÜXEN, OBERDORFER (or others belonging to the same school), which were written during the third, fourth, fifth, and sixth decades of this century.

The result of this development was a progressive narrowing of the association concept. The increase in number of associations corresponded to a decrease in number of character species that remained useful for identifying these small associations. The number of species with narrow amplitudes is limited in all regions. For example, many of TÜXEN's (1950a) associations have no character species of their own. They are recognized by character species of alliances that are combined with differential species.

Such small associations have many advantages to a specialist, just as finer species divisions have advantages to a taxonomist interested in a special group. However, comprehension of such classifications becomes very difficult as a result of too much splitting. Particularly difficult to follow is the repeated changing and renewed defining of associations, alliances, or higher units after they have been defined once, because this requires a reevaluation of the species used for diagnostic purposes.

Another reason for repeated reevaluation of the different hierarchical units was related to the extension of the vegetation studies over increasingly larger areas. As long as an investigator is concerned only with a small geographic area, he can use many character and differential species with locally valid diagnostic value. However, with increasing geographic range, he will find more and more instances where a species that locally may show a strong correlation to a particular association may be present also in other associations that may not even occur in the smaller geographic area.

For example, Bromus erectus is restricted to the alliance Mesobromion in Northwest Germany. In Southwest Germany, Bromus erectus is found in this alliance and also in the Xerobromion. The latter alliance has many exclusive character species, and it occupies much drier and warmer habitats than exist in the subatlantic climate of Northwest Germany. There is even a relatively large number of floristically different communities with Bromus erectus as major component in South-Central Europe. Because of their high number, they have been grouped into a special order (Brometalia).

Another example will further illustrate this point. The Cirsium group consisting of *Cirsium oleraceum, Deschampsia caespitosa, Angelica silvestris,* and others (TABLE 9.9) can be considered as strongly associated character species of the moist meadows only in the geographically limited area of South Würtemberg. Here the moist meadows are grouped into one association, the *Cirsium oleraceum* association. The species of the *Cirsium* group extend here only as differential species into the moistest segment of the otherwise drier *Arrhenatherum* meadows, and they do not occur in any other grassland type, because there are only fertilized and neutral-to-weakly acid, moist meadow-habitats in this area.

However, if one considers a larger geographic area that includes also unfertilized, extremely acid moist meadows, for example, by extending the vegetation classification to the Rhine-plain of North Baden, one observes that all species of the *Cirsium* group are found also in the *Molinia*-meadows (*Molinietum*). *Cirsium oleraceum* may still be considered a favorably associated character species of the above *Cirsium oleraceum* association with *Arrhenatherum.* But, the other group members, *Filipendula ulmaria, Deschampsia caespitosa, Angelica silvestris,* and many others are equally common in the *Molinia* meadows. Most of these can be considered character species for both the *Molinia-* and *Cirsium oleraceum*-meadows. Therefore, they become useful as character species of an alliance when enlarging the geographic outline of the vegetation study.

If the geographic area is further enlarged to include areas with predominantly acid soils where even drier soils are acid (for example, Northwest Germany) one will notice still other types of *Cirsium oleraceum-* and *Molinia* meadows. In this area, the *Cirsium oleraceum-* and *Molinia*-meadow types can be subdivided into several units in relation to increasing soil moisture. Therefore, it appeared useful to distinguish several associations within each meadow type. Consequently, all fertilized moist meadows were grouped into an alliance (which today is called *Bromion racemosi*) and the *Molinia* meadows were grouped into another alliance, *Molinion.* Another consequence is the raising of *Angelica silvestris, Deschampsia caespitosa,* and other species into the rank of character species of an order (*Molinietalia*).

Finally, it should be mentioned that *Deschampsia caespitosa* extends its amplitude beyond the frame of this order in East and Southeast Europe, where it does not occur only in meadows, but also in wetland pastures and forests. Consequently, it may hardly be called a character species of a single vegetation unit at all. Therefore, a species like *Deschampsia caespitosa* can be considered in progression as character species for an association, an alliance, an order, or a class, or it may even lose its diagnostic value entirely. This depends only on the size of the geographic area considered and on the current state of knowledge.

Similar examples could be cited for other geographic areas and also for all formations, particularly for forests. A perfectly clear vegetation classification of a relatively small geographic area, which may show floristically well-defined vegetation units, becomes increasingly diffuse when applied to successively larger areas.

A compromise for this dilemma was found by many authors in returning, so to speak, to the starting point of phytosociological research; that is, by limiting the diagnostic validity of character species geographically. On this basis, the following types of character species have been distinguished:

1.   Local character species; with a closely limited range of diagnostic validity (e.g., applicable to a mountain valley, an island, or the surroundings of a village).

2.   Territorial character species; which have validity for larger, naturally defined regions (e.g., the Vienna basin, the Northwest German lowland or Ireland).

3.   Absolute character species; which have diagnostic validity without any geographic limitation.

The latter type is practically nonexistent, according to our present state of knowledge. A few rare exceptions are geographically limited endemics or ecologically strongly specialized plants on extreme habitats, for example, certain alpine species, or species of serpentine soils.

The local or territorial limitation of the character species concept, however, leads to three rather undesirable consequences. First, one arrives at a large number of basic units with different names, although many of these may be quite similar. Also, the same species may be used as character species in several units (TÜXEN 1950a). Thus, synoptic comprehension of such vegetation units becomes even more difficult than it was through establishing many small associations. Second, it proved to be impossible, especially in extensive level terrain, to draw the boundaries between the geographic ranges of associations without some arbitrary decision. The third conclusion held the greatest weight— local and territorial units could no longer be claimed as universally applicable basic units. Yet, the associations were intended for just this purpose according to the original concept of BRAUN-BLANQUET.

However, it is indeed questionable whether it is possible to recognize basic units in vegetation science that are so closely defined and universally applicable as is the species concept in taxonomy. This question must be answered negatively, because of the nonorganismic attributes of plant communities as previously discussed. Similarities among communities are more or less gradually distributed along geographic gradients. While similarities and differences of communities can be validly established for restricted geographic areas, it is not possible to do this in the form of a super-regional hierarchy of universal application (see also WALTER and WALTER 1953, SCAMONI 1954).

Other important phenomena, which point in the same direction, are the findings made from transplant studies of widely distributed species. For example, the classical genecological study by CLAUSEN, KECK and HIESEY (1948) of *Achillea borealis* in California has shown that this species is separated into distinct ecological races or ecotypes. Differences in temperature and rainfall seasonality at different geographic locations in California were genetically manifested in physiologically and morphologically quite distinct populations. It is highly probable that geographically widespread species, such as *Bromus erectus* and *Deschampsia caespitosa* are comprised of several genetical races with somewhat differing environmental adaptations.

Therefore, we may say that three important observations run contrary to the validity of the character species concept as originally intended.

1. The spectrum of available habitats may vary from one geographical region to another, particularly where agricultural modifications, such as regular fertilization of meadow communities, are superimposed. This can open new niches, or eliminate others, for the same species occurring in different geographic regions.

2. The same species becomes associated with different competitors from one geographic region to the next. This may bring about different environmental relationships for the same species.

3. Different geographic regions are associated with macroclimatic variations that may be genetically manifested in the same species. This would separate the same species into ecological races or ecotypes with differing physiological adaptations.

The three factors will act together in bringing about differing sociological relationships for any species that has a wide geographic distribution.

9.7   VALUE OF THE ALLIANCE CONCEPT

For orientation purposes we believe it more useful to distinguish relatively large and easily identified vegetation units, than a large number of narrowly defined and overlapping associations. According to the present state of European phytosociology, such relatively large vegetation units are the alliances of the BRAUN-BLANQUET system. They correspond approximately to the earlier association concept of the BRAUN-BLANQUET school, but also to type concepts of earlier authors, for example, to the "meadow types" of STEBLER and SCHRÖTER, to the "forest types" of CAJANDER and to the ecological formations of WARMING. Also, a few climax associations of CLEMENTS have a comparable content.

This parallelism in concept of many ecologists is not accidental, because alliances are in general vegetation units that recur in similar forms in many areas. They are as a rule structurally and physiognomically quite distinct. Their habitats can usually be recognized independently of the vegetation cover and without special measurements. In addition they are well separated floristically, either by many differential species or even by several character species that have diagnostic significance in large geographic areas. Therefore, they fulfill largely the requirement originally set forth for the association concept at the International Botanical Congress at Brussels in 1910 (see SECTION 8.92). As examples of alliances may be named, among others, the *Arrhenatherum* meadows (*Arrhenatherion*), the *Phragmites* water grass communities (*Phragmition*) and the moist alder forest communities (*Alnion glutinosae*) and the therophytic *Salicornia* marshland communities (*Salicornion europaeae*). In their order of magnitude alliances are comparable to the North American concept of cover type, although alliances are not always dominance-communities. For example, the alliance *Arrhenatherion* does not imply the presence of *Arrhenatherum elatius* as dominant grass in each relevé (see TABLE 9.7). The alliance is therefore more a qualitative than a quantitative vegetation type concept.

Alliances can be identified rather easily by several criteria. Thus they are more or less natural units. However, our intention is not to suggest these as rigid, basic units, but rather as orientation points. As such, they can be used for establishing relationships to the many, already described local and territorial associations and to the equally numerous sociations. Also, vegetation units derived from local studies, whose systematic status has not yet been decided, can be related to alliances. Moreover, we consider it useful to maintain an unsystematic status for abstract vegetation communities in all cases where the emphasis is on intensive local vegetation studies. However, a hierarchical scheme becomes very desirable where the emphasis lies on developing a vegetation synopsis at a more extensive geographical scale.

Mathematical Treatment of
Vegetation Data

10 The mathematical approach to grouping and ordering vegetation samples has received a great deal of attention. The basic supposition is that mathematical synthesis of data leads to a more objective presentation of results. More objective in this context means that mathematical definitions of steps in the analysis and synthesis of data will permit exact repetition of the procedure by anyone who follows the prescription.

Therefore, mathematical methods as objective treatments have a special value as prescriptions for the novice, or the person with a limited knowledge or familiarity with the data.

The latter, of course, applies to some extent to any investigator who has completed a field study of a number of vegetation samples. He may know a great deal about the floristic composition of his samples, the distribution of the species therein, and the ecological relationships. He can subsequently find out a great deal more about his data by ordinary, nonmathematical synthesis techniques as described in the preceding chapter. Yet, a mathematical treatment may help put his subsequent findings on a more creditable basis for some readers. It may give further insights into similarity relationships and may aid in making a better judgment, or preventing erroneous ones.

For example, the ordering or ordination of species and stands is a necessary step prior to any good floristic classification. However, the ways of ordering this material may range from crude ranking to the most sophisticated factor analysis.

Closer examination of mathematical methods and their application, particularly of the more complicated ones, shows that familiarity with the kind of vegetation to be processed is required to come up with a reasonable interpretation. Therefore, as a rule, the successful application of mathematical methods requires a number of decisions and adjustments. Decisions and adjustments always involve judgment, and any judgment is a subjective act.

It is not the purpose of this book to treat the most recent developments in mathematical techniques. To these, this chapter may be considered an introduction. Instead, only the more widely used and simpler mathematical approaches will be discussed.

10.1 SIMILARITY COEFFICIENTS OF PLANT COMMUNITIES

It was emphasized before (SECTIONS 9.5 and 9.6) that the grouping of communities into an association and the assigning of unit categories for abstract plant communities was a matter of judgment on the part of the investigator. The reason is that different investigators may have different opinions about the degree of similarity allowable in combining individual communities into an association. It is difficult to establish a generally acceptable degree of similarity or dissimilarity. However, similarity relations can be expressed mathematically, and one can set arbitrary limits on a mathematical basis.

These mathematical expressions of community similarity are variously referred to as indices of similarity or community coefficients.

10.11 Index of Similarity According to JACCARD. A very simple mathematical expression for the similarity of plant communities is the community coefficient of JACCARD (1901, 1912, 1928). It is based on the presence-absence relationship between the number of species common to two areas (or communities) and the total number of species. Therefore, the coefficient expresses the ratio of the common species to all species found in two vegetation segments. Thus,

$$\frac{\text{common species}}{\text{all species}} \times 100$$

The coefficient was used originally only for comparing the floras of larger areas. However, it can be used also for comparing vegetation relevés or sample stands. Moreover, the community coefficient can be applied in two different ways. It can be used to compare two communities simply on the basis of their species lists for presence and absence of common and unique species. It can also be expanded with information on the amount or quantity of each species present. The quantitative modification was first applied by GLEASON (1920).

For calculating these coefficients we will use the first two relevés shown in TABLE 9.7 (running Nos. 1 and 2, releves 4 and 10).

The simplest community coefficient is concerned only with the presence of species in two relevés. The presence concept does not involve quantity. The presence-community coefficient or index of similarity (IS) is equal to the number of common species as expressed in percent of total number of species in both plots. It is calculated from the following formula (Index J = JACCARD):

$$IS_J = \frac{c}{a+b+c} \times 100$$

where c is the number of common species, a is the number of species unique to the first relevé, and b is the number unique to the second. The values can be obtained simply by counting the species in the table. In our example, the index of similarity based on presence is:

$$IS_J = \frac{18}{10+11+18} \times 100 = 46 \text{ percent}$$

Therefore, in giving equal weight to presence and absence of all species, the first two releve analyses are not as similar as one could assume on first view.

JACCARD's presence-community coefficient is often written as:

$$IS_J = \frac{c}{A+B-c} \times 100$$

In this formula, A stands for presence of the total number of species in relevé A and B stands for the total number of species in relevé B, while c has the same meaning as in the first formula. The result is the same as shown by application to the same data (relevés 1 and 2, TABLE 9.7):

$$IS_J = \frac{18}{28 + 29 - 18} \times 100 = 46 \text{ percent}$$

However, when using the first given interpretation, where a and b refer to the species unique in each relevé, the counting of individuals from the table is a little faster and simpler.

10.12   Index of Similarity According to SØRENSEN. A number of other community coefficients or similarity indices have been proposed. However, most of these are very similar and derived from JACCARD's coefficient. One of the more widely applied similarity indices is the one by SØRENSEN (1948), which reads (Index S = SØRENSEN):

$$IS_S = \frac{c}{1/2\,(A + B)} \times 100 \text{ or } \frac{2c}{A + B} \times 100$$

Here, c = number of species common to two releves

A = total number of species in releve A
B = total number of species in releve B

When applied to the same data (releves 1 and 2 in TABLE 9.7), SØRENSEN's presence community coefficient results in a greater similarity than JACCARD's:

$$IS_S = \frac{18 \times\ 2}{28 + 29} \times 100 = 63 \text{ percent}$$

SØRENSEN's main reasoning for his modification was that in JAC-CARD's index both the numerator and the denominator change simultaneously, while in SØRENSEN's formula the denominator is independent of the numerator. He points out that theoretically each species has an equal chance of being present in two areas implying that any one species can either occur in the two communities under comparison or in only one. Thus, the expression $(A + B)\,/2$ or $\frac{1}{2}(A + B)$ represents the sum of the theoretically realizable coinciding occurrences, while the numerator c is an expression of the actually encountered coinciding occurrences. Therefore, SØRENSEN's index expresses the actually measured coinciding species occurrences against the theoretically possible ones. This may be mathematically more satisfactory as it includes a statistical probability term.

The difference between JACCARD's (IS_J) and SØRENSEN's (IS_S) indices can also be described as the first measuring the ratio of common (c) to total species in two samples and the second measuring the ratio of the common to the average number of species in two samples, i.e.,

$$IS_J = \frac{c}{\text{total number of species}} \times 100$$

$$IS_S = \frac{c}{\text{average number of species}} \times 100$$

This also shows that SORENSEN's index gives greater weight than JACCARD's to the species that recur in the two test areas than to those that are unique to either area.

10.13 Similarity Relations Involving Species Quantities. The similarity of two communities is not only a function of the number of common and unique species, but also of the amount of each species present. Of course, the similarity concept could be further expanded to include life form composition and other criteria. This depends on the objectives.

The same indices can be used with the amount of each species incorporated. By including, for example, the percent plant biomass (M) of each species (as recorded in TABLES 9.2 to 9.7) in JACCARD's community coefficient, the formula can be rewritten as modified by ELLENBERG (1956) (Index E = ELLENBERG):

$$IS_E = \frac{Mc:2}{Ma + Mb + Mc:2} \times 100$$

Where Mc is the sum of the percent biomass values of the species common to both stands, Ma is the sum of the biomass values of the species restricted to the first stand, and Mb is the corresponding sum for the species restricted to the second stand. (Species whose biomass was rated +, were here evaluated with 0.1 percent.) Application of the quantitative modification of JACCARD's index to our example (relevés 1 and 2, TABLE 9.7) has the effect of increasing the similarity of the first two communities. TABLE 10.1 shows the complete example of computation. The IS_E of the two relevés is 79 percent, because several of the common species have high biomass percentages.

TABLE 10.1 Calculation of JACCARD's Index of Similarity Based on Percent Biomass. Example of Two Vegetation Samples (Relevés 4 and 10, TABLE 9.7). Further Explanation in Text.

SPECIES	Ma	Mc		Mb
Bromus erectus		74	21	
Scabiosa columbaria		1	0.1	
Thymus serpyllum		0.1	2	
Salvia pratensis				5
Koeleria pyramidata		2	3	
Festuca ovina				2
Campanula glomerata	1			
Viola hirta				3
Briza media				1
Linum catharticum	0.1			
Arrhenatherum elatius		2	8	
Dactylis glomerata		5	6	
Galium mollugo		1	12	
Poa pratensis		5	8	
Plantago lanceolata		1	1	
Festuca pratensis				2
Chrysanthemum leucanthemum		0.1	5	
Ranunculus acer				3
Veronica chamaedrys		0.1	1	
Achillea millefolium		1	8	
Daucus carota		0.1	1	

The division by 2 of the sum of biomass values for the common species is a modification introduced by ELLENBERG (1956). The reason for this is that the common species represent two sets of values when their biomass values are used, but in terms of presence they represent only a single set.

GLEASON (1920) applied quantitative values directly to JACCARD's formula without modification. According to GLEASON's application the values in this example would become

$$IS_G = \frac{177.1}{5.7 + 18.3 + 177.1} \times 100 = 88 \text{ percent}$$

In this application, double weight is given to the common species relative to the unique ones.

Other quantitative values can be similarly treated. For example, a

TABLE 10.1 (Continued)

SPECIES	Ma	Mc		Mb
Rumex acetosa	0.1			
Taraxacum officinale		0.1	0.1	
Trifolium pratense	0.1			
Medicago lupulina		0.1	0.1	
Helictotrichon pubescens				0.1
Plantago media	0.1			
Campanula rotundifolia		1	1	
Festuca rubra		2	4	
Cerastium caespitosum	0.1			
Prunella vulgaris				0.1
Bellis perennis				1
Vicia sepium		0.1	0.1	
Carex flacca	3			
Vicia cracca				0.1
Dianthus superbus	1			
Galium boreale	0.1			
Potentilla reptans				1
Pimpinella saxifraga	0.1			
Total	5.7	177.1		18.3
Mc:2		88.6		

Calculation according to the formula

$$IS_E = \frac{88.6}{5.7 + 18.3 + 88.6} \times 100 = \frac{88.6}{112.6} \times 100 = 79 \text{ percent}$$

"frequency-coefficient of community similarity" can be calculated in the same way, by simply using frequency values in place of percent biomass values. For determining the BRAUN-BLANQUET "cover-abundance coefficient of community similarity," it is necessary to first convert the cover-abundance values into mean cover percentages, because the scale values correspond to unequal ranges of cover percentages (see TABLE 8.1, SECTION 8.3).

10.14 Application of Similarity Indices for Ordering Relevés. The community coefficients or similarity indices provide for a mathematical check on the tabular arrangement of relevés, which one may want to arrange in order of their similarity in floristic composition. For this purpose, the first and last relevés in TABLE 9.7 are used as standards of comparison for the other relevés.

The result of such a comparison is shown graphically in FIGURE

10.1. The species lists of the relevés with running Nos. 2 to 7 in TABLE 9.7 show great similarities to that of relevé 1. The others are less similar. In relation to the last relevé in TABLE 9.7, the trend is approximately reversed. However, a few irregularities are shown in the curves. For example, running No. 16 falls out of the sequence by particularly low values. Therefore, the elimination of this relevé (which is original No. 19) in TABLE 9.7, is supported by this calculation. Improvements in the similarity trend would result from exchanging relevé positions. But such repositioning is significant only where the objective is to arrive at an optimal stand ordination (see SECTION 10.7).

In using these mathematical tests it is always important to consider their possibilities and limitations. Their advantage is the evaluation of all species in each vegetation sample. At the same time, this is in some respects also a disadvantage, because the species with narrower amplitudies, i.e., the character and differential species, receive equal weight with those having wide amplitudes, or with those that may occur only temporarily or by accident in the community.

The application of mathematical methods in vegetation ecology becomes problematic where the initial stand choice is based on subjective judgment. In such cases the basis for application of mathematical tests is inexact in the strict sense. The most exact mathematical evaluation does not render the initial values more exact. The uncritical reader is often deceived about this fact. Moreover, the effort involved in calcu-

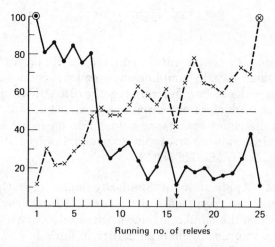

FIGURE 10.1.   Biomass-percent coefficients (JACCARD–ELLENBERG) of the meadow communities from TABLE 9.7. The values are related to relevés 1 (dots and solid lines) and 25 (crosses and dashed lines).

lating quantitative similarity relations is often not warranted for classification purposes in which the amplitude or constancy of species is considered more critical than the overall floristic similarity. They are of value, however, where the objective is community ordination (see SECTION 10.7), which in its simplest form relates to the orderly arrangement of plots or sample stands in a linear quantitative-floristic sequence of similarity as crudely shown in FIGURE 10.1. Of course, the limitations with regard to initial plot choice apply here as well.

With the generally increased availability of computer facilities, similarity comparisons are often made not only between one or two reference stands and the remaining stand. Instead each stand is compared with each other. For example, applied to the 25 relevés in TABLE 9.7 this means that not only $n-1=24$ comparisons are made as shown for each reference stand in FIGURE 10.1, but also relevés 2, 3, 4, etc. are compared to all other releves. This requires $n \times (n-1)/2$ comparisons, which comes to 300 comparisons for 25 relevés.

The quantitative similarity index most widely used for ordination purposes is SØRENSEN's community coefficient in its quantitative modification as applied first apparently by MOTYKA, DOBRZANSKI and ZAWADSKI (1950) (Index Mo = MOTYKA):

$$IS_{Mo} = \frac{2Mw}{MA+MB} \times 100$$

Here Mw refers to the sum of the smaller quantitative values of the species common to two relevés, and not to the sum of both values. The reason for this is that the smaller value is contained in both the larger and the smaller values of the common species. MA is the sum of the quantitative values of all species in one of the two plots, and MB is the sum of the quantitative values of all species in the other plot.

Applied to the first two stands (relevés 4 and 10) in TABLE 9.7, the values are calculated as follows:

$$Mw = 21+0.1+0.1+2+2+ \quad 5+\cdots+0.1 = \ 41.8 \ (n=18)$$

$$MA = 74+1+0.1+2+1+0.1+\cdots+0.1 = 101.4 \ (n=28)$$

$$MB = 12+0.1+ 2+5+3+2+\cdots+ \quad 1 = \ 99.7 \ (n=29)$$

$$IS_{Mo} = \frac{41.8 \times 2}{101.4+99.7} \times 100 = 41.6 \text{ percent}$$

(The complete example is shown in TABLE 10.13 where it is discussed in detail for stand ordination.)

10.15 Comparison of Seven Similarity Indices. BRAY and CURTIS (1957) introduced a computational simplification, which was recently also suggested by DAGET and POISSONET (1969). This provides for substituting the value for MA by 100 and adjusting the value for MB to 100 as well, so that $MA + MB = 200$. In this case the similarity index reduces to (Index BC = BRAY and CURTIS):

$$IS_{BC} = \frac{2\,Mw}{200} \times 100 \qquad \text{or} \qquad IS_{BC} = Mw \text{ (percent)}$$

In the application of this formula one has to use percent values for each individual quantity, because the sum of the quantitative values must come to 100 for each stand. However, since the quantitative values for each species in TABLE 9.7 are already percent values, namely percent plant biomass in the form of dry weight estimates, the column-sum of each relevé analysis gives a close approximation to 100 as indicated by the MA and MB values above of 101.4 and 99.7, respectively. Therefore, for the first two relevés (4 and 10) the IS value is 41.8 percent. It is of some interest to note that for the 25 *Arrhenatherum* relevés in TABLE 9.7 the IS_{BC} values based on $MA + MB = 200$ are in all comparisons nearly equivalent to the IS values based on the actual quantitative totals for MA and MB. This is so, because the ratio of $2\,Mc/(MA + MB)$ is almost a constant, namely 2/200 or 1/100. Therefore, in this example the unique species can be ignored and the degree of similarity is sufficiently explained by the number and quantity of the common species. This does not likely apply to stand comparisons in which the quantitative values are other than percent values.

SØRENSEN (1948) applied his index in both the qualitative and quantitative sense. For the latter he used degree of constancy, which essentially is the same as frequency. One may wonder why it is sufficient to use merely the sum of the smaller quantitative values of the common species.

SØRENSEN explains that the smaller value is the only value present twice, because it is contained in the larger value of the same species occurring in the second stand under comparison. Therefore, it seems sufficient to add only the smaller values and to then multiply this sum by 2. The procedure is analogous to the application of this index to presence and absence data, because the presence of a species occurring in both relevés is also counted only as 1 and multiplied by 2.

The quantitative modification applied by ELLENBERG (1956) to JACCARD's index involves adding the smaller and greater value for each common species. The sum is then divided by 2 resulting in an average quantitative value for the species common to two stands. This gives a somewhat higher value for the common species than SØREN-SEN's method. This procedure also is analogous to the treatment of presence and absence data, because in the qualitative application of JACCARD's index the presence of a species occurring in both relevés is counted only once. This is the same as adding the two occurrences and dividing them by 2.

Recently, SPATZ (1970) developed another modification of JAC-CARD's index. This index consists of two distinct components. The first component is an expression of the relative similarity (R) of the two stands being compared. The second component is the straight quantitative application of JACCARD's index as introduced by GLEA-SON (1920). This second component is multiplied with the first. The index is written as follows (Index Sp = SPATZ):

$$IS_{Sp} = R \times \frac{Mc}{Ma + Mb + Mc} \times 100$$

To obtain the first component, the smaller quantitative value of the species common to the two stands is divided by the greater quantitative value. The resulting fractions, whose number is equal to the number of common species, are added up and the sum is divided by the total number of species in the two stands.

SPATZ gives the following example relating to assumed plant bio-mass values in percent.

SAMPLE STAND, SPECIES	A	B
1	50	35
2	30	55
3	10	10
4	8	·
5	2	·

$$R = \frac{35:50 + 30:55 + 10:10}{5}$$

$$= \frac{0.7 + 0.6 + 1.0}{5} = 0.46$$

$$IS_{Sp} = 0.46 \times \frac{190}{200} \times 100 = 43.7 \text{ percent}$$

This index of similarity requires more computational work than any other index discussed so far. But where electronic computer facilities are available, this difference becomes unimportant.

There are several other community indices published in the literature, but JACCARD's and SØRENSEN's indices and their quantitative modifications are the most widely used.

The community indices discussed are summarized in TABLE 10.2. For providing a better indication of how they differ in their expression of similarity relationships, a set of hypothetical relevé data is applied to each from TABLE 10.3.

The differences in results for the same relevé pairs are rather striking. The first relevé pair (TABLE 10.3), which has all species in common, shows 100 percent similarity with the first three forms of JACCARD's index and with SØRENSEN's presence index. These indices reflect the qualitative likeness of this relevé pair, but they do not indicate the great quantitative difference. The latter is reflected by SPATZ's modification of JACCARD's index and by the two quantitative modifications of SØRENSEN's index. Among these three indices, SPATZ's index takes better cognizance of the qualitative similarity of relevé pair 1 by showing a higher value, 17.8 percent as opposed to 8 percent (see TABLE 10.2). Relevé pair 4 (TABLE 10.3), which has only 5 species in common out of 15, shows almost the same low similarity result (6 percent) with SØRENSEN's quantitative indices as for releve pair 1 (8 percent), which, however, has all 15 species in common. In contrast, SPATZ's index reflects this difference by showing only 1.7 percent similarity for relevé pair 4. Relevé pair 2 has 10 species in common. These are the decidedly most abundant ones in both stands. The simpler quantitative indices all reflect this as very high similarity of 99 percent or more. SPATZ's index takes cognizance of the five species unique to relevé A_2, which shows again the greater sensitivity of SPATZ's index to this qualitative difference. In this comparison the result is almost the same (SPATZ 62.7 percent) as that of the purely qualitative application of JACCARD's index (with 66.7 percent similarity) (see TABLE 10.2).

Therefore, one may conclude that SPATZ's quantitative modification of JACCARD's index has achieved the best compromise in reflecting both the quantitative and qualitative differences of the hypothetical relevés here compared. However, one cannot conclude automatically that one similarity index is superior to another. The value of an index depends on the properties that one intends to emphasize. It is important, however, that one is aware of the diagnostic capacity of a given index.

10.16   Threshhold Values for Classifying Associations.   For the recognition of associations, similarity in species combinations (i.e. presence or absence of particular species) is often considered more important than the quantitative contribution of each species (WILLIAMS and LAMBERT 1959). Even for subjectively chosen sample stands, it is quite valid to set arbitrarily certain extreme limits for classifying given relevés into the same or a different association. For this purpose, we may use JACCARD's presence-community coefficient. As an example, we will select the most typical or central stand as standard, to then relate all other stands to this one standard relevé by computing their similarity indices.

In our series of *Arrhenatherum* meadows from the Danube lowland south of Ulm, we may consider relevé 20 (running No. 12, TABLE 9.7) as the most typical *Arrhenatherum* relevé, because this community has been sorted into the central position on TABLE 9.7. It is characterized by a very high proportion of *Arrhenatherum*, also by the presence of *Crepis biennis*, *Heracleum spondylium*, and *Trisetum flavescens*, while few moisture-indicating plants are represented. In using this relevé as standard and by calculating JACCARD's presence-community coefficient one obtains the values shown in FIGURE 10.2.

Experience has shown that there is rarely a JACCARD *IS* value based on presence which exceeds 50 or 60 percent. Even the species lists of two similar, neighboring communities have rarely more than ⅔ of the

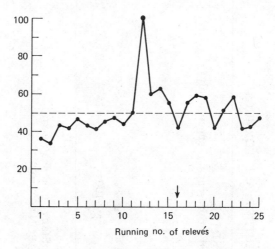

FIGURE 10.2.   Presence-community coefficients (JACCARD) of the meadow communities shown in TABLE 9.7 as calculated with reference to the most typical community (running No. 12).

TABLE 10.2 Summary of Community-Similarity Indices and Their Results (After SPATZ 1970, Modified). Numerical Data from TABLE 10.3.

AUTHOR AND FORMULA	RELEVÉ PAIR (SEE TABLE 10.3)	CALCULATION	RESULT (PERCENT)
JACCARD (=J) (based on presence of species only) $$IS_J = \frac{c}{a+b+c} \times 100$$ or $$IS_J = \frac{c}{A+B-c} \times 100$$	1	$\dfrac{15\times100}{0+0+15}$ or $\dfrac{15\times100}{15+15-15}$	100
	2	$\dfrac{10\times100}{5+0+10}$ or $\dfrac{10\times100}{15+10-10}$	66.7
	3	$\dfrac{5\times100}{5+5+5}$ or $\dfrac{5\times100}{10+10-5}$	33.3
	4	$\dfrac{5\times100}{5+5+5}$ or $\dfrac{5\times100}{10+10-5}$	33.3
JACCARD (quantitative modification by GLEASON 1920 = G) $$IS_G = \frac{Mc}{Ma+Mb+Mc} \times 100$$	1	$\dfrac{100+100}{0+0+200}\times100$	100
	2	$\dfrac{100+99}{1+0+199}\times100$	99.5
	3	$\dfrac{89+89}{11+11+178}\times100$	89
The same as MATUSKIEWICZ (fide SPATZ 1970) $$IS_G = \frac{Mc}{MA+MB}\times100$$	4	$\dfrac{11+11}{89+89+22}\times100$	11

JACCARD
(quantitative modification by
ELLENBERG 1956 = E)

$$IS_E = \frac{Mc : 2}{Ma + Mb + Mc : 2} \times 100$$

1	$\dfrac{200 : 2}{0 + 0 + 200 : 2} \times 100$	100
2	$\dfrac{199 : 2}{1 + 0 + 199 : 2} \times 100$	99
3	$\dfrac{178 : 2}{11 + 11 + 178 : 2} \times 100$	80.2
4	$\dfrac{22 : 2}{89 + 89 + 22 : 2} \times 100$	5.8

JACCARD
(quantitative modification by
SPATZ 1970 = Sp)

$$IS_{sp} = \frac{\Sigma (Mw : Mg)}{a + b + c} \times \frac{Mc}{Ma + Mb + Mc} \times 100$$

1	$\dfrac{0.178 \times 200}{200} \times 100$	17.8
2	$\dfrac{0.63 \times 199}{200} \times 100$	62.7
3	$\dfrac{0.33 \times 178}{200} \times 100$	29.4
4	$\dfrac{0.16 \times 21}{200} \times 100$	1.7

AUTHOR AND FORMULA	RELEVE PAIR (SEE TABLE 10.3)	CALCULATION	RESULT (PERCENT)
SØRENSEN (= S) (based on presence of species only) $$IS_s = \frac{2c}{A+B} \times 100$$	1	$\dfrac{2 \times 15}{15+15} \times 100$	100
	2	$\dfrac{2 \times 10}{15+10} \times 100$	80
	3	$\dfrac{2 \times 5}{10+10} \times 100$	50
	4	$\dfrac{2 \times 5}{10+10} \times 100$	50
SØRENSEN (quantitative modification, originally applied to constancy, but also introduced apparently for other quantities by MOTYKA et al.= Mo) $$IS_{Mo} = \frac{2\,\Sigma\,Mw}{MA+MB} \times 100$$ In the American literature usually written without summation sign and without symbol M	1	$\dfrac{2 \times 8}{100+100} \times 100$	8
	2	$\dfrac{2 \times 99}{100+100} \times 100$	99
	3	$\dfrac{2 \times 89}{100+100} \times 100$	89
	4	$\dfrac{2 \times 6}{100+100} \times 100$	6

SØRENSEN

(quantitative modification based on relative values, where the sum of the quantitative values of all species per stand is 100; introduced by BRAY and CURTIS 1957 = BC)

1	$0.2+0.2+0.2+0.2+0.2$ $+1+1+2+1+1$ $+0.2+0.2+0.2+0.2+0.2$	8
2	$44+20+10+8+7$ $+4+2+2+1+1$	99
3	$44+20+10+8+7$	89

$IS_{BC} = \Sigma Mw \times 100$

usually written without summation sign and without symbol M

4	$1+1+2+1+1$	6

Explanation of symbols in TABLE 10.2: IS=any index of similarity, a=number of species occurring in stand A only, b=number of species occurring in stand B only, c=number of species common to both stands A and B, A=total number of species in stand A, B=total number of species in stand B, Mc=sum of plant biomass or other quantitative values of the species common to stands A and B, Ma=sum of quantitative values of the species unique to stand A, Mb=sum of quantitative values of the species unique to stand B, Mw=smaller quantitative value of a species common to stands A and B, Mg=greater quantitative value of a species common to stands A and B, MA=sum of quantitative values of all species in stand A, MB=sum of quantitative values of all species in stand B.

TABLE 10.3   *Example of Hypothetical Releve Data in Terms of Plant Biomass Percentages (from SPATZ 1970).*

RELEVÉ PAIR	1		2		3		4	
RELEVÉ	A_1	B_1	A_2	B_2	A_3	B_3	A_4	B_4
Species								
1	44	$+^a$	44	44	44	44	44	·
2	20	+	20	20	20	20	20	·
3	10	+	10	10	10	10	10	·
4	8	+	8	8	8	8	8	·
5	7	+	7	7	7	7	7	·
6	4	1	4	4	5	·	5	1
7	2	1	2	2	·	5	2	1
8	2	2	2	2	2	·	2	2
9	1	2	1	2	·	2	1	2
10	1	4	1	1	2	·	1	5
11	+	7	+	·	·	2	·	7
12	+	8	+	·	1	·	·	8
13	+	10	+	·	·	1	·	10
14	+	20	+	·	1	·	·	20
15	+	44	+	·	·	1	·	44

[a] The symbol + is here assessed with 0.2 percent plant biomass.

species in common. Therefore, it can be concluded that the IS_J values of all relevés in TABLE 9.7 are relatively large, in spite of their being only between 33 and 66 percent (as shown in FIG. 10.2).

For deciding whether or not a community can be considered part of the same association, we suggest from our experience the following presence-community coefficients as threshhold values:

IS_J more than 25 percent, but less than 50 percent (based on presence only)

One can hardly speak of "similar floristic composition"—as requested in the definition of the term association—if the IS_J value is less than 25 percent. If, on the other hand, the IS_J value exceeds 50 percent, the similarity is so great that it is recommended to either distinguish subunits beneath the rank of association or to ignore further grouping, as the number of classes may become too many (with too few individual relevés) to result in ecologically meaningful generalizations. Of course, these criteria should only be taken as a guide. They should not be made an absolute requirement.

10.2   CORRELATION BETWEEN SPECIES

In the same way as one can use coefficients for evaluating similarity relations between two communities, one can use coefficients for evaluating the similarity between two species with regard to their distribution among different plant communities. One may calculate for this purpose the correlation between species by an "index of association" (*IA*). A number of association indices and their relative information values have been reviewed by COLE (1949, 1957), DAGNELIE (1960), and SOKAL and SNEATH (1963). However, at this point we will give primary consideration to JACCARD's coefficient, because of its simplicity in application.

JACCARD's index of community similarity can be applied as an index of species association in three ways:

1.   For evaluating the correlation in presence and absence of species between sample plots, quadrats, or relevés.

2.   For comparing the correspondence in quantitative values or amounts of each species present.

3.   To correlate the constancy of species among vegetation types.

The index of species association (*IA*) based on presence (p) is written in this form for comparing relevés:

$$IA_p = \frac{c}{a+b+c} \times 100$$

However, here c stands for the number of relevés in which the two species under comparison occur together; a refers to the number of relevés in which one of the two species occurs alone, and b refers to the number of relevés in which the other species is found alone. For example, for *Bromus erectus* and *Salvia pratensis* in TABLE 9.7 the presence is $c=5$, $a=2$ and $b=0$, resulting in an *IA* value of:

$$IA_p = \frac{5}{2+0+5} \times 100 = \frac{7}{5} \times 100 = 72 \text{ percent}$$

The *IA* values between *Bromus* and all other species in TABLE 9.7 can be calculated in the same way. The result is shown in TABLE 10.4. The *IA* values in this table provide for a measure of the degree of association of all other species (shown in TABLE 9.7) with *Bromus erectus*. TABLE 10.4 lists only those species whose distribution overlaps to some degree with that of *Bromus erectus*.

TABLE 10.4   Indices of Species Association (IA_p values in Percent) Based on Presence for Bromus erectus as Shared with the Other Species in TABLE 9.7.

Bromus erectus	100	Arrhenatherum elatius	28
		Knautia arvensis	25
Salvia pratensis	72	Cerastium caespitosum	25
Viola hirta	63	Ranunculus acer	25
		Trifolium pratense	25
Koeleria pyramidata	57	Centaurea jacea	24
Campanula glomerata	56	Trisetum flavescens	22
Campanula rotundifolia	54	Lotus corniculatus	22
		Dianthus superbus	22
Scabiosa columbaria	50	Galium boreale	21
Thymus serpyllum	50	Helictotrichon pubescens	21
Linum catharticum	50		
		Festuca pratensis	20
Vicia sepium	40	Bellis perennis	19
Festuca rubra	38	Prunella vulgaris	19
Festuca ovina	37	Senecio jacobaea	18
Briza media	33	Trifolium repens	17
Achillea millefolium	33	Heracleum sphondylium	17
Medicago lupulina	33	Brachypodium pinnatum	14
Vicia cracca	33	Alchemilla vulgaris	14
Veronica chamaedrys	32	Rumex acetosa	13
Plantago media	31	Potentilla reptans	11
		Silene inflata	11
Chrysanthemum leucanthemum	30		
Taraxacum officinale	30	Leontodon hispidus	10
Carex flacca	30	Crepis biennis	10
Daucus carota	29	Anthriscus sylvestris	9
Pimpinella saxifraga	29	Tragopogon pratensis	8
Galium verum	29	Ajuga reptans	6
Galium mollugo	28	Lathyrus pratensis	6
Poa pratensis	28		
Plantago lanceolata	28	All other species	0
Dactylis glomerata	28		

The IA shared with *Arrhenatherum elatius* or with *Cirsium oleraceum* or with any other species can be established in the same way.

Calculation of the species association index based on quantities (IA_M, here M for biomass) involves a little more work. For this purpose, we may also use *Bromus* and *Salvia* as examples (see TABLE 9.7):

The percent similarity in associated quantity between the two species

$$Mc = 21+5+47+2+37+1+35+4+10+4 = 166$$
$$Ma = 0$$
$$Mb = 74+50 = 124$$
$$IA_M = \frac{166:2}{124+0+166:2} \times 100 = \frac{83}{207} \times 100 = 40 \text{ percent}$$

is much less than the similarity based on presence alone, because *Bromus* is always present in great quantity, while *Salvia*, though closely associated with *Bromus*, is present always in low quantity.

Two species with similar quantitative values, but less closely associated occurrences may show the same quantitative association index as two closely associated species with unequal abundance. Therefore, the inclusion of species quantities may obscure this coefficient as an index of association.

Moreover, two species may be positively correlated in occurrence, but negatively correlated in quantity. Such an inverse quantitative relationship may indicate differing degrees of competitive relations along an environmental gradient. They are thus of ecological interest. Quantitative relationships of this sort are tested statistically by the so-called product-moment correlation coefficient (r), which is discussed in any statistical handbook. An example of application to two negatively associated species is given by KERSHAW (1964:78).

TABLE 10.5 shows a number of *IA* values which were calculated from the summary table of Northwest German plant communities (TABLE 9.10). The *IA* values are based on constancy (C) and were related once to *Arrhenatherum elatius* and once to *Bromus erectus*.

For example, the index of association (*IA*) based on constancy (C) between *Arrhenatherum* and *Tragopogon* in TABLE 9.10 was calculated as follows:

$$IA_C = \frac{Cc:2}{Ca+Cb+Cc:2} \times 100$$

where

$Cc = 18+9+9+18+82+86+85+60+64+18+17+4+20$
$+20 = 510$ (i.e., seven common occurrences)

$Ca = 22+20+8 = 50$ (i.e., three unique occurrences of *Arrhenatherum*)

$Cb = 0$ (i.e., no unique occurrence of *Tragopogan*)

Thus,

$$IA_C = \frac{255 \times 100}{305} = 83.6 \text{ or } 84 \text{ percent}$$

TABLE 10.5 *Indices of Association Based on Constancy* (IA$_C$) *for* Arrhenatherum elatius *and* Bromus erectus *Shared with Some Meadow Plants in Northwest Germany. Calculated from the Constancy Values in TABLE 9.10.*

| | IA$_C$ RELATED TO | |
	ARRHENATHERUM	BROMUS
Arrhenatherum elatius	100	—
Tragopogon pratensis	84	—
Crepis biennis	80	—
Pastinaca sativa	74	—
Anthriscus sylvestris	71 (86)[a]	—
Daucus carota	69	6
Heracleum sphondylium	60 (68)	—
Trisetum flavescens	58	—
Galium mollugo	57	3
Dactylis glomerata	40 (59)	—
Veronica chamaedrys	51	—
Chrysanthemum leucanthemum	48	—
Lysimachia nummularia	44	—
Bellis perennis	41	2
Taraxacum officinale	33 (46)	1
Achillea millefolium	33 (40)	3
Poa pratensis	37	2
Alopecurus pratensis	35	—
Lolium perenne	32 (39)	—
Festuca pratensis	33	—
Plantago lanceolata	31 (35)	2
Trifolium repens	29 (32)	—
Holcus lanatus	28 (31)	—
Cirsium oleraceum	27 (32)	—
Deschampsia caespitosa	19 (31)	—
Angelica sylvestris	17 (22)	—
Thymus serpyllum	15	14
Geum rivale	12 (13)	—
Festuca ovina	9 (11)	14
Scabiosa columbaria	6	59
Koeleria pyramidata	5	56
Salvia pratensis	—	100
Bromus erectus	—	100

[a] Indices in parentheses show the within-grassland associations for those species that occur also outside the grassland community.

The values in TABLE 10.5 portray very clearly the close sociological and ecological relationship of *Arrhenatherum*, *Tragopogon*, *Crepis*, *Pastinaca* and *Anthriscus*. Moreover, *Arrhenatherum* has high association indices with many other meadow plants. In contrast, the *Bromus erectus–Salvia–Koeleria–Scabiosa* group (on bottom of TABLE 10.5) is quite unrelated to the other species and shows thereby that it is representative of a rather different vegetation unit.

Therefore, the species association indices offer a means for expressing naturally occurring associations of species in figures. They can be used also as tests for evaluating species groups derived through other methods. However, group formation or association among species can be assessed statistically only if one uses a large number of randomly selected samples.

10.3 SPECIES ORDINATION

DE VRIES (1954) calculated, in a similar way, the correlations of the more common grassland species in the Netherlands from 1000 randomly located 1 m² quadrats. He showed his results graphically in a form which clearly indicates the ecological relations of the species groups (FIG. 10.3). Following a suggestion of G. HAMMING, he included the quadrats in which the species-pair was absent. In this mathematical manner, DE VRIES arrived at species groups which coincided in general with those derived by the BRAUN-BLANQUET method. The following groups can be found in FIGURE 10.3:

	Group on upper left		**Group on upper right**
M	Molinia coerulea	Arr	Arrhenatherum elatius
S	Sieglingia decumbens	Tri	Trisetus flavescens
Pe	Potentilla erecta	D	Dactylis glomerata
Cp	Carex panicea		
Ci	Cirsium disectum		

	Group on left		**Group in center left**
Cal	Caltha palustris	Ao	Anthoxanthum odoratum
LFc	Lychnis flos cuculi	Ru	Rumex acetosa
Gm	Glyceria maxima	Hl	Holcus lanatus

	Group on lower left		**Group on lower right**
Rr	Ranunculus repens	Lp	Lolium perenne
Ag	Alopecurus geniculatus	Cy	Cynosurus cristatus
Gf	Glyceria fluitans	Tr	Trifolium pratense
		Phl	Phleum pratense

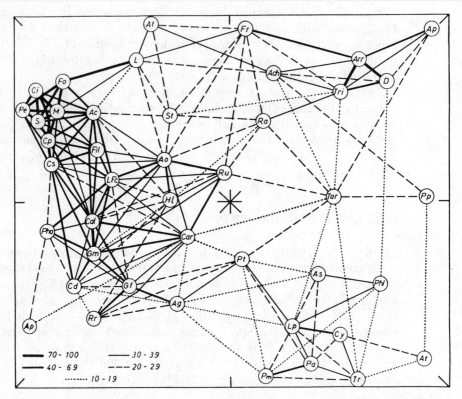

FIGURE 10.3. Correlation between some species in grassland communities in the Netherlands (after DE VRIES). The representation is intended to give a three-dimensional relationship. Legend: Ac = Agrostis canina; Ach = Achillea millefolium; Ag = Alopecurus geniculatus; Ao = Anthoxanthum odoratum; Ap = Alopecurus pratensis; Arr = Arrhenatherum elatius; As = Agrostis stolonifera; At = Agrostis tenuis; Cal = Caltha palustris; Car = Cardamine pratensis; Cd = Carex disticha; Ci = Cirsium dissectum; Cp = Carex panicea; Cs = Carex stolonifera (=fusca); Cy = Cynosurus cristatus; D = Dactylis glomerata; Fil = Filipendula ulmaria; Fo = Festuca ovina; Fr = Festuca rubra; Gf = Glyceria fluitans; Gm = Glyceria maxima; Hl = Holcus lanatus; L = Luzula campestris; LFc = Lychnis flos-cuculi; Lp = Lolium perenne; M = Molinia coerulea; Pa = Poa annua; Pe = Potentilla erecta; Pha = Phalaris arundinacea; Phl = Phleum pratense; Pm = Plantago major; Pp = Poa pratensis; Pt = Poa trivialis; Ra = Ranunculus acer; Rr = Ranunculus repens; Ru = Rumex acetosa; S = Sieglingia decumbens; St = Stellaria graminea; Tar = Taraxacum officinale; Tr = Trifolium pratense; Tri = Trisetum flavescens.

The coordinates on FIGURE 10.3 divide the species from the center upward into hay meadow plants, from the center downwards into pasture plants. Plants indicating moist substrates are the lower left groups, plants indicating nutritionally poor substrates are at the upper left side and plants typical of well-drained hay meadows are on the upper right. Lime indicators are on the right hand side, while acid soil indicators are on the left hand side.

The geographic variation in species associations are mappable and the association indices can be transferred on such a map.

It is obvious also that none of the groups form distinct entities. Instead they are variously interconnected. For example, the *Caltha palustris* group on the left is closely related to the *Ranunculus repens* group on the lower left through the interconnecting species *Glyceria maxima*.

The two-dimensional representation of species correlations on FIGURE 10.3 required some approximations because the correlations are in fact three-dimensional, and DE VRIES' model would appropriately be a sphere. Yet, the diagram clearly shows a continuum of stronger and weaker correlations among species, in brief, a "varying continuum." DE VRIES' diagram is clearly an ordination of species. Group separation, or distinction of units (i.e., classification) is a further step in abstraction, which is most appropriately done by separation along the weaker correlations.

However, the diagram also shows a number of discontinuities. For example, *Phleum pratense* and *Cynosurus* are both correlated with *Lolium perenne* and *Trifolium repens* (lower right on diagram), but *Phleum* and *Cynosurus* are not positively correlated with each other. DE VRIES points out that this fact pleads in favor of a division of the *Lolieto-Cynosuretum* association into a "fat" (*Trifolium*) and a "poor" (*Lolium*) variant.

The species correlations on FIGURE 10.3 are valid only for the study site; that is, for a restricted region with particularly uniform climate and floristics. The more extensive the area and the more variable the habitats, the less reliable the species association indices become. Thus, the same difficulties apply that were found in tabular comparisons of the plant communities. The reason for this is the varying continuum of vegetation. Plant communities can be arranged into a logical order only locally or within certain geographic limits. This basic phenomenon cannot be evaded by mathematical treatment.

10.31 Application of the 2×2 Contingency Table. As mentioned before, DE VRIES included in his calculation a fourth value, namely the number of quadrats in which both species were absent. He used a statistical correlation coefficient (r). A value of $r = 0$ implied that the two test species occurred together as frequently as expected by chance, $r = +1$ implied that the two test species occurred always together, $r = -1$ implied that they never occurred together. In his diagram, however, only positive species correlations are clearly identified.

The four combinations of two species are conveniently recorded in a 2×2 contingency table. This table serves as the data arrangement for the chi-square (χ^2) test.

A 2×2 continguency table for a two-species correlation can be written as follows:

Species A

		+	−	
Species B	+	a	b	$a+b$
	−	c	d	$c+d$
		$a+c$	$b+d$	$n=a+b+c+d$

Where

a = observed number of quadrats (or relevés) containing both species

b = observed number of quadrats containing only species B

c = observed number of quadrats containing only species A

d = observed number of quadrats without any of the two species

The values for a, b, and c, were used in the previous application of JACCARD's formula. Value d is the additional parameter included in DE VRIES' calculations of species correlations.

Incorporation of a value for joint absences (d) is important for statistical evaluations. A coefficient that compares in simplicity with JACCARD's and which incorporates joint-absences, is the so-called "simple matching coefficient" (SOKAL and SNEATH 1963).

As an index of association, the simple matching coefficient (IA_{SM}) relates the sum of quadrats with joint-occurrences and joint-absences of species to the total number of quadrats sampled. It is written as follows:

$$IA_{SM} = \frac{a+d}{n}$$

Apparently, this coefficient has not been used very much in vegetation ecology. A recent paper that employed the simple matching coefficient as an index of species association is the one by MOORE et al. (1970).

The tabular arrangement of the 2×2 contingency table permits a comparison between the number of quadrats actually occupied by the species in question with those expected under the hypothesis of a random dispersion of the plants on the ground. This hypothesis postulates that the number of occurrences of all combinations follows a predictable pattern as mathematically calculated by the χ^2 distribution. χ^2 tests are used to see if the proportion of samples containing one species is significantly correlated with the occurrence of a second species. The test does not require a random distribution of the test species, but the quadrats should really be randomized. This renders it particularly suitable for application to plant species, because they usually occur in more or less clumped distribution patterns. However, in a large number of samples, a certain number of joint species occurrences and joint absences are expected by chance. These have to be discounted. But their percentage is low, about 5 percent (GOODALL 1953a, GREIG-SMITH 1964).

The expected number of quadrats (occurrences or frequences of events) for any table combination is calculated by multiplying the marginal totals and by dividing this product by the total number of quadrats.

For example, the expected number of quadrats containing:

1. Both species A and B is calculated as

$$\frac{(a+b)\,(a+c)}{n}$$

2. Only species B is calculated as

$$\frac{(a+b)\,(b+d)}{n} \quad \text{or} \quad (a+c) - \frac{(a+b)\,(a+c)}{n}$$

3. Only species A is calculated as

$$\frac{(a+c)\,(c+d)}{n} \quad \text{or} \quad (a+b) - \frac{(a+b)\,(a+c)}{n}$$

4. Neither of the two species is calculated as

$$\frac{(b+d)\,(c+d)}{n}$$

The observed (O) and expected (E) number of quadrats can then be compared. This is done by calculating the χ^2 value. The χ^2 value is the sum of ratios of the squared deviations to the expected values. This can be written as (MORONEY 1954):

$$\chi^2 = \Sigma \frac{(O-E)^2}{E}$$

As an example let us compare the occurrence of *Bromus erectus* with *Scabiosa columbaria* in TABLE 9.7.

OBSERVED VALUES

Bromus erectus

		+	−	
Scabiosa columbaria	+	4	1	5
	−	3	17	20
		7	18	25

EXPECTED VALUES

Bromus erectus

		+	−	
Scabiosa columbaria	+	1.4	3.6	5
	−	5.6	14.4	20
		7	18	25

$$\chi^2 = \frac{2.6^2}{1.4} + \frac{2.6^2}{3.6} + \frac{2.6^2}{5.6} + \frac{2.6^2}{14.4}$$

$$\chi^2 = 4.83 + 1.88 + 1.21 + 0.47 = 8.39$$

The χ^2 value is usually calculated from the formula

$$\chi^2 = \frac{(ad-bc)^2 \times n}{(a+b)\ (c+d)\ (a+c)\ (b+d)}$$

Thus, for the same data we obtain

$$\chi^2 = \frac{[(4 \times 17) - (1 \times 3)]^2 \times 25}{5 \times 20 \times 7 \times 18} = \frac{105,625}{12,600} = 8.38$$

The second formula eliminates the need for calculating the expected values and their deviations from the observed values.

The χ^2 value is looked up in a χ^2 table. Such tables are published in any statistical handbook. The χ^2 value is entered with the appropriate degrees of freedom, which are always one less than the number of items compared. Here, the use of the 2×2 contingency table renders the degree of freedom to be 1.

Inspection of such a χ^2 table shows that the value of 8.38 obtained for the comparison of *Bromus erectus* with *Scabiosa columbaria* is significant at the 0.5 percent level. The χ^2 table gives a value of 7.88 at the 0.5 percent level, which is somewhat exceeded by our value. Chi-square cannot be fully relied on as the original relevés were not chosen at random.

Moreover, statistical handbooks (e.g., MORONEY 1954) inform us that YATES' correction should be applied when the test is used for small numbers of samples. This correction consists of subtracting 0.5 from an observed value that exceeds the expected value and of adding 0.5 to an observed value that is less than the expected value. This has the effect of lowering the χ^2 value markedly in small numbers of samples, less so in larger numbers of samples.

A simpler calculation of YATES' correction is built into the modified formula below

$$\chi^2 \text{ (with YATES' correction)} = \frac{(ad - bc - n/2)^2 \times n}{(a+b)\,(c+d)\,(a+c)\,(b+d)}$$

Applied to our example the corrected value becomes

$$\chi^2 = \frac{[(4 \times 17) - (1 \times 3) - 12.5]^2 \times 25}{5 \times 20 \times 7 \times 18} = \frac{68,906.25}{12,600} = 5.47$$

Reinspection of the χ^2 table shows that a value of 5.02 is just significant at the 2.5 percent level of probability. Thus, our corrected χ^2 value indicates still a significant association of *Bromus erectus* and *Scabiosa columbaria*.

GREIG-SMITH (1964) recommends the application of YATES' correction where any expected value is less than 500. Such large numbers are rarely sampled. Therefore, in χ^2 tests of species associations it seems almost always desirable to use the corrected χ^2 formula.

The χ^2 test is not considered sufficiently reliable if the expected value in any one cell of the contingency table is less than 5. This arbitrary level is suggested in statistical handbooks (e.g., MORONEY 1954); however it is not a rigid requirement. A low expected value may produce disproportionately small deviations. Thus, the application of the χ^2 test to *Bromus erectus* and *Scabiosa columbaria* is statistically not entirely valid, because two of the expected values are below 5.

GREIG-SMITH (1964) further points out that in 100 species-pair comparisons, 5 can be expected to yield a significant level simply by chance alone. This means that 5 percent in any set of significant χ^2 values may have to be discounted. Therefore, isolated tests are of questionable value. Moreover, a series of χ^2 tests becomes meaningful only if based on a large number of samples, preferably in excess of 100 (GREIG-SMITH 1964).

The observed numbers of quadrats containing any of the four possible combinations of two species are frequency values, because the information relates merely to presence and absence of the species in a limited area (the quadrat). Such frequency values are dependent upon quadrat size. Therefore, the significance of a two-species association as shown by χ^2 depends also on the size of the quadrat used. For example, a very small quadrat may fail to detect positive species correlations simply because its small size excludes the correlated species. A large quadrat may yield a correspondingly larger number of species and the result may be a greater number of correlated species-pairs than would be realistic in terms of their dispersion pattern in a community (KERSHAW 1964).

Therefore, species correlations derived in this way cannot be claimed as absolute. They are absolute only in relation to a given quadrat or plot size. However, the plot size is based on subjective choice.

10.32 Two Literature Examples. Huber (1955) studied bryophyte communities on limeless (acid) soils in forests near Basel (Switzerland) by application of the 2×2 contingency technique. He shows a list of 23 moss and 4 herb species that occurred in at least 10 of the 233 releves studied. His releves were selected not for typical species combinations, but for homogeneity, and they were located on all known limeless forest sites. (These are rare in this area.) His intention, like that of DE VRIES, was to mathematically test the species combinations that occur on these sites. His releve sizes varied from a few square meters to about 100 m².

He explains that random samples of equal size would have been preferable, but the moss communities did not cover homogeneous areas large enough for random sampling. Moreover, random sampling would have required exact knowledge of the extent of limeless substrates. This knowledge was only gradually forthcoming during the course of the study.

HUBER points out that because of his unequal releve sizes he could not establish constancy. But, he says, one can nevertheless calculate if two species are found more commonly together than can be expected from independent distribution of the two species. However, this point is debatable.

It is probable that 1 m² samples satisfied the minimal area requirement in his bryophyte communities. In that case an increase in sample plot size should not yield any great increase in species number. Since the moss communities were selected for their homogeneity, a larger sample should therefore not be significantly richer in species. Moreover, HUBER eliminated all species that occurred in less than 12 out of his 233 releves. This resulted in the elimination of about 40 rare moss species, which may have been the only contributors to any increased species richness in the larger releves.

He applied the χ^2 test to 24 species (including a few vascular plants). This required the calculation of 276 χ^2 values, namely n $\times$ (n–1)/2 = 24 $\times$ 11.5. This resulted in a large number (35) of two-species combinations that occurred more frequently together than could be expected if their distributions were entirely independent. He also found but a small number (13) of two-species combinations that occurred together less frequently than would be expected if these species were independently distributed. The latter species therefore showed a pattern of avoidance, which was indicated by a significantly negative χ^2 value.

The first group of positively associated species combinations was portrayed in a species diagram (FIG. 10.4) similar to that of DE VRIES (FIG. 10.3). In HUBER's diagram the position of each species name was placed such that a minimum of lines would cross each other. The connecting lines indicate significant positive correlations. However, since the samples were not chosen at random, the testing of statistical significance is of questionable value. The strength of association between species is shown by the number of parallel connecting lines. One line refers to a positive χ^2 value significant between the 1 and 0.5 percent level, two parallel lines denote a significance of between 0.5 and 0.05 percent, three parallel lines a significance in excess of the 0.05 percent level.

Two species groups were identified by HUBER, a *Pleurozium* group which is shown on the upper left in the diagram and an *Isopterygium* group which is shown on the lower left. At least a third group can be recognized on the lower right in the diagram. Those related to mineral soil are found on the lower part. A tendency to lime-tolerance is shown by the *Fissidens* group on the lower right. However, the ecological amplitude of the species is not indicated.

A similar study was done by AGNEW (1961). He used 99 quadrats of 1 m^2 size to test species associations in a series of communities that contained *Juncus effusus*. He eliminated the rare species that occurred with a constancy of less than 5 percent. This left 53 species for which he calculated the observed and the expected joint occurrences, isolated occurrences, and joint absences from 2 $\times$ 2 contingency tables. The deviations between observed and expected values were then tested for significance by χ^2 values.

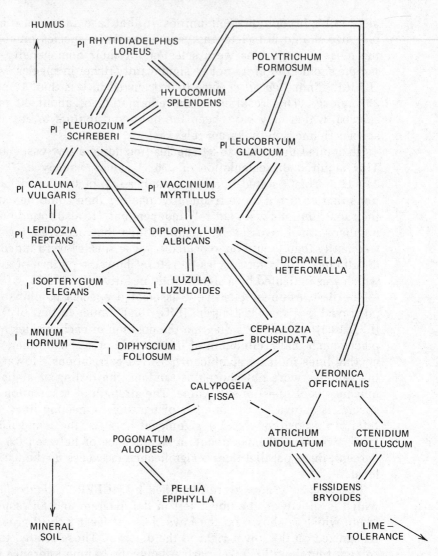

FIGURE 10.4   *Species diagram showing the association of mass species on forest soils near Basel, Switzerland. Species connected by lines are positively correlated as determined by application of the chi-square test. Legend: Pl = Pleurozium group; I = Isopterygium group. (After HUBER 1955.)*

AGNEW shows the whole matrix of χ^2 values for the 53 species sampled in the 99 quadrats. This required $53 \times 52/2 = 1378$ χ^2 value calculations. But he only indicates the significant positive and negative correlations on this matrix.

He then constructed a diagram, similar to HUBER's, in which the positively associated species are positioned close together. In addition, the distance between species-pairs is based on the reciprocal of the χ^2 values. Therefore, the greater the χ^2 value, the closer together the species-pair. However, since the similarity relationships between a great number of species are always three-dimensional, the portrayal in a two-dimensional diagram required some approximation. Significant correlations are indicated by connecting lines between the respective species-pairs, just as in HUBER's diagram.

From AGNEW's diagram, which is shown also by KERSHAW (1964:142) and GREIG-SMITH (1964:197) and is not repeated here, three groups of associated species are quite evident. Two of these are similar to the upper left (*Molinia coerulea*) and lower right (*Lolium perenne*) groups on DE VRIES' diagram (FIG. 10.3). A third group includes only mosses in the most poorly drained segment of the *Juncus effusus* communities. AGNEW concludes that his associated species groups correspond to associations in the sense of BRAUN-BLANQUET.

AGNEW's form of representing species relationships, like that of DE VRIES and HUBER, is an ordination, because the uniqueness of each species is shown by its position in a diagrammatic model.

The species correlations were derived in both DE VRIES' and AGNEW's study from random quadrats. Therefore, they were derived objectively, except for the determination of quadrat size.

Classification of associated species groups from such models, however, still requires judgment and therefore remains subjective.

To a mathematician, this may still be a deficiency. However, the species arrangement and strengths of correlations indicate where class boundaries are meaningful. Moreover, any separation of a species assemblage into classes must be related to a set of purposes. These may vary for the same vegetation allowing for different classifications that are all correct. Subjective decisions must even precede the so-called objective classifications, which require a great deal more mathematical manipulation.

10.4   OBJECTIVE CLASSIFICATION

Objective procedures for this final step, the classification of associated species groups, were apparently first introduced to vegetation ecology by GOODALL (1953) and subsequently by WILLIAMS and LAMBERT (1959, 1960, 1961). Objective classification requires an exorbitant amount of computation. Its practicability therefore rests on the availability of sufficient electronic computer facilities.

The methods again are primarily concerned only with small frequency rather than large minimal area quadrats. GOODALL (1953*a*) used 5 m² quadrats in a low-stature *Eucalyptus* forest in Australia in which he enumerated the presence of herbaceous and woody plants without stratification into life forms. No quantitative data was obtained. WILLIAMS and LAMBERT (1959) also used only small quadrats (1 m²) in several *Calluna* heath stands in England and recorded merely presence and absence data. Of course, the record of species presence and absence in a number of quadrats results in frequency, which can be considered a quantitative parameter. Frequency is fundamentally the same as constancy or degree of presence in a number of releves. But frequency applies to an arbitrary set of quadrats, constancy to a set determined by kind.

Objective classification methods are concerned essentially with the qualitative problem of species composition of plant communities. In this respect their underlying aim is closely related to the releve method of BRAUN-BLANQUET. However, small frequency quadrats do not necessarily contain 90 percent of the species of the community at each placement as do the minimal area or releve plots. Therefore, computer classification based on small frequency quadrats may deviate from a classification based on the releve method.

GOODALL's (1953) method of objective classification is thoroughly discussed by GREIG-SMITH (1964:162) and by KERSHAW (1964:147).

WILLIAMS and LAMBERT's work is basically a modification of GOODALL's technique. One important reason for their modification was the time-consuming computation work involved in the method recommended by GOODALL.

The method GOODALL recommends is to group species on the most frequent (or constant) species involved in positive correlations determined as significant by χ^2, and to pool the remaining species at each stage. This requires an exorbitant amount of calculation time and labor.

WILLAMS and LAMBERT devised a more efficient method which makes use of both positive and negative associations. They consider the use of significant negative species correlations just as important in classification as the use of positive associations. They emphasize this point particularly, since the failure to make use of negative correlations tends to overstress continuity in distribution. Certainly, absence of an otherwise constant species can be just as critical for the delimitation of vegetation units as the presence of constant species.

Moreover, WILLIAMS and LAMBERT (1959) criticized GOODALL's classification procedure for its tendency to separate units of ecologically questionable value.

GOODALL defines a species group as a homogeneous unit if there is no significant positive correlation left that would connect this unit with another unit, or if the χ^2 value cannot be calculated. WILLIAMS and LAMBERT (1959) point out that indeterminacy of the χ^2 value is not the same as nonsignificance. They prefer to separate groups on the basis of nonsignificance only, but include both positively and negatively nonsignificant correlations as criteria. Their criteria have a tendency to distinguish fewer but more significant units.

Their procedure consists of selecting a dominance-community (e.g., *Calluna* heath) of a given size. In the first example, they selected a very simple six-species heath community (WILLIAMS and LAMBERT 1959) in which they sampled an area of 70×220 m. In this area 616 m² quadrats were laid out in a systematic order, each quadrat 5 m apart. It is of interest to note here that the authors did not consider a random layout of their quadrats to be a necessary prerequisite for a subsequent statistical analysis by chi-square. They argue that they could use a systematic layout because they were dealing only with presence-absence data, not with estimates of density, and because they were concerned with the pattern of an area as a whole. They further state that the chance of their quadrat grid to "resonate" with the vegetation was extremely remote, and that they were prepared, therefore, to take the risk in view of the great practical advantages associated with a systematic sampling layout.

The presence and absence data were then analyzed in 2×2 contingency tables. Joint occurrences, joint absences, and single occurrences were compared to the statistically expected values for each combination. The deviations were tested for significance by chi-square values, which were calculated for each species combination encountered. The χ^2 values were written into a separate matrix for all levels of species interaction.

For example, a separate chi-square matrix was established for the number of species and for the first-order classes. The number of species used in the first matrix was only five, because one species was excluded as too rare, occurring in only 2 percent of the quadrats. For five species, the number of combinations is $n \times (n{-}1)/2 = 5 \times 2 = 10$. But each combination was entered twice into the matrix to obtain an equal number of entries for each column. Only the statistically signicant χ^2 values were entered with their numerical values. The χ^2 values were then summed up for each column in the matrix regardless of whether they showed positive or negative correlations.

The species with the greatest sum chi-square value was selected for classifying the quadrats into two groups. Then each of the two groups was reexamined and further subdivided on the species with the next highest sum χ^2 value. This procedure was continued until no significant correlations were left within a group. At this point the group was designated as a final class.

WILLIAMS and LAMBERT call this separating procedure "efficient" subdivision, because each subdivision results in a maximum reduction of heterogeneity of the two resulting subunits. It involves subdivision on that species which in the two resulting subclasses produces the smallest number of residual significant correlations. This species, which has the highest sum χ^2 value, can be interpreted as the most strongly correlated species. This procedure is quite different from GOODALL's, which separates on the most frequent species. In GOODALL's procedure only one efficient subdivision arises at each stage. This is the one containing the most frequent species. The other subdivision is left rather heterogeneous. Several such heterogeneous subdivisions are recombined or pooled before they are newly subdivided. GOODALL's system is likely to result in similar final classes, but these are obtained through a much longer process of computation.

After experimenting with different forms of the sum χ^2 value, WILLIAMS and LAMBERT (1960) came to the conclusion that the sum $\sqrt{(\chi^2/N)}$ was the most sensitive association index. However, this more complicated parameter can only be applied efficiently, where computer facilities are available. The authors wrote a computer program, which was successfully tested on four different vegetation covers.

Three of these were *Calluna* heath communities with relatively few species, up to 29. The sampling was done in all of these by a rather dense network of 1 m^2 quadrats, not more than 10 m apart. The 10 m distance gave a somewhat superficial pattern, which was rectified by closer spacing at 2 m intervals in another community. The areas sampled varied from 2000 m^2 to about 4 ha. Two communities showed marked species associations in relation to more obvious burning patterns and one, which extended from bog over wet mineral soil to well drained soil, showed an almost perfect relationship to this soil moisture pattern by significant association patterns.

A fourth analysis was concerned with a grazed marsh containing 72 species and extending over a rather larger area of 133 ha. This vegetation was surveyed earlier by LAMBERT (1948). At that time only 115 one–m^2 quadrats were placed subjectively at representative points, probably into typical communities with a fair geographic spread. The computer analysis resulted in units of much greater heterogeneity than in the densely sampled, species-poor *Calluna* communities. This would of course be expected from such a limited sample. Nineteen final classes appeared. But 11 of these contained less than 8 quadrats. Eight quadrats were considered too few for valid unit distinction. The remaining 8 units were further reduced to 5 units, which reflected genuine ecological entities in relation to grazing and soil variations.

An accompanying map (WILLIAMS and LAMBERT 1960:207) shows a certain clustering of the sample quadrats. The statistical test supported the original subjective classification. An objective layout of sample quadrats would probably have required a much greater sampling density to give the same result.

In a third paper of the same series, WILLIAMS and LAMBERT (1961) distinguish as a "new" type of association analysis, the so-called "inverse" analysis. The procedure, so far discussed, refers to the sorting of quadrats by species pairs and is called "normal" analysis. The important aspect in normal analysis is that quadrats, not species, are classified.

In contrast, inverse association analysis relates to the classification of species into groups by manipulation of the quadrats in which they occur. The correlation of quadrat pairs is tested in all combinations.

Normal analysis, or what is sometimes referred to as an R-technique (CATTELL 1952, SOKAL and SNEATH 1963, GREIG-SMITH 1964:199), is analogous to the sorting of species-pairs in all combinations according to their similarity or difference in occurrence. This can be accomplished by application of a species correlation coefficient or by indices of association (see SECTION 10.2). In this case, the species are sorted into groups of similar amplitude. Their amplitude is indicated by the relevés in which they occur and these are moved together horizontally to result in the differential synthesis table (TABLE 9.7). Inverse analysis or what is sometimes called Q-technique is analogous to sorting of quadrats, relevés or stands into similarity groups. Both sorting aspects are combined in the BRAUN-BLANQUET tabulation technique.

The same computer analysis using $\sqrt{(x^2/N)}$ as association index has been applied in inverse analysis, and both types of analysis have been combined and applied to the same community data.

The result is a statistically derived differential table. In layout and ecological interpretation it is identical to the differential table derived from the tabulation technique (TABLE 9.7). Vertical separation lines distinguish the major community types by relevés or quadrats. Horizontal blocks distinguish the species of comparable amplitude (LAMBERT and WILLIAMS 1962).

10.5   RELEVE SYNTHESIS BY PUNCH CARDS WITH SIGHTING HOLES

In modifying a method originally suggested by ELLENBERG and CRISTOFOLINI (1964), ELLENBERG (1968a) used 26.5 × 28.5 cm large punch cards for transferring relevé data by punching holes for each species present. These cards, one for each relevé, were then used for sorting stands into similarity groups.

Each punch card shows a major divisioning into tree (B), shrub (S), herb (K) and moss or lichen (M) species (FIG. 10.5, See insert in back pocket). Within each of these life form groups, species are further grouped according to known community types. The herbaceous species are further subdivided according to their known ecological indicator value.

Therefore, the species arrangement on each punch card is based on previous knowledge and familiarity with the vegetation. This was done to facilitate comprehension of the data from the start. However, the same method can be used with any other species order.

The species are transferred from a field-relevé sheet to the punch card by a number code. The number code and the species name are printed on a transparent plastic sheet that matches the species positions on the punch card. FIG. 10.6, See insert in back pocket). By super-positioning the transparent species sheet, each punch card can imme-diately be decodified and read like a normal relevé sheet.

In addition to its presence, the quantity of each species is recorded by a 5-scale rating. Five holes punched across the space for a species name indicate that this species is covering 75 percent or more of the plot area, one hole punched at the position of the species name indi-cates presence of few individuals with less than 5 percent cover. The other values lie in between and correspond to the cover-abundance scale of BRAUN-BLANQUET (SECTION 5.42.1).

The punched relevé cards are then sorted according to their floristic similarity.

A card is picked randomly from the file and placed on an illuminator. A second card is placed over the first one and its degree of similarity is immediately seen. The greater the number of common punched holes, the greater the floristic similarity. This corresponds to a visual calcu-lation of floristic similarity. A mathematical calculation using JAC-CARD's similarity index would involve counting the common holes plus the semi-illuminated holes. The latter show the species unique to either relevé.

In this way all cards are sorted in relation to the first randomly picked card according to their varying similarity in species composition. It is inconsequential which card is taken as the first reference card. All others will depart in similarity from this first one, from most similar (largest number of species in common) to least similar (no species in common).

On this basis five stacks of cards are established. They are:
- Stack 1 relevés with many species in common.

- Stack 2 relevés with several species in common.
- Stack 3 relevés with few species in common.
- Stack 4 relevés with an occasional species in common.
- Stack 5 relevés with no species in common.

This first sorting process is only a presorting. For 1000 cards it may take approximately 1 hour.

In the second sorting process a card is picked randomly from stack 1. Then three other cards are searched from stack 1 that give the closest match to the first picked card. In this way four cards are matched up that show the closest resemblance. The reason for using four cards is that the illuminator allows the investigator to spot species with 50 percent constancy or more. This means that the similarity is based first on species with 100 percent constancy (present in all four relevés), second on species with 75 percent constancy (present in three of the four relevés) and third on species with 50 percent constancy (present in two of the four relevés).

After matching the first four most similar cards from stack 1, a next card is picked randomly from stack 1. Three other cards are searched out that give the closest match to this secondly picked card.

Each set of four cards forms the core of a community type (without rank). But for holding the number of community types within reasonable limits, a standard may be set. For example, for the hardwood forests of Switzerland, ELLENBERG found it convenient to restrict the community cores to those four cards that had ten or more species in common. In this way 27 groups of four cards emerged which had each ten or more species in common.

For speeding up this process one "search card" was prepared from each of the 27 core groups. The search card was punched by transferring only the species from 50 to 100 percent constancy into a new card.

Then any card was taken from the card file, placed on the illuminator, and one search card after the other was placed over the top. The card was then grouped to the search card with which it showed the closest similarity. All cards were subjected to the same treatment until they were grouped into 27 classes.

If one wants to extract groups of differential species from the relevé cards, one can prepare a "differential species card" in the way the search card was prepared. The differential species card serves the same purpose as the "partial table," which is used for grouping relevés that have certain species in common.

Species known to be common over a restricted amplitude form important differentiating species. These are punched into a new card, which then becomes the differential species card. Relevé cards can then be sorted according to their similarity with the differential species.

This sorting process is no longer based on total floristic similarity, as was the first sorting into 27 classes. Instead, the sorting for a close match with the differential species card emphasizes a diagnostic similarity, meaning that the relevé similarity is being weighted in favor of the differential species.

Several other sorting processes can be accomplished with the relevé punch cards, such as sorting according to the pressure of certain life forms, ecological species groups, or species groups with affinities to certain floristic provinces.

The relevé punch card with sighting holes represents a development based on the synthesis table technique. Each relevé punch card resembles a column in a synthesis table. However, the difference is that the species order, which is at first mixed in the synthesis table, is already prearranged in well-ordered form on the plot punch card. The prearrangement of the species order is based on previous knowledge gained from work with synthesis tables of a particular regional vegetation.

The plot punch card method has considerable time advantage over the table technique, when a large number of relevés need to be compared. ELLENBERG and KLÖTZLI (1972) treated over 3000 relevés of Switzerland by this method, and could accomplish the sorting into 27 groups in 3.5 to 4 hours. For relevés up to 200 in number, the plot punch card method has no time advantage.

The relevé punch card method has an advantage over the individual species card method in so far as the relevé information is kept together in a readily accessible form. A special advantage is in the ready possibility of comparing any individual relevé with any other in the set. This is not possible in the table-technique after several relevés have been grouped into one column in a summary table (see, for example, TABLE 9.9, SECTION 9.5).

The punching process for relevé cards requires much less time than for individual species cards. However, once the time in preparing individual species cards for each relevé analysis has been invested, their subsequent analysis can be done by the electronic computer. Considerable time can be saved at this stage. Therefore, the future in processing of relevé-data lies with the computer methods, such as will be described in the following sections.

10.6   RELEVÉ SYNTHESIS BY ELECTRONIC COMPUTER

With development of electronic computer facilities, particularly within the last 10 years, many vegetation ecologists have attempted to adapt the tabular comparison technique for automatic processing by computer. Among a number of more or less successful methods, we will discuss here a few that appeared interesting to us—particularly those that made use of the 1956 vegetation analysis example of ELLENBERG, which was used here to explain the basic BRAUN-BLANQUET table technique (CHAP. 9).

SPATZ (1969, 1970, 1972) used an electronic computer for simulating the synthesis table technique.

The species of a relevé analysis are transferred on standard computer cards with 80 columns, one card for each species. The species are identified by a code number, and the appropriate number of the species is punched on each species card. The species code number is the first value shown on each card. A second value identifies the relevé number, a third value the quantity of the species. For the purpose of having the computer print out the whole species name on the synthesis tables, a second card is prepared for each species that bears the name of the species written out in full. This card can further be supplied with information about the life form of the species, its ecological indicator value, and other useful data.

The result of data analysis in the form of a differentiated synthesis table is achieved in two separate programs.

10.61 Sorting of Species by Constancy. In the first program both sets of species cards are fed into the computer. The computer is programmed to calculate for each species the constancy value both in absolute terms (i.e., number of times a species is present in the total number of relevés compared) and in percent. At this point the computer can print out a completed raw table, which shows the species names on the left side of the table in numerical order of the species code. The relevés are listed in numerical order at the head of the table. The species quantities are shown in horizontal rows for each species and in vertical columns for each relevé. The two constancy values are shown at the right end of the table.

The 25 *Arrhenatherum* grass cover stands of TABLE 9.2 were subjected to computer analysis by this method. The computer-printed raw table is shown as TABLE 10.6.

The next step involves a resorting of the species order from high to low constancy. The relevé order remains the same. The result of this species rearrangement is printed out as the constancy table (TABLE 10.7).

10.62 Sorting of Relevés in Similarity Groups. In a third step of the first program, the similarity between relevés is calculated. This can be done with any similarity index.

The calculation in this example was done with the quantitative modification of JACCARD's index as developed by SPATZ (SECTION 10.15).

TABLE 10.6   Computer-Printed Raw Table of the 25 Arrhenatherum Relevés Shown in TABLE 9.2. From SPATZ 1970.

Species, Species No.	Species No.	1	2	3	4	5	6	7	8	9	10	11	12	13	14	15	16	17	18	19	20	21	22	23	24	25	Absol.	Percent
Arrhenatherum elatius	1	+	5	2	2	4	4	4	4	5	8	9	10	10	12	15	15	15	22	22	24	25	25	26	30	35	25	100.0
Dactylis glomerata	2	5	5	15	5	12	12	4	10	2	6	12	32	15	10	6	15	15	18	1	5	18	8	8	8	18	25	100.0
Helictotrichon pubescens	3	1	1	1	·	20	8	3	·	4	+	·	1	4	·	·	13	4	4	28	28	+	2	·	10	·	16	64.0
Bromus erectus	4	50	·	35	74	·	·	·	·	47	21	·	·	·	·	·	·	·	·	·	·	·	·	·	·	·	7	28.0
Festuca ovina	5	2	1	·	·	·	·	·	·	1	2	·	·	·	·	37	·	·	·	·	·	·	·	·	·	·	4	16.0
Poa pratensis	6	4	74	10	5	4	2	3	4	10	8	6	25	2	5	15	10	5	10	1	6	1	9	16	20	10	25	100.0
Briza media	7	1	1	·	·	·	·	·	·	2	1	·	·	·	·	·	2	·	·	·	·	·	·	·	·	·	5	20.0
Koeleria pyramidata	8	3	·	·	·	·	·	·	·	3	3	·	·	1	·	·	·	·	·	·	·	·	·	·	·	·	4	16.0
Festuca rubra	9	15	·	·	2	+	·	2	+	3	2	2	2	1	·	6	·	2	12	+	2	15	15	+	2	·	15	60.0
Festuca pratensis	10	·	5	3	20	3	·	2	8	5	10	10	·	6	·	6	5	28	12	10	8	15	16	4	2	3	23	92.0
Trisetum flavescens	11	2	2	5	·	8	·	·	·	3	·	·	·	·	·	4	10	5	5	5	8	8	16	·	·	·	15	60.0
Alopecurus pratensis	12	·	·	·	·	2	9	·	10	·	·	6	·	4	·	·	·	·	2	·	·	15	10	·	·	1	9	36.0
Holcus lanatus	13	·	·	·	·	1	1	1	2	1	·	2	·	1	2	·	10	2	5	+	2	2	·	+	·	15	12	48.0
Deschampsia caespitosa	14	·	·	·	·	·	·	11	2	·	·	28	·	·	1	·	·	·	·	·	·	5	·	·	·	·	7	28.0
Poa trivialis	15	·	·	·	·	·	·	2	·	·	·	·	·	·	·	·	·	·	·	·	·	·	·	·	·	·	1	4.0
Phleum pratense	16	·	·	·	·	·	·	·	·	·	·	·	+	·	·	·	·	·	·	·	·	·	·	·	·	·	1	4.0
Festuca arundinacea	17	·	·	·	·	·	·	·	·	·	·	·	·	·	·	·	·	·	·	·	·	·	·	·	·	·	1	4.0
Lolium perenne	18	·	·	·	·	·	·	·	·	·	·	·	·	·	·	·	·	·	2	·	·	·	·	·	·	·	1	4.0
Glyceria fluitans	19	·	·	·	·	·	·	·	·	·	·	·	·	·	·	·	·	·	·	20	·	·	·	·	·	·	1	4.0
Phalaris arundinacea	20	·	·	·	·	·	·	·	·	·	·	·	·	·	·	·	·	·	·	·	28	·	·	·	·	·	1	4.0
Phragmites communis	21	·	·	·	·	·	·	·	·	·	·	·	·	·	·	·	·	·	·	·	·	+	·	·	·	·	1	4.0
Brachypodium pinnatum	22	·	·	·	·	·	·	·	·	·	·	·	·	·	·	·	·	·	·	·	·	·	·	5	·	·	1	4.0
Carex flacca	23	2	2	·	3	·	·	·	·	·	·	·	·	·	·	2	1	3	·	2	·	1	·	·	·	2	6	24.0
Carex acutiformis	24	·	·	·	·	·	·	10	·	·	·	4	·	·	·	·	·	·	·	2	·	·	·	·	·	·	5	20.0
Carex hirta	25	·	·	·	·	·	·	·	·	·	·	6	·	·	·	·	·	·	·	·	·	·	·	·	·	·	2	8.0
Carex panicea	26	·	·	·	·	·	·	·	·	·	·	·	·	·	·	·	·	2	·	·	·	·	·	·	·	·	1	4.0
Carex gracilis	27	·	·	·	·	·	·	·	·	·	·	·	·	·	·	·	1	·	·	4	·	·	·	·	·	1	2	8.0
Trifolium pratense	28	+	+	1	+	4	1	1	·	+	·	·	·	·	·	+	1	·	·	4	+	+	1	2	2	1	18	72.0
Trifolium repens	29	+	·	+	+	2	+	+	·	1	·	·	·	·	·	+	·	2	+	+	+	+	1	·	·	·	14	56.0
Medicago lupulina	30	1	1	·	+	2	+	+	+	+	·	·	·	·	·	+	·	2	+	·	·	·	1	1	·	·	17	68.0

TABLE 10.6 (Continued)

			%
31	Vicia sepium	7	28,0
32	Lotus corniculatus	4	16,0
33	Lathyrus pratensis	10	40,0
34	Vicia cracca	5	20,0
35	Achillea millefolium	21	84,0
36	Daucus carota	20	80,0
37	Campanula rotundifolia	13	52,0
38	Plantago lanceolata	25	100,0
39	Heracleum sphondylium	14	56,0
40	Galium mollugo	25	100,0
41	Chrysanthemum leucanthemum	23	92,0
42	Scabiosa columbaria	5	20,0
43	Linum catharticum	5	20,0
44	Rumex acetosa	19	80,0
45	Ranunculus acer	23	92,0
46	Thymus serpyllum	5	20,0
47	Cerastium caespitosum	12	48,0
48	Centaurea jacea	14	56,0
49	Taraxacum officinale	18	72,0
50	Campanula glomerata	7	28,0
51	Veronica chamaedrys	22	88,0
52	Plantago media	14	56,0
53	Silene inflata	3	12,0
54	Leontodon hispidus	4	16,0
55	Crepis biennis	15	60,0
56	Myosotis arvensis	2	8,0
57	Ajuga reptans	16	64,0
58	Salvia pratensis	5	20,0
59	Knautia arvensis	3	12,0
60	Viola hirta	6	24,0
61	Bellis perennis	11	44,0
62	Dianthus superbus	4	16,0
63	Pimpinella saxifraga	2	8,0
64	Galium boreale	4	16,0
65	Cirsium oleraceum	14	56,0

TABLE 10.6 (Continued)

Species, Species No.	Running No. / Relevé No.	Constancy Absol.	Percent
Tragopogon pratensis	66	7	28.0
Glechoma hederacea	67	6	24.0
Anthriscus silvestris	68	5	20.0
Filipendula ulmaria	69	4	16.0
Geum rivale	70	14	56.0
Melandrium diurnum	71	11	44.0
Angelica silvestris	72	6	24.0
Lysimachia nummularia	73	8	32.0
Prunella vulgaris	74	12	48.0
Pimpinella magna	75	4	16.0
Polygonum bistorta	76	4	16.0
Lychnis flos-cuculi	77	7	28.0
Senecio jacobaea	78	6	24.0
Potentilla reptans	79	3	12.0
Cardamine pratensis	80	2	8.0
Myosotis palustris	81	2	8.0
Geranium pratense	82	1	4.0
Pastinaca sativa	83	3	12.0
Galium uliginosum	84	1	4.0
Sanguisorba officinalis	85	1	4.0
Galium verum	86	2	8.0
Silaus pratensis	87	3	12.0
Ranunculus repens	88	1	4.0
Euphrasia odontites	89	2	8.0
Lamium album	90	1	4.0
Rumex crispus	91	2	4.0
Polygonum convolvulus	92	1	4.0
Chenopodium album	93	1	4.0
Alchemilla vulgaris	94	1	4.0

Relevé columns: 1 2 3 4 5 6 7 8 9 10 11 12 13 14 15 16 17 18 19 20 21 22 23 24 25

Sum per relevé: 98 100 99 100 100 100 100 100 100 100 100 100 100 99 100 99 99 100 100 99 100 100 99 100 100

One relevé is chosen as a reference stand. This is a sample stand considered extreme within the group of stands under comparison, but related enough to the group so as not to represent an entirely different type. This stand choice can be brought on a more objective basis by using, for example, the stand choice criterion proposed by SWAN and DIX (1966) (see SECTION 10.72.1).

For this program relevé 10 was chosen as it fulfilled the above requirements. It may be noted that relevé 10 appears as second stand in the hand-prepared differential table (TABLE 9.7).

The third step in the first program does not only involve the calculation of the floristic similarity among the relevés. It involves at the same time a linear ordination of the relevés. This results in presorting the stands for classification. This process will be explained using the computer printed correlation test table shown as TABLE 10.8.

The first line on TABLE 10.8 lists the selected reference stand (relevé 10) and the time used for computation of the correlation test table.

All relevés are then compared to the reference stand by calculation of their similarity indices shared with the reference stand. For the 25 relevés this resulted in $n-1=24$ indices of similarity (*IS* values). The 24 stand comparisons with relevé 10 are shown with their respective *IS* values in the next four lines of TABLE 10.8. The *IS* values are computed to the fourth decimal place.

The computer was then programmed to scan these 24 *IS* values and to select those stands that showed a similarity of at least 20 percent with the first reference stand. Six stands fulfilled this criterion. These are printed out in the sixth line on the correlation test table, and are recognizable at once by their *IS* values that are printed only to the nearest decimal place. The six stands form a similarity group together with the reference stand, which is the seventh stand. An *IS* value of 20 percent was considered to reflect a fairly high degree of floristic similarity. In SPATZ's quantitative modification of JACCARD's index, 20 percent implies a floristic similarity which would be greater than 50 percent when using SØRENSEN's index in its quantitative application (TABLE 10.2). The 20 percent *IS* value was used as a breakoff point for a similarity group. It is, of course, an arbitrary value. But, it is based on field experience with grass cover communities and their quantitative evaluation by plant biomass values.

After selecting the first similarity group from the $(n-1)$ *IS* values, the computer selects next the stand most similar to the first reference stand. This is relevé 9 with 28.1 percent similarity to relevé 10. The computer lists relevé 9 as the second stand in the seventh line on TABLE 10.8. This also means relevé 9 becomes the second stand in the linear ordination.

TABLE 10.7 Computer-Printed Constancy Table. From SPATZ 1970.

Species, Species No.	Relevé No.	1	2	3	4	5	6	7	8	9	10	11	12	13	14	15	16	17	18	19	20	21	22	23	24	25	Absol.	Percent
Arrhenatherum elatius	1	+	1	2	2	4	4	4	4	5	8	9	10	10	12	15	15	15	22	22	24	25	25	26	30	35	25	100,0
Dactylis glomerata	2	5	5	15	5	12	12	4	10	2	6	12	32	15	10	6	15	15	18	1	5	18	8	8	8	18	25	100,0
Poa pratensis	6	4	74	10	5	4	2	3	4	10	8	6	25	2	5	15	10	5	10	1	6	1	9	16	20	10	25	100,0
Plantago lanceolata	38	1	1	1	1	2	2	6	4	1	1	+	1	4	5	1	2	+	5	+	1	1	2	8	+	+	25	100,0
Galium mollugo	40	3	2	7	1	3	24	6	8	2	12	5	6	10	14	3	5	+	6	2	5	15	3	12	3	2	25	100,0
Festuca pratensis	10	.	5	3	.	20	3	2	2	5	2	10	10	2	2	1	5	1	12	10	2	15	15	2	2	3	23	92,0
Chrysanthemum leucanthemum	41	1	+	3	+	+	3	2	1	2	5	3	3	2	2	2	1	1	+	+	3	+	1	1	6	3	23	92,0
Ranunculus acer	45	+	+	+	+	+	1	2	+	2	3	+	+	2	1	1	1	+	+	.	1	+	+	+	.	.	23	92,0
Veronica chamaedrys	51	+	1	.	1	+	2	1	+	1	1	+	1	1	+	+	+	2	1	1	1	1	4	2	2	1	22	88,0
Achillea millefolium	35	6	1	3	1	2	1	1	1	2	8	+	+	4	16	2	5	+	12	.	6	4	4	+	3	+	21	84,0
Daucus carota	36	1	1	1	+	+	.	1	5	.	1	.	2	2	1	1	1	+	1	.	1	.	2	2	.	+	20	80,0
Rumex acetosa	44	+	+	+	+	+	.	1	3	.	.	2	.	.	2	+	1	+	1	.	+	1	+	.	.	1	20	80,0
Trifolium pratense	28	+	.	1	4	4	1	1	1	.	.	.	.	2	.	6	+	2	5	5	+	+	+	2	2	1	18	72,0
Taraxacum officinale	49	+	+	+	1	3	1	3	1	1	.	.	2	.	.	+	+	+	5	1	+	+	4	4	+	+	18	72,0
Medicago lupulina	30	1	.	+	1	2	.	.	.	1	1	1	.	.	.	.	.	+	4	+	.	1	+	.	1	+	17	68,0
Helictotrichon pubescens	3	1	1	.	.	20	8	.	.	4	.	+	4	+	.	.	13	4	4	.	28	.	2	+	1	+	16	64,0
Ajuga reptans	57	+	+	+	+	+	.	1	3	.	.	1	1	+	.	+	.	.	+	.	.	1	+	.	+	.	16	64,0
Festuca rubra	9	15	.	.	2	.	.	.	.	.	4	+	.	2	.	6	.	.	.	.	+	1	.	.	.	.	15	60,0
Trisetum flavescens	11	.	2	5	.	8	.	.	+	3	.	3	1	6	18	4	10	5	5	5	8	8	16	6	2	3	15	60,0
Crepis biennis	55	.	.	+	1	2	1	3	8	1	.	3	.	.	.	1	.	5	1	.	1	1	1	4	.	.	15	60,0
Trifolium repens	29	+	.	.	+	2	.	4	.	1	1	.	1	.	2	+	2	1	+	1	2	+	.	1	+	.	14	56,0
Heracleum sphondylium	39	+	1	1	3	3	2	1	+	.	.	.	.	26	4	.	.	+	.	.	+	.	+	1	+	+	14	56,0
Centaurea jacea	48	.	1	6	.	.	2	.	2	.	.	.	.	.	.	1	2	2	.	2	2	.	.	.	.	.	14	56,0
Plantago media	52	+	+	+	+	+	3	1	1	2	+	+	1	.	.	.	.	.	.	+	.	2	.	+	+	3	14	56,0
Cirsium oleraceum	65	.	.	.	.	+	12	20	+	.	.	3	.	.	18	.	.	2	.	.	.	.	.	1	.	.	14	56,0
Geum rivale	70	.	.	.	.	.	2	1	5	.	.	3	.	.	.	.	.	.	.	.	.	1	.	+	+	+	14	56,0
Campanula rotundifolia	37	1	.	.	.	1	.	1	2	.	.	2	1	1	2	.	.	.	.	.	.	2	.	.	.	.	13	52,0
Holcus lanatus	13	.	.	.	.	.	.	.	.	.	.	.	.	.	.	.	.	.	.	.	.	.	.	.	15	.	12	48,0
Cerastium caespitosum	47	+	.	+	+	+	+	+	+	+	+	+	+	.	.	.	+	+	+	+	.	.	.	+	+	+	12	48,0
Prunella vulgaris	74	.	.	+	.	+	.	+	2	+	1	.	1	.	.	.	.	2	.	.	.	+	1	1	.	.	12	48,0
Bellis perennis	61	.	.	+	.	+	.	.	+	.	+	.	.	.	.	.	.	.	+	.	.	.	.	.	.	.	11	44,0

TABLE 10.7 (Continued)

No.	Species	Number	Percent
71	Melandrium diurnum	11	44,0
33	Lathyrus pratensis	10	40,0
12	Alopecurus pratensis	9	36,0
73	Lysimachia nummularia	8	32,0
4	Bromus erectus	7	28,0
14	Deschampsia caespitosa	7	28,0
31	Vicia sepium	7	28,0
50	Campanula glomerata	7	28,0
66	Tragopogon pratensis	7	28,0
77	Lychnis flos-cuculi	6	24,0
23	Carex flacca	6	24,0
60	Viola hirta	6	24,0
67	Glechoma hederacea	6	24,0
72	Angelica silvestris	6	24,0
78	Senecio jacobaea	5	20,0
7	Briza media	5	20,0
24	Carex acutiformis	5	20,0
34	Vicia cracca	5	20,0
42	Scabiosa columbaria	5	20,0
43	Linum catharticum	5	20,0
46	Thymus serpyllum	5	20,0
58	Salvia pratensis	5	20,0
68	Anthriscus silvestris	5	20,0
5	Festuca ovina	4	16,0
8	Koeleria pyramidata	4	16,0
32	Lotus corniculatus	4	16,0
54	Leontodon hispidus	4	16,0
62	Dianthus superbus	4	16,0
64	Galium boreale	4	16,0
69	Filipendula ulmaria	4	16,0
75	Pimpinella magna	4	16,0
76	Polygonum bistorta	4	16,0
53	Silene inflata	3	12,0
59	Knautia arvensis	3	12,0
79	Potentilla reptans	3	12,0

TABLE 10.7 (Continued)

Species, Species No.		1	2	3	4	5	6	7	8	9	10	11	12	13	14	15	16	17	18	19	20	21	22	23	24	25	Constancy Absol.	Percent
Pastinaca sativa	83	·	·	·	·	·	·	·	·	·	·	·	1	·	+	·	·	·	3	·	·	·	·	·	·	·	3	12.0
Silaus pratensis	87	·	·	·	·	·	·	·	·	·	·	·	·	·	·	·	1	+	·	·	+	·	·	·	·	·	3	12.0
Carex hirta	25	·	·	·	·	·	·	·	·	·	·	6	·	·	·	·	·	2	·	·	·	·	·	·	·	·	2	8.0
Carex gracilis	27	·	·	·	·	·	·	·	·	·	·	·	·	·	·	·	·	·	·	4	·	·	·	·	·	1	2	8.0
Myosotis arvensis	56	·	+	·	·	·	·	+	·	·	·	·	·	·	·	·	·	·	·	·	·	·	·	·	·	·	2	8.0
Pimpinella saxifraga	63	·	·	·	+	·	·	·	·	·	·	·	·	·	·	·	+	·	·	·	·	·	·	·	·	·	2	8.0
Cardamine pratensis	80	·	·	·	·	·	·	·	·	+	·	·	+	·	·	·	·	·	·	·	·	·	·	·	·	·	2	8.0
Myosotis palustris	81	·	·	·	·	·	·	·	·	·	·	·	+	·	·	·	·	·	·	·	·	+	·	·	·	·	2	8.0
Galium verum	86	·	·	·	·	·	·	·	·	·	·	·	·	·	·	1	·	·	·	·	·	·	·	·	·	·	2	8.0
Euphrasia odontites	89	·	·	·	·	·	·	·	·	·	·	·	+	·	·	·	·	·	·	·	·	·	·	·	·	·	2	8.0
Rumex crispus	91	·	·	·	·	·	·	2	·	·	·	·	·	·	·	·	·	·	·	·	·	·	·	·	·	·	2	8.0
Poa trivialis	15	·	·	·	·	·	·	·	·	·	·	·	·	·	1	·	·	·	·	·	·	·	·	·	·	·	1	4.0
Phleum pratense	16	·	·	·	·	·	·	·	·	·	·	·	·	·	·	·	·	·	·	·	·	·	·	·	·	·	1	4.0
Festuca arundinacea	17	·	·	·	·	·	·	·	·	·	·	·	·	1	·	·	·	·	·	·	·	·	·	·	·	·	1	4.0
Lolium perenne	18	·	·	·	·	·	·	·	·	·	·	·	·	·	·	·	·	·	2	·	·	·	·	·	·	·	1	4.0
Glyceria fluitans	19	·	·	·	·	·	·	·	·	·	·	·	·	·	·	·	·	·	·	20	·	·	·	·	·	·	1	4.0
Phalaris arundinacea	20	·	·	·	·	·	·	·	·	·	·	·	·	·	·	·	·	·	·	28	·	·	·	·	·	·	1	4.0
Phragmites communis	21	·	·	·	·	·	·	·	·	·	·	·	·	·	·	·	·	·	·	·	·	·	·	·	·	·	1	4.0
Brachypodium pinnatum	22	·	·	·	·	·	·	·	·	·	·	·	·	·	·	·	·	·	·	·	·	·	·	·	5	·	1	4.0
Carex panicea	26	·	·	·	·	·	·	·	·	·	·	·	·	·	·	·	2	·	·	·	·	·	·	·	·	·	1	4.0
Geranium pratense	82	·	·	·	·	·	·	·	·	·	·	·	+	·	·	·	·	·	·	·	·	·	·	·	·	·	1	4.0
Galium uliginosum	84	·	·	·	·	·	·	·	·	·	·	·	·	·	+	·	·	·	·	·	·	·	·	·	·	·	1	4.0
Sanguisorba officinalis	85	·	·	·	·	·	·	·	·	·	·	·	·	·	+	·	·	·	·	·	·	·	·	·	·	·	1	4.0
Ranunculus repens	88	·	·	·	·	·	·	·	·	·	·	·	·	·	·	·	·	+	·	·	·	·	·	·	·	·	1	4.0
Lanium album	90	·	·	·	·	·	·	·	·	·	·	·	·	·	·	·	·	·	·	·	·	·	·	·	+	·	1	4.0
Polygonum convolvulus	92	·	·	·	·	·	·	·	·	·	·	·	·	·	·	·	·	·	·	·	·	+	·	·	·	·	1	4.0
Chenopodium album	93	·	·	·	·	·	·	·	·	·	·	·	·	·	·	·	·	·	·	·	·	·	·	·	+	·	1	4.0
Alchemilla vulgaris	94	·	·	·	·	·	·	·	·	·	·	·	·	·	·	·	·	·	·	·	·	·	·	·	·	+	1	4.0
Sum per relevé		98	100	99	100	100	100	100	100	100	100	100	100	100	99	100	99	99	100	100	100	99	100	100	100	100		

As the next step, the remaining five stands of the first similarity group are compared to relevé 9 by calculation of a new set of similarity indices. Their *IS* values shared with relevé 9 are printed out in the eighth line. Following this, the computer again scans these values for the highest. The highest *IS* value is 29.1 which relevé 9 shares with releve 3. Stand 3 then becomes the third relevé in the linear ordination.

The remaining four stands of the first similarity group are then compared to relevé 3. The most similar one (relevé 24 with a 35.7 similarity to relevé 3) is selected by the computer, and the fourth stand is found for the linear ordination. This process is continued until all stands of the first similarity group are ordinated. The selected stand order shown is relevé 10, 9, 3, 24, 15, 1, and 4. Relevé 4 is the last stand of the first similarity group.

The computer then scans once more the first set of *IS* values that the relevés shared with the first reference stand, releve 10. It now selects the stand that shared an *IS* value with relevé 10 that was next below the 20 percent threshhold value of similarity. This is relevé 20, which shares an *IS* value of 15.65 percent with relevé 10. Relevé 20 then becomes the second reference stand of a second similarity group. In the correlation test table, relevé 20 is listed as the eighth stand in the linear ordination. Now all stands not already placed are compared by the computer to relevé 20. This results in the printout of $25-8=17$ new *IS* values. These are easily spotted on the correlation test table as the second set of *IS* values computed to the fourth decimal place.

Now the process is repeated with the 17 *IS* values as described before for the initial 24 *IS* values. That is, the computer selects the second similarity group from the 17 *IS* comparisons by scanning these values for the relevés with at least 20 percent similarity to the second reference stand, relevé 20. This includes nine relevés. They are easily spotted in the correlation table under relevé 20 as those stands whose *IS* values are computed only to the nearest decimal place. Of these nine relevés, the one most similar to relevé 20 becomes the ninth stand in the linear ordination. This is relevé 22. Again, the remaining stands of this similarity group are compared to relevé 22, and the ordination process is continued until all nine stands of the second similarity group are placed in the proper sequence.

Following this, a third reference stand and similarity group is selected by the same criteria. The third reference stand is the one that shared an *IS* value next below the 20 percent threshold value with the second reference stand. The choice goes to relevé 23. However, it may be noted in the correlation test table that none of the remaining seven relevés shared an *IS* value of 20 percent or greater with the third reference stand. Therefore, there is no similarity group established. The next

TABLE 10.8 Correlation Test Table and Printout of the Similarity Indices (IS Values) of the 25 Arrhenatherum Relevés. From SPATZ 1970.

```
Correlation test

1.  Selected relevé  = 10                                         Time = 415 Sec.
10   1  27,6400   10   2  10,4642   10   3  21,7299    4  20,4219   10   5   6,1694    6   9,3549
10   7   7,5261    8   4,8181        9  28,1280       11   4,9303   10  12  15,4463   13  10,8711
10  14  12,3390   15  24,8092       16  11,9514       17   8,8293       18  12,6139   19   1,3149
10  20  15,6504   21   4,9940       22  10,1286       23  12,4427   10  24  20,0096   25   7,1391
10   1  27,6       3  21,7           4  20,4       10          28,1        24  20,0          24,8

2.  Relevé No. =  9
 9   1  27,8       3  29,1           4  24,1        9          27,6        24  20,4

3.  Relevé No. =  3
 3   1  21,2       4  27,0           3          35,7           30,4

4.  Relevé No. = 24
24   1  21,6       4  13,1          24          21,6       15  27,9

5.  Relevé No. = 15
15   1  37,2       4  32,6          15          37,2

6.  Relevé No. =  1
 1   4  40,8

7.  Relevé No. =  4

                                              Greatest IS value, under 20

8.  Relevé No. = 20
20   2  18,5410  20   5  24,0421  20   6  28,9570    7  15,6250  20   8  10,3761   11  10,2276
20  12  23,6054  13  26,0037       14  14,4213       16  21,2357  20  17  29,3566   18  25,1658
20  19   6,8421  21  22,3390       22  32,0178       23  18,7638  20  25  13,7427    5  24,0
20   6  29,0     12  23,6          13  26,0          16  26,0     20      29,4       18  25,2
20  21  22,3     22  32,0          20  21,2          13      29,4        20  29,4       25,2

9.  Relevé No. = 22
22   5  32,8       6  21,6         22  38,9       22  23,0           16  28,6
22  18  30,5      21  24,9         13      24,6           27,2       17

10. Relevé No. = 13
13   5  34,8       6  32,1         13  16,6       13  24,6           17  30,8
```

TABLE 10.8 (Continued)

```
                13
                21  ,  19.9
11.                                                                          Relevé No. = 5
    5    6    21.6    5   12   14.0    5   16   18.0    5   17   21.8    5   18   21.9    5   21   18.3
12.                                                                          Relevé No. = 18
   18    6    19.6   18   12   18.2   18   16   21.7   18   17   20.0   18   21   17.0
13.                                                                          Relevé No. = 16
   16    6    20.1   16   12   19.1   16   17   23.9   16   21   15.3
14.                                                                          Relevé No. = 17
   17    6    18.4   17   12   23.6   17   21   30.2
15.                                                                          Relevé No. = 21
   21    6    17.9   21   12   22.0
16.                                                                          Relevé No. = 12
   12    6    19.0
17.                                                                          Relevé No. = 6

Greatest IS value, under 20 = 18.
   23    2   13.9018   23    7   13.0520    8   17.9766   23   11   5.4171   23   14   14.9445    8.4435
   23   25   16.4824

Greatest IS value, under 20 = 19.
    8    2   10.3028    8    7   15.8776    8   11   29.0355    8   14   18.7209    8   19   11.3207   14.3455
    8   11   29.0

20.                                                                          Relevé No. = 11

Greatest IS value, under 20 = 21.
   14    2    9.1193   14    7   14.8333   14   19    4.3796   14   25   13.9800

Greatest IS value, under 20 = 22.
    7    2    9.0766    7   19    6.9907    7   25   10.5423

Greatest IS value, under 20 = 23.
   25   19    9.1554   25   19   16.2152

24.                                                                          Relevé No. = 19
   19    2    9.3531

25.                                                                          Relevé No. = 2

Greatest IS value, under 20
```

stand following relevé 23 is the one most similar, even though not reaching the 20 percent minimum similarity. This is relevé 8 which shares an *IS* value of only 17.97 percent with relevé 23. Thereafter, the remaining six stands are compared to relevé 8. Only one of these, relevé 11 shows a similarity greater than 20 percent to stand 8. Thus, the third similarity group consists only of two stands, relevé 8 and 11.

This process is continued until all 25 relevés are ordinated according to these criteria. The computer then prints out an ordinated table (TABLE 10.9), which shows the new relevé order established by the similarity group criteria. The group separation lines are not printed by the computer. They were drawn in subsequently from the group segregation established on the correlation test table (TABLE 10.8).

It may be noted that the computer achieved only a preliminary classification. Two large similarity groups and a small group were segregated. The first consisting of seven relevés, the second of ten, the third of only two relevés (i.e., relevé 8 and 11). This leaves six stands ungrouped. The reason for this is that these six remaining relevés share similarity indices of less than the stipulated threshold value of 20 percent. In other words, they represent separate groups or classes by themselves, and one could say that the computer established $3 + 6 = 9$ separate classes.

However, nine classes would hardly appear a meaningful breakdown of the sample of 25 *Arrhenatherum* releves. In TABLE 9.7, the original data was separated into three classes.

Yet, a close comparison of the relevé order and separation lines indicates that the computer achieved a close approximation.

The following tabulation of the two rows of figures matches the sequential positions of original relevé numbers in the head of TABLE 9.7 (here the upper row) with the computer-derived sequence (here the lower row) as recorded in the head of TABLE 10.9. The computer derived similarity groups are underlined in the lower row. The letters *A, B* and *C* in the upper row stand for the three vegetation units recognized in TABLE 9.7.

	A			*B*			*C*	
4, 10, 1,	9, 15, 3, 24,	2	16, 18, 12, 20,	5, 22	23, 19, 13, 17, 6, 11,	8, 21, 25, 14, 7
10, 9, 3, 24,	15, 1,	4	20, 22, 13,	5, 18, 16, 17, 21, 12,	6	23, 8, 11, 14,	7, 25, 19, 2

The first computer-selected group coincides well with the *Arrhenatherum-Bromus erectus* group (*A*). However, relevé 2 was sorted at the end of the computer series because it shares a low *IS* value with relevé 10 (10.5 percent). The second computer-selected group contains all six relevés of *Arrhenatherum* type *B*, but in addition also four relevés of type *C*. The relevés sorted at the end of the series by the computer are all sample stands of the *Arrhenatherum–Circium oleraceum* type *C*, except for relevé 2. Relevé 2 was also discarded by the computer, from the second similarity group, because its *IS* value shared with relevé 20 was just below 20 percent, namely 18.5 percent.

However, it may be noted that the *IS* value shared with relevé 20 is the highest for relevé 2 on the correlation test table. The *IS* value shared with the neighboring relevé (19) is much lower, only 9.4 percent.

Therefore, at this point an adjustment was made by placing relevé 2 forward next before relevé 20. A similar adjustment was made for relevé 25 by placing it forward next following relevé 23, because these two relevés shared a higher *IS* value (16.5 percent) than the neighboring relevé 7 with 25 (10.5 percent) in the computer ordinated series.

SPATZ's method of linear stand ordination through similarity groups with threshold values therefore results in a good presorting, but final adjustments of the stand sequence may be necessary. The same difficulty would arise if all stands had been compared to each other resulting in $n \times (n-1)/2$ or 300 *IS* values for 25 stands. Sorting stands next to each other that show the closest similarity often separates fairly similar stands quite far apart on a linear ordination. The reason for this seemingly contradictory observation is that the floritic similarity relationships are actually multi-dimensional. A representation in a linear sequence of one dimension is only a crude approximation. However, this approximation has the great advantage that all species are shown with their stand to stand variation in one table. This is not possible in a multidimensional ordination.

As in any ordination, classification still remains a matter of judgment of the investigator. Following the two relevé order adjustments discussed above, SPATZ divided the linear relevé sequence into three (instead of the nine computer-indicated) classes. The first and second computer-established classes were used, and a third class was assigned to the remaining relevés.

10.63 Selection of Differential Species. The second computer program is initiated after establishing the classes or community types. The species cards are fed into the computer once more. The computer is now programmed to select from all species those that are preferentially associated with either one or two of the three community types just established. These are the species that permit us to recognize and differentiate vegetation samples in the field as belonging to any one of the established classes. For this reason they are usually called differential or differentiating species.

Two differential species types were established. The criteria for a dif-

TABLE 10.9 Computer-Printed Ordinated Relevé Table. The Species Order from High to Low Constancy Is the Same as in TABLE 10.7. From SPATZ 1970.

Running No.	1	2	3	4	5	6	7	8	9	10	11	12	13	14	15	16	17	18	19	20	21	22	23	24	25	Constancy Absol.	Percent
Relevé No.	10	9	3	24	15	1	4	20	22	13	5	18	16	17	21	12	6	23	8	11	14	7	25	19	2		
Species, Species No.																											
Arrhenatherum elatius 1	8	5	2	30	15	+	2	24	25	10	4	22	15	15	25	10	4	26	4	9	12	4	35	22	1	25	100,0
Dactylis glomerata 2	6	2	15	8	6	5	5	5	8	15	12	18	10	15	18	32	12	8	10	12	10	18	18	1	5	25	100,0
Poa pratensis 6	8	10	10	20	15	5	5	6	9	15	4	10	10	5	1	25	+	16	4	6	5	3	10	1	74	25	100,0
Plantago lanceolata 38	1	1	1	+	3	+	1	1	2	2	2	2	1	+	1	+	4	4	8	4	5	4	+	+	1	25	100,0
Galium mollugo 40	12	2	7	3	3	5	1	5	3	10	20	6	5	3	1	6	24	12	4	5	14	6	2	2	+	25	100,0
Festuca pratensis 10	2	5	3	2	6	1	1	2	15	2	12	5	5	28	1	10	3	2	1	10	2	2	3	10	5	23	92,0
Chrysanthemum leucanthemum 41	5	1	+	+	2	+	+	3	1	1	20	+	+	+	15	+	4	1	8	10	2	2	3	.	5	23	92,0
Ranunculus acer 45	3	2	+	+	+	1	+	1	2	1	1	2	1	+	+	3	2	1	1	+	2	2	+	+	5	23	92,0
Veronica chamaedrys 51	8	2	3	+	+	.	+	1	4	+	1	+	1	1	1	1	1	2	+	1	1	1	+	+	+	22	88,0
Achillea millefolium 35	.	1	+	3	6	6	+	6	4	4	12	1	5	5	+	+	1	2	.	.	16	1	+	+	+	21	84,0
Daucus carota 36	1	1	+	+	+	1	+	+	+	2	1	1	+	+	+	+	2	2	5	+	2	1	+	1	+	20	80,0
Rumex acetosa 44	.	.	+	+	+	+	+	+	1	1	+	+	+	+	1	1	+	+	3	1	2	1	+	+	1	20	80,0
Trifolium pratense 28	+	.	1	2	.	.	.	+	+	+	8	1	1	+	+	+	1	4	+	+	2	1	5	.	+	18	72,0
Taraxacum officinale 49	+	1	.	.	.	+	1	+	1	2	2	+	+	+	.	+	2	+	4	+	+	3	.	.	.	18	72,0
Medicago lupulina 30	+	+	.	1	1	1	+	.	1	4	3	+	+	.	+	.	4	1	+	+	.	.	.	.	.	17	68,0
Helictotrichon pubescens 3	.	4	.	.	.	.	.	28	2	4	20	4	13	4	.	+	8	+	.	.	.	.	.	.	.	16	64,0
Ajuga reptans 57	.	1	.	+	2	.	+	2	1	+	+	1	+	+	+	+	+	+	3	+	1	1	+	1	+	16	64,0
Festuca rubra 9	4	3	5	2	6	15	2	2	.	1	5	5	.	5	.	2	+	4	.	.	.	.	.	.	2	15	60,0
Trisetum flavescens 11	.	3	5	2	4	.	.	8	16	6	8	5	10	5	8	1	1	6	8	.	.	.	5	.	2	15	60,0
Crepis biennis 55	.	.	.	.	.	+	1	1	1	.	2	2	.	.	.	+	+	1	+	+	2	+	+	+	.	15	60,0
Trifolium repens 29	+	.	+	+	1	1	1	+	+	26	3	6	.	2	.	+	1	+	+	4	4	3	+	+	.	14	56,0
Heracleum sphondylium 39	.	.	6	2	.	.	.	1	.	.	.	.	2	2	.	.	1	1	.	.	4	4	.	.	1	14	56,0
Centaurea jacea 48	.	.	.	.	.	.	.	2	.	.	.	2	.	2	+	1	+	+	+	.	.	3	+	.	1	14	56,0
Plantago media 52	.	.	.	.	.	.	.	+	+	+	+	+	1	2	2	+	.	.	.	.	.	.	+	.	.	14	56,0
Cirsium oleraceum 65	.	.	.	.	.	.	.	+	.	.	.	1	+	.	2	1	12	1	20	3	.	1	.	.	.	14	56,0
Geum rivale 70	.	.	.	.	.	.	.	+	+	+	.	+	.	.	2	+	2	.	5	18	.	20	+	+	.	14	56,0
Campanula rotundifolia 37	1	.	.	.	.	1	1	.	.	1	.	.	1	.	.	+	.	.	+	3	1	1	+	1	.	13	52,0
Holcus lanatus 13	.	.	.	+	1	1	.	.	1	.	.	2	.	2	2	1	.	6	+	2	2	2	15	.	.	12	48,0
Cerastium caespitosum 47	.	+	.	+	.	1	.	.	.	.	.	+	.	.	1	.	.	1	+	2	2	1	+	+	.	12	48,0
Prunella vulgaris 74	+	+	.	+	+	.	1	.	.	+	.	.	.	.	.	.	.	1	2	.	.	+	.	.	.	12	48,0

TABLE 10.9 (Continued)

Species	No.																								Count	%
Bellis perennis	61	1	+							+									+	+	+	1			11	44,0
Melandrium diurnum	71			+						+	1							+	+	2	1	1	4	+		11	44,0
Lathyrus pratensis	33		+					+		1	+							1	1					+		10	40,0
Alopecurus pratensis	12				10		1			4					6	10		8	+				1	+		9	36,0
Lysimachia nummularia	73			35	10	37	50	74		+						1		+			+					8	32,0
Bromus erectus	4	21	47							+			15		+											7	28,0
Deschampsia caespitosa	14												5			28		10								7	28,0
Vicia sepium	31	+		+	+		+			1		1		2		1		1	+				+			7	28,0
Campanula glomerata	50		1	1		1				+								+	2							7	28,0
Tragopogon pratensis	66						+			1	+	1		1				+	1							7	28,0
Lychnis flos-cuculi	77									+		1	+					+	+							7	28,0
Carex flacca	23		2		1	2	3				2	3	+		+	+	2	+								6	24,0
Viola hirta	60	3	2	+	+	+				+								+								6	24,0
Glechoma hederacea	67									+								+					1	+		6	24,0
Angelica silvestris	72			+	+	1				+		+			2			+								6	24,0
Senecio jacobaea	78	+		+	+					+					+			+								6	24,0
Briza media	7	1	2	+	1		1			2								+								5	20,0
Carex acutiformis	24												1		4	10	1									5	20,0
Vicia cracca	34	+								1	+	1				1		+								5	20,0
Scabiosa columbaria	42	+		+	1					+							2	+								5	20,0
Linum catharticum	43			+	+	1	+			+								+								5	20,0
Thymus serpyllum	46	2		+	1					+								+								5	20,0
Salvia pratensis	58	5	2	4						+	2							+								5	20,0
Anthriscus silvestris	68	+		1		+				+		2						+	+				1			5	20,0
Festuca ovina	5	2	1		2																					4	16,0
Koeleria pyramidata	8	3	3		3	2																				4	16,0
Lotus corniculatus	32			+	+	+				+		+						+								4	16,0
Leontodon hispidus	54			+		1				1				4				+								4	16,0
Dianthus superbus	62	+		+				1		+																4	16,0
Galium boreale	64	1				+						1						1								4	16,0
Filipendula ulmaria	69									+				3					+							4	16,0
Pimpinella magna	75									+					2											4	16,0
Polygonum bistorta	76									+	1							+	1		4	1				4	16,0
Silene inflata	53		+		+	+				+															2	3	12,0
Knautia arvensis	59		1							+																3	12,0

TABLE 10.9 (Continued)

Running No.		1	2	3	4	5	6	7	8	9	10	11	12	13	14	15	16	17	18	19	20	21	22	23	24	25	Constancy	
Species, Species No.	Releve No.	10	9	3	24	15	1	4	20	22	13	5	18	16	17	21	12	8	23	8	11	14	7	25	19	2	Absol.	Percent
Potentilla reptans	79	1	.	.	.	.	.	.	1	.	.	.	.	.	.	.	.	.	.	.	.	.	.	.	.	.	3	12.0
Pastinaca sativa	83	.	.	.	.	.	.	.	.	.	1	.	3	.	.	.	.	.	.	.	.	+	.	.	.	.	3	12.0
Silaus pratensis	87	.	.	.	.	.	.	.	+	.	1	.	.	1	.	.	.	.	.	.	.	+	.	.	.	.	3	12.0
Carex hirta	25	.	.	.	.	.	.	.	.	.	.	.	.	.	.	.	.	.	.	.	6	.	.	.	.	.	2	8.0
Carex gracilis	27	.	.	.	.	.	.	.	.	.	.	.	.	.	2	.	.	.	.	.	.	.	.	1	4	.	2	8.0
Myosotis arvensis	56	.	.	.	.	.	.	.	.	.	.	.	.	.	.	.	.	.	.	.	.	.	+	.	.	+	2	8.0
Pimpinella saxifrage	63	.	+	.	.	.	.	+	.	.	.	.	.	.	.	.	.	.	.	.	.	.	.	.	.	.	2	8.0
Cardamine pratensis	80	.	.	.	.	.	.	.	.	.	.	.	.	+	.	.	+	.	.	.	.	.	.	.	.	.	2	8.0
Myosotis palustris	81	.	.	.	.	.	.	.	.	.	.	.	.	.	.	+	.	.	+	.	.	.	.	.	.	.	2	8.0
Galium verum	86	.	.	.	+	1	.	.	.	.	.	.	.	.	.	.	.	.	.	.	.	.	.	.	.	.	2	8.0
Euphrasia odontites	89	.	.	.	.	.	.	.	.	.	.	+	+	.	.	.	.	.	.	.	.	.	.	.	.	.	2	8.0
Rumex crispus	91	.	.	.	.	.	.	.	.	.	.	.	.	.	.	+	.	.	.	.	.	+	.	.	.	.	2	8.0
Poa trivialis	15	.	.	.	.	.	.	.	.	.	.	.	.	.	.	.	.	.	.	.	.	.	**2**	.	.	.	1	4.0
Phleum pratense	16	.	.	.	.	.	.	.	.	.	.	.	.	.	.	.	.	.	.	.	.	1	.	.	.	.	1	4.0
Festuca arundinacea	17	.	.	.	.	.	.	.	.	.	.	.	.	.	.	.	.	.	+	.	.	.	.	.	.	.	1	4.0
Lolium perenne	18	.	.	.	.	.	.	.	.	.	.	.	2	.	.	.	.	.	.	.	.	.	.	.	.	.	1	4.0
Glyceria fluitans	19	.	.	.	.	.	.	.	.	.	.	.	.	.	.	.	.	.	.	.	.	.	.	20	.	.	1	4.0
Phalaris arundinacea	20	.	.	.	.	.	.	.	.	.	.	.	.	.	.	.	.	.	.	.	.	.	.	28	.	.	1	4.0
Phragmites communis	21	.	.	.	.	.	.	.	.	.	.	.	.	.	.	.	.	.	.	.	.	.	.	.	.	.	1	4.0
Brachypodium pinnatum	22	.	.	.	5	.	.	.	.	.	.	.	.	.	.	.	.	.	.	.	.	.	.	.	.	.	1	4.0
Carex panicea	26	.	.	.	.	.	.	.	.	.	.	.	.	2	.	.	.	.	.	.	.	.	.	.	.	.	1	4.0
Geranium pratense	82	.	.	.	.	.	.	.	.	.	.	.	.	.	.	.	.	.	.	.	.	.	.	.	.	.	1	4.0
Galium uliginosum	84	.	.	.	.	.	.	.	.	.	.	.	.	.	.	.	.	.	.	.	.	.	.	.	.	.	1	4.0
Sanguisorba officinalis	85	.	.	.	.	.	.	.	.	.	.	.	.	.	.	.	+	.	.	.	.	.	.	.	.	.	1	4.0
Ranunculus repens	88	.	.	.	.	.	.	.	.	.	.	.	.	.	+	.	.	.	.	.	.	.	.	.	.	.	1	4.0
Lamium album	90	.	.	.	+	.	.	.	.	.	.	.	.	.	.	.	.	.	.	.	.	.	.	.	.	.	1	4.0
Polygonum convolvulus	92	.	.	.	.	.	.	.	.	.	.	.	.	.	.	.	.	.	.	.	.	.	.	+	.	.	1	4.0
Chenopodium album	93	.	.	.	.	.	.	.	.	.	.	.	.	.	.	.	.	.	.	.	.	.	.	+	.	.	1	4.0
Alchemilla vulgaris	94	.	.	.	+	.	.	.	.	.	.	.	.	.	.	.	.	.	.	.	.	.	.	.	.	.	1	4.0
Sum per relevé		100	100	99	100	100	98	100	100	100	100	100	100	99	99	99	100	100	100	100	99	99	100	100	100	100		

ferential species of type 1 was to occur with at least 50 percent constancy in one or two of the community types but with not more than 10 percent outside. A differential species of type 2 was to occur with at least 50 percent constancy within, but with not more than 20 percent outside the designated community types. The species are sorted first by total constancy. Then, the computer calculates the constancy and mean biomass values of each species for each class or community type.

The differential species are then selected by the computer according to their class-constancy criteria; first for each community type and then for any combination of two community types. The differential species are then printed out by the computer (TABLE 10.10). TABLE 10.10 shows that the computer selected 21 differential species, of which twelve are of the stronger differential type 1, and nine are of the weaker differential type 2. The numbers at the right on TABLE 10.10 are the species code numbers. Community types 1, 2, and 3 correspond to community types *A*, *B*, and *C* on TABLE 9.7.

The computer then prints out a summary table (TABLE 10.11) which shows first the 21 differential species with their constancy in percent and their mean biomass value for each community type. The differential species are followed by all other species arranged in order of high to low constancy. The summary table is followed by a differential table (TABLE 10.12). In this table the species order is the same as in the summary table (TABLE 10.11). The relevé order is the same as decided at the beginning of the second program. But the table clearly shows the amplitude of each species, which cannot be visualized from the summary table. This is particularly important with regard to the 21 differentiating species listed first. The blocks have been drawn subsequently. Solid lines identify the stronger differential species of type 1, dashed lines identify the weaker differential species of type 2. Shown at the right on the table are the total constancy values in absolute terms and in percent. The computer-printed differential table is a close approximation of the hand-manipulated differential table (TABLE 9.7).

The same method employing the two programs in succession was used by SPATZ to ordinate, classify, and differentiate 171 alpine pasture relevés in South Germany. Computerization becomes useful when large numbers of relevés have to be processed. Moreover, errors in transcribing tables are avoided.

Even though SPATZ's procedure is an attempt to simulate the original table technique as described in CHAPTER 9, there is one principal difference. The relevé order in TABLE 9.7 was established from use of a partial table (TABLE 9.4). Here, the species with high total constancy of more than 60 percent and with low total constancy of less than 10 percent were eliminated. Therefore, the relevé order in TABLE 9.7 is based on partial floristic similarity, not on total floristic similarity as in the

TABLE 10.10   Computer-Selected Differential Species for Community Types 1 and 3 and for the Community Type Combinations 1+2, 1+3, and 2+3. From SPATZ 1970. Community Types 1, 2, and 3 Correspond to A, B, and C in TABLE 9.7.

Community type 1

Differential species type 1

1.Species	=	Bromus erectus	4
2.Species	=	Salvia pratensis	58
3.Species	=	Viola hirta	60
4.Species	=	Koeleria pyramidaia	8
5.Species	=	Thymus serpyllum	46
6.Species	=	Scabiosa columbaria	42
7.Species	=	Linum catharticum	43

Community type 1

Differential species type 2

8.Species	=	Campanula glomerata	50
9.Species	=	Vicia sepium	31

Community type 3

Differential species type 1

10.Species	=	Carex acutiformis	24

Community type 3

Differential species type 2

11.Species	=	Deschampsia caespitosa	14
12.Species	=	Angelica silvestris	72

Community type combination 1 + 2

Differential species type 2

13.Species	=	Helictotrichon pubescens	3
14.Species	=	Trisetum flavescens	11

Community type combination 1 + 3

Differential species type 2

15.Species	=	Campanula rotundifolia	37

Community type combination 2 + 3

Differential species type 1

16.Species	=	Geum rivale	70
17.Species	=	Cirsium oleraceum	65
18.Species	=	Melandrium diurnum	71
19.Species	=	Alopecurus pratensis	12

Community type combination 2 + 3

Differential species type 2

20.Species	=	Ajuga reptans	57
21.Species	=	Lathyrus pratensis	33

computer tables. The use of partial floristic similarity established through species of intermediate total constancy may lead to a still more satisfactory relevé ordination and classification. The order and grouping is then not influenced by sporadic or accidental species, nor is it influenced by the omnipresent ones, which have no value for differentiating finer, more homogeneous units.

Computer programs that group relevés on the basis of species with intermediate constancy were described by LIETH and MOORE (1971) and by CESKA and ROEMER (1971). The initiating step in these programs is the isolation of species that occur together in certain relevés and that are mutually absent in others. This step is followed by a moving-together of those relevés that contain the mutually present species. The first step, the isolation of differential species, is again what WILLIAMS and LAMBERT (1961) called "normal" analysis, and the second step is their so-called "inverse" analysis. Alternatively, the first step is referred to now usually in computer analyses as R-technique and the second step as Q-technique (VAN DER MAAREL 1969).

A successive alternation between the two types of analyses was accomplished by the "Travelling Salesman" program (or "Lin's algorithm") that was used by LIETH and MOORE (1971). These authors based their procedure on an earlier program by MOORE, BENNINGHOFF, and DWYER (1967). The final printout is a crude approximation of a differential table, wherein computer-selected species define by their amplitude the extremes of a floristic gardient. Classes have to be established subsequently by the investigator.

CESKA and ROEMER (1971) describe their program as employing two rules as an algorithm. Rule I requires the computer to scan the data set for the species describing a particular group of relevés (i.e., an R-technique). Rule II requires the computer to establish relevé groups through the use of species (i.e., a Q-technique). The procedure works in form of a successive alternation of rules I and II until all diagnostic or differential species are extracted. The authors applied their program to the 1956 data set of ELLENBERG and the final outcome is a fairly close approximation of the hand-derived, original differential table (TABLE 9.7).

10.7 SAMPLE STAND ORDINATION

Ordination has been defined by GOODALL (1954) as "an arrangement of units in a uni- or multidimensional order." The emphasis is on the arrangement of sample units (species or stands) by individual values rather than by group values. In contrast, an arrangement by group values or within a range of values would result in classification. Ordi-

TABLE 10.11   Computer-Printed Summary Table Showing Percent Constancy and Mean Plant Biomass Values for Each Species by Community Type.

Species, Species No.	Group No.	1 Constancy (%)	Mean yield (%)	2 Constancy (%)	Mean yield (%)	3 Constancy (%)	Mean yield (%)
Bromus erectus	4	100.0	039.1	•	•	•	•
Salvia pratensis	58	71.4	003.2	•	•	•	•
Viola hirta	60	71.4	001.1	9.1	000.2	•	•
Koeleria pyramidata	8	57.1	002.8	•	•	•	•
Thymus serpyllum	46	57.1	000.9	9.1	000.2	•	•
Scabiosa columbaria	42	57.1	000.4	9.1	002.0	•	•
Linum catharticum	43	57.1	000.2	9.1	000.2	•	•
Campanula glomerata	50	71.4	000.7	18.2	000.6	•	•
Vicia sepium	31	57.1	000.2	18.2	000.6	16.7	000.2
Carex acutiformis	24	•	•	9.1	001.0	50.0	005.3
Deschampsia caespitosa	14	•	•	18.2	007.5	66.7	010.5
Angelica silvestris	72	•	•	18.2	000.6	66.7	001.1
Helictotrichon pubescens	3	57.1	001.6	100.0	007.7	16.7	003.0
Trisetum flavescens	11	57.1	003.5	81.8	007.6	16.7	004.0
Campanula rotundifolia	37	100.0	000.7	18.2	000.6	66.7	000.6
Geum rivale	70	•	•	81.8	000.6	66.7	002.3
Cirsium oleraceum	65	•	•	63.6	002.4	100.0	010.8
Melandrium diurnum	71	•	•	63.6	000.6	50.0	002.0
Alopecurus pratensis	12	•	•	54.5	006.8	50.0	005.7
Ajuga reptans	57	14.3	001.0	90.9	000.4	83.3	001.1
Lathyrus pratensis	33	14.3	000.2	54.5	000.5	50.0	000.2
Arrhenatherum elatius	1	100.0	008.9	100.0	014.1	100.0	015.0
Dactylis glomerata	2	100.0	006.7	100.0	014.1	100.0	010.3
Poa pratensis	6	100.0	010.3	100.0	013.5	100.0	007.3
Plantago lanceolata	38	100.0	000.9	100.0	001.5	100.0	003.1
Galium mollugo	40	100.0	004.4	100.0	006.2	100.0	007.2
Festuca pratensis	10	71.4	003.6	100.0	010.6	100.0	004.5
Chrysanthemum leucanthemum	41	100.0	002.6	90.9	001.5	83.3	001.2
Ranunculus acer	45	85.7	000.7	90.9	000.9	100.0	000.8
Veronica chamaedrys	51	100.0	000.3	81.8	000.6	83.3	000.9
Achillea millefolium	35	100.0	003.6	90.9	003.5	66.7	004.4
Daucus carota	36	85.7	000.7	100.0	000.7	50.0	002.7
Rumex acetosa	44	42.9	000.2	81.8	000.8	100.0	001.5
Trifolium pratense	28	71.4	000.7	81.8	000.8	50.0	001.1
Taraxacum officinale	49	85.7	000.2	63.6	001.1	83.3	001.7
Medicago lupulina	30	85.7	000.3	72.7	000.5	50.0	000.5
Festuca rubra	9	85.7	005.3	45.5	001.4	50.0	000.8
Crepis biennis	55	28.6	000.2	81.8	000.8	66.7	003.6
Trifolium repens	29	42.9	000.5	63.6	000.5	66.7	000.4
Heracleum sphondylium	39	42.9	000.5	72.7	004.0	50.0	001.7
Centaurea jacea	48	57.1	002.1	63.6	001.5	50.0	002.4
Plantago media	52	71.4	000.2	54.5	000.3	50.0	000.5
Holcus lanatus	13	•	•	45.5	001.4	100.0	003.7
Cerastium caespitosum	47	57.1	000.2	36.4	000.2	50.0	000.2
Prunella vulgaris	74	42.9	000.2	54.5	000.3	50.0	001.1
Bellis perennis	61	42.9	000.5	63.6	000.3	16.7	000.2
Lysimachia nummularia	73	•	•	45.5	000.2	50.0	000.5

TABLE 10.11 (Continued)   The Differential Species Are Followed by the "Other" Species. From SPATZ 1970. This TABLE Is Analogous to TABLE 9.9.

Species, Species No.	Group No.	1 Constancy (%)	Mean yield (%)	2 Constancy (%)	Mean yield (%)	3 Constancy (%)	Mean yield (%)
Tragopogon pratensis	66	14.3	000.2	45.5	001.0	16.7	000.2
Lychnis flos-cuculi	77	.	.	27.3	000.5	50.0	000.2
Carex flacca	23	42.9	002.0	18.2	002.5	16.7	002.0
Glechoma hederacea	67	.	.	27.3	000.2	50.0	000.2
Senecio jacobaea	78	28.6	000.2	27.3	000.2	16.7	000.2
Briza media	7	42.9	001.3	18.2	001.5	.	.
Vicia cracca	34	42.9	000.2	9.1	000.2	16.7	001.0
Anthriscus silvestris	68	14.3	000.2	18.2	001.1	33.3	000.2
Festuca ovina	5	42.9	001.7	9.1	001.0	.	.
Lotus corniculatus	32	28.6	000.2	.	.	33.3	000.2
Leontodon hispidus	54	14.3	000.2	18.2	001.0	16.7	004.0
Dianthus superbus	62	28.6	000.6	9.1	000.2	.	.
Galium boreale	64	28.6	000.6	.	.	33.3	001.0
Filipendula ulmaria	69	.	.	18.2	001.6	33.3	001.1
Pimpinella magna	75	.	.	27.3	000.7	16.7	000.2
Polygonum bistorta	76	.	.	9.1	000.2	33.3	002.5
Silene inflata	53	14.3	000.2	18.2	001.1	.	.
Knautia arvensis	59	28.6	000.6	9.1	000.2	.	.
Potentilla reptans	79	14.3	001.0	9.1	001.0	16.7	000.2
Pastinaca sativa	83	.	.	18.2	002.0	16.7	000.2
Silaus pratensis	87	.	.	27.3	000.5	.	.
Carex hirta	25	.	.	9.1	002.0	16.7	006.0
Carex gracilis	27	.	.	.	.	16.7	001.0
Myosotis arvensis	56	.	.	9.1	000.2	16.7	000.2
Pimpinella saxifraga	63	28.6	000.2	.	.	.	.
Cardamine pratensis	80	.	.	9.1	000.2	16.7	000.2
Myosotis palustris	81	.	.	18.2	000.2	.	.
Galium verum	86	28.6	000.6	.	.	.	.
Euphrasia odontites	89	.	.	9.1	000.2	16.7	000.2
Rumex crispus	91	.	.	.	.	16.7	000.2
Poa trivialis	15	.	.	.	.	16.7	002.0
Phleum pratense	16	.	.	9.1	000.2	.	.
Festuca arundinacea	17	.	.	.	.	16.7	001.0
Lolium perenne	18	.	.	9.1	002.0	.	.
Glyceria fluitans	19	.	.	.	.	.	.
Phalaris arundinacea	20	.	.	.	.	.	.
Phragmites communis	21	.	.	9.1	000.2	.	.
Brachypodium pinnatum	22	14.3	005.0	.	.	.	.
Carex panicea	26	.	.	9.1	002.0	.	.
Geranium pratense	82	.	.	9.1	000.2	.	.
Galium uliginosum	84	.	.	.	.	16.7	000.2
Sanguisorba officinalis	85	.	.	.	.	16.7	000.2
Ranunculus repens	88	.	.	9.1	000.2	.	.
Lamium album	90	.	.	.	.	.	.
Polygonum convolvulus	92	.	.	.	.	.	.
Chenopodium album	93	.	.	.	.	.	.
Alchemilla vulgaris	94	14.3	000.2	.	.	.	.

TABLE 10.12   Computer-Printed Differential Table. From SPATZ 1970. This Table is Analogous to TABLE 9.7.

Vegetation type					1									2											3			Constancy	
Running No.																													Absol.	Percent
Relevé No.	10	9	3	24	15	1	4	2	20	22	13	5	18	16	17	21	12	6	23	19	25	8	11	14	7			
Species, Species No.																														
Bromus erectus 4	21	47	35	10	37	50	74	2																					7	28,0
Salvia pratensis 58	5	2	2	4	1																								5	20,0
Viola hirta 60	3	2	+		2																								6	24,0
Koeleria pyramidata 8	3	3	+		2																								4	16,0
Thymus serpyllum 46	+		+		1	1	2																						5	20,0
Scabiosa columbaria 42	+	+	+		2																								5	20,0
Linum catharticum 43			+	+	1	1																							5	20,0
Campanula glomerata 50	+		1	1	1	+																							7	28,0
Vicia sepium 31	+		+	1	1	+																							7	28,0
Carex acutiformis 24																2				4	10	2							5	20,0
Deschampsia caespitosa 14													10		5	1			28	1	2								7	28,0
Angelica silvestris 72						+	1	1			1	1	1			+	1	1	1	2	+								6	24,0
Helictotrichon pubescens 3	+	4			1		1	1	2	20	4		8	13	4	5			1	1	3								16	64,0
Trisetum flavescens 11	3	5	2	4	1	2	8	16	8		6	8		10	5	8		4					5						15	60,0
Campanula rotundifolia 37	1	+	+		1	1			+	+		+	1	+	1	1	1	1	1	1		+	1						13	52,0
Geum rivale 70												+	2	1	+	1		12	1	1	3	5	3	18	1				14	56,0
Cirsium oleraceum 65						+	+	+	+	+	+	+	1		2	2		12	1	3	20	3	3	18	1				14	56,0
Melandrium diurnum 71							+	+	+	10	4	+	2			+	1	2	+	4		6	3	1	1				11	44,0
Alopecurus pratensis 12							+	+	10	+	2	2	8		15	15	1	8	+	1	1	10	6	1	1				9	36,0
Ajuga reptans 57	1	+	+	+	+	+	+	+	+	+	+	+	1		1	1	1	1	+	3	3	3	1	1	1				16	64,0
Lathyrus pratensis 33		+				8		+	1	+	+	+	+	+	+	+	+	+	+	+	+	1	+	+	+				10	40,0
Arrhenatherum elatius 1	8	5	2		15	6	2	+	24	25	10	22	15	15	15	25	32	12	8	26	35	4	9	12	4	22			25	100,0
Dactylis glomerata 2	6	2	15	8	8	2	15	5	5	8	15	12	18	15	15	18	12	8	8	18	18	10	12	10	4	1			25	100,0
Poa pratensis 6	8	10	10	20	15	4	5	74	6	9	2	4	2	10	5	5	1	25	2	16	10	4	6	5	3	1			25	100,0
Plantago lanceolata 38	1	1	1	3	1	1	1	1	1	2	2	4	1	2	1	1	1	1	1	8	4	4	6	2	4	+			25	100,0
Galium mollugo 40	12	2	7	3	3	3	3	+	5	3	3	2	6	5	3	3	6	24	12	2	5	2	5	14	6	2			25	100,0
Festuca pratensis 10	2	5	2	6	1	2	5	5	2	15	2	20	12	5	28	15	1	3	1	3	10	8	10	2	2	10			23	92,0
Chrysanthemum leucanthemum 41	5	2	3	1	1	+	1	3	1	2	1	1	5	1	1	1	6	4	1	1	1	1	1	2	2	+			23	92,0
Ranunculus acer 45	3	3	+		+	+	1	2	2	2	1	1	1	1	1	15	3	2	1	3		+	1	2	2	+			23	92,0

TABLE 10.12 (Continued)

Species	No.	Count	%
Veronica chamaedrys	51	22	88.0
Achillea millefolium	35	21	84.0
Daucus carota	36	20	80.0
Rumex acetosa	44	20	80.0
Trifolium pratense	28	18	72.0
Taraxacum officinale	49	18	72.0
Medicago lupulina	30	17	68.0
Festuca rubra	9	15	60.0
Crepis biennis	55	15	60.0
Trifolium repens	29	14	56.0
Heracleum sphondylium	39	14	56.0
Centaurea jacea	48	14	56.0
Plantago media	52	14	56.0
Holcus lanatus	13	12	48.0
Cerastium caespitosum	47	12	48.0
Prunella vulgaris	74	12	48.0
Bellis perennis	61	11	44.0
Lysimachia nummularia	73	8	32.0
Tragopogon pratensis	66	7	28.0
Lychnis flos-cuculi	77	7	28.0
Carex flacca	23	6	24.0
Glechoma hederacea	67	6	24.0
Senecio jacobaea	78	6	24.0
Briza media	7	5	20.0
Vicia cracca	34	5	20.0
Anthriscus silvestris	68	5	20.0
Festuca ovina	5	4	16.0
Lotus corniculatus	32	4	16.0
Leontodon hispidus	54	4	16.0
Dianthus superbus	62	4	16.0
Galium boreale	64	4	16.0
Filipendula ulmaria	69	4	16.0
Pimpinella magna	75	4	16.0
Polygonum bistorta	76	4	16.0
Silene inflata	53	3	12.0

TABLE 10.12 (Continued)

Vegetation type			1										2										3			Constancy	
Running No.	10	9	3	24	15	1	4	2	20	22	13	5	18	16	17	21	12	6	23	25	8	11	14	7	19		
Relevé No.	1	2	3	4	5	6	7	8	9	10	11	12	13	14	15	16	17	1R	19	20	21	22	23	24	25		
Species, Species No.																										Absol.	Percent
Knautia arvensis 59	.	1	.	.	.	.	.	.	.	.	.	.	.	.	.	.	.	.	.	.	.	.	.	.	.	3	12,0
Potentilla reptans 79	1	.	.	.	.	.	.	.	.	.	.	.	.	.	.	.	.	.	.	.	.	.	+	.	.	3	12,0
Pastinaca sativa 83	.	.	.	.	.	.	.	.	.	.	.	.	.	.	.	.	.	.	.	.	.	+	+	.	.	3	12,0
Silaus pratensis 87	.	.	.	.	.	+	.	.	+	.	1	.	.	.	.	.	.	.	.	.	.	.	.	.	.	3	12,0
Carex hirta 25	.	.	.	.	.	.	.	.	.	6	.	.	.	.	2	.	.	.	.	.	.	.	.	.	.	2	8,0
Carex gracilis 27	.	.	.	.	.	.	.	.	1	.	.	.	.	.	.	.	.	.	.	.	.	.	.	.	4	2	8,0
Myosotis arvensis 56	.	.	.	.	.	.	+	.	.	.	.	.	+	.	.	.	.	.	.	.	.	.	.	.	.	2	8,0
Pimpinella saxifraga 63	.	+	.	.	.	.	+	.	.	.	.	.	.	.	.	.	.	.	.	.	.	.	.	.	.	2	8,0
Cardamine pratensis 80	.	.	.	.	.	.	.	.	.	+	.	.	.	.	.	.	+	.	.	.	.	.	.	.	.	2	8,0
Myosotis palustris 81	.	.	.	.	.	.	.	.	.	.	.	+	.	.	.	.	.	.	.	.	.	.	.	.	.	2	8,0
Galium verum 86	.	.	.	.	1	.	.	.	.	.	.	.	+	.	.	.	.	.	.	.	.	.	.	.	.	2	8,0
Euphrasia odontites 89	.	.	.	.	.	.	.	.	.	.	.	.	+	.	.	.	.	.	+	.	.	.	.	.	.	2	8,0
Rumex crispus 91	.	.	.	.	.	.	.	.	+	.	.	.	.	.	.	.	.	.	.	+	.	.	.	.	.	2	8,0
Poa trivialis 15	.	.	.	.	.	.	.	.	.	.	.	.	.	.	.	.	.	2	.	.	.	.	.	2	.	2	8,0
Phleum pratense 16	.	.	.	.	.	.	.	.	.	.	.	.	.	.	.	.	.	.	.	.	.	.	.	.	.	1	4,0
Festuca arundinacea 17	.	.	.	.	.	.	.	.	.	.	.	.	.	1	.	.	.	.	.	.	.	.	.	.	.	1	4,0
Lolium perenne 18	.	.	.	.	.	.	.	.	.	.	.	.	2	.	.	.	.	.	.	.	.	.	.	.	.	1	4,0
Glyceria fluitans 19	.	.	.	.	.	.	.	.	.	.	.	.	.	.	.	.	.	.	.	.	20	.	.	.	.	1	4,0
Phalaris arundinacea 20	.	.	.	.	.	.	.	.	.	.	.	.	.	.	.	.	.	.	.	.	28	.	.	.	.	1	4,0
Phragmites communis 21	.	.	.	.	.	.	.	.	.	.	.	.	+	.	.	.	.	.	.	.	.	.	.	.	.	1	4,0
Brachypodium pinnatum 22	.	.	.	5	.	.	.	.	.	.	.	.	.	.	.	.	.	.	.	.	.	.	.	.	.	1	4,0
Carex panicea 26	.	.	.	.	.	.	.	.	.	.	.	.	.	2	.	.	.	.	.	.	.	.	.	.	.	1	4,0
Geranium pratense 82	.	.	.	.	.	.	.	.	.	.	.	.	.	.	.	.	+	.	.	.	.	.	.	.	.	1	4,0
Galium uliginosum 84	.	.	.	.	.	.	.	.	.	.	.	.	.	.	.	+	.	.	.	.	.	.	.	.	.	1	4,0
Sanguisorba officinalis 85	.	.	.	.	.	.	.	.	.	.	.	.	.	.	.	+	.	.	.	.	.	.	.	?	.	1	4,0
Ranunculus repens 88	.	.	.	.	+	.	.	.	.	.	.	.	.	.	.	.	.	.	.	.	.	.	.	.	.	1	4,0
Lamium album 90	.	.	.	.	.	.	.	.	.	.	.	.	.	.	.	.	.	.	.	.	.	.	.	+	.	1	4,0
Polygonum convolvulus 92	.	.	.	.	.	.	.	.	.	.	.	.	.	.	.	.	.	.	.	.	.	.	.	+	.	1	4,0
Chenopodium album 93	.	.	.	.	.	.	.	.	.	.	.	.	.	.	.	.	.	.	.	.	.	.	+	.	.	1	4,0
Alchemilla vulgaris 94	.	.	.	.	.	.	.	.	.	+	.	.	.	.	.	.	.	.	.	.	.	.	.	.	.	1	4,0

nation is sometimes considered an alternative to classification. This is based on the idea that grouping stands into classes may display a discreteness that does not really exist in nature. Instead, an ordination of sample stands will expose the relative continuity or discontinuity among them. In this respect, we may, therefore, say that ordination and classification are two quite separate but intimately linked processes, whereby ordination is the step prior to classification.

10.71 Uni-dimensional Ordination. A uni-dimensional ordination is simply a linear ranking of values from one extreme to another. For example, TABLE 9.7 shows at least two ordinations. One links the relevés together horizontally by the amplitudes of the differential species groups. Here, the sequence of relevés is one from dry to moist (from left to right). That this relevé sequence across the table follows a ranked order is further supported by the plotting of the floristic similarity indices shown in FIGURES 10.1 and 10.2.

In these figures the linear ordination is not perfect with regard to total floristic similarity among relevés as shown by the zigzag shape of the curves. It would be a simple matter to smooth the curves by repositioning some of the relevés so that they form a perfect linear ordination. However, in that case total floristic similarity would be given a higher emphasis than diagnostic floristic similarity. The latter is displayed by certain key plants that form the differential species groups. Therefore, a regrouping was not attempted at this point.

A second obvious linear ordination in TABLE 9.7 is displayed by the group of "other species" that are written down in a ranked order from high to low constancy. Further, in referring back to TABLE 9.3, we can see at once another simple, but important principle. In the ranked order of species constancy there are self-evident classes, namely three species have the same constancy of 15, six species a constancy of 14, and so on. Therefore, this ranking shows a step-wise change in contrast to a continuous change from high to low. The latter is often shown when stands are ranked by floristic similarity indices.

It was also a simple linear or uni-dimensional ordination that was employed by CURTIS and McINTOSH (1951) to demonstrate a continuum of sample stands for the forests in Wisconsin. CURTIS and McINTOSH's paper gave rise to the continuum school (DANSEREAU et al. 1968) and the subsequent development of their multidimensional ordination techniques (BRAY and CURTIS 1957) that are often believed to give a better abstract representation of vegetation as it really exists than any form of classification.

In their 1951 paper, CURTIS and McINTOSH did not apply a standard floristic similarity index for linear ranking of their sample stands, but they used instead a synthetic index, which they called the "continuum index." First, the stands were arranged in order of increasing abundance of one tree species, *Acer saccharum*. This species was considered to be indicative of climax stands in the Wisconsin forest. The abundance measure in this case was the "importance value," which represented the sum of three abundance parameters (see SECTION 7.67).

This stand arrangement was interpreted as a linear ordination from earlier to later successional stages. The other twenty-two tree species were plotted with their importance values on this stand arrangement. *Acer saccharum* was then given a rating of 10 as the best developed climax species. A rating from 1 to 9 was assigned to the other tree species according to the distance of their peak importance value from the *Acer saccharum* peak in the ordination graph. In this way, each tree species received a relative number from 1 to 10, called "climax adaptation number." A high climax adaptation number meant good adaptation to all of the environmental factors (particularly shade) present in terminal stands.

The climax adaptation number of each tree species was then multiplied by its importance value. This product was called "vegetational continuum index," because of the smooth curvilinear relationship that resulted from the graphical representation of these indices. The continuum indices (i.e., the products of importance value and climax adaptation number) were added for each tree species in each stand. The resultant weighted total was used as a basis of placing each stand in relation to any other stand sampled. This gave a second uni-dimensional ordination. The method utilized all of the tree species in a stand to express the position of that stand along a successional gradient. However, no shrubs and herbs were used in the continuum index. Therefore, the stand arrangement was based only on partial floristic similarity.

In this respect the ordination was similar to the relevé order in TABLE 9.7, which is also based only on partial floristic similarity of certain key species. However, in TABLE 9.7 the floristic similarity is one of common diagnostic species groups. In CURTIS and McINTOSH's (1951) study the floristic similarity was based merely on the tree species that were arranged in order of tree species change in an assumed chronosequence towards climax. In this arrangement, each stand showed a unique position. This was interpreted as a continuum.

In principle, the same can be said about the relevé order in TABLE 9.7. Each stand has a unique position in a linear ordination as attested by the plotting of relevés by floristic similarity indices in FIGURES 10.1 and 10.2.

The important question is only whether there are relatively greater and lesser differences among stands that make it possible to recognize clusters or groups. The solid curve in FIGURE 10.1 permits immediate recognition of at least two groups. The first seven relevés are rather similar to one another, differing by a rather large difference from the rest. The same was shown by the computer arrangement based on similarity indices as discussed in SECTION 10.62. Of course, one can interpret the same sequence as a continuum. This depends on viewpoint and common sense. As an aid, one can establish arbitrary guide lines.

10.72   Multidimensional Ordination. Subsequently, BRAY and CURTIS (1957) developed a multidimensional technique for the ordination of sample stands. This was a major breakthrough, because with this technique it was possible to demonstrate similarity relations geometrically.

Instead of an equidistant arrangement of stands along a linear axis as used in the uni-dimensional ordinations, stands are placed so that their positions approximate the degree of similarity directly. This has the advantage that clustering tendencies of stands can be shown geometrically. However, the recognition of groups or classes remains still a matter of personal judgment. In addition to cluster recognition, ordinations can serve to identify trends in vegetation variation, which may lead to explanations in terms of environmental gradients.

Two-dimensional ordinations of species have already been discussed (see SECTION 10.3). But, in these techniques, the spacing of species to one another was not perfected geometrically to the degree developed for the multidimensional ordination of stands. Since BRAY and CURTIS' 1957 paper, a number of modifications have been proposed, but the use of the original model still continues with little change.

The first step in multidimensional stand ordination is the calculation of a community coefficient or similarity index for each stand-pair of the sample series. This results in $n \times (n-1)/2$ similarity indices. For example, for a multidimensional ordination of the 25 relevés in TABLE 9.7, 300 similarity indices have to be calculated. The most widely applied index of similarity (IS) is the one by SØRENSEN, which is usually applied to the quantities of each species present in the stand (see SECTION 10.13). It is written as

$$IS = \frac{2w}{A+B} \times 100 \text{ (percent)}$$

where w is the sum of the smaller of the two quantitative values of the species that are common to two sample stands or relevés, A is the sum of all quantitative values in one of the stands, and B is the sum of all quantitative values in the other stand. The IS value can be expressed as a fraction or in percent.

If the quantitative field data are recorded in percent of plant biomass values, the sum of these values for all species in a stand comes to 100 percent (or approximately so). This applies to the relevés in TABLE 9.7 (see also SECTION 10.15). In this case the *IS* values reduce automatically to *w*. The same can be achieved by expressing any quantitative value as a fraction of the total for the stand. This modification was used by BRAY and CURTIS (1957) to save time and labor in hand computation. However, a change from absolute to relative quantitative values may also change the similarity relations among stands.

As an example of the Wisconsin multidimensional ordination technique, the 25 *Arrhenatherum* grass cover relevés (TABLE 9.7) are used. Three hundred *IS* = *w* values were calculated by an electronic computer from cards punched by G. SPATZ.*

The indices of similarity (*IS* values) were transcribed into a correlation matrix, TABLE 10.13. The 300 *IS* values are in the triangle on the lower left side of TABLE 10.13. In addition, 300 indices of dissimilarity (*ID* values) are shown in the upper right triangle of the correlation matrix. The *ID* values were obtained by subtracting the respective similarity indices from 100 (*ID* = 100 − *IS*).

The *ID* values are used for determining the spatial distance between each stand along the ordination axes. As mentioned before, this distance arrangement is the essential difference from a simple linear stand ordination as shown in FIGURES 10.1 and 10.2 and as used by CURTIS to demonstrate a vegetational continuum (CURTIS and McINTOSH 1951, CURTIS 1955). (See also WHITTAKER 1953, DAUBENMIRE 1966).

Theoretically, the similarity indices could vary from zero to 100 percent, i.e., from complete dissimilarity to perfect similarity. In practice, however, even the most similar replicate stands will hardly ever show 100 percent similarity. A more realistic index of dissimilarity for each stand-pair would result from subtracting their similarity indices from a replicate stand sample of the most typical or central stand in the group. The similarity of two replicate stands may be nearer to 75 percent than to 100 percent. The most typical or central stand among the 25 releves is, on a mathematical basis, relevé 18, because it has the highest

---

* A slight variation from the original data was introduced by translating the + ratings into 0.2 percent plant biomass values (instead of 0.1 percent as used in the example given in SECTION 10.14). Moreover, in the comparison of relevés 4 and 10 (running Nos. 1 and 2) TABLE 9.7, *Galium boreale* occurs only in relevé 4 whereas in SPATZ's data it appears also in releve 10. This results in a similarity between relevés 4 and 10 of 42.8 percent instead of 41.8 percent as shown in the former computation example (SECTION 10.14). However, this is a minor discrepancy that changes little in the overall result.

similarity to all remaining relevés. This is shown by the fact that the sum of *IS* values is the highest (with 1139) among all relevés recorded in TABLE 10.13. No effort was made during field analysis of the *Arrhenatherum* grassland samples to establish a close replicate stand. The most similar stand, relevé 22, was only 70.6 percent similar to relevé 18. The highest similarity among the *Arrhenatherum* stands was only 73.8 percent (see relevé-pair 9 and 15, TABLE 10.13). BEALS (1960) found an average replicate bird-community similarity of 85 percent. However, this value is difficult to establish objectively. The difference does not appear to be significant for ordination purposes, because the dissimilarity among stands as shown by geometric distance is only a relative distance. For this reason it has become common practice to use *ID* = 100 − *IS* as index of dissimilarity.

10.72.1 x Axis Construction.   For demonstrating the similarity relationships geometrically among the 25 relevés, the most dissimilar stands must first be determined. These become the terminal or reference stands of the first ordination or x axis. This first axis is in principle nothing else than any simple linear sequence or ordination of stands as may be appearing in the table heads of any differential table.

The most dissimilar stand-pair in TABLE 10.13 is the relevé-pair 1 and 19. These relevés are only 5.8 percent similar. They would be selected according to BRAY and CURTIS' 1957 criteria. However, since relevé 19 was crossed out on TABLE 9.7, we may discard it as too unrelated to the rest of the relevés to form the end points of an ordination axis. Instead, we may leave relevé 19 in the group to be ordinated with the rest. In this way relevé 19 does not influence the ordination of the other relevés. One reason for eliminating relevé 19, as discussed in CHAPTER 9, was its relatively low number of species. It has only 20 in comparison to the other relevés that showed between 25 and 35 species (TABLE 9.7). Another reason was the dominance (here high plant biomass) of two wetsite species (*Glyceria fluitans* and *Phalaris arundinacea*). The presence of the two species, which did not occur in any of the other relevés, indicated that this site, in contrast to all others, was subject to periodic flooding.

The problem associated with using as reference stands those that are too unrelated to the rest of the group led to a modification of the original reference stand selection criterion of BRAY and CURTIS (1957). BEALS (1960) selected as the first reference stand the one that shows the lowest sum of similarity indices. This would be relevé 2 in TABLE 10.13. The other terminal stand would then be the one most dissimilar to relevé 2, which is relevé 19 again. Since this was already eliminated as terminal stand, relevé 4 would become the other terminal stand since

TABLE 10.13   *Matrix of Indices of Dissimilarity (ID) and Similarity (IS) in Percent for the 25 Relevés Shown in TABLE 9.7. Further Explanation in Text.*

ID values

		1	2	3	4 (x, z ↓)	5	6	7	8	9	10	11	12 (z ↓)	13
	1	X	82.0	44.4	28.2	80.4	84.0	83.2	82.8	29.2	44.6	84.4	79.0	78.8
	2	18.0	X	72.4	84.6	77.2	81.8	81.6	79.6	72.0	76.8	79.8	57.6	81.6
	3	55.6	27.6	X	46.6	64.0	64.6	71.8	72.2	32.6	40.2	70.8	55.2	58.6
	4	71.8	15.4	53.4	X	83.8	87.6	86.4	84.8	34.4	57.2	83.8	79.6	85.2
	5	19.6	22.8	36.0	16.2	X	57.0	67.0	60.2	70.2	75.0	60.8	60.4	50.2
	6	16.0	18.2	35.4	12.4	43.0	X	50.2	43.4	77.6	64.2	57.4	61.8	50.4
	7	16.8	18.4	28.2	13.6	33.0	49.8	X	48.6	77.2	72.2	58.0	69.2	59.6
	8	17.2	20.4	27.8	15.2	39.8	56.6	51.4	X	77.2	74.6	48.8	62.6	61.2
	9	70.8	28.0	67.4	65.6	29.8	22.4	22.8	22.8	X	40.6	75.4	65.4	72.2
	10	55.4	23.2	59.8	42.8	25.0	35.8	27.8	25.4	59.4	X	69.2	59.0	60.6
	11	15.6	20.2	29.2	16.2	39.2	42.6	42.0	51.2	24.6	30.8	X	51.8	60.2
	12	21.0	42.4	44.8	20.4	39.6	38.2	30.8	37.4	34.6	41.0	48.2	X	56.6
	13	21.2	18.4	41.4	14.8	49.8	49.6	40.4	38.8	27.8	39.4	39.8	43.4	X
IS	14	24.6	18.2	37.2	19.2	30.8	56.2	51.4	52.8	22.2	47.8	41.2	41.2	46.0
values	15	63.6	33.0	70.4	58.2	32.8	22.2	21.0	27.2	73.8	59.8	33.6	48.0	34.2
	16	25.8	31.2	50.4	20.2	57.0	45.2	34.2	34.0	39.0	39.6	40.4	54.0	55.4
	17	20.2	23.4	38.8	18.6	56.8	36.8	38.4	38.8	30.8	29.0	59.2	50.8	48.0
	18	22.4	29.0	48.0	16.2	52.8	38.4	27.4	37.4	36.4	41.4	45.2	59.6	55.8
	19	5.8	13.0	15.2	6.6	25.0	14.6	18.6	22.0	18.2	15.0	30.8	26.6	24.2
	20	25.6	23.2	36.8	18.0	53.8	38.6	32.2	25.2	33.0	42.2	32.0	40.4	45.6
	21	10.4	18.0	29.6	11.4	47.6	37.6	27.4	46.4	21.2	21.2	53.2	45.8	45.2
	22	21.6	27.4	36.6	16.4	55.0	38.2	27.4	44.2	33.4	36.2	46.0	48.0	49.4
	23	19.4	31.4	39.8	17.0	39.2	38.4	34.4	39.8	29.4	42.4	32.8	48.8	48.8
	24	31.2	35.4	52.2	27.6	29.8	28.2	24.4	24.2	43.4	52.8	31.0	52.6	34.6
	25	13.0	21.8	33.4	14.2	29.2	33.8	26.6	32.0	23.4	28.2	42.4	47.0	36.6

	1	2	3	4	5	6	7	8	9	10	11	12	13

IS values

Σ 578.0   Σ 601.4

it is next dissimilar to relevé 2 with 15.4 percent similarity. SWAN and DIX (1966) and NEWSOME and DIX (1968) used a still more discriminatory criterion by requiring that all reference stands must have at least three stand comparisons with similarity indices greater than 50 percent. Their criterion was followed in the present example.

The stand with the lowest sum of similarity indices, but including at least three comparisons with similarity indices greater than 50 percent is relevé 4, TABLE 10.13. Note that relevé 4 is also the first stand in TABLE 9.7.

ID values

y↓							x↓			y↓			
14	15	16	17	18	19	20	21	22	23	24	25		
75.4	36.4	74.2	79.8	77.6	94.2	74.4	89.6	78.4	80.6	68.8	87.0	1	
81.8	67.0	68.8	76.6	71.0	87.0	76.8	82.0	72.6	68.6	64.6	78 2	2	
62.8	29.6	49.6	61.2	52.0	84.8	63.2	70.4	63.4	60.2	47.8	66.6	3	
80.8	41.8	79.8	81.4	83.8	93.4	82.0	88.6	83.6	83.0	72.4	85.8	4	←x,z
69.2	67.2	43.0	43.2	47.2	75.0	46.2	52.4	45.0	60.8	70.2	70.8	5	
43.8	77.8	54.8	63.2	61.6	85.4	61.4	62.4	61.8	61.6	71.8	66.2	6	
48.6	79.0	65.8	61.6	72.6	81.4	67.8	72.6	72.6	65.6	75.6	73.4	7	
47.2	72.8	66.0	61.2	62.6	78.0	74.8	53.6	55.8	60.2	75.8	68.0	8	
77.8	26.2	61.0	69.2	63.6	81.8	67.0	78.8	66.6	70.6	56.6	76.6	9	
52.2	41.2	61.4	71.0	58.6	85.0	57.8	78.8	63.8	57.6	47.2	71.8	10	ID
58.8	66.4	59.6	40.8	54.8	69.2	68.0	46.8	54.0	67.2	69.0	57.6	11	values
58.8	52.0	46.0	49.2	40.4	73.4	59.6	54.2	52.0	51.2	47.4	58.0	12	←z
54.0	66.8	44.6	52.0	44.2	75.8	54.4	54.8	50.6	51.2	65.4	63.4	13	
X	64.4	51.0	56.2	50.8	78.6	57.8	66.0	59.8	53.2	61.6	60.4	14	←y
35.6	X	49.6	55.2	51.0	69.4	57.6	63.8	51.6	49.6	38.8	62.4	15	
49.0	50.4	X	40.2	33.2	67.4	34.4	49.6	38.2	48.8	50.2	50.4	16	
43.8	44.8	59.8	X	38.8	61.8	52.6	35.6	43.0	57.4	57.4	53.6	17	
49.2	49.0	66.8	61.2	X	58.4	41.0	36.0	29.4	42.4	47.4	42.6	18	
21.4	30.6	32.6	38.2	41.6	X	65.2	53.8	56.4	66.2	68.8	64.2	19	
42.2	42.4	65.6	47.4	59.0	34.8	X	54.0	40.2	46.4	45.8	58.0	20	
34.0	36.2	50.4	64.4	64.0	46.2	46.0	X	28.0	53.0	59.0	43.0	21	←x
40.2	48.4	61.8	57.0	70.6	43.6	59.8	72.0	X	42.0	44.4	49.0	22	
46.8	50.4	51.2	42.6	57.6	33.8	53.6	47.0	58.0	X	37.6	47.8	23	
38.4	61.2	49.8	42.6	52.6	31.2	54.2	41.0	55.6	62.4	X	46.6	24	←y
39.6	37.6	49.6	46.4	57.4	35.8	42.0	57.0	51.0	52.2	53.4	X	25	
14	15	16	17	18	19	20	21	22	23	24	25		

IS values

Σ 1139.0

Relevé 4 therefore, becomes reference stand A, forming the zero point on the first ordination axis, the x axis. The stand most dissimilar to 4 (excepting 19) is 21. Since relevé 21 also conforms to the requirement of reasonable similarity with the rest of the stands by showing at least three comparisons in which the similarity indices exceed 50 percent, it becomes the second terminal stand, B, of the x axis.

The length (L) of the x axis is given by the dissimilarity between relevés 4 and 21 (ID = 88.6 = L). The remaining stands are then positioned along the x axis by their dissimilarity to the two terminal stands. However, if the two dissimilarity values that each stand shares with the terminal stands were plotted directly on the x axis, each stand would have two positions. The results would be two different linear ordinations similar to the two curves shown in FIGURE 10.1.

Plotting only one correct position for each stand is accomplished as follows. In FIGURE 10.7 P' represents any stand that is to be positioned between the reference stands A and B. An arc is plotted from A whose radius is the ID value that the stand in question (P') shares with the reference stand A (in the matrix TABLE 10.13). This ID value is equivalent to the distance (dA) of A to P'. A second arc, dB, is plotted around B. The intersection of the two arcs (dA and dB) locates P' above and below the x axis. The intersecting arcs below the axis are not shown. The x axis position of any stand is obtained by the vertical projection of P' to P.

The same can be accomplished by calculation using the Pythagorean theorem, which states that $dA^2 = e^2 + x^2$ and $dB^2 = e^2 + (L-x)^2$. By subtracting one equation from the other, e^2 is eliminated. The formula then becomes

$$x = \frac{L^2 + (dA)^2 - (dB)^2}{2L}$$

This formula was used by BEALS (1960). It is now most commonly applied for positioning of stands along ordination axes.

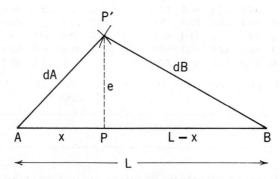

FIGURE 10.7. Triangle to show stand placing technique for ordinating stands geometrically along one axis. Legend: A=first terminal stand; B= second terminal stand; P'=any other stand in the group under comparison; dA=the ID value or distance (d) of stand P' to stand A; dB=the ID value or distance of stand P' to stand B; P=the position of any stand (P') on the x axis; e=the height of the intersecting arcs (dA and dB) above (or below) the x axis; AB=the x axis=L, the dissimilarity value shared by reference stand A and reference stand B; x=the distance of P from A along the x axis; L−x=the distance of P from B along the x axis.

As an example, we will show the calculation for positioning relevé 1 on the x axis which is formed by relevés 4 and 21 as terminal stands. For relevé 1 the *dA* value is 28.2 (see TABLE 10.13, *ID* value for relevés 1 and 4), the *dB* value is 89.6 (*ID* value for relevés 1 and 21), L = 88.6 (the *ID* value for the reference stands 4 and 21). By substitution of these values in BEALS' formula, the position of relevé 1 along the x axis becomes

$$x = \frac{7849.96 + 795.24 - 8028.16}{177.2} = 3.5$$

This calculation is carried through until all stands are positioned along the x axis. FIGURE 10.8 shows the 25 relevés at their proper position. TABLE 10.14 gives the ordination values. This x axis ordination does not yet fulfill the objective of indicating by stand position the similarity of the stands to one another. Many relevés that are close together on the x axis are not really very similar to one another. This is particularly true of the stands in the middle part of the axis. For example, relevés 6 and 20 occupy the same position on the x axis, but their similarity is only 38.6 percent (*IS* value, TABLE 10.13).

10.72.2 y Axis Construction.   The next step involves separating the stands in a second dimension for a better geometric approximation of their similarity relations. This requires construction of a y axis. First one has to select two terminal stands for the y axis.

It is logical that the first reference stand on the y axis should be a stand that is particularly ill fitted to the x axis. Such a poorly fitted stand would be one that shares especially high *ID* values with the

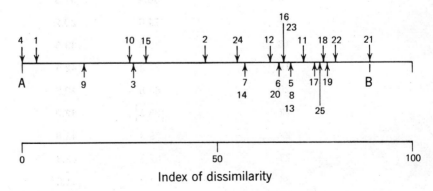

FIGURE 10.8 Ordination of the 25 Arrhenatherum relevés (TABLE 9.7) along the x axis with reference to stand 4 and 21 as A and B terminal stands, respectively.

TABLE 10.14 Ordination Values of 25 Arrhenatherum Relevés for x, y, and z Ordination Axes. Blocked Out Values Represent Those of the Terminal Relevés for Each Axis.

RELEVE NUMBER	x	y	z
1	3.5	38.5	74.0
2	46.7	51.2	15.7
3	28.6	44.3	45.3
4	$\boxed{0}$	41.2	$\boxed{79.6}$
5	68.4	29.7	18.6
6	65.6	4.5	15.6
7	56.7	3.6	7.1
8	68.7	2.2	19.2
9	16.0	53.9	59.2
10	27.7	34.8	41.1
11	71.6	20.2	12.5
12	63.5	40.6	$\boxed{0}$
13	68.3	19.6	14.3
14	56.6	$\boxed{0}$	20.5
15	31.2	52.2	45.8
16	66.4	31.5	13.1
17	74.5	29.7	13.4
18	76.6	33.5	5.9
19	77.2	42.5	18.8
20	65.8	40.7	19.9
21	$\boxed{88.6}$	37.9	8.9
22	79.3	43.8	12.9
23	67.3	42.3	13.0
24	54.2	$\boxed{61.6}$	21.0
25	75.4	42.8	14.7

terminal stands. The x axis fit of each stand is indicated by "e" on FIGURE 10.7. Therefore, the fit of any stand can be calculated from the formula $e^2 = (dA)^2 - x^2$.

As an example, the e^2 value for relevé 1 is $e^2 = 28.2^2 - 3.5^2 = 795.24 - 12.25 = 782.99$. The e^2 values for all 25 relevés are shown in TABLE 10.15. The stand with the highest e^2 value is relevé 2. But the y axis reference stand should also have at least three IS values over 50 percent, which relevé 2 does not have (see TABLE 10.13). The stand that fits this criterion is relevé 14, which has the fourth highest e^2 value. According to NEWSOME and DIX (1968) the stand should also lie within the mid-50 percent of the first ordination axis. This requirement is fulfilled by relevé 14, which therefore becomes the first reference stand (A') of the y axis. The second reference stand (B') should be within close proximity to the first along the x axis so that the y axis will be as nearly perpendicular to the x axis as possible. The arbitrary position limit is given as within 10 percent of the total x axis length. In this example, B' should therefore be within ±8.9 units from A'. The x value of A' (relevé 14) is 56.6 (TABLE 10.14). There are three stands within the range of 47.7 and 65.5, namely relevés 7, 12 and 24. Of these, relevé 24 shows the greatest difference from relevé 14. Thus, relevé 24 becomes B'. The stand also shows more than three similarity indices that exceed 50 percent, and therefore fulfills the requirement of sufficient similarity to all the stands under comparison.

The next step involves the computation of y values for each relevé. This is again done by BEALS' formula.

$$y = \frac{(L')^2 + (dA')^2 - (dB')^2}{2L'}$$

The y axis values are shown in TABLE 10.14.

Now, it is a simple matter to plot each relevé position by its x and y value into a graph (FIG. 10.9). The relevés are here identified as they were previously classified in TABLE 9.7. Three groups were recognized, $A = Arrhenatherum–Bromus$ $erectus$ relevés, $B = Arrhenatherum$ $typicum$ relevés, and $C = Arrhenatherum–Cirsium$ $oleraceum$ relevés. The three groups coincide also with dry, mesic, and moist soil moisture regimes, respectively. The two-axis ordination graph shows the seven relevés of group A well separated from groups B and C. Group C relevés are exclusively on the lower right of the graph, while the group B reléves are clustered above group C. Obviously, the B and C relevés form a vegetational continuum on this basis of total floristic comparison. Two relevés of group C (23 and 25) are even intermingled in group B. The distribution pattern also shows that the A relevés, though flor-

TABLE 10.15 The e^2 Values That Indicate the Fit of Each Sample Stand to the First Ordination Axis Formed by Relevés 4 and 21; $e^2 = (dA)^2 - x^2$.

RELEVÉ NUMBER	e^2 VALUE	RELEVÉ NUMBER	e^2 VALUE
1	782.99	14	3325.08
2	4976.27	15	773.80
3	1353.60	16	1959.08
4	0	17	1075.71
5	2343.88	18	1154.88
6	3370.40	19	2763.72
7	4250.07	20	2394.36
8	2471.35	21	0
9	927.36	22	700.47
10	2504.55	23	2359.71
11	1895.88	24	2304.12
12	2303.91	25	1676.48
13	2594.15		

istically clearly distinct from the other two groups, are in general more dissimilar among each other than the six plots of group B, which are not very far apart. Relevé 19, which was eliminated in TABLE 9.7 for its floristic abnormality, lies here in the B group cluster. This is probably due to the fact that the similarity index ($IS = w$) sorts the relevés only on the basis of their common species, while the unique ones are ignored.

10.72.3   z Axis Construction.   It may be of interest to see whether the B and C group relevés will further separate by aligning the relevés in a third dimension. This can be done by constructing a z axis.

The procedure is almost the same as in the construction of the y axis. However, selection of the first reference stand (A'') of the z axis involves scanning both the x and y axes for stands that lie in the mid-50 percent of their ranges (NEWSOME and DIX 1968).

The length of the x axis is 88.6. Its mid-50 percent range is 44.3 ± 22.2. Stands with x ordination values between 22.1 and 66.5 include relevés 2, 3, 6, 7, 10, 12, 14, 15, 16, 20, and 24 (TABLE 10.14). But A'' should also lie in the mid-50 percent of the y axis. The y axis length is 61.6. The mid-50 percent range, therefore, is 30.8 ± 15.4. There are five relevés that have y values between 15.4 and 46.2 and at the same time lie in the mid-50 percent range of the x axis. These are relevés 3, 10, 12, 16, and 20. One of these will be A''.

The first reference stand (A'') should be the relevé most poorly fitted to both axes. The poorest fit is indicated by the highest sum of $e_x^2 + e_y^2$. A third requirement imposed is that the stand is sufficiently related to the rest of the relevés. Therefore, it should also show three IS values greater than 50 percent.

Relevés selected for A'' candidacy on the z axis:

RELEVÉ NUMBER	e_x^2	e_y^2	$(e_x^2 + e_y^2)$
3	1353.60	1981.35	3334.95
10	2504.55	1513.8	4018.35
12	2303.91	1809.08	4112.99
16	1959.08	1608.75	3567.83
20	2394.36	1684.35	4078.71

Where

$$e_x^2 = (dA)^2 - x^2$$
$$e_y^2 = (dA')^2 - y^2$$

Example releve 3 $x^2 = 28.9^2$ $(dA')^2 = 46.6^2$
$y^2 = 44.3^2$ $(dA')^2 = 62.8^2$

Relevé 12 becomes A'', because it has the highest sum $(e_x^2 + e_y^2)$ value, and also shows at least three IS values greater than 50 percent. In fact, it shows four. The stand most dissimilar to relevé 12 is relevé 4. Its similarity to relevé 4 is only 20.4 percent (TABLE 10.13). Relevé 4 also satisfies the overall group similarity requirement of having at least three IS values in excess of 50 percent. Therefore, the two z axis terminal stands are relevé $12 = A''$ and relevé $4 = B''$.

The ordination values of the z axis are calculated as before with BEALS' formula as

$$z = \frac{(L'')^2 + (dA'')^2 - (dB'')^2}{2L''}$$

The z axis values are recorded in TABLE 10.14. FIGURE 10.10 shows the z ordination values plotted over the x ordination values. A three-dimensional ordination can be obtained by combining the two graphs, the y/x ordination (FIG. 10.9) and the z/x ordination (FIG. 10.10). The points that are plotted in a two-dimensional plane on the y/x ordination can be raised vertically to a height given by the z axis. However,

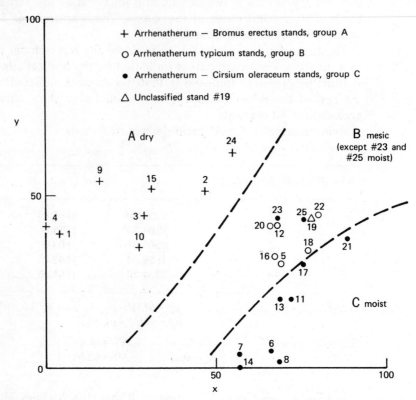

FIGURE 10.9.   *Ordination of y/x values of the 25 relevés shown on TABLE 9.7.*

a complicated three-dimensional drawing would not add to a pictorial clarification in this case. The z/x ordination shows that the *A* group relevés are lifted to the highest levels, except for relevé 2. The separation between the *A* and *B* group relevés is maintained. However, the z/x ordination brings out a more continuous connection of the *A* and *C* group relevés. Moreover, the *B* group relevés do not sort out in an intermediate position as they did in the y/x ordination. Instead, the *B* and *C* group relevés form one cluster.

This can be explained by the fact that relevé 12, which became the first reference stand of the z axis, is truly a central or *Arrhenatherum typicum* stand. The z axis ordination therefore, is not an ordination from one floristic extreme to another, but an ordination from an average or typical stand to an extreme stand. Relevé 4, which became the second reference stand on the z axis, served also as the first reference stand on the x axis. We may, therefore, conclude that the y/x ordina-

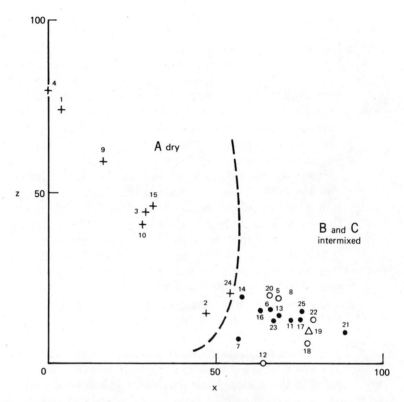

FIGURE 10.10. Ordination of z/x values of the 25 Arrhenatherum grass cover stands shown on TABLE 9.7. Symbols as on FIGURE 10.9. A three-dimensional ordination can be constructed by combining this graph with FIGURE 10.9. The stand positions along the x axis are the same in both.

tion shows quite sufficiently the important trends of total floristic variation among the stands. However, a certain distortion is to be expected in any scatter diagram that attempts to portray in two dimensions similarity relationships which are in reality multidimensional.

10.72.4 Modification of Procedure. Of greater immediate interest than the portrayal of the relevé relationships in a third dimension is the question, whether other reference stands may not result in a different two-dimensional ordination. The selection of reference stands is a very critical aspect in ordination and several papers have been published on this problem (ORLOCI 1966, AUSTIN and ORLOCI 1966, SWAN, DIX and WEHRHAN 1969, VAN DER MAAREL 1969). The essential objective is to find the axes that define the stand relationships by their maximal vegetational variation. This is not necessarily identical to defining the axes by extreme stands, since extreme stands may be quite unrelated to the major variation among the rest of the stands. Such an unrelated extreme stand is, for example, relevé 19.

The mathematical procedures for finding the axes that best define the maximal vegetational variations in a series of stands are rather complicated. An explanation of these improved ordination techniques, known as factor analysis (DAGNELIE 1960), principal components analysis, and position vectors technique (ORLOCI 1966), would go beyond the scope of this book.

However, we will apply to our data a simple alternative in reference stand selection. We may assume that the axes showing maximal variation in our stand series are defined by the floristically extreme stands. But we will exclude as reference stand releve 19, because this stand is indeed too unrelated to the rest for reasons given earlier.

Therefore, instead of using the criterion of SWAN and DIX (1966) that all stands show at least three similarity indices of 50 percent or greater, we will apply the greatest-difference-criterion as used by BRAY and CURTIS (1957). The stands with the greatest ID value in TABLE 10.13 (excluding relevé 19) are relevés 1 and 21. We may use relevé 1 as zero point of the x axis and relevé 21 as the second terminal point. The x axis positions of each remaining stand are calculated as before. The stand with the highest e^2 value is relevé 2, which becomes the zero point of the y axis. Most dissimilar to relevé 2, and within 10 percent of the x axis length of relevé 2, is relevé 14. Thus, relevé 14 becomes the second terminal stand of the y axis.

FIGURE 10.11 shows this new ordination which still has relevé 21 as second x axis reference stand in common with the first ordination (FIG. 10.9).

It is interesting that the change in reference stands has resulted in a 90° rotation of the axes in comparison to FIGURE 10.9. The overall constellation of relevé positions has not changed very much. However, one important difference is noteworthy. Relevé 25 is no longer intermingled in group B, but now is clearly associated with the relevés of group C. In that respect, this second ordination can be considered an improvement over the first.

DAGNELIE (1960) subjected the same 25 relevés to factor analysis and came up with a very similar relevé constellation (FIG. 10.12). The mathematically much more involved and sophisticated technique of factor analysis will not be discussed here. It requires about four times as much computation work as the simple ordination method discussed here. DAGNELIE's ordination shows a somewhat clearer clustering of relevé groups A, B, and C. Relevé 25 is well in the center of relevé

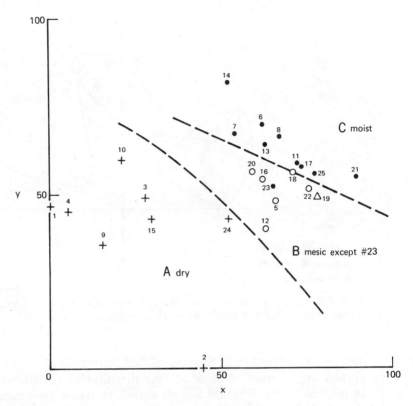

FIGURE 10.11.   Ordination of y/x values of the 25 Arrhenatherum relevés based on different reference stands. Symbols as on FIGURE 10.9. Explanation in text.

group C on DAGNELIE's graph, thereby showing its similarity to this group even more so than on FIGURE 10.11. Relevés 13 and 17 indicate a somewhat intermediate relationship to group B and C on DAG-NELIE's graph. This is also true for the simple ordination on FIGURE 10.11. However, relevé 11 is the most extreme C stand in DAGNELIE's ordination. In FIGURE 10.11 this relevé has a position close to relevé 18, which can be considered a typical stand of group B, because it shows the greatest similarity to all other stands (TABLE 10.13).

Relevé 23, which is classified in the differential table (TABLE 9.7) as a C stand, (because it contains two species of the *Cirsium oleraceum* group), also fell in DAGNELIE's ordination outside the C stand cluster. On the basis of total floristics it has a closer similarity with group B. As seen on TABLE 9.7, relevé 23 lacks several differential species that are found in the other C stands.

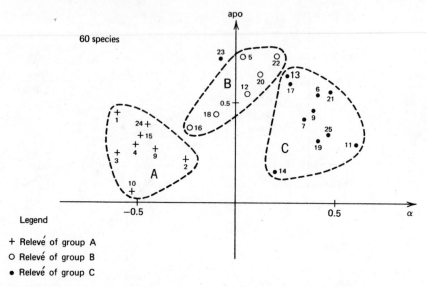

FIGURE 10.12. Ordination of the 25 Arrhenatherum relevés by Q-type factor analysis. (From DAGNELIE 1960, somewhat modified.)

10.72.5 Distribution of Species in Ordination Graph. The ordination graphs can be used to illustrate the distribution of individual species in relation to the different sample stands in which they occur. In addition, their quantitative representation can be indicated by dividing the quantitative parameters into a limited number of classes that can be identified by symbols on the graphs.

As an example, we may illustrate in the first y/x ordination (FIG. 10.9) the quantitative distribution of the four species, *Arrhenatherum elatius, Bromus erectus, Geum rivale,* and *Cirsium oleraceum* (FIG. 10.13). For this purpose, the plant biomass percentages (shown in TABLES 9.2–9.7 are divided into five classes as shown in FIGURE 10.13. The pattern of species distribution shows certain tendencies quite clearly. *Arrhenatherum elatius* is present in all stands, but it develops its greatest biomass on the mesic soils of the grass cover vegetation. *Bromus erectus* is clearly restricted to dry soils, where it assumes considerable dominance. It is either present in large quantities or not at all. Both *Geum rivale* and *Cirsium oleraceum* are present only in the mesic and moist section of the grass cover habitats. *Cirsium* grows in greater quantities (attaining up to one quarter of the plant biomass of a stand) only in moist habitats. Yet by presence, the two species differ little. They occupy much the same habitat segment.

This information, of course, can also be read directly from the differential table (TABLE 9.7). The plotting of species with their quantitative values on an ordination graph is, therefore, analogous to showing

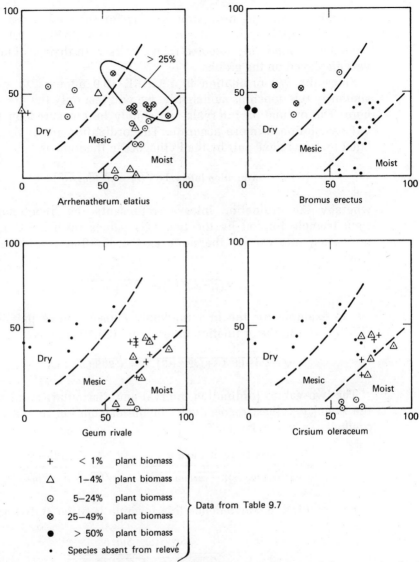

FIGURE 10.13. *Showing the quantitative distribution of four species in the 25 Arrhenatherum releves as ordinated in FIGURE 10.9.*

the species distribution in a differential table. The latter is more comprehensive, since all species can be compared at once and in greater detail.

10.72.6 Statistical Test of an Ordination. It must be understood that the ordination graphs are only a geometric approximation for portraying the dissimilarity values (of TABLE 10.13). For this reason it becomes necessary to test the ordination graphs for their efficiency. This was done by comparing dissimilarity values (from TABLE 10.13) between a number of randomly selected relevé pairs with their ordination intervals displayed on the graphs.

Since the y/x ordination in FIGURE 10.9 appears to explain the similarity relationships sufficiently, we will test only the y/x ordination here. The ordination intervals can simply be measured on the graph, but calculation is more accurate. The ordination intervals are calculated for any relevé pair by the Pythagorean theorem as

$$\text{ordination interval} = \sqrt{(x_1 - x_2)^2 + (y_1 - y_2)^2}$$

whereby the ordination interval represents the hypotenuse of the right triangle formed by the two axis values (in TABLE 10.14). The ordination interval for the three axis ordination would be obtained from

$$\sqrt{(x_1 - x_2)^2 + (y_1 - y_2)^2 + (z_1 - z_2)^2}$$

As an example for the first randomly chosen relevé pair 16 and 25 (TABLE 10.13), the ordination interval in FIGURE 10.9 is

$$\sqrt{(75.4 - 66.4)^2 + (42.8 - 31.5)^2} = \sqrt{208.69} = 14.4$$

The two values (ordination interval and dissimilarity index) become the x and y values, respectively, in the calculation of the correlation coefficient r, which is

$$r = \frac{\text{covariance of } x \text{ and } y}{\sqrt{[\text{variance } (x)] \cdot [\text{variance } (y)]}} = \frac{(x - x)\,(y - y)}{\sqrt{[\Sigma(x - x)^2 \cdot \Sigma(y - y)^2]}}$$

For calculating convenience this formula is often stated as

$$r = \frac{Sxy - (Sx \cdot Sy)/n}{\sqrt{(Sx^2 - (Sx)^2/n)\,(Sy^2 - (Sy)^2/n)}}$$

A table is first prepared for all expressions in the above formula (TABLE 10.16). Ten comparisons of ordination intervals and dissimilarity indices were selected randomly by using a random table from a statistical handbook (QUENOUILLE 1953).

For $n - 2 = 8$ degrees of freedom we find in a t table (QUENOUILLE 1950) the t value of 3.83 at the 0.5 percent level and $t = 5.04$ at the 0.1 percent level of significance. Thus, our t value of 4.53 shows that the

RELEVÉ PAIR	ORDINATION INTERVAL x	ID VALUE y	xy	x^2	y^2
16 : 25	14.4	50.4	725.76	207.36	2,540.16
21 : 2	44.0	82.0	3,608.00	1,936.00	6,724.00
22 :17	14.9	43.0	640.70	222.01	1,849.00
23 : 6	37.8	61.6	2,328.48	1,428.84	3,794.56
18 : 4	77.0	83.8	6,452.60	5,929.00	7,022.44
24 : 8	61.1	75.8	4,631.38	3,733.21	5,745.64
13 : 3	46.8	58.6	2,742.48	2,190.24	3,433.96
12 : 1	60.0	79.0	4,740.00	3,600.00	6,241.00
14 : 7	3.6	48.6	174.96	12.96	2,361.96
9 :15	15.3	26.2	400.86	234.09	686.44
Sum	374.9	609.0	26,445.22	19,493.71	40,399.16

Calculation of numerator:

$$Sxy = 26,445.22$$

$$\frac{Sx \cdot Sy}{n} = \frac{374.9 \times 609.0}{10} = 22,831.41$$

$$\text{numerator} = 26,445.22 - 22,831.41 = 3613.81$$

Calculation of denominator:

$$Sx^2 = 19,493.71$$

$$\frac{(Sx)^2}{n} = \frac{374.9^2}{10} = 14,055.00$$

$$19,493.71 - 14,055.00 = 5438.71$$

$$Sy^2 = 40,399.16$$

$$\frac{(Sy)^2}{n} = \frac{609^2}{10} = 37,088.10$$

$$40,399.16 - 37,088.10 = 3311.06$$

$$\text{denominator} = \sqrt{5438.71 \times 3311.06} = \sqrt{18,007,895.13} = 4243.57$$

$$r = \frac{3613.81}{4243.57} = 0.85$$

We may now ask whether the correlation coefficient r=0.85 is significant. For this, one can use Student's t-test and the formula (MORONEY 1954)

$$t = \frac{r\sqrt{n-2}}{\sqrt{1-r^2}} = \frac{0.85 \sqrt{8}}{\sqrt{1-(0.85)^2}} = \frac{0.85 \times 2.829}{\sqrt{1-0.72}}$$

$$= \frac{2.4}{\sqrt{0.28}} = \frac{2.4}{0.53} = 4.53$$

correlation between dissimilarity indices and ordination intervals is highly significant, exceeding the 0.5 percent level of probability.

It is useful to plot the *ID* values over the ordination intervals into a scatter diagram (FIG. 10.14), and to then fit a straight line to these data. This will permit a gross check on the calculation and provide a visual idea of the correlation of the two dissimilarity measures (x and y).

The straight line is expressed by the formula

$$y = a + bx$$

where x = any ordination interval that corresponds to y = the *ID* values (as above), a = the point of intersection on the y axis, b = the slope of the line. A short-cut calculation for the straight line is as follows (MORONEY 1954):

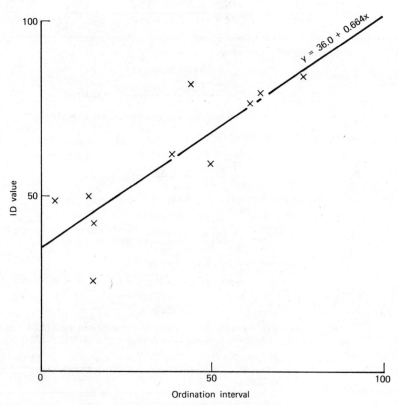

FIGURE 10.14. Scatter diagram of indices of dissimilarity (ID values) over ordination intervals and the straight line showing the trend of these values; x and y values from TABLE 10.16.

Substitue the x and y values (from TABLE 10.16) in the formula:

50.4 = $a +$ **14.4**b multiply by 14.4   **725.76 =** **14.4**$a +$ **207.36**b

82.0 = $a +$ **44.0**b multiply by 44.00   **3608.00 =** **44.0**$a +$ **1936.00**b

.   .   .   .

.   etc.   .   etc.   .   .   etc.   .

.   .   .   .

Sum   **609.0 = 10**a **+ 374.9**b Sum   **26,445.22 = 374.9**a **+ 19493.71**b

Eliminate b by combining both equations,

$$609.0 = 10a + 374.9b \quad x\ 51.9971$$

$$26{,}445.22 - 374.9a + 19{,}493.71b$$

$$31{,}666.23 = 520.0a + 19{,}493.71b$$

$$26{,}445.22 = 374.9a + 19{,}493.71b \quad -$$

$$5221.01 = 145.1a$$

$$a = 36.0$$

$$609 = 10a + 374.9b$$

$$609 = 360 + 374.9b$$

$$b = \frac{249}{374.9} = 0.664$$

$$y = 36.0 + 0.664x$$

For checking: Since sum x = 374.9 and sum y = 609 for n = 10, we have x = 37.5 and y = 60.9, and

$$y = a + bx$$

$$60.9 = 36.0 + 24.9$$

To draw the straight line on the scatter diagram (FIG. 10.14), use x = 37.5 and y = 60.9 as one point. For the second point use x − 0 in the formula; thus y = 36.0.

10.8 THE DENDROGRAM METHOD

Within the last decade, dendrograms have become popular for presenting the results of computer-derived classifications. These classification techniques are also known as cluster analyses. Like an ordination graph, a dendrogram is a graphical tool to represent the relationships in a similarity matrix. Yet, in contrast to a two-dimensional ordination, stands (or species) are not shown as individual points in a geometric space; instead they are shown as clusters of pairs linked together at certain levels of similarity. The clusters of pairs are further combined into more inclusive or generalized clusters that form a hierarchical arrangement.

There are a number of cluster analysis techniques that can make use of dendrograms for data representation. Several of these are reviewed by SOKAL and SNEATH (1963). GOWER (1967) distinguishes between divisive and agglomerative methods. Many computer-derived classification techniques (e.g., GOODALL 1953a, WILLIAMS and LAMBERT 1959) are divisive, meaning the cluster analysis proceeds from above by subdivisioning. Such a cluster analysis is analogous, for example, to the structural classification of vegetation by FOSBERG which proceeds from the general to the detailed with the help of a key (see SECTION 8.73). Another feature of divisive cluster analysis techniques is that they are often monothetic, meaning that the division is made on a single character at each stage. In contrast, a classification can be polythetic, meaning that the division into groups is based on several characters. An agglomerative cluster analysis proceeds from below; that is, from the detailed to the more general level of inclusivenss. Such a system is that of BRAUN-BLANQUET, which builds hierarchical units from comparison of individual stands and differential species groups (CHAP. 9). Moreover, the method is polythetic, because groups of species (differential groups) rather than single species are used.

10.81 Construction of a Dendrogram by Cluster Analysis. The dendrogram shown in FIGURE 10.15 was constructed by SPATZ and SIEGMUND (1973) for ELLENBERG's (1956) data pertaining to the *Arrhenatherum* relevés in TABLE 9.7. SPATZ and SIEGMUND used an agglomerative cluster analysis similar to the "weighted pair-group method" as described by SOKAL and MICHENER (1958, SOKAL and SNEATH 1963) and also similar to the "centroid sorting method" as described by DUCKER, WILLIAMS and LANCE (1965).

The computations leading to the construction of the dendrogram (FIG. 10.15) are as follows:

1. A similarity matrix is established, which compares all ($n = 25$) relevés to one another by similarity indices (i.e., $n \times (n-1)/2 = 300$ *IS* values). The result can be tabulated as, for example, shown in TABLE 10.13. At this point of data reduction, an ordination graph (as discussed in the previous sections) or a dendrogram can be prepared. For the

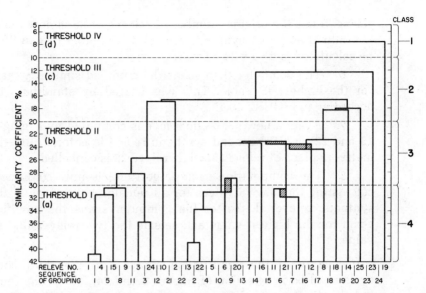

FIGURE 10.15. Dendrogram derived from cluster analysis applied to the vegetation data in TABLE 9.2 (from SPATZ 1972). Two major clusters are shown in class 2 near 17–18 percent similarity. The first cluster of eight relevés (releves 1 through 2) coincides exactly with vegetation unit A on TABLE 9.7. The second cluster (relevés 13 through 25) includes both units, B and C. The dendrogram "stems" (vertical connecting lines) that join the cross-linkages (horizontal connecting lines) are based on "synthetic" relevés. These were derived from averaging the quantitative species values in the relevés connected by the cross-linkages.

dendrogram in FIGURE 10.15, SPATZ and SIEGMUND (1973) used SPATZ's index of similarity, which was described in SECTION 10.15 as one of the modifications of JACCARD's coefficient.

2. The computer scans the established matrix of similarity indices for the highest *IS* value and then isolates the stand-pair that shares this value. In FIGURE 10.15, the stand-pair with the highest *IS* value is relevés 1 and 4. This information is stored.

3. Then, the two first selected stands are combined into a synthetic relevé. This is done by averaging the quantitative values of each species in the two relevés.

4. The synthetic or mean relevé is returned to the *IS* matrix in place of one of the two individual stands. The place arbitrarily chosen is that of the relevé with the smaller number (1), while the position of the relevé with the larger number (4) is cancelled in the matrix.

5. Now, a new series of *IS* values is calculated for the synthetic

relevé. Since the original number of relevés of 25 is now reduced to 24 (23 individual plus 1 synthetic relevé), the number of new *IS* values to be calculated is $n-1 = 24-1 = 23$.

6.   The computer then scans the modified similarity matrix again for the highest *IS* value. This was shared by stand-pair 13 and 22 according to FIGURE 10.15.

7.   Again a new synthetic relevé is computed by averaging stands 13 and 22, and the second synthetic relevé is entered into the second matrix in place of relevé 13, while relevé 22 is cancelled.

8.   The computation cycle (algorithm) is now continued by first calculating the new *IS* values for all remaining stands with the second synthetic relevé; thereafter the computer scans the modified matrix again for the highest value and selects the two relevés that share this value.

Inspection of the dendrogram shows that the third computation cycle resulted in selection of relevés 3 and 24. But, in the fourth cycle, the situation occurred for the first time that the highest *IS* value was shared by a synthetic and an individual relevé. The synthetic relevé is made up of 13 and 22, the individual relevé that shared an *IS* value of near 34 percent with the combined relevé is 5.

At this point a new synthetic relevé was calculated that consisted of three original relevés (13, 22 and 5). In this averaging procedure, the synthetic relevé (labelled 13) receives double weight because it consisted of two individual relevés. Now a new synthetic relevé is created that consists of three individual relevés. The *IS* matrix is again recalculated for all combinations with the remaining stands. According to FIGURE 10.15 the fifth computation cycle also linked an individual releve (15) with a synthetic relevé (1, comprised of 1 and 4).

Combining two synthetic relevés did not occur until the eleventh computation cycle, when the highest *IS* value was shared by synthetic relevé 1 (consisting of individual relevés 1, 4, 15, and 9) and synthetic relevé 3 (consisting of individual relevés 3 and 24).

As seen in FIGURE 10.15, the successive computation cycles yielded a ranked sequence of *IS* values from high to low.

The relevé sequence from left to right on the abscissa resulted from the linkage of the "stems" of the first relevé-pair in the dendrogram. The individual relevé most similar to the synthetic relevé (consisting of 1 and 4) was stand 15. Therefore, 15 was positioned as the third relevé next to 1 and 4 on the abscissa. Following 15 was 9, because the latter relevé was most similar to the synthetic relevé 1 (comprised in this linkage of 1, 4 and 15). The finally derived relevé sequence along the abscissa is a linear or uni-dimensional ordination. As such, it is subject to a certain amount of distortion because, as we have seen earlier (SECTION 10.72), even a two-dimensional ordination would not allow the positioning of the stands without distortion of their true similarities. Nevertheless, the relevé sequence arrived at is an optimal compromise as will be shown next.

10.82   Classification of Dendrogram Clusters.   Since the paired and grouped relevé clusters in the dendrogram are the result of linkages at various levels of similarity, an ecologically meaningful classification is not automatically indicated. But, classification can be automated by the setting of more or less arbitrary threshold values. This was done in the program that led to the dendrogram in FIGURE 10.15. Threshold values were set at 30, 20, and 10 percent. These are indicated by the horizontal dashed lines on the dendrogram. In the application of SPATZ's similarity index, a similarity over 30 percent is very high. Classification at this level of similarity would result in too many units with limited, generalizable information value. The clusters most similar to the class-units A, B, and C recognized on the differentiated synthesis table (TABLE 9.7) and in the ordination graphs (e.g., FIG. 10.11) are found in class 2 (between 10 and 20 percent similarity) or in class 3 (between 20 and 30 percent similarity).

In class 2 there are two main clusters; one that links together the first eight relevés (1 through 2) at about 17 percent similarity, and a second one that joins the next 15 relevés (13 through 25) at about 18 percent similarity. The two remaining stands, 23 and 19, are so dissimilar to the others that they remain rather isolated. It is worth noting here that relevé 19, which was eliminated as "untypical" on TABLE 9.7, turned out also as the most dissimilar stand (with less than 8 percent similarity to the others) in the cluster analysis on the dendrogram.

In class 3, there are also two main clusters, one that links together relevés 1 through 10 (7 stands) at about 26 percent similarity and a second one that combines relevés 13 through 18 (13 stands) at about 23 percent similarity. This leaves five stands unclassified as relatively dissimilar to the rest, i.e., 2, 14, 25, 23, and 19.

The clustering in class 2 on the dendrogram coincides closely with the relevé grouping in TABLE 9.7 and the result of the ordination technique as shown in FIGURES 10.9 and 10.11. Just how closely the dendrogram clusters coincide with the relevé groups in TABLE 9.7 is shown in the following contingency table.

The contingency table shows that the first cluster in class 2 on the dendrogram coincides exactly with vegetation type A in TABLE 9.7. The second cluster on the dendrogram was separated on TABLE 9.7 into two subunits, B and C. This separation is not clearly indicated on the dendrogram. This implies that the floristic variation among the

Units on Table 9.7

	A	B	C	Σ
1	8	0	0	8
2	0	6	9	15
3[a]	0	0	2	2
Σ	8	6	11	25

Class 2
dendrogram
clusters
(on FIG. 10.15)

[a] Relevés 23 and 19

releves in units B and C on TABLE 9.7 is continuous. The same was shown in the ordination graphs (FIGS. 10.9, 10.10 and 10.11).

Thus, the three techniques—the tabulation method, the ordination, and cluster analysis—all give essentially the same result. However, the two mathematical methods permit a statement on the degree of variation among the units, which is not evident from mere inspection of the association table (TABLE 9.7).

PART
IV

Spatial and Temporal
Explanations of Vegetation
Patterns

Vegetation-Environment Correlation Studies

11 In CHAPTERS 9 and 10, we discussed methods of ordering vegetation data. The implicit purpose of these ordering methods is to first establish a meaningful framework of vegetation patterns to which we may subsequently relate environmental parameters and patterns. This approach from vegetation patterns to environmental explanations is the tradition of vegetation science.

One may also study vegetation from its response to known variations in environment. This approach is a reversal in direction. Instead of beginning with an analysis of vegetation patterns, one begins with an analysis of the environmental patterns. From the environmental patterns one proceeds to observations of the vegetation response. The ultimate objective of both approaches is to explain the underlying causes of the vegetation patterns.

The environment-to-vegetation approach was particularly empha- sized by WARMING (1909) and others (see SECTION 2.3). It has led to the understanding of plant ecology as defined in all English speaking countries. We will discuss the principles of this approach under three headings; (a) the concept and derivation of ecological species groups, (b) ecological land classification, and (c) environmental gradient analysis.

11.1   ECOLOGICAL SPECIES GROUPS

11.11   The Concept.   Treatment of vegetation samples through tabular comparisons or mathematical correlations independently leads to the conclusion that no two plant species have exactly the same ecological and sociological amplitude. However, these treatments also show that some species are so similar in their distribution that they can be com- bined into groups. This grouping is the first and surest step to an orderly arrangement of plant communities with respect to species compo- sition.

Beside tabular comparison and mathematical treatments of vegeta- tion samples, there is still a third method for establishing species groups; namely, through study of the environmental response of the species. This method, however, requires a considerable knowledge of the environmental relations of all plant species. Therefore, it cannot be applied during the first stages of vegetation research of an area. How- ever, with progressing ecological knowledge, the method gains impor- tance, because it supports the search for the causes of plant distribu- tion and pattern, it simplifies the definition of vegetation units, and thereby facilitates practical application of vegetation information.

Similarly, as vegetation units can be distinguished by groups of differential and character species, they can be distinguished also by the presence of the so-called "ecological groups." All those species can be combined into an ecological group, which show similar "ecological behavior." This means that the species forming an ecological group must show similar relationships to the more important site factors. The groups are then named after a species that represents the character- istics of the group particularly well.

It is an important implication of the concept that the species of an ecological group must also be closely similar in their life forms. There- fore, they represent at the same time a synusia. However, a synusia is not quite the same as an ecological group. A synusia can be distin- guished in the field without further study. Ecological groups can only be defined after the response of species to known environmental gradients has been established.

In general, an ecological group is defined more narrowly than a synusia. A synusia may include several ecological groups, and a greater number of species which may show quite different correlations to environmental factors. WHITTAKER (1967) defines "ecological groups" as species with closely similar distribution modes. This is quite a different concept as it does not restrict the life forms. When speaking of ecological groups, we here refer only to species of closely similar life forms.

The ecological group concept was first presented by WARMING, who referred to "ecological unions." KELLER (1923) and ALECHIN (1926) later used the term "consumer associations" (Genossenschaften) and KLEOPOW talked about "eco-elements." (This term appears suitable for international communication.) However, systematic use of ecological groups in characterizing vegetation units was made only later, by SCHLENKER (1950) in forest communities and by ELLENBERG (1950) in agricultural weed communities. The method can be applied in principle to all plant communities.

The method is based on the premise that plant communities are not organisms, which can only be treated as wholes, but that communities are combinations of plant species, whose composition is dependent upon local environment. Plant communities are composed in part by species, which complement each other rather well in growth form and site requirements, for example, trees and shrubs, deep-rooting and shallow-rooting herbs, spring geophytes and late-summer perennials, etc. Nearly all of the plant types mentioned are represented also by several species, and these can be very similar in gross-morphology (life form) and ecological behavior. Such species form the ecological groups.

Before explaining the method of deriving ecological groups, we intend to give a few examples to further explain this little known concept.

An especially impressive example is the *Corydalis* group in the Central European hardwood forests (see ELLENBERG 1939). It is comprised of *Corydalis cava, C. solida, Allium ursinum, Gagea lutea, G. spathacea, Leucoium vernum, Anemone ranunculoides,* and similar species. These are all spring geophytes. They are almost exclusively restricted to shady hardwood forests, since they cannot compete with the meadow plants that are adapted to grow among dense grasses. The spring geophytes have relatively deep roots and they prefer a deep, loose, fresh to moist, well aerated soil, rich in bases and nutrients. Therefore, they populate the most productive hardwood forest soils, but they are not dependent upon a particular tree species or restricted to one association only. They are found, for example, in *Fagus sylvatica* forests on calcareous soils in microclimatically humid positions; that is, predominantly on northerly and easterly slopes. However, they are characteristic also for mixed hardwood forests on colluvially enriched (fine textured) lower-slope soils; that is, on deep, mild loams. Moreover, they are abundant in oak- or elm-alluvial bottomland forests, where beech is absent because of periodic flooding. The species of the *Corydalis* group are likewise sensitive to water logging and poor soil aeration, but they do not extend their roots as deeply as the beech trees and thus are able to maintain themselves on higher local positions on the alluvial bottomlands. Here they indicate rather clearly the normal height of the flood-water level.

Thus, the species of the *Corydalis* group are rather similar in their ecological behavior. The group is well defined with respect to several important site factors. The individual members also show a relatively similar response to temperature by avoiding, as a group, the Mediterranean as well as the subalpine and alpine climates. However, the species are not entirely ecologically alike, which is already evident from the fact that their geographical distributions are not the same. Yet, within their common range of distribution, there are usually several members of the *Corydalis* group in the above described habitats.

Another well-investigated ecological group is the *Gnaphalium uliginosum* group (occurring among weed communities of European fields and gardens), which, besides *Gnaphalium*, includes *Sagina procumbens*, *S. apetala*, *Plantago intermedia*, and to a certain extent also *Polygonum hydropiper*. These species require wet conditions for germination. As shallow-rooted plants they are very sensitive to soil-surface drying, but they cannot tolerate soil wetness when correlated with lack of oxygen. On agricultural fields they form a well-defined group of surface soil moisture indicators. They are relatively indifferent to the level of soil nutrients. Further examples of ecological groups are given, among others, by HAUFF, SCHLENKER and KRAUSS (1950), SCHÖNHAR (1954) and ELLENBERG (1950, 1952, 1963).

Some of these ecological groups are more or less identical to floristically determined groups of differential and character species. For example, TÜXEN (1937) has used members of the *Corydalis* group in one case as differential species for the *Querceto–Carpinetum corydaletosum* and in another for the *Fagetum allietosum ursinae*.

The *Arrhenatherum* group with *Arrhenatherum elatius*, *Crepis biennis*, *Tragopogon pratensis*, and *Geranium pratense* can be used as an example of an ecological group whose members were evaluated largely as character species of an alliance. Here, therefore, is a case where the systematic-floristic grouping is identical to the ecological grouping. Such coincidence seems to occur particularly where the character species have a strong competitive capacity and where they react in a similar way to a controlling site factor.

A contrasting example is shown by the character species' of the Northwest German *Querceto–Carpinetum* (TÜXEN 1937). This is not an ecological group since it is composed of trees (*Carpinus*), shrubs (*Evonymus europaea*), herbs and mosses (*Catharinaea undulata*), i.e., of very different life forms.

These examples may explain the ecological group concept. Ecological species groups represent, so to speak, within-community components at a level above the species. These species group components play an important role in the fine-structure of a plant community. The recognition of ecological groups is, as we will see, independent of the method of locating releves and independent of any floristically derived vegetation units. However, the interpretation of ecological species groups may require reevaluation with increasing ecological knowledge.

11.12 Derivation of Ecological Groups. How can we recognize ecological groups? Since there is not a single species with entirely the same ecological relationships, ecological groups can be established only through abstraction. Which site factors are emphasized and which are neglected is a matter of judgment. This depends on the questions asked or on the purposes of the study. The choice of environmental factors does not preclude a rather objective derivation of the ecological groups, as their evaluation can always be checked by measurements.

There are three principal ways in which ecological groups can be established. (a) One may conduct field observations in such a way that one looks immediately for groups of species with similar ecological relationships. (b) One may study the ecological behavior of many individual species in relation to single site factors until the ecological relationships of these species can be expressed in figures. The species are then sorted into ecological groups subsequently. (c) The third approach makes use of tabular comparisons or mathematical correlation methods for species groups by investigating which of the differential or character species (or which species with high correlation coefficients) may form ecological groups. The latter method makes use of the experience accumulated in vegetation science. This third approach is therefore particularly suited to establish the link between the floristic pattern and the environmental pattern approach to vegetation ecology.

The first approach yields results quickly but it is also the least exact one. SCHLENKER and his collaborators used this approach because their objective was to develop as rapidly as possible, without many vegetation samples, forest site maps that provided a habitat inventory sufficient for forestry purposes. The tentatively established ecological groups were later investigated in more detail and, where necessary, were redefined on the basis of measurements of pH and other soil factors (SCHÖNHAR 1954). However, these groups may not be considered as final since several of them include a rather large number of species, and these are not all of the same life forms.

ELLENBERG (1950, 1952) used the inductive, second approach in the study of agricultural weed communities. He began by recording the response of many species to a selected single environmental gradient (e.g., to various degrees of soil acidity). He then classified all investigated species into six pH related groups. He used for this his own pH measurements and supplementary pH values from the literature. The six groups were defined as:

1. Species, that can grow under very unfavorable conditions occurring almost exclusively at very low values of the environmental gradient (e.g., plants restricted to very acid conditions = low pH indicators).

2. Species of similar occurrrence, but of wider amplitude (e.g., plants generally correlated with acid conditions).

3. Species which are particularly common at the medium range of the gradient (e.g., plants correlated with moderately acid conditions).

4. Species of similar response as group 5, but of wider amplitude (e.g., plants generally correlated with calcareous conditions).

5. Species, which occur almost exclusively at the upper end of the gradient; that is, under generally very favorable or suprafavorable conditions (e.g., high pH or lime-indicators).

6. Species that are indifferent to the gradient investigated; that is, they have a very wide amplitude.

A useful way of presenting such species groups is in an "ecological series table." This is a two-way table, which shows the species amplitudes and quantities in relation to the environmental gradients investigated. For this purpose, vegetation samples are first transferred into a raw table. But, the samples are ordinated right away in relation to the increasing or decreasing intensity of the environmental factor investigated, e.g., to a gradient of decreasing soil acidity. If this gradient has a wide enough amplitude, one may recognize species that are more or less restricted to certain pH ranges. All species can then be rearranged according to their amplitude in the ecological series table. Only the very rare ("accidental") species cannot be properly evaluated. On the other extreme, the species with high constancy must be considered as indifferent to the factor investigated.

OLSEN (1923) published the first series tables in which the species were ranked according to soil pH. TABLE 11.1 shows species from ELLENBERG's (1950) example to explain the work process. The vertical columns refer to relevés of agricultural weed communities that occurred in the surroundings of Stuttgart. Their soils were tested for pH several times during 1948. By ignoring certain irregularities, it is quite possible to divide the weed species into the five "reaction groups," as sorted out in TABLE 11.1.

Such an order applies initially only to the area investigated, that is, for a narrowly limited geographic area. However, if such series tables are prepared also for other areas, it will be shown that most species respond similarly. Certain species may deviate more or less strongly from the pattern shown here. Such was the case with *Convallaria majalis* in Central European hardwood forests (not on TABLE 11.1). This species was found in certain geographic districts predominantly on acid soils; in others, it appeared to be restricted to calcareous soils. In reality, the species is relatively indifferent and responds much more strongly to other site factors, for example, to light, to soil water, and soil air relations.

If the investigation is extended over very large geographic areas (e.g., over the entire area of Europe), then many species which locally show a definite response to a narrow range of pH would need to be classified as indifferent. Therefore, it is advisable to restrict the investigation right away to a limited geographic area (e.g., to an area the size of South Germany or to that of the Northwest German lowland). Then, the separation of species into races or ecotypes—which are often hard to distinguish morphologically—presents a lesser problem also.

Similar series tables can be established for other environmental gradients, such as soil moisture, soil nitrogen, temperature, etc., and one may combine species with similar responses into groups. However, most other site factors are more difficult to measure than soil pH and certain factors can hardly be measured at all. In such cases one may have to be satisfied with relative differences. For example, to establish six groups in relation to soil moisture, relevés may be placed along a transect in such a way that they represent an ecological series of decreasing depth to a water table. Or, to establish ecological groups in grassland communities or among agricultural weeds, one can utilize fertilization experiments relating to differently fertilized grasslands or crop fields which occur side by side on the same soil.

After a sufficiently large number of species has been studied and arranged in relation to the most important site factors, one can proceed to establish the ecological species groups. The division into always six categories allows for a schematic and easily comprehensible

TABLE 11.1 Extract of a Series Table of Weed Communities Arranged in Order of Increasing Soil pH (After ELLENBERG 1950:37). The Figures Following Each Species Always Refer to the Highest Cover Value (BRAUN-BLANQUET scale) That Was Attained During the Year 1948.

Running Number	1	2	3	4	5	6	7	8	9	10	11	12	13	14	15	16	17	18	19	20	21	22	23	24	25
pH mode for 1948 fraction of	4.				5.						6.								7.						
pH value	5	6	6	8	1	2	3	5	7	8	1	3	4	5	5	5	7	9	0	2	2	2	2	3	3
R 1																									
Rumex acetosella	1	1																							
Scleranthus annuus	1	2	+		1	2																			
Spergula arvensis	2	2			1	2																			
R 2																									
Alchemilla arvensis			2	+					+				+												
Raphanus raph.	2	2				2		2			3		+		2										
R 3																									
Matricaria cham.						2		2		2	+		2	2	2	2									
Apera spica venti	1	2	2	2		+	2	+	1			3	+	1	+	+	1								
Poa annua	1	1	+	+			2	1	1	2	1		1	1	+	1	1		+						
R 4																									
Sinapis arvensis						1		2		1	2		1	2	2	1	2	1	2	1	1	3	1	2	1
Papaver rhoeas						+				1		+	+	+	3	+	2		3	2	1	1	+	+	
Fumaria officinalis										1			1	1	2						+	1	1	1	
Sonchus oleraceus												+		1		1	+		+	+	+	+	1	1	1
R 5																									
Caucalis lappula																			1	3		3	1	3	1
Delphinium cons.																			1	1	3	1	+	+	+
Galium tricorne																				+	+	2	2	1	
Anagallis coerulea																				+		+	1	1	1

summary of the ecological behavior of species. ELLENBERG (1950) applied this system, for the first time, to agricultural weeds, whereby he arrived at 25 ecological groups. (TABLE 11.2 gives two examples of these.)

TABLE 11.2 clarifies at a single glance the ecological relations that are considered important for the species combined into groups and those that are considered less important.

In deriving ecological groups, it is important to consider the behavior of the same species separately for meadows, agricultural fields, heathlands and forests or other formations. This is necessary, because the same species may grow in several formations. In that case they grow in combination with quite different competitors, and competition exerts a great influence on the ecological behavior of plants.

We have described the procedure for deriving ecological groups in some detail. One of the purposes was to show in what way this method necessitates subjective decisions and in what way the procedure is objective. The choice of the sample locations requires some experience. It is, however, largely objective. This applies also to the tabular treatment of the data. The final step, namely the combining of several species into a group by abstraction, is also a matter of personal judgment.

11.13   Plant Communities as Combinations of Ecological Groups.   In any several-layered community there are always several ecological groups that participate in the structure of such a community. But, also in single-layered communities, we usually find a few species groups that complement each other in their ecological adaptation, provided that the habitat is not so extreme, as for example, the marine *Salicornia* mud flats (*Salicornietum strictae*).

The combination of ecological groups is rather a simple means for characterizing plant communities. TABLE 11.3 shows examples for agricultural weed communities. The relative magnitude of each group in a certain plant community is shown in the table by symbols.

A special advantage of describing communities by the combination of ecological groups is that each individual community can be evaluated at once, without prior floristic ordering, through the environmental and floristic relations of its ecological groups. The groups provide a certain idea of the more important factor combinations operative in the respective habitats. The group names and their ecological implications become quickly familiar with frequent use.

Just as unranked vegetation units can be named after groups of differential species, they can be named also after ecological species groups. This allows for an adaptable and yet precise nomenclature

TABLE 11.2   *Examples of Two Ecological Groups and Their Response
to Four Site Factors. The Groups Were Established Through the Response of Each Species to Individual Site Factors. The Controlling
Factors Are Emphasized by Enframing. The Numbers and Names of the
Groups Correspond to the Summary on TABLE 11.3.*

	LIFE FORM	RELATION TO SITE FACTORS[a] T	W	R	N
2. Coringia group					
Coringia orientalis	T	5	1	5	1–2
Caucalis latifolia	T	5	1–2	5	2 ?
Galeopsis angustifolia	T	3	1	5	2 ?
Ajuga chamaepytis	T	5	1	5	2
Bunias orientalis?	G	3	1 ?	5	1–2
13. Ranunculus repens group					
Ranunculus repens	H	1	5	0	0
Agrostis prorepens	Ch–H	1	5	0	3 ?
Potentilla anserina	H	1	5	4	4
Poa trivialis	H	0	5	0–3	4
Mentha arvensis	G–H	3	5	0–3	4–3
Stachys palustris	G–H	2	5	4	4
Equisetum silvaticum?	G	1	5	0	2 ?

[a] T=temperature; W=soil water; R=soil reaction; N=nitrogen. The figures correspond to the six categories explained on p. 310 (W1 means occurring on loose,
permeable soils that are often dry; W5 means occurring on interchangeably moist,
often water-logged and poorly aerated soils.)

of unranked vegetation units. The only rule here is to emphasize the
dominant ecological group in the community by putting it at the end
of the community name, e.g.,

Tussilago farfara–Sinapis arvensis **community**
(or briefly: *Tussilago–Sinapis* field)

Ranunculus repens–Gnaphalium uliginosum–Sinapis arvensis **community**
(or briefly: *Ranunculus–Gnaphalium–Sinapis* field)

Corydalis cava–Fagus silvatica **community**
(or briefly: *Corydalis–Fagus* forest)

Such communities are floristically, ecologically, and structurally
defined vegetation units. Therefore, they comply with the association

concept as proposed at the Third International Congress in Brussels. Vegetation units that are designated through ecological groups are independent of any hierarchical unit designation, but it is not difficult to relate them to the unit hierarchy of BRAUN-BLANQUET.

11.14 Ecological Group Spectra. The contribution of ecological groups to the structure of a plant community can be shown graphically in form of "ecological group spectra." In agricultural weed communities it is satisfactory for this purpose to represent the number of species for each group as shown in FIGURE 11.1.

The numbering system and the division of the groups follow the detailed representation of these ecological groups after ELLENBERG (1950). Here, we will give only a summary with a brief ecological characterization of the groups (TABLE 11.4). The relevés that were used in calculating the two spectra on FIGURE 11.1 are found also in ELLENBERG (1950:121, 122).

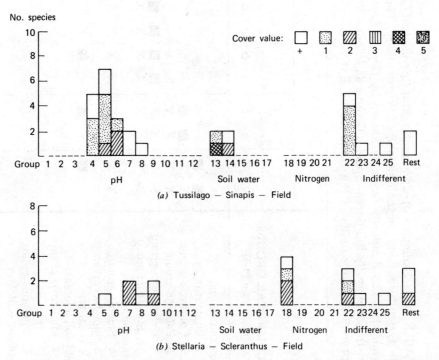

FIGURE 11.1. *Ecological group spectra of two agricultural weed communities. The numbers of the groups are explained in TABLE 11.4.*

TABLE 11.3 Examples of Combinations of "Ecological Groups" as They Occur in Some Weed Communities on Agricultural Fields in the Lowland of Württemberg. The Vegetation Units Are Ranked According to Their Similarity. The Similarity Order Is Shown by Capital Letters (A–C), Numbers (1–3), and Lowercase Letters (a–d). Ecologically Indifferent Species Groups and Those Not Present in the Area Are Not Shown in the Table. Forms Transitory Between the Units Occur, but Are Not Represented Here for Reasons of Simplification.

ECOLOGICAL GROUPS (see Table 11.4)	A 1 a	A 1 b	A 2 a	A 2 b	A 3 a	A 3 b	A 3 c	A 3 d	B 1 a	B 1 b	B 1 c	B 1 d	B 2 a	B 2 b	B 2 c	B 2 d	C 1 a	C 1 b	C 1 c	C 1 d
Lime indicators:																				
1. Bupleurum-group	■	X	—																	
2. Conringia-group	X	X	■	■	—															
3. Falcaria-group	X	V	X	V	—	V														
4. Delphinium-group	V	V	X	X	■	■	■	■	■	■	■	■	■	V	V	V				
4a. Sherardia-group	V	V	X	X	X	X	X	X	V	V	V	V	X	V	V	V				
Lime tolerant:																				
5. Sinapis-group	X	V	X	X	X	X	X	X	■	■	■	■								
6. Sonchus arvensis-group			—		—				X	X	X	X	X	X	X	X	—	—	—	—
Acid-soil tolerant:																				
7. Matricaria cham.-group	■		—	■	—	■	■	X												
8. Raphanus-group													V	V	V	V	V	V	V	V
													X	X	X	X	X	X	X	X
Acid-soil indicator:																				
9. Scleranthus-group																	■	■	■	■

Tolerant to water-logging:
13. Ranunc. repens.-group
14. Tussilago-group

Surface soil-moisture
related:
15. Gnaphalium-group
16. Juncus bufonius-group
17. Riccia-group

Nitrogen indicators:
18. Stellaria media-group
19. Panicum-group
20. Euph. peplus-group

Group															
13. Ranunc. repens.-group	—	—		■	■	—		■	■	—		■	■	V	■	■
14. Tussilago-group	—	—		—	—	—		—	—	—		V	V	V	V	
15. Gnaphalium-group	—	—		■	■	■		■	■	■		■	■	■	■	■
16. Juncus bufonius-group												V	V	V	V	V
17. Riccia-group														V	V	V
18. Stellaria media-group	—	—		V	V	V		V	V	V		V	V	V	—	—
19. Panicum-group	—	—		—	—	—		—	—	—		—	—	—	—	—
20. Euph. peplus-group				V	V	V		V	V	V		—	—	—	—	—

■ =dominant group; X=frequently represented group; V=sparsely represented group; —=group represented only in few of the communities.

TABLE 11.4   *Summary of Ecological Groups of Agricultural Weeds in Southwest Germany (After ELLENBERG 1950).*

I. Controlling factor: soil reaction

(a) Lime indicators (R^a5)
     tolerating droughtiness (W^a1)
        adapted to low nitrogen levels
           (N^a1)_ _ _ _ _ _ _ _ _ _ _ _ _1. *Bupleurum* group
        requiring moderate amount of
           nitrogen (N2)_ _ _ _ _ _ _ _ _2. *Coringia* group
        deeply rooting   3. *Falcaria* group
        preferring fresh conditions (W2)_ _ _4. *Delphinium* group
        plants with a tendency to group 5
           (R4–5)_ _ _ _ _ _ _ _ _ _ _ _4a. *Sherardia* group

(b) Lime-preferring (R4)
     preferring fresh conditions (W2–3)_ _5. *Sinapis* group
     tolerating wet conditions (W4)_ _ _ _6. *Sonchus arvensis* group

(c) Acid-preferring (R3 and R2)
     tolerating wet conditions (W4)_ _ _ _7. *Matricaria chamomilla* group
     moisture indifferent (W0–2)_ _ _ _ _8. *Raphanus* group

(d) Acid (=low pH) indicators (R1)
     moisture indifferent (W0–3)_ _ _ _ _ 9. *Scleranthus* group
     tolerating droughtiness (W1–2)_ _ _10. *Ornithopus* group
     preferring fresh conditions (W2–3;
     atlantic-subatlantic)
        requiring little nitrogen (N1–2)_ _11. *Arnoseris* group
        requiring much nitrogen (N3–4)_ _12. *Galeopsis segetum* group

---

The spectra give an indication of the respective constellation of site factors. Species of the *Tussilago* and *Ranunculus repens* group dominate in relevé A. They indicate moisture in the subsoil. In addition, the *Delphinium*- and the *Sinapis arvensis* groups indicate clay. These species have a high competitive capacity at weakly acid to alkaline soil reactions. High nitrogen requiring species are little represented. Relevé B shows an entirely different spectrum, in which acid-tolerating and nitrogen-preferring species are dominating.

As seen from the spectra in FIGURE 11.1, most weed species of Southwest Germany can be classified into one of the 25 different ecological groups. Therefore, nearly the entire species list of a plant community can be evaluated with the aid of ecological groups. In contrast, character and differential species usually encompass only a fraction of the total species list of a community.

TABLE 11.4   (Continued)

II. Controlling factor: water

    (e) Tolerating stagnating water (W5)

        shallow rooting_ _ _ _ _ _ _ _13. *Ranunculus repens* group

        deeply rooting _ _ _ _ _ _ _ _ 14. *Tussilago* group

    (f) Preferring surface soil moisture
    (W5s)

        indifferent (R0)_ _ _ _ _ _ _ _15. *Gnaphalium* group

        tolerating acid conditions (R2)_ _16. *Juncus bufonius* group

        liverworts _ _ _ _ _ _ _ _ _ 17. *Riccia* group

III. Controlling factor: high nitrogen level

    (g) Nitrogen indicators (N4–5)
    indifferent to pH (R0)

        tolerating cool temperatures

        (T^a0–2)_ _ _ _ _ _ _ _ _ 18. *Stellaria media* group

        preferring warm temperatures

        (T3–4)_ _ _ _ _ _ _ _ _ _19. *Panicum crus-galli* group

    lime-preferring (R4–5)

        tolerating cool temperatures

        (T1–3)_ _ _ _ _ _ _ _ _ 20. *Euphorbia peplus* group

        preferring warm temperatures

        (T4–5)_ _ _ _ _ _ _ _ 21. *Mercurialis* group

IV. Indifferent species

    (h) Indifferent annuals_ _ _ _ _ _22. *Myosotis arvensis* group

    (i) Indifferent perennials _ _ _ _ _ 23. *Taraxacum* group

V. Special groups

    (k) Seed-weeds_ _ _ _ _ _ _ _ _24. *Agrostemma* group

    (l) Indicating trampling (=compaction)_25. *Plantago major* group

[a] N=Nitrogen group, R=Reaction group, T=Temperature group, W=Water group.

11.2 ECOLOGICAL LAND CLASSIFICATION

11.21 Environmental Factors and Land Use Purposes. The environmental factors investigated in the previous example of ecological group derivation were soil reaction, soil nitrogen, surface soil water, and surface soil temperature. The plants used as indicators of these factors were agricultural weeds. Most of them are annuals, some are biennials, geophytes, and hemicrytophytes. But all are more or less restricted with their root systems to the surface soil or plough-zone.

The environmental factors can be translated for practical application into agricultural use-purposes for which these factors are critical. Such agricultural use-purposes are, for example:

- Crop choice (pH, surface water, soil aeration).
- Crop productivity (soil water, soil nitrogen).
- Germination chance or survival (surface temperature, surface water).
- Ploughing practice and timing and other mechanical treatments (surface water).

The four environmental factors are critical in different proportions for all of these use-aspects.

The same principle of isolating basic environmental factors through a study of their indicators can be applied to forestry. A classification of forest habitat types for southeastern Manitoba (MUELLER-DOMBOIS 1964) was done on this basis. The investigation was specifically for practical application in forest research and management. The main guiding use-purposes for defining basic habitat difference were:

- Productive capacity for trees.
- Natural regeneration chance.
- Choice of trees for planting.
- Melioration aspects, such as fertilization requirements for seed production.
- Engineering properties of the land (for example, the suitability of digging water holes for fire protection and wildlife management, the chance of locating gravel deposits for road construction).

These use-aspects were translated into basic environmental factors that influence tree growth and tree species composition in the area. Therefore, the study was approached in the following order:

1. Tree measurements were made to establish height/age relationships in the form of site indices. (Site index determination involves a standard forest mensurational technique, whereby the height of selected dominant trees in forest stands is related to a reference age (usually 50 years) on the basis of previously established height/age curves for each tree species).

2. Environmental factors were studied and identified that were believed to be responsible for the site index differences found, in particular, soil moisture–and nutrient regimes. These included the more obvious factor of topographic or physiographic position of each stand.

3. Responses to these factors other than tree height and age relations were studied and identified. These included undergrowth plant distribution, abundance, soil profile characteristics, and stand climate.

The approach from tree observations and practical use-purposes to the identification of growth-controlling environmental factors and their indicators led to the derivation of a generalized ecological series of forest habitats.

11.22   Derivation of Ecological Series. An ecological series is here defined as a group of two or more habitats along a transect which differ from one another by a different stage of intensity of a major environmental control factor. The concept is similar to the well-known catena concept used in soil science. This refers to soils comprised of the same parent material that differ in soil water relations along a transect. The concept is also similar to the environmental gradient concept, except that the latter does not require recognition of habitats or communities. The ecological series concept was apparently introduced by Russian authors and used, for example, by SUKACHEV (1928). He used it for a group of communities along transects, but also implied a successional meaning, which is *not* implied in the ecological series concept as defined here. Topographic series of communities differing by seepage water relations were also used by KRAJINA (1969) in British Columbia. This is a special kind of ecological series.

A generalized ecological series is understood to mean the same as an abstract ecological series, which results from grouping habitat or community samples taken along specific environmental transects.

Reconnaissance of the approximately 3000 square mile area of southeastern Manitoba was carried out with the aid of air photographs. Field trips were directed first to those areas that looked unusual or extreme on the air photos, then to those that showed the more general patterns. In this way the area was traversed from one extreme habitat to the other with transects going across the more optimal segments of the forested areas.

It soon became apparent that one extreme forested site was represented by inland dunes. These were often occupied by more open-grown, very branchy and poorly growing *Pinus banksiana* (jack pine) trees. Another extreme jack pine habitat was found on sandy soils that bordered on boggy areas which were occupied by *Picea mariana* (black spruce stands). These boggy areas in themselves formed variations that ranged from shallow organic peat deposits on mineral soils (classified as "half bogs") to deep deposits in floating bogs.

Other extreme habitats with mineral soil at the surface were found in low-lying areas where snow-melt water accumulates in the spring resulting in several weeks of inundation. Here, broad-leaved tree species such as *Populus balsamifera, P. tremuloides* and others often grew mixed with coniferous species such as *Picea glauca* (white spruce).

The orientation from dry-extreme sites on sandy dunes to moist-extreme sites resulted in the recognition of soil water regime differences. These were imprinted in the soil profiles by the presence or absence of gley horizons, by the degree, color, and depth of mottling, by fragipans (semipermeable clay lenses), underlying gravel strata, the amount of leaching as expressed by the depth and development of a podzolized A_2 horizon, the depth of overlying humus, presence of free water at certain depths (in relation to time of the year), etc. The soil profile studies led to the recognition of three soil water regimes. These were, ground water, vadose water, and flood water regimes and their combinations in very moist habitats.

Ground water soils were defined as sandy soils that showed a permanent water table in the subsoil and within the tree root zone. The water table was maintained by an underlying impervious clay pan.

Vadose water soils were defined as soils in which the water supply for tree growth resulted from the water storage capacity in the rooted soil portion. Their storage capacity was related to the presence of fragipans (in sandy soils) or increases in the fine soil fraction. Thus, finer textured soils (sandy loams, loams, and clay-loams) were in this category.

Flood water soils were defined as soils that were flooded for several weeks after snow melt in the spring. This included clay soils in low lying areas, sandy soils forming very thin caps on claypans and alluvial bottomlands and stream banks.

From measurement of pH and organic matter in the soil profiles, it became apparent that the three soil water regimes were closely correlated with soil nutrient regimes. The ground water soils were nutritionally poor (oligotrophic) by showing low pH and organic carbon values; the flood water soils were nutritionally enriched (eutrophic) by showing circum-neutral soil reactions and an enrichment with humus colloids mixed into the mineral soil. The vadose water soils showed pH values similar to those of the flood water soils but less enrichment with humus colloids. They were defined as mesotrophic.

Within each of these soil water and nutrient regimes, habitat types were defined by stages of intensity of these factors as expressed in each profile. This led to the recognition of ecological series.

However, for a more rapid evaluation and for field mapping, it was important to translate the soil water and nutrient regime variations into more readily recognizable surface features. For this, physiographic position, soil surface characteristics (humus type, depth and degree of A horizon development) and undergrowth plant distribution were used (MUELLER-DOMBOIS 1965).

FIGURE 11.2 shows the generalized ecological series of fourteen forest habitat types that were recognized from this study. In the sense of SUKACHEV (1928) one can recognize on this diagram five ecological se-

TREELESS — eutrophic

MIXED-WOOD — mesotrophic

JACK PINE — oligotrophic

BLACK SPRUCE — mesotrophic eutrophic mesotrophic

JACK PINE — oligotrophic

MIXED-WOOD — mesotrophic eutrophic — **HARDWOOD**

MIXED-WOOD — mesotrophic

Col	Type description
a	semi-aquatic Typha-Phragmites marsh
b	very wet Salix-Carex swamp
c	11 wet Cornus-Carex -Caltha type mainly tamarack and/or aspen
d	6 very moist Cornus st.-Petasites palmatus type white spruce, cedar, aspen balsam-poplar, ash, birch, black spruce
e	4 moist Ledum-Rubus type jack pine SI 54±4 with commonly black spruce understorey; sometimes aspen
f	3 fresh Arctostaphylos-Linnaea type jack pine SI 51±6
g	2 dry Arctostaphylos-Cladonia type jack pine SI 46±4
h	3 fresh Arctostaphylos-Linnaea type jack pine SI 51±6
i	4 moist Ledum-Rubus type jack pine SI 54±4 w/commonly black spruce understorey
j	5 very moist Feathermoss-Sphagnum type black spruce, a type variant with cedar occurs
k	13 wet Feathermoss-Sphagnum sink hole type black spruce some tamarack
l	12 very wet Betula glandulosa-Carex-Caltha-Potentilla type black spruce some tamarack
m	13 wet Feathermoss-Sphagnum sink hole type black spruce some tamarack
n	14 wet ± Sphagnum-Feathermoss type black spruce SI 24±4
o	5 very moist Feathermoss-Sphagnum type black spruce SI 38±4
p	4 moist Ledum-Rubus type jack pine SI 54±4 with commonly black spruce understorey
q	3 fresh Arctostaphylos-Linnaea type jack pine SI 51±6
r	2 dry Arctostaphylos-Cladonia type jack pine SI 46±4
s	1 very dry Cladonia-nudum type jack pine SI 40±4
t	2 dry Arctostaphylos-Cladonia type jack pine SI 46±4
u	7 fresh Corylus-Linnaea type jack pine SI 50±4 as a rule mixed with birch and/or aspen
v	9 very fresh Cornus st-Corylus-Petasites palmatus type white spruce, balsam fir; usually stocked dominantly by hardwoods
w	10 very moist Ulmus-Matteuccia type elm, green and black ash, balsam poplar, aspen Manitoba maple
x	9 v. fresh Cornus st.-Corylus-Petasites p. type often covered w/thickets of tall shrubs
y	8 moist Cornus stolonifera Corylus-Petasites type white spruce, balsam poplar, aspen, birch, ash, with usually abundant ash regeneration
z	7 fresh Corylus-Linnaea type jack pine SI 50±4 usually mixed with birch or aspen, occasionally pure stands; seldom red pine, which shows however very good growth

Profile substrate notes: firm sand, sorted (gravel) substrate · calcareous clay – loam · some sorting · firm sand · loose sand · unsorted · calcareous clay – loam substrate · Bt organic enriched · organic enriched · loam – clay-loam occasional boulders · loamy sand

Col	Soil description
a	Sapropel over marl
b	mucky peat >6 inches, water table at or near surface
c	muck <6 inches on α-gley and Stagno-gley soils
d	twin humus (root mor over mull), α-gley soil
e	mycelial mor, β-gley Podzol
f	insect to mycelial mor, low β- and γ-gley Podzols
g	shallow insect mor, Minimal Podzol
h	insect to mycelial mor, β-gley Podzol
i	mycelial mor, β-gley Podzol
j	fibrous peat (6-20 inches) Peaty Gleysol
k	fibrous peat hummocks and muck-filled depressions
l	mucky peat >6 inches water table at or near surface
m	fibrous peat hummocks and muck-filled depressions
n	fibrous peat deeper than 20 inches
o	fibrous peat (6-20 inches), Peaty Gleysol
p	mycelial mor, β-gley Podzol
q	insect to mycelial mor, low β- and γ-gley Podzol
r	shallow insect mor, Minimal Podzol
s	very thin insect crust mor, Aeolian Regosol
t	shallow insect mor, Minimal Podzol
u	insect mor to duff mull Bisequa Podzols and Grey Wooded soils
v	mull, Alluvial Grey Wooded, Chernozemic Dark Grey soils
w	earth mull, Alluvial Grey Wooded soil
x	mull, Alluvial Grey Wooded soil
y	root mor to duff mull, Pseudo-gley soil
z	insect mor to duff mull, Bisequa Podzols Bisequa Grey Wooded and Orthic Grey Wooded soils

LACUSTRINE DEPOSITS | **BEACH DEPOSITS** | **MUSKEG OVER LACUSTRINE DEPOSITS TILL PLAINS OR GROUND MORAINES OCCASIONAL SAND BANKS** | **GLACIAL OUTWASH AND SANDY GROUND MORAINES** | **DUNES** | **RECESSIONAL MORAINES** | **RECENT ALLUVIUM** | **TILL PLAINS AND GROUND MORAINES** | **OUTWASH OVER TILL GRAVELLY BEACH DEPOSITS**

FIGURE 11.2—Forest habitat types in SE Manitoba (generalized ecological series).

ries, namely (a) a treeless, (b) a mixed-wood, (c) a jack pine, (d) a black spruce, and (e) a hardwood series of communities. On a land form basis one can recognize at least six ecological series of habitats, namely (a) on lacustrine deposits, (b) on beach deposits, (c) on muskeg (i.e., organic soil), (d) on sandy moraines (including dunes), (e) on recent alluvium, and (f) on glacial till.

TABLE 11.5 links the same habitat types into a schematic grouping identifying them by their moisture and nutrient regimes and other important ecological qualities. Each of these ecological qualities forms the basis for a specific ecological series as here defined in the habitat sense. Eight such series are identified on TABLE 11.2. These are (a) an oligotrophic, (b) a mesotrophic, (c) a eutrophic, (d) a ground water, (e) a vadose water, (f) a flood water, (g) a half-bog, and (h) a bog series. It does not matter in the ecological series concept that there is considerable overlap among the nutrient and water regime series. The habitats within each of these eight series occur along transect segments as shown in FIGURE 11.2.

11.23 Undergrowth Species as Ecological Indicators. A specific ecological series is formed, for example, by the oligotrophic ground water habitats shown as habitats 1 through 5 in FIGURE 11.2 and TABLE 11.5. Tree growth on habitats 1 and 2 is not normally influenced by ground water, because where present it was found to be too low to supply capillary water effectively. However, habitats 1 and 2 are on oligotrophic sands, which form the rooting medium also for habitats 3 to 4. In habitat 5 the acid moss peat forms the main rooting medium for *Picea mariana* stands, but tree roots often extend into the sandy, gleyed mineral soil below.

Therefore, the habitat sequence from 1 through 5 forms a sequence of decreasing water table depth. The depth extends from below the reach of tree roots in the spring (habitat 1) to the surface of the mineral soil (habitat 5). In terms of tree growth, this water table depth sequence is reflected in an increase in site index for *Pinus banksiana* from 46 (habitat 1) to 54 (habitat 4). The site index value refers to the height in feet that *Pinus banksiana* trees can grow in 50 years. The change from habitat 4 to 5 is responded to by a change in forest cover type to *Picea mariana*, and a reduction in site index to 38.

Undergrowth species were important in defining the limits of each habitat type. A correlation was shown in some species by restricted amplitudes and quantitative differences across this water table depth gradient.

The species are listed in TABLE 11.6. They are ordered into six groups similarly as shown before for agricultural weed species in their

TABLE 11.5 Forest Habitat Types in Southeast Manitoba and their Major Ecological Characteristics (after MUELLER-DOMBOIS 1964).

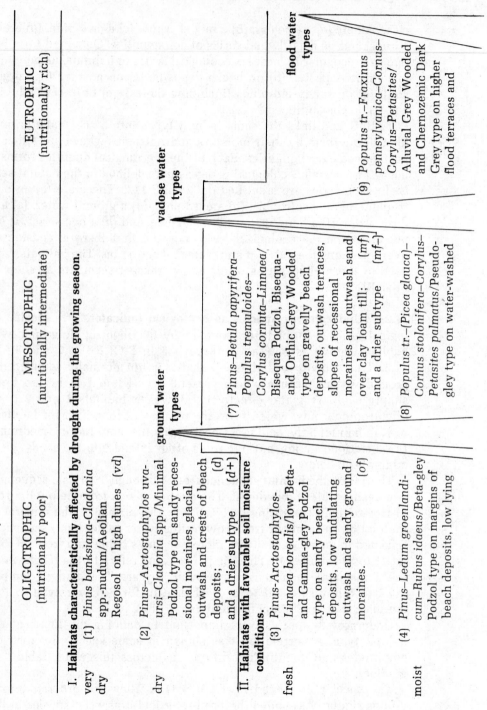

	OLIGOTROPHIC (nutritionally poor)	MESOTROPHIC (nutritionally intermediate)	EUTROPHIC (nutritionally rich)
I. Habitats characteristically affected by drought during the growing season.			
very dry	(1) Pinus banksiana-Cladonia spp.-nudum/Aeolian Regosol on high dunes (vd)		
dry	(2) Pinus–Arctostaphylos uva-ursi-Cladonia spp./Minimal Podzol type on sandy recessional moraines, glacial outwash and crests of beach deposits; (d) and a drier subtype (d+)		
II. Habitats with favorable soil moisture conditions.		ground water types	vadose water types / flood water types
fresh	(3) Pinus-Arctostaphylos-Linnaea borealis/low Beta- and Gamma-gley Podzol type on sandy beach deposits, low undulating outwash and sandy ground moraines. (of)	(7) Pinus–Betula papyrifera–Populus tremuloides–Corylus cornuta–Linnaea/Corylus cornuta–Linnaea/Bisequa Podzol, Bisequa- and Orthic Grey Wooded type on gravelly beach deposits, outwash terraces, slopes of recessional moraines and outwash sand over clay loam till; (mf) and a drier subtype (mf–)	(9) Populus tr.–Fraxinus pennsylvanica–Cornus–Corylus–Petasites/Alluvial Grey Wooded and Chernozemic Dark Grey type on higher flood terraces and
moist	(4) Pinus–Ledum groenlandicum–Rubus idaeus/Beta-gley Podzol type on margins of beach deposits, low lying	(8) Populus tr.–(Picea glauca)–Cornus stolonifera–Corylus–Petasites palmatus/Pseudogley type on water-washed	

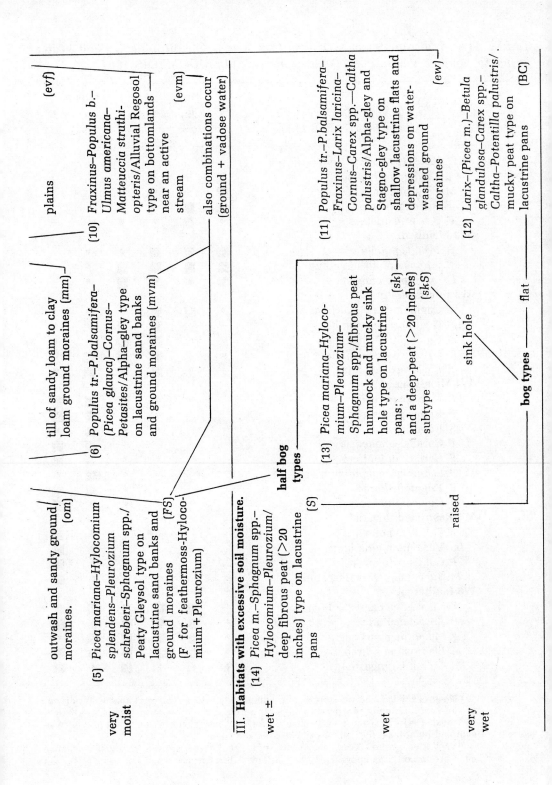

plains (evf)

till of sandy loam to clay loam ground moraines (mm)

outwash and sandy ground moraines (om)

(10) Fraxinus–Populus b.–Ulmus americana–Matteuccia struthiopteris/Alluvial Regosol type on bottomlands near an active stream (evm)

(6) Populus tr.–P.balsamifera–(Picea glauca)–Cornus–Petasites/Alpha-gley type on lacustrine sand banks and ground moraines (mvm)

(5) Picea mariana–Hylocomium splendens–Pleurozium schreberi–Sphagnum spp./Peaty Gleysol type on lacustrine sand banks and ground moraines (FS) (F for feathermoss–Hylocomium+Pleurozium)

also combinations occur (ground + vadose water)

(11) Populus tr.–P.balsamifera–Fraxinus–Larix laricina–Cornus–Carex spp.—Caltha palustris/Alpha-gley and Stagno-gley type on shallow lacustrine flats and depressions on water-washed ground moraines (ew)

(13) Picea mariana–Hylocomium–Pleurozium–Sphagnum spp./fibrous peat hummock and mucky sink hole type on lacustrine pans; (sk) and a deep-peat (>20 inches) subtype (skS)

(12) Larix–(Picea m.)–Betula glandulosa–Carex spp.—Caltha–Potentilla palustris/. muckv peat type on lacustrine pans (BC)

half bog types

bog types —— flat

sink hole

raised

III. **Habitats with excessive soil moisture.**

(14) Picea m.–Sphagnum spp.–Hylocomium–Pleurozium/deep fibrous peat (>20 inches) type on lacustrine pans (S)

very moist

wet ±

wet

very wet

TABLE 11.6   Indicator Species Distribution Across Oligotrophic Ground Water Series in Southeast Manitoba. (Data from MUELLER-DOMBOIS 1964)[a].

		DRY		MOIST		
HABITAT TYPE	1	(2)	2	3	4	5
(Symbol as in TABLE 11.5)	vd	d+	d	f	m	vm
W1[b] **Restricted to dry**						
L Cladonia silvatica	■	+	•	•		
M Polytrichum piliferum	■	+	•			
F Solidago nemoralis	■	■	+	+		
G Koeleria cristata	•	■	•	•		
W2 Mainly dry						
S Prunus pumila	■	■	■	•	•	
LS Arctostaphylos uva-ursi	■	■	■	■	•	
G Oryzopsis pungens	■	■	■	■	•	
W3 Intermediate						
F Campanula rotundifolia	•	■	■	+		
LS Gaultheria procumbens	+	+	■	■	+	•
LS Chimaphila umbellata	•	•	■	■	•	•
G Oryzopsis asperifolia	•	+	■	■	■	
F Fragaria virginiana		•	■	■	■	•
W4 Mainly moist						
LS Linnaea borealis		+	•	■		
F Rubus pubescens			•	•	■	
S Rubus idaeus			•	•	■	•
F Cornus canadensis				+	■	
S Cornus stolonifera		•	•	•	■	
S Ledum groenlandicum				•	■	+
F Trientalis borealis				•	■	+
M Hylocomium splendens		•	•	•	•	
W5 Restricted to moist						
F Petasites palmatus				•	+	
LS Vaccinium vitis-idaea				•	•	
F Coptis trifolia				•	•	
M Sphagnum warnstorfianum					•	
W6 Indifferent						
S Vaccinium angustifolium	+	■	■	■	■	
S Rosa acicularis	•	■	■	■	■	
F Maianthemum canadense	•	■	■	■	■	•
M Pleurozium schreberi	•	■	■	■	■	
M Dicranum rugosum	+	+	+	■	+	

[a] ■=present in >80 percent of the sample stands, +=present in 60–80 percent of the sample stands, •=present in <60 percent of the sample stands, L=lichen, M=moss, F=forb, G=grass, S=shrub, LS=very low shrub. Each habitat type was sampled by at least five relevés of 800 m² size.

[b] W=Water group. Note, these are not "ecological species groups," because each group contains species of quite different life forms.

relation to soil pH. However, it should be noted that these species groups are not ecological groups in the proper sense because they are comprised of quite different life forms. For example, in the first group, there are at least two ecological groups, one formed by the lichen and moss which are merely attached to the surface of the soil, and the other perhaps by the forb and grass, which are both deep-rooted hemicryptophytes. Therefore, in the groups shown here, several of the species are drawing their soil moisture and nutrient supply from different depths in the soil profile. They are thus complementary rather than directly competitive in their habitat utilization.

11.3   ENVIRONMENTAL GRADIENT ANALYSIS

11.31   The Concept   The concept of environmental gradient analysis does not differ in principle from the previously discussed approaches of ecological species group derivation and ecological classification. All three relate to the analysis of species and community distribution along known environmental gradients. The approach from environment to floristic analysis has been distinguished in a review paper by WHITTAKER (1967) as *direct* gradient analysis. The approach through patterns formed by the vegetation itself is called, by WHITTAKER, *indirect* gradient analysis. The latter does not differ in principle from the community classification by means of differential species groups or the ordination of vegetation samples by community correlation coefficients or similarity indices. In the latter approach, the floristic pattern evaluation comes first. Thereafter, an attempt is made to interpret the floristic patterns in terms of environmental relations. In *direct* gradient analysis, the environmental variation is known at the start. The floristic variation is then correlated to find out to what extent the known variation of environment coincides with the floristic variation encountered.

Environmental gradient analysis contributes greatly to elucidation of the underlying causes of plant and community distribution. However, cause and effect relations at finer levels of plant distribution are difficult to establish through field observations because plant to plant interactions, animal browsing or grazing, fire, and other superimposed factors often obscure direct relations to known environmental factor intensities. The cause of these finer patterns can only be determined through experimentation (see CHAPTER 12).

11.32   The Technique.   The basic question in environmental gradient analysis asked by WHITTAKER (1967, 1970) is, "How are species populations distributed in relation to one another and [to] communities along an environmental gradient?"

As an example, consider an altitudinal gradient upslope along a mountain side that shows more or less clear vegetation belts or zones. WHITTAKER's approach to vegetation sampling was to purposely ignore the vegetation boundaries and to subdivide the altitudinal gradient more or less arbitrarily into 300 m elevation belts. Within these belts he sampled the vegetation along a subgradient, the so-called "topographic moisture gradient." The latter gradient relates to a series of topographic sites that by their position indicate a gradient from moist sites (valley bottoms or ravines) to dry sites (ridge crests or south and southwest facing slopes). Such topographic variations occur in almost any physiographically aged mountain system. An exception would be a recent volcanic mountain, or a smooth mountain slope that shows no important variation in topographic positions and associated moisture conditions. On such a smooth slope, positional variations in moisture may be ignored in an attempt to sample species and community distributions in relation to the elevation gradient.

In his monographic vegetation study of the Siskiyou Mountains, WHITTAKER (1960) used approximately 60 vegetation samples of each 50 x 20 m plot size to sample a topographic moisture gradient within a 300 m elevation belt. In each plot he counted all trees and shrubs by species. Herbs and low-shrubs were assessed in terms of frequency by twenty-five placements of a 1 m² frequency frame per 1000 m² plot.

The criterion for the layout of the vegetation samples was that the site had to be uniform over the 1000 m² plot area, i.e., areas with uneven surfaces or changing slope directions were avoided. Also, stands with signs of recent disturbances were excluded. In addition, WHITTAKER tried to accomplish a "random" coverage of the vegetation within each elevation belt. "Random" here implies a plot layout that aims at a geographically representative sampling across the elevation belt. Areas along mountain trails that met the site-uniformity criterion were sampled. Thus, WHITTAKER's sampling was not random in the statistical sense since it was limited by access. However, access limitations and the need for a geographically or spatially representative coverage often determine the layout in vegetation sampling.

In his earlier monographic study in the Smoky Mountains, WHITTAKER (1956) used two sampling designs, so-called "field transects" and "site samples." The "field transects" involved a layout of plots along predetermined intervals of a gradient (either a predetermined horizontal distance or a predetermined elevational interval). Since the predetermined sample location would not always satisfy the "uniform-site" criterion, he relaxed his more objective "field transect" sampling into the "site sample" layout which is the above described procedure of "random" coverage.

Depending on the site-moisture range within an elevational belt, WHITTAKER divided the topographic moisture gradient into six to ten "steps" and tried to sample each step with about five vegetation samples, which may be described as approximate replicate samples. A "step" was defined by a topographic position-type, for example, a ravine bottom position, lower slope position, N-facing slope position, etc. These were arranged into a step-wise series of topographic positions or site-types that would be judged by experience to vary from moist to dry within the defined altitudinal belt. No environmental factor was measured, but WHITTAKER points out that the moisture gradient was the overriding site factor and that other factors, such as temperature, wind exposure, and sun duration would change concomitantly along a topographic moisture gradient, which he also called a "complex gradient" for this reason.

The purpose of sampling within each elevational belt along a topographic moisture gradient was obviously to apply the fundamental experimental technique of varying a factor (or factor complex) whose influence one wishes to investigate while holding all other factors constant. In this way WHITTAKER accomplished sampling the effect of spatial changes in moisture on population and community changes within a more or less constant or uniform air-temperature regime that may be said to exist within a narrow (300 m) altitudinal belt. Conversely, he was able to group samples within a topographic moisture regime along an altitudinal range crossing several sequential 300 m elevation belts. In the Siskiyou Mountains. WHITTAKER applied the same controlled sampling scheme to a gradient across major parent materials (from diorite, gabbro to serpentine) and also to a major climatic gradient from a humid-maritime climate (Pacific Coast) to a more xeric continental climate (at the California-Oregon border).

For the presentation of results, WHITTAKER used two-dimensional ordinations of stands and species. For stand ordination he introduced a "weighted average" technique. This method requires first a classification of species into ecological amplitude or gradient-distribution groups (as applied by ELLENBERG, SECTION 11.12), except WHITTAKER's amplitude classes are less detailed. Along the topographic moisture gradient, tree species are classified into four amplitude-groups: (a) mesic (those restricted to the moist end of the gradient), (b) submesic (species of wider amplitude, but extending over the moist half of the gradient), (c) subxeric (species distributed over the dryer half of the gradient), and (d) xeric (species limited to the dry end of the topographic moisture gradient). An arbitrary value is assigned to each amplitude class (0 for mesic, 1 for submesic, 2 for subxeric, and 3 for xeric). These arbitrary values are used as multipliers to rank or ordinate the sample stands along the moisture gradient axis. For example, a ravine forest sample may include 100 trees. Of these, 60 individuals may belong to mesic species, 30 to submesic, 10 to subxeric and none to xeric. For this stand the moisture index would be (60 $\times$ 0 + 30 $\times$ 1 + 10 $\times$ 2 + 0 $\times$ 3)/100 = 0.5. For each forest stand sample two such "weighted averages" or stand moisture indices were calculated, one for the tree and tall-shrub stratum combined, and a second one for the herb and low-shrub stratum combined. As mentioned before, these smaller plants were quantitatively evaluated through frequency values, and the same technique of weighing is applied to their frequency values as shown above for the density values of trees and tall-shrubs. After computation, the two "weighted averages" for each stand are plotted into a two-dimensional graph. The y axis is used to plot the weighted indices of the tree and tall-shrub layer, and the x axis is used for the weighted indices of the herb- and low-shrub layer. Thus, each stand receives a position by two coordinates. The length of each axis is determined by the range of indices encountered for each layer in the series of sample stands. WHITTAKER (1967) calls this technique a "double weighted-average ordination." Stands ranked in this way form a more or less continuous, but scattered alignment along the diagonal of the two axes. This was interpreted by WHITTAKER as proof for the continuous variation of vegetation along environmental gradients, since in this technique the plot points or stands form no distinct clusters with intermittent breaks.

For the ordination of individual species, WHITTAKER uses as the abscissa the environmental gradient directly (i.e., the topographic moisture gradient, altitudinal gradient, or other). The spatial ranges of selected species are entered along this x axis. On the ordinate he uses as y value the quantitative value obtained for the species in the stand at the sampling position along the gradient. The quantitative species values are, however, averages obtained from usually five composite stand samples (which may be viewed as approximating replicate samples). The quantitative species values are expressed as relative values—for tree and tall-shrub species they are relative densities and for herb and low-shrub species relative frequencies.

The diagrams produced in this way show a series of bell-shaped species curves. The characteristics of these species curves are that they show each an individual position with regard to curve peak (mode) and range. The species modes differ from one another by position along the x axis (the gradient) and by height above the x axis (quantitative importance). None of the species portrayed by WHITTAKER show closely similar distribution ranges. Moreover, hardly any of the species show closely coinciding positional modes along the elevational axes. These results are in contrast to those relating to ecological groups (SECTION 11.12).

11.33   Conclusions.   From his studies of species distributions along environmental gradients done in this manner, WHITTAKER (1970: 37) draws two conclusions (quote):

1. Each species is distributed in its own way, according to its own genetic, physiological, and life-cycle characteristics and its way of relating to both physical environment and interactions with other species; hence no two species are alike in distribution.

2. The broad overlap and scattered centers of species populations along a gradient imply that most communities intergrade continuously along environmental gradients, rather than forming distinct, clearly separated zones. (Either environmental discontinuity or disturbance by fire, logging, and so on, can, of course, produce discontinuities between communities.)

With this conclusion, WHITTAKER clearly supports and reemphasizes the concept of species and community individuality of RAMENSKY (1924, 1930) and GLEASON (1926). Moreover, he has come to the same conclusion as CURTIS and his collaborators of the Wisconsin school for the continuum. WHITTAKER arrived at this conclusion through direct gradient analysis, the Wisconsin workers primarily through indirect gradient analysis. The variance of these conclusions with the ecological group concept may be related to the differences in plant life forms admitted to the tests.

Moreover, certain questions remain unresolved. For example:

1.   How is it possible that so many vegetation ecologists have come to the conclusion that species groups can be recognized that have closely similar distributions in nature?

2.   How can one explain the often abrupt change in vegetation structure in an altitudinal zonation from forest to scrub, the often marked zonation of herb-shrub-trees on a uniform coastal dune substrate, or the marked zonation on salt flats or other habitats where the environmental gradient does not change abruptly, but the vegetation structure does?

The existence of the phenomena mentioned under the second point is well documented in many studies, and was reemphasized by DAUBENMIRE (1966, 1968:18). WHITTAKER and others (e.g., McINTOSH 1970) apparently deny this phenomenon.

The first question is answered in part by WHITTAKER himself. He says that groups of species (or communities) can only be recognized arbitrarily. This is true. But the answer is unsatisfactory, because it evades the real question. "Arbitrary" is too general a word for the phenomenon that some species have more closely coinciding distribution ranges (though not exactly coinciding ones) along specific environmental gradients than others. It does not matter that the degree of similarity or dissimilarity of distribution ranges is continuous in principle.

WHITTAKER (1967) defines ecological groups of species as exhibiting closely coinciding modes along a specific environmental gradient. Therefore, their ranges may be broadly overlapping. Also, WHITTAKER's "ecological groups" may be species of different life forms. Therefore, trees, shrubs, and herbs may form one ecological group. This is in clear contrast to the ecological group concept of ELLENBERG, who defines them as species of closely similar life forms growing together in the same habitat and showing closely similar distribution ranges with regard to a specific environmental gradient.

It is, of course, easier to find species with closely similar distribution ranges by permitting combinations of different life forms. The clearest cases of such closely coinciding distributions are host-parasite relationships. But the degree of complementation and interdependency among species should, in principle, vary continuously among different life forms and species.

It is hard to understand how one viewpoint of noncorrelation of species ranges should answer the question of species and community distribution in nature. Certainly it would be equally dogmatic to claim correlation of species ranges as the only true expression of community development in nature. However, such close correlations can be expected also, and there seems to be no good reason why the two forms of species distributions cannot coexist in different situations or in the same situation. It is for future research to decide what patterns of species correlations apply to particular groups of organisms and life forms and to particular areas and environmental gradients.

11.34 Four Hypotheses. WHITTAKER (1970:35) presents four distinct hypotheses to this problem. These are (quote):

1. Competing species, including dominant plants, exclude one another along sharp boundaries. Other species evolve toward close association with the dominants and toward adaptation for living with one another. There thus develop distinct zones along the gradient, each zone having its own assemblage of species adapted to one another, and giving way at a sharp boundary to another assemblage of species adapted to one another.

2. Competing species exclude one another along sharp boundaries, but do not become organized into groups with parallel distributions.

3. Competition does not, for the most part, result in sharp boun-

daries between species populations. Evolution of species toward adaptation to one another will, however, result in the appearance of groups of species with similar distributions.

4. Competition does not usually produce sharp boundaries between species populations, and evolution of species in relation to one another does not produce well-defined groups of species with similar distributions. Centers and boundaries of species populations are scattered along the environmental gradient.

These four hypotheses were quoted in full because they are perhaps the clearest statements recently presented on this problem.

All four seem possible: The first in species-poor areas, where each synusia in a community is represented by only few species, perhaps by only one, and where the parallel distribution ranges relate to species of different life forms or different sociological value (*sensu* DAUBENMIRE 1968). Here, one such integrated unit (community) competes with another such integrated unit of complementary life forms. Each species of the same life-form type excludes the other by competition. This situation of all synusiae replacing each other at the same point along an environmental gradient, is, however, unlikely to recur at several points with such precision as suggested in hypothesis 1.

The second hypothesis may be possible also in species-poor areas where the synusiae consist of only a few or one species. But here the different life forms or synusiae are not well integrated. Each synusia has its individuality of distribution. For example, an undergrowth fern synusia may extend beyond the limits of the forest overstory into a neighbouring scrub community. Where the fern synusia meets a grass synusia there may be a sharp boundary along the gradient. Such distributions have been observed also.

The third hypothesis postulates correlated modes of species distribution ("ecological groups," *sensu* WHITTAKER 1967), but overlapping ranges. The correlated modes are most likely to be found among species of different life forms that do not compete with one another for the same general niche. Such distributions have been observed by many investigators.

The fourth hypothesis is the one of noncorrelation in species modes and distribution ranges. WHITTAKER has demonstrated this "individualistic" pattern of species distribution with data from several areas. Therefore, he holds to the view that this is the only realistic pattern in nature. His ecological group concept is thus clearly only a marginal idea, as it is only explained by the third hypothesis.

11.35 Application to Evolution.   It is clear that the question of species and community distribution is not yet closed, with an acknowledgement of only one of the four hypotheses as portraying a realistic trend. Further research into this question is important as it has a great bearing on the evolution of species interactions, the evolution of communities, and ecosystems. WHITTAKER (1967, 1970) holds that the highest degree of integration is accomplished by a high "beta-diversity." This implies accommodation of a high number of species with restricted distributions along an environmental gradient, as opposed to few wide-ranging species on the same gradient. The term *beta-diversity* is defined by WHITTAKER (1970:39) as "the degree of change in species composition of communities along a gradient" and contrasted by him to *alpha-diversity*, which refers to the number of species in a given community. The distinction of these two kinds of diversity appears to have useful application.

WHITTAKER believes that the trend in natural selection and plant evolution has gone away from, rather than towards, the formation of species with parallel or closely similar distributions. This is an interesting thought. But an increase in floristic richness on a given environmental gradient leading to narrower distribution ranges of most species involved, may also be accompanied by life form complementation and thus may not necessarily result in a tendency toward greater overlap in distribution ranges. In other words, it does seem equally possible that another trend in natural selection and plant evolution has been towards the formation of species with parallel or closely similar distributions. Moreover, species with parallel distributions do not need to belong to diverse life forms only, but also can be expected among species of closely similar life forms as explained in the ecological group concept (SECTION 11.11).

To clarify once more, the argument here presented is not against the findings of WHITTAKER and those of the Wisconsin school that species ranges overlap along a gradient; but it is against the idea that distribution overlap is the only true manner of species distribution in nature and that evolution always results in distribution overlap. Instead, it is argued that both trends, distribution overlap and synchronized distributions among species, can be observed in nature and that both trends may occur in the course of evolution.

Causal-Analytical Inquiries into the Origin of Plant Communities

12

12.1 SCOPE OF THE CAUSAL-ANALYTICAL APPROACH

Any thorough study of plant communities leads to the observation that species compositions show certain responses to their environments. Proof for this interpretation can, however, only be obtained from measurements and experimentation.

The causal-analytical approach to vegetation ecology is only in its beginnings and represents neither methodically nor by its subject matter a clearly defined field of science. It is closely interlinked with other branches of botany and biology (biogeography, palynology, physiology, genetics, biochemistry) and with many related disciplines.

However, here only those questions shall be considered that relate specifically to the sociological interactions of plants growing in natural or seminatural communities. Under this viewpoint, questions about allelopathy, for example, are considered part of causal-analytical vegetation research only if they relate to species that grow together in their natural habitats, where they may compete with one another. An important complementary research area is the field of genecology as defined by BAKER (1952), which is concerned with experimental studies into the nature of species populations (HIESEY, NOBS and BJÖRKMAN 1971).

The basic question asked from a causal-analytical viewpoint is: How did the species composition of a certain plant community originate? This question cannot be given a general answer, because the constellation of reasons for the origin of a plant community is different at any given point on the earth's surface.

12.2 CAUSES OF COMMUNITY FORMATION

Basic causal factors or factor-complexes for the development of a community are the following:

1.   The *flora* (f) of a given area, which provides for the basic material, the kinds of plants of various taxonomic ranks here referred to as species, for simplicity.

2.   The ability of a given species to reach the habitat in question. This depends primarily on the distribution of the species in the neighborhood, their dispersal mechanisms and establishment chances. This can be called the *accessibility factor* (a).

3.   The properties of the species themselves, particularly their life forms, their physiological requirements and other characteristics that have an influence on their ability to compete with each other once they have become established in the habitat. This can be called the *ecological plant properties* (e).

4.   The *habitat*, (h). This is the sum total of environmental factors operative at the particular locality in question.

5.   The *time*, (t) that has passed following a major historical event that initiated vegetation invasion or a change of one or another habitat factor.

The five factor-complexes effective in community formation can be expressed in the form of a function ($\int$):

$$\text{plant community} = \int (f, a, e, h, t)$$

where

f = flora
a = accessibility factor
e = ecological plant properties
h = habitat
t = time

This function differs little from the five-factor approach of JENNY (1941) to soil formation that MAJOR (1951) applied to vegetation or community formation. The latter reads:

$$\text{plant community} = f(o, c, p, r, t)$$

where

o = organisms
c = climate
p = parent soil material $\Big\}$ = habitat
r = relief or topography
t = time

There is only a difference of emphasis between the two equations.

In MAJOR's function, the factor complex "organisms" is very broadly stated. In the function here presented the same organism complex is further specified into three, namely flora, accessibility, and ecological plant properties, which are very important factors in causal vegetation ecology. Among the ecological plant properties, it is particularly the competitive capacity of the available species that determines community patterns in addition to the habitat factors. Competition is discussed in some detail in SECTION 12.5.

In MAJOR's function, the "habitat" factor complex is separated into three, namely climate, parent soil material, and topography. This is merely a question of emphasis. The present discussion emphasizes the totality and combined effect of all environmental factors and therefore, one symbol, (h), appears satisfactory. Of course, the concept of habitat can be detailed even far beyond the three components (climate, soil, and topography) into differing intensities and combinations of the basic environmental parameters needed for community existence (energy, water, nutrients). Moreover, our concept of habitat includes also such important factors as fire, animal grazing and browsing, seed predation (JANZEN 1971), and all forms of human intervention as long as these factors exert a continuous or regularly repeated influence.

Time as used in the above equation is a complex of at least three separately definable elements, (a) time as "rate of change," (b) time as "community age" and (c) time as any significant event that affected a community in the past in the form of a "historical perturbation." As pointed out by BILLINGS (1965), time is not a factor like temperature or water. Instead, he describes time as an overriding dimension. Time, most certainly, is an important element in community formation. But the importance of time is fundamental to all processes, because time is the underlying scale for the rate of change. Community formation can be viewed with respect to different scales of time. On a short-time scale, we may view the process as succession. This time scale may involve decades in secondary succession and centuries in primary succession (for definitions see CHAPTER 13). On a long-time scale we may view the process as evolution (SECTION 12.7).

As a rule, the five factor-complexes (f, a, e, h, t) interact in various ways. For example, the phylogenetic development of species and eco-types is influenced by local environment. On the other hand, the habitat becomes changed by the species occupying it. Such changes may provide for better living conditions of plants other than those that caused the habitat modification. This results in a time-chain reaction and a string of complex cause-effect relations that are difficult to untangle.

Such a complicated cause-and-effect system can only be understood step by step. In most cases, several of the factors can be eliminated right away in specialized investigations. For example, in short-term and locally restricted investigations, it is often possible to consider the flora and properties of the species as given constants. The ability of the species to reach a habitat will play an important role only in studies of plant invasion on new substrates, on denuded soils, in areas where introduced species have not yet become naturalized or where species occur in small relict populations. Locally, the accessibility factor loses importance with time. Time is an important factor only in those cases, where the plant community has not yet attained an equilibrium with the prevailing factors of the environment. This varies among regions. But, in many regions—even in cultivated landscapes—more or less self-perpetuating or relatively balanced vegetations can be found. Under those conditions, the habitat factors and the ecological properties become usually the core research activity in causal analysis.

12.3 IMPORTANCE OF FLORA IN COMMUNITY FORMATION

The influence of the flora on the vegetation can only be appreciated properly if one travels in a wider geographic area and then attempts a comparison of different parts of the world. One will find that similar habitats may be occupied by quite different plant communities, wherever these habitats occur in different floristic regions.

Good examples of this are the evergreen scrub and forest formations of the Mediterranean region when one compares these to four other world-regions with Mediterranean climates—the South African Cape Province, parts of South Australia, and certain sections of the west coasts of North and South America (in southern California and Chile). These areas have similar climates and, as a consequence, similar life forms dominate in the natural vegetation. However, the evergreen formations of these five regions are floristically entirely different. Less different are the forest communities in the glaciated parts of North America and North Europe, wherever they occur in similar climates. The plants occupying similar positions in the communities are, in this comparison, partly related species. However, the North American tree flora is richer in species.

12.31 Access to Habitat. From a regional flora, only those species are available for occupation of a habitat and subsequent competition that are able to reach the particular locality. The accessibility of a habitat for different species may be quite variable depending upon the condition of the vegetation itself, on the characteristics of the surrounding terrain, and on the current location of potential invaders.

So far, this factor has received little attention. HEIMANNS (1954) and EGLER (1954) emphasized accessibility as an important aspect in causal analysis. Particularly, the "contact communities" that adjoin the habitat in question play a great role.

Where, as described by ELLENBERG (1956), a *Calluna* heath is bordering seed-producing pine stands, one can soon observe pine seedlings in the heath community. Moreover, the frequency of the young pines differs greatly with distance of the seed trees and with prevailing wind direction. One can often use frequency along transects leading away from the seed source as a relative measure of access to a given sample area. Similar observations can be made where a strip of sand-birch trees (*Betula pendula*) traverses through heathland. It can usually be seen that the mean access-range, and therefore also the accessibility of equidistant points from the seed source, is greater for birch seeds than for those of pine. However, larger distances can usually not be bridged even by birch. For this reason, one finds, for example, in the neighborhood of swamp-birch stands (*Betula pubescens*) nearly exclusively swamp-birch as forest pioneers on neighboring heath and cutover land, while near sand-birch trees, the latter are more dominant as pioneers.

These examples show that the access to a growth location depends mainly on three factors: (*a*) on the specific mechanism for distribution of the plant species under consideration, (b) on the age and developmental stage, the vitality or height, and reproductive capacity of representatives of this species, and (c) on the abundance of the seed source of the species, its distance and position.

The importance of plant-access for the formation of plant communities can be studied very well in newly afforested areas, i.e., areas that were without forest cover for several decades. The list of species growing under the planted trees may initially have very little similarity with the species lists of neighboring forests. On the afforested land, it may show greater resemblance with those of heath-communities, grassland, or agricultural weed communities, depending on the former use of the land. However, typical forest undergrowth plants soon appear, and often show a superior competitive ability under the conditions changed by afforestation, for example, the reduced light availability. Yet, it usually takes a long time until those species of the forest arrive that lack efficient means for distribution, or that are among the quantitatively less abundant members of the typical undergrowth communities.

Growth locations in alpine valleys are commonly accessible for plants of all alpine vegetation belts, particularly if these are formed by anemochorous species, e.g., conifers. The high accessibility of alpine valley habitats for alpine plants may be one of the reasons why the course of succession, following invasion of raw substrates, is particularly rapid and regular in its pattern in spite of the harsh climate. Access plays an important role also in the foothill area of the mountains. The foothills are easily reached by seeds and fruits of plants from the high-alpine and subalpine belts. However, they usually cannot compete successfully under the less severe climatic conditions. Yet, these extrazonal elements can often survive for long periods on more extreme substrates such as alluvial gravel banks, on rock slides, and on land scars and artificially cleared soils.

If a growth location is of easy access for a particular species, this species may even become a permanent resident, in spite of not reproducing under the given environmental conditions, or competing in vegetative growth. In such cases, an abundant and sustained supply of disseminules can maintain the presence of this species. This applies, for example, to Pinus sylvestris in bogs where this species does not develop seed. Particularly effective are vegetative means of distribution as, for example, in Phragmites communis. This plant can even occur on dry soils by means of far-reaching runners, provided that the runners originate from a rhizome located in a moist habitat.

12.32   Latently Present Plant Species.   Any habitat is accessible to a much greater number of species than are growing on it. This is already shown by the fact that there are usually many living seeds visible in the soil, which belong to species that do not become seedlings or mature plants in the community. The number of latent seeds in the soil is surprisingly great in most investigated examples. BARBOUR and LANGE (1967) cite a number of studies concerned with this subject. Calculated on the basis of one square meter surface of agricultural soil, WEHSARG (1954) found a content of 25,000 to 150,000 seeds that he induced to germinate by repeated turning of the soil.

The most impressive example of the importance of latent seeds is the spontaneous appearance of new species on cut-over forest land. These are suddenly exposed to direct sunlight and oxygen and thus become activated. Similar observations can be made on ploughed grassland soils. Annual weeds appear soon after ploughing. These include even species whose seeds are not produced at that time of the year. Therefore, their seeds must have been brought to the grassland habitat at an earlier time, either with the fertilizer, by wind, or by other means.

The observation that there are often more species present in a habitat than are actually found growing at a given time reveals the presence of selective forces. The causes for the selection of a specific combination of currently growing species are to be found in part among the habitat factors and in part among the plants themselves that affect each other through competition or allelopathy.

12.4   VEGETATION AND HABITAT

12.41   Qualitative Analysis of the Habitat.   It is not the aim of causal-analytical vegetation research to study the reaction of plants to individual site factors, but to analyse their reaction to the combination of all factors. In particular, it is important to recognize the factors that are primarily responsible for the control of the species combination of the plant community under study.

It must be emphasized, however, that an understanding of the dependancy of plant life on individual site factors is a prerequisite to a proper evaluation of factor interaction and plant response. A detailed approach to individual factor analysis and plant response is given by DAUBENMIRE (1962) or in German by WALTER (1960) It is, therefore, not our intention to elaborate on this more elementary aspect of plant ecology.

Since the number of factors and their possible combinations are very great, it is not advisable to begin at once with the measurement of certain factors. A broad qualitative analysis should precede any quantitative analysis, as is done in chemistry before one attempts to analyze an unknown substance. Such a preinvestigation helps to save time and effort and may focus attention on factors that were initially not considered.

It was shown, for example, in a comparison of weed communities on cereal and hoed-vegetable fields that the dependency of certain weed groups on their cover crops was not related primarily to the cover plants themselves. The same combinations of weed species can also be produced without the respective cover crops (ELLENBERG 1950, 1963, 1974). This is so, provided the soil is treated in the same way and at the same time of the year for both kinds of crops. Yet, the similarity of treatments is not the only reason either. Several weeds typically associated with hoed vegetable, e.g., *Solanum nigrum* and *Chenopodium polyspermum* are more sensitive to frequent hoeing than *Matricaria chamomilla, Scleranthus annuus* and other cereal weeds. The hoed-vegetable weeds develop best, if hoeing is carried out only once, provided this soil treatment is done in the summer or late spring. If, on the other hand, the single hoeing treatment is applied in early spring or late fall, only cereal weeds are found to germinate in addition to the many indifferent species that occur under both cover crops. Thus, important for germination is the time of the year, or, more directly, the soil temperature at that time. Adequate moisture of the seedbed is required also by the cereal weeds. Therefore, in this case, quantitative measurements are best applied to germination temperatures and moisture conditions. In fact, LAUER (1953) found that most of the cereal weeds were found to germinate at cool temperatures and most of the hoed-vegetable weeds at warm temperatures. However, this investigation does not yet constitute a complete causal analysis of the two weed communities. It represents merely an explanation for the presence and absence of two characteristic species groups.

Further examples for the qualitative analysis of habitat relationships of plant communities are given by ELLENBERG (1963, 1974). In most cases the relations are not as simple as in the above example.

12.42 Quantitative Analysis of the Habitat Factors.   After recognition of the main controlling factors (i.e., the factors believed to be decisive), the selection of the method suitable for their measurement follows. Such methods are discussed by WALTER (1960), and in a manual on quantitative environmental methods in physiological plant ecology published by STEUBING (1965).

As shown already by the detailed discussions of BRAUN-BLANQUET (1928, 1951), the methods applied to the habitat of plant individuals do not differ basically from those used for the plant community.

12.5 HERBIVORY

Quantitative techniques to assess the effects of herbivory are usually not discussed in plant ecological methods books, even though herbivory may be a main controlling factor. A probable reason is that browsing and grazing effects are often subtle and difficult to recognize in natural communities. A widely applicable technique in the study of herbivory is experimental displacement of the animal in question by the use of exclosures. For example, it was possible with this technique to demonstrate that feral goats in Hawaii have a controlling influence on the spatial pattern of *Acacia koa* (koa) trees in the mountain parkland ecosystem on Mauna Loa (SPATZ and MUELLER-DOMBOIS 1973). Browsing of small herbaceous koa suckers causes profuse resprouting of roots of adult trees. When goat pressure is released long enough locally for the sprouts to grow with their crowns above the upper browse-level of goats (approx. $\geq$ 2 m height), the suckers develop into dense tree colonies. If, however, koa suckers of sapling size (0.5–2.0 m tall) are girdled by goats, the suckers do not resprout—they die. This contributes to the maintenance of tree islands in the parkland ecosystem by preventing the formation of continuous forest. Very infrequent fires are another factor (MUELLER-DOMBOIS and LAMOUREUX 1967). The goat-browsing effect was evaluated quantitatively through several population-structure analyses as described in SECTION 7.31. Similarly, with the help of exclosures, goat-grazing effects were evaluated on a tropical lowland grass community in Hawaii (MUELLER-DOMBOIS and SPATZ 1972). Here goats have a controlling effect on the species composition. Release of goat pressure in an exclosure resulted in very rapid displacement of the dominant annual grass (*Eragrostis tenella*) by perennial bunchgrasses (*Rhynchelytrum repens, Sporobolus africanus, Heteropogon contortus,* and others). After 18 months, woody chamaephytes (*Waltheria americana, Cassia leschenaultiana, Indigofera fruticosa*) and an endemic herbaceous vine (*Canavalia kauensis*) were well established. The vine covered 50 percent of the 7×100 m exclosure. Surprisingly, the vine turned out to be a new endemic species, never before described in Hawaii (ST. JOHN 1972). The causality in plant-species replacement upon release of goat pressure was related to the adaptive strategy of the different plant life forms involved. Annual grasses and creeping (stoloniferous) perennial grasses (in this case *Cynodon dactylon* and *Chrysopogon aciculatus*) are adapted to a frequently repeated mechanical disturbance from clipping, tearing, and trampling. The annual grass (*Eragrostis tenella*) responds by rapid growth to maturity in a few weeks under temporarily favorable soil moisture conditions and by subsequent production of abundant seed

This, coupled with a high capacity to survive as seed in drought seasons and rapid germination upon return of favorable moisture, guarantees its success under high grazing pressure. As soon as this pressure is released, these properties are inadequate. Instead, species which can outgrow the annuals and creeping perennials in biomass, such as the perennial bunchgrasses, find more favorable conditions. In turn, the success of the native vine was attributable to its capacity to overgrow even the bunchgrasses and in part the woody chamaephytes. For quantitative evaluation of the outside- and inside-exclosure variations, vegetation analysis methods were used as described in CHAPTERS 5 and 6.

Exclosures are not the only way of studying animal influences on the vegetation. They can also be evaluated through animal activity manifestations. For example, crown distortion by elephants (MUELLER-DOMBOIS 1972) was evaluated through a quantitative survey of defined crown damages, using the quantitative plot technique (SECTION 7.3). It was found that the structure of certain woody vegetations (deciduous tall-scrub with scattered evergreen emergents and *Feronia limonia* forest-scrub on grumusols) was strongly affected by repeated past elephant feeding (i.e., as manifested in crown distortion), while the evergreen low-stature monsoon forest (climax) showed much less activity and thus feeding preference for elephant-browse.

12.6   COMPETITION

Measurement of the most important habitat factors and their correlation to plant response is only the beginning of a full causal-analytical investigation. The real problems begin at this point, because one knows only the beginning and the end of the causal chain, which runs from the given combination of habitat factors to the present plant community.

As a next step one needs to explore the reasons for the present species combination of the community, because usually many other species can also reach the habitat and most of these may potentially be able to grow on it. This question cannot be answered so simply and generally as is often done.

12.61 Physiological and Ecological Response. Most communities contain plants which grow much better under cultivated conditions (such as in botanical gardens) than in their natural habitats. Therefore, plants do not as a rule obtain a physiologically optimal supply of environmental resources in their natural habitat. And yet, the performance

at these natural habitats may be the best over their entire natural distribution range.

The reason for this is plant competition. Plants of low and slow growth are commonly displaced from their physiologically optimal habitat, because other plants with faster growth rates and greater height development may also find their optimal conditions on the same habitat.

CAJANDER (1925) has drawn attention to these relationships by citing the example of pine. *Pinus sylvestris,* when protected by the forester against competing tree species, attains its optimal biomass and seed production on deep, fresh, weakly acid loamy sands and sandy loams. As a pioneer species, however, it is pushed out in the process of natural competition on such fertile soils by shade-tolerant species, in particular by spruce and beech. In natural landscapes, pine can maintain a position only on extreme or poor soils, where it is superior because of its tolerance to the poor habitat conditions. For example, pine is found on very dry and acid soils, such as formed on siliceous dunes. It occurs also on limestone soils, which are similarly very dry, but which have an alkaline reaction. Yet in Europe the same pine species dominates also on very wet habitats if these are acid and poor in nutrients and even on certain bogs. It would be incorrect to interpret this distribution as a "preference" of pine to such habitats. Pine is able to grow under these extreme conditions, because it has a greater ecological tolerance or amplitude with respect to moisture, soil pH, and soil nitrogen than most of the other forest trees, but it does not "like" these conditions.

Since CAJANDER's observations, a similar response has been observed for many other plants. The first experimental proof was given by the competition experiments of DE VRIES (1934) and ELLENBERG (1953). The latter has shown that a mixed cultivation of two species can be sufficient to displace their physiological optima in a way similar to those found under natural conditions.

As shown in FIGURE 12.1, the pH optima of *Sinapis arvensis* and *Raphanus raphanistrum* are close together when each species is grown in monoculture. In mixed culture the two species push each other away from their physiological optima. *Raphanus* then shows the greatest dry matter production at pH 4 and *Sinapis* at an alkaline reaction. *Spergula arvensis* is displaced completely from its optimum by the stronger competitor *Raphanus*. Both have similar optima when grown alone. In competition, *Spergula* can best develop only on strongly acid soils, where its light requirements are satisfied. In nature, these weeds are in fact similarly distributed on agricultural crop fields. Further examples are reviewed by ELLENBERG (1954, 1963, 1974) and KNAPP (1954, 1971).

A very impressive example of the influence of different competitors

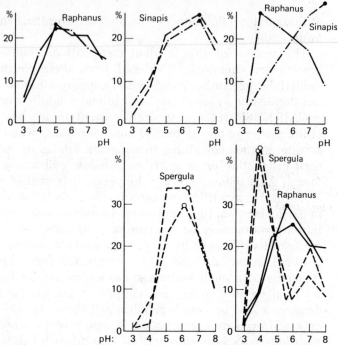

FIGURE 12.1. *Monocultures and mixed cultures in relation to a soil pH gradient. Above: monocultures of* Raphanus raphanistrum *and* Sinapis arvensis *(2 replicates each), and mixed cultures of both species. Below: monocultures of* Spergula arvensis *and mixed cultures of* Spergula *and* Raphanus, *with replicates in each case. Y axis values are expressed in percent biomass for each species.*

on the pH amplitude of the same plant species can be seen from data of OLSEN (1923), who, however, did not give this interpretation himself. OLSEN, in his classical investigation, determined the pH values of the habitats of different plant communities in which always the same species occurred. He also cultivated this species without competitors on the same soils as found at the natural habitats. His commonly cited conclusion from these experiments, namely that the response of plants in nature would coincide with that of pot cultures of gradually differing pH values is not quite correct in a strict sense.

The pH amplitude of *Avenella (Deschampsia) flexuosa* which was thoroughly investigated by OLSEN, is much wider in monoculture than under natural conditions (FIG. 12.2). The physiological pH optimum of this species is near 5. Therefore, the optimum is on the acid side, but not in the extremely acid range. On cutover land *Avenella* has a much narrower amplitude and its optimum is still more in the acid range. This

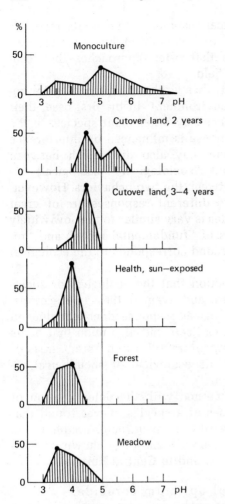

FIGURE 12.2.   *Optimum curves of Aven-ella (=Deschampsia) flexuosa in relation to soil pH in the rooting-zone—in mono-culture and in different plant communi-ties. (Based on data of OLSEN 1923.)*

optimum displacement is more pronounced on older cutover areas, which contain more competitors than the less densely stocked recent cutover areas. The optimum of *Avenella* is still further displaced into the acid pH range, when it occurs in heath. communities. A similar optimum is found on forest soils or even in relatively tall and dense meadows, where *Avenella* has to compete with many perennating species. Thus, the competitors are primarily responsible for the ascribed role of *Avenella* as a characteristic indicator of the most acid and raw humus soils.

ELLENBERG (1953) suggested distinguishing the response of a species in single culture or in monoculture as its "physiological response" and contrasting this to its "ecological response," that is, its response under natural conditions inclusive of competition by other species. Mixed cultures exhibit a response that often approaches the "ecological" response as observed in the field.

One may argue that plants must always show a physiological response, regardless of their growth environment. In the first case, their physiological response is not influenced by other plant species. In the second case, their physiological response is influenced by interference with other plant species. Thus, one may also distinguish between "individualistic" (physiological) and "sociological" (ecological) responses to indicate the difference between the two behaviors. However, the important point is that the two different responses are of great significance in causal ecology. The idea is very similar to the now widely used concepts in ecological genetics of "fundamental niche" and "realized niche" (MILLER 1967). The second corresponds to the ecological or sociological response.

We may emphasize in this connection that the ecological or sociological response of the same species, and even of the same ecotype, cannot be a constant one. The ecological response does not depend only on the constitution of the species itself, but also upon that of its competitors. Wherever the main competitors differ in different areas of the distribution range of a species, its ecological or sociological response is expected to be different also.

In all previous investigations concerned with the determination of ecological and physiological responses of a species, it was found that these were rarely identical. With regard to a given factor-gradient, one can distinguish the following responses (schematically shown on FIGURE 12.3; all examples refer to soil pH and to Central Europe).

1. Physiological and ecological optimum coincide.
    a.   The ecological amplitude is only somewhat more restricted (e.g., *Fagus sylvatica*).
    b.   The ecological amplitude is strongly restricted on both sides (e.g., *Senecio sylvaticus*).

2. The ecological optimum is displaced to one side in relation to the physiological optimum.
    a.   Very close to the low-extreme end of the physiological amplitude (e.g., *Avenella flexuosa* and *Spergula arvensis*).
    b.   Only little in direction of the low-extreme end of the physiological amplitude (e.g., *Alopecurus pratensis* and *Raphanus raphanistrum*).
    c.   Only little in direction of the high-extreme end of the physiological amplitude (e.g., *Arrhenatherum elatius* and *Sinapis arvensis*).

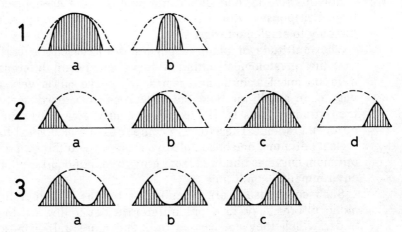

FIGURE 12.3. *Schematic representation of types of ecological responses in relation to a given factor-gradient. Physiological optimum curves dashed, ecological optimum curves solid. The latter show the response of a species under competition. Y axis values are expressed in percent biomass. Further explanation in text.*

    d. Very close to the high-extreme end of the physiological amplitude (e.g., *Tussilago farfara*).

3. Two ecological optima are found that occur near the two physiological extremes.

    a. The ecological optimum near the physiological low-extreme is more pronounced (e.g., *Festuca rubra* var. *fallax*).

    b. The two ecological optima are approximately equal (e.g., *Pinus silvestris*).

    c. The ecological optimum near the physiological high-extreme is more pronounced (e.g., *Convallaria majalis, Legousia speculum-veneris*).

The same species can show quite a different type of ecological response to different factor-gradients. For example, *Bromus erectus, Briza media,* and *Arrhenatherum elatius* belong to the following groups of the above outline:

FACTOR-GRADIENT:	pH	WATER	NITROGEN
Bromus erectus	2d	3a	2a
Briza media	3c	3b	2a
Arrhenatherum elatius	2c	2c	2c

For this reason, one should not compare the ecological and physiological responses with regard to the total factor complex of a habitat, but only to single factor-gradients.

The multitude of plant responses is increased further by the fact that the physiological optimum curve can be of different shapes for different species. In some species, the curve has a very narrow or a very wide amplitude, and the curve peak can be distorted to one of the extremes. This is, for example, nearly aways true in case of the nitrogen gradient. Therefore, in comparing the ecological to the physiological optimum, one has to always relate this to the actual shape of the optimum curve as obtained from growth in monoculture under the same environmental conditions.

Such comparisons are complicated by the fact that the results of monocultures may turn out differently depending on the conditions under which they are carried out. The more unlike these are to the natural conditions, the less reliable becomes the comparison. Therefore, nutrient solution cultures are useless for this purpose. Pot cultures in the open or field experiments are suggested instead. Here, the requirement is not to keep all other factors constant except the one to be investigated. Rather, the requirement is that the investigated plant species are grown under the same environmental conditions in mono- and mixed cultures.

A further complication is the fact that most wide-ranging species are comprised of more than one ecotype (HIESEY, NOBS and BJÖRKMAN 1971). This implies that different segments of the same species populations may have evolved different physiological requirements and adaptations. It is thus important not to view a wide-ranging species as an entity with the same physiological response throughout its range. Therefore, for valid experimental interpretations of the causality in community formation, the test species must be from the same local seed source or from vegetatively reproduced material.

The plant-parameter used for the above discussed comparisons of plant development was in each case the yield (dry shoot weight) produced in a vegetation period. This parameter is more indicative than a measurement of cover or plant height, of leaf size, or of other parameters that are measurable and that can be obtained without disturbance of the plant. However, in perennial plants, it is often more important to study their response over a period of several years than to obtain exact yield figures.

In annuals, however, seed production is often more important in competition than their individual growth capacity. Therefore, field experiments with annuals should be carried out over several years wherever possible to assess their reproductive success over a longer period of time. For example, DE VRIES (1934) sowed several weed species in rows which extended parallel across seed beds that represented different levels of a pH gradient. In the first year, the species did not yet show any interaction and they developed nearly equally well on all seed beds at all pH values. In the following years, *Stellaria media* became denser in the less acid range and even spread across the rows, while in the strongly acid range, *Rumex acetosella* and other species known as "acid indicators" were more successful than *Stellaria media*.

It is understandable, in view of the difficulties and relatively long time required for experimental investigations, that only few plants are as yet classifiable into the above outlined competition groups. There is still much work to be done in the causal-analytical direction of vegetation research.

However, a general insight can certainly be considered as definitive, namely that one has to be extremely careful with a characterization of the physiological requirements of certain species or ecotypes from field observations. One can certainly not conclude from the fact that a species occurs predominantly or exclusively on acid soils that this species is acidophilous. This is true only if it belongs to the types 1a and 1b. Most "low-pH indicators" belong to types 2a, 2b, or 3a. It is therefore better in such cases to speak merely of species that are tolerant to acid soils. Similarly problematic are such concepts as "nitrophilous," "xerophilous" or "hydrophilous." For example, so far no really "xerophilous" plant has ever been found. All species that were grown in absence of competitors developed best on soils with intermediate moisture regimes (see ELLENBERG 1963, 1974). BARBOUR (1970) came to a similar conclusion in a literature review on halophytes. He states that in all cases investigated so far there is no evidence for obligatory halophytes (i.e., "halophilous" plants) and that instead one can refer to some plants on saline soils as facultative halophytes.

It may be of interest to determine, in different floristic regions and major communities, the proportion of plants that show specialized physiological adaptation (i.e., real "preference") in terms of genetic fixation to their natural habitats. Where the proportion of such community members is high, one may speak perhaps of an evolutionary species equilibrium as proposed by WILSON (See SECTION 13.54).

12.62 Mechanisms in Competition. What means do the plants possess to succeed in competition? This question also belongs in the area of causal analysis, since it is concerned with the reasons for the grouping phenomena of plants in nature. Undoubtedly, in most cases, this is related to a modification of the environment by one plant influencing the other. The successful competitor withdraws more light, water, and macro- or micro- nutrients from the local environment to the disadvantage of the associated plant, or, it may influence other important environmental factors. Even the mere occupation of growing space may play a certain role.

For example, preempting of such a space through disturbance of the soil may provide a competitive advantage to a species with a superior propagation ability. This is the case with the European grass *Holcus lanatus* whose expansion is readily facilitated in the native *Deschampsia australis* grassland on Mauna Loa, Hawaii by the soil rooting activity of feral pigs (SPATZ and MUELLER-DOMBOIS 1972). Similarly, the advance of exotic *Bromus tectorum* (cheatgrass) into *Agropyron spicatum* grass communities, native in the Inter-Mountain Region of the Pacific Northwest, has been attributed to a competitive advantage of cheatgrass gained from various forms of ground disturbance (HARRIS 1967).

In seedlings that germinated at the same time, an initial advantage may become a lasting superiority in growth. The neighboring plants become a little overshaded and attain only a small yield. As a result, the superiority of the advantaged competitor increases even more.

Plants that are already established can only be outdone by competitors that are able to grow in their shade, or that can grow at a faster rate through openings in the canopy and then attain a greater height, or that may be more competitive for still other reasons.

Generally speaking, the competitive ability of a species depends on its genetic potential which is manifested in its morphological structure and physiological requirements. The following properties can be considered particularly important. Each of these may be especially decisive when others are equal:

1. *Morphological Structure* (largely expressed in the life form, see SECTION 8.2):

    a. Germination and growth rate in the early stages of development.

    b. Ontogenetic rhythm (duration of photosynthesis). Species with the same rhythms are strong competitors, species with different rhythms are more or less "complementary."

    c. Height. The final height, according to BOYSEN-JENSEN (1949), is the most important property in the competition struggle. The final stage in vegetation development is usually marked by the tallest plants. Smaller plants can succeed only if they can grow in the shade of the taller ones.

    d. Longevity. Longer living plants succeed by their "lasting ability" (see KNAPP and KNAPP 1954).

    e. Root system. In particular density, depth and morphology of the water- and nutrient-absorbing roots.

f. Means of reproduction. Reproduction from seed favors the migration into other communities, while vegetative reproduction is favorable for the maintenance and enlargement of an already established growth position. Vegetatively spreading herbaceous plants with a dense or closed growth habit succeed by "lateral exclusion" (e.g., *Arrhenatherum, Dactylis,* see KNAPP 1954, 1967), plants with a loose or open growth habit succeed by "penetration" (e.g., *Phragmites, Ranunculus repens*).

g. Regenerative capacity of the shoot system. This is of particular importance after temporary suppression (e.g., *Melica uniflora* in cutover vegetation) and upon mechanical disturbance (by logging, fire, mowing, grazing, trampling, etc.).

2. *Physiological Requirements,* i.e., the requirements for particular quantities and combinations of environmental resources and the response to these resources. The most important properties are:

  a. Light requirements.

  b. Heat requirements.

  c. Water requirements.

  d. Nutrient requirements and response to other chemical influences.

  e. Response to mechanical influences.

Since the competitive capacity of a species varies with habitat factors, it is not a constant species property. It follows that the competitive capacity cannot be determined through the plant qualities alone, but must be related to a specific environment and community. Therefore, experiments and field trials for determining the competitive ability of a species, as done for example by CAPUTA (1948) and others, are applicable only to the experimental conditions under which the tests were made.

For example, based on *Arrhenatherum elatius,* the relative yields of associated grasses were very different at different depths to water table (ELLENBERG 1954):

Relative Yield in Percent

	Depth to water table		
	140 cm	80 cm	5 cm
Arrhenatherum elatius	100	100	100
Alopecurus pratensis	18	15	1430
Dactylis glomerata	15	27	960
Festuca pratensis	12	21	485
Poa palustris	4	11	1125

12.63 Importance of Life Form in Spatial Competition. Spatial displacement of species optima along environmental gradients appears to be largely a function of similarity of life forms of the competing species.

ELLENBERG (1954) used an artificial slope with a controlled water table to produce a habitat gradient from dry to wet (see also WALTER 1971). Among other plants, ELLENBERG tested on this slope the behavior of three grass species that under natural conditions are associated with edaphically dry sites (*Bromus erectus*), with edaphically moist sites (*Alopecurus pratensis*), and with edaphically intermediate sites (*Arrhenatherum elatius*). Their growth in mixture resulted in sorting out of species optima similar to that found under natural conditions.

MUELLER-DOMBOIS and SIMS (1966) came to a result that appeared in variance on first view, but on closer examination supported the findings of ELLENBERG. They used a similar experimental model and three grasses that show a similar ecological behavior in southeast Manitoba. These were *Koeleria cristata*, a dry-site grass; *Calamagrostis canadensis*, a moist-site grass; and *Andropogon gerardi*, a prairie grass that shows some indifference to soil moisture regimes on sandy cutover sites. The expected shifts in optima did not occur during the 13 months this experiment was run. All three species ranged from dry to wet sites on the two soil slopes used in this experiment and their yields were highest in the middle sections.

When the experiment was dismantled, it became clear from the root structures that the three species did not compete with one another during the running time of this experiment. *Calamagrostis canadensis* had developed roots that formed a dense, but shallow lateral network beneath the soil surface. *Koeleria cristata* had formed deep-reaching fibrous roots that were everywhere in contact with the water table. *Andropogon gerardi* showed more or less single tap-roots that reached into the capillary fringe but not into the water table.

The experiment had been carried out in a greenhouse with artificial surface watering. Thus, *Calamagrostis canadensis* was kept alive, because of repeated surface watering. However, in contrast to the other two grasses, it had not extended its roots down to the ground water. Therefore, each species occupied a different niche below the ground. A similar structural separation was shown by their shoot systems. *Koeleria cristata* occurred as a caespitose bunch grass. *Andropogon gerardi* had produced more or less single, upright stems with lateral blades. *Calamagrostis canadensis* assumed a low-matted shoot habit with only occasional upright stems in the wetter portion of the tanks.

All three species were perennial grasses as were the three species ELLENBERG had used. However, the three European grasses are all caespitose and deep-rooting life forms and thus, are morphologically more alike.

The only significant competition produced in the Canadian experiment occurred within *Calamagrostis canadensis* itself. This species showed bi-modal density curves on each of the two artificial slopes. Density peaks were found in the wet and dry sections of each tank, with the lowest number of individuals occurring in the middle sections. The middle sections had the most favorable water supply because the root zone was kept near field capacity throughout the experiment. This apparently was the area of greatest intra-specific competition which resulted in a reduction of number of individuals per unit area associated with an increase in yield. This trend was further supported by the fact that the number of *Calamagrostis* individuals was much greater on the poor sandy soil as compared to the nutritionally richer loamy sand. The yield was much greater on the richer soil.

The individuals of *Calamagrostis* competed strongly with one another because they were close together in each tank and drew their environmental resources (in this case, soil water and nutrients) from the same narrow surface soil portion as indicated by their root structure.

Of course, morphologically the individuals of one species are very similar. They are forced to use the same habitat segment or portion ("specific" niche) of the environmental resources available at a site. As individuals they are the strongest spatial competitors.

It follows that the next strongest spatial competitors are species that are most similar in their morphological structure. This would apply to species of the same life form. Anatomical structure and physiological requirements may often be closely related to morphological structure. Perennial grasses are, of course, already a life form group by themselves. But within perennial grasses, one can distinguish several more specialized life forms, such as rhizomatous, stoloniferous, narrow-caespitose, and broad-caespitose (bunch) grasses, among others.

Thus, for producing spatial displacement from physiological optima under experimental conditions within a relatively short time, it is necessary to select species having closely similar life forms.

This experiment also brings to light the functional importance of the synusia concept. This relates to structural subunits in any plant community. These structural subunits or synusiae are composed of different species of closely similar life form (see SECTION 8.75). It is in these synusiae that one may expect species that compete strongly for a similar portion of the community environment or for the same "general" niche.

However, the emphasis on interspecific competition within the same life-form group growing in the same habitat is not meant to underestimate the competition that may occur between different life forms. For example, WALTER (1971) has interpreted the coexistence of scattered trees and dense grasses in some African savannas as a competitive relationship in the use of soil water, which results in a dynamic equilibrium. Moreover, competition between life forms is a fundamental process in autogenic succession (see CHAP. 13).

The ecological behavior of a plant species, i.e., the total response of a plant to its natural environment, is a rather complex function. It is the result of both its morphological structure and physiological requirements as matched with the environmental resources available in the habitat and modified by the competing plant species. Therefore, the ecological behavior of a species varies with differing environmental and sociological relationships as does its competitive ability.

12.7 ALLELOPATHY

Among the factors that potentially influence plant competition, we have not yet discussed the allelopathic ones. Allelopathy is usually understood as a biochemical inhibition among higher plants under natural conditions that is caused by the release of metabolic substances. (GRÜMMER 1953, RADEMACHER 1959, EVENARI 1961, MULLER 1969, BÖRNER 1971, GRODZINSKIJ 1973). The question of whether or not allelopathy plays a significant role in controlling plant distribution under natural conditions has not yet been fully clarified.

New emphasis has recently been given to chemical interactions between species as possible mechanisms of population control (WHITTAKER and FEENEY 1971), but the evidence brought forth for plant to plant interactions is mostly circumstantial. As correctly emphasized by GRÜMMER (1953), it is not sufficient to prove the existence of toxic leaf or root excretions. The evidence must be supplied that the substances are present in sufficient concentration under the described conditions and that their effect is not eliminated by soil adsorbtion or by interaction with microorganisms. Up to now, full evidence has not been established in any example. On the contrary, upon closer investigation, it has always been found that toxic substances (which are undoubtedly present in many situations and whose chemical structure is already known in many cases) lose their effectiveness in the soil. The initially allelopathically interpreted phenomena could in most cases be traced to other causes.

For example, the toxic substance produced by *Encelia farinosa* was shown by MULLER (1953, MULLER and MULLER 1956) to have no influence on the distribution of herbs growing among chaparral shrubs contrary to the earlier opinions of WENT (1942) and BONNER (1950).

The toxic substance loses its poisoning effect on the herbaceous plants in the soil, and the presence or absence of certain associated plants depends on the amount of accumulated wind-blown soil under shrubs, on the kind of humus formation, and on other non-allelopathic factors.

In subsequent publications, however, MULLER (1966, 1971) recognized allelopathy as the primary cause for the absence of certain desert-annuals around chaparral shrubs in California. He explained the barren zones surrounding some of the shrubs as caused by accumulation of toxic leachates from the shrub canopy and litter. Root competition from shrubs for soil water and nutrients to the disadvantage of annuals was ruled out as factor because soil analyses from under the shrubs through the barren zones and into adjacent perennial grass cover showed non-significant differences. Shading effects were ruled out by comparative light measurements. Grazing and trampling effects were ruled out by similarly sparse and stunted herb growth observed in small exclosures in the unprotected barren zones. However, BARTHOLOMEW (1970, 1971) found that increased rodent activity around the shrubs (determined through seed removal experiments, feces counts, and exclosure studies) and edaphic differences (harder soil surfaces, less organic enrichment in the bare zones and thus somewhat different water relations than under shrubs) are sufficient to explain the bare zones. Apparently the animal influences had not been fully eliminated in MULLER's earlier experiments. HANAWALT (1971) transferred soil blocks from the barren zones and from adjacent herb-covered areas into the greenhouse and found that poor germination and stunted growth of desert-annuals continued on the soil blocks from the barren zones upon watering with distilled water. However, after several months in the greenhouse, the soil blocks from the barren zones lost their inhibitory effect, presumably because the toxic substances had been leached out by watering. Also, artificial soil heating was found to result in significantly increased germination of annuals (McPHERSON and MULLER 1969) indicating release of an inhibitory substance. Such soil heating occurs under natural conditions from fires.

Therefore, allelopathic effects on germination and growth of desert-annuals from certain chaparral shrubs seem fairly well established in this case. However, it was also found that the barren zones disappear in moist years and that they are best expressed only on certain clay soils (MULLER 1971). Allelopathy under the described conditions is thus a temporary, though recurring, phenomenon that influences plant distribution in these environments with a pronounced drought season. Yet, it is only a contributory and not a sole cause for the bare zones (MULLER and DEL MORAL 1971).

BÖRNER (1971) suggested enlargement of the definition of allelopathy to include inhibiting biochemical effects of higher plants on soil microorganisms. This enlarged meaning would give more general significance to allelopathy. For example, RICE (1964) demonstrated that nitrogen-fixing and nitrifying bacteria were inhibited by plant extracts and leachates from sand cultures stocked with pioneer grasses and other herbaceous plants of abandoned fields in Oklahoma. He concluded that the annual grass stage in succession was prolonged because the grasses created a low-nitrogen environment to which they themselves were tolerant. While the reduction of these bacteria was not demonstrated under natural conditions, effects of higher plants on changing the soil microflora are most certainly expected. The higher plants provide the litter and humus, and the microorganism community can be expected to adjust to these conditions. Therefore, such an expanded meaning of allelopathy circumvents the issue.

More relevant to the basic question are autotoxic effects, i.e., allelopathy within the same species of higher plants. For example, McNAUGHTON (1968) showed that *Typha latifolia* germination was completely inhibited when seeds were treated with aqueous *Typha*-leaf extracts. *Typha* seedlings were strongly inhibited by water squeezed from *Typha* soils. While the latter could also be explained as a result of differential absorption of nutrients, it explains perhaps why *Typha* stands normally lack seedlings.

A significant investigation suggesting autotoxicity in the tree species *Grevillea robusta* was made by WEBB, TRACEY and HAYDOCK (1967). The authors observed absence of *Grevillea* seedlings in *Grevillea* plantations in Queensland, where *Grevillea* seedlings were abundant in a neighboring plantation of *Araucaria cunninghamii*. Seeding and planting of *Grevillea* under the *Grevillea* stand failed. The failure was not due to lack of light. Root competition for water and perhaps nutrients were ruled out as factors, by providing sufficient water and by transferring soil from the *Grevillea* seedling-rich, neighboring *Araucaria* stand into small excavated plots under the *Grevillea* stand. Some of the plots were trenched so as to eliminate root competition by tall *Grevilleas*. Seedlings survived and grew until roots of the mature *Grevilleas* penetrated into the trenched plots. Then the leaves of the young plants turned black and soon thereafter, they died.

In addition, sand cultures with *Grevillea* seedlings were watered with Hoagland solution and the leachate was recirculated. Growth was inhibited in the leachate treatment and the seedlings died after a certain time. This treatment, if the Hoagland solution was adjusted for nutrient loss, rules out root competition for nutrients. However, *Grevillea* seedlings watered with Hoagland solution collected as leachate from pots containing *Araucaria* plants died also after a certain time. This implies that recycling soil solutions from species not showing toxicity under natural conditions may also produce toxic effects. The treatment may result in accumulation of metabolic waste products, and certainly this sort of recycling does not occur in nature.

The authors suggested as cause for *Grevillea* seedling mortality a water-transferable toxic root exudate that is detrimental to the seedlings of the same species. The substance may act by changing the microflora in the soil, thus resulting in an antibiotic reaction specific to *Grevillea*. Root protuberances, typically found in the family Proteaceae, were suggested as the possible organs that may excrete the substance. The substance itself was not yet isolated.

The authors suggest that autotoxicity may well be the factor that causes wide-spaced distribution of so-called "nongregarious" species in species-rich tropical forests. They found that other naturally nongregarious tree species failed to produce vigorous mature individuals when planted in gregarious stands.

The study of WEBB, TRACEY and HAYDOCK is perhaps the only well-documented work of allelopathy under nonextreme edaphic conditions. Their study is not yet totally conclusive, since possible superior, absorption of nutrients by adult roots has not yet been eliminated as a factor. Moreover, the toxic substance has not yet been isolated and the mechanism of transfer in the soil has not yet been explained. We therefore believe it premature to offer any prediction about the general ecological significance of allelopathy.

12.8 IMPORTANCE OF HISTORY AND EVOLUTION IN CAUSAL ANALYSIS

As pointed out in SECTION 12.2, time is not only important as the underlying rate of all processes. In community formation, time becomes important in the sense of historical events that may induce a new course of community development.

Any organism is the product of history and evolution. The same can be said for species, communities, and ecosystems. While usually a considerable part of community structure and pattern can be explained from an analysis of current environmental factors and competitive relations among the species, another part cannot be understood without an appreciation of certain significant events that took place in the past. Such historical events, which may appear in the form of perturbations, differ from region to region and from locality to locality.

12.81 Historical Cause and Effect Relations. The more common historical factors may be listed as the impacts of man, fire, storms, animal influences, floods, diseases, etc. Most of these have mechanical effects on the vegetation, which are drastic in contrast to the more subtle effects of the interacting habitat factors of radiation, water, and nutrients. Therefore, we normally refer to such historical factors as disturbances or perturbations. However, historical causes can also act in a gradual way, such as the gradual change of a habitat factor or the invasion or introduction of a new species. These more gradual changes may also be recognized as perturbations on a telescopic scale of history.

Studies of the access of species to a given region (i.e., arrival stages of a regional flora, its origin, degrees of endemism and distribution) form specific inquiries of historical vegetation research. (GOOD 1964, WALTER and STRAKA 1970). The tools are in general fossil records (in paleobotany) or pollen analysis (in palynology). In this context, historical vegetation research deals with the first two factors of community formation—the flora and its regional access. Historical vegetation research may also give information on past changes in climate. Thus, also to some extent, changes in habitat factors are studied by means of historical floristic records.

History can be defined as a series of significant events and an interpretation of their causes. In vegetation ecology, however, we are primarily interested in the manifestation of such past events and in the identification of their causes.

A manifestation can be considered temporary or permanent. For example, a forest may be cut, and the subsequent vegetation development proceeds from the herbaceous cutover vegetation through a new immature tree stand to another mature forest. This is called secondary succession, and this phenomenon is essentially repeatable and temporary. Succession will be discussed in the next chapter. In secondary succession, the underlying time scale of the vegetation changes is usually measured in years or decades only. In primary succession, which originates, by definition, on newly exposed substrates or land-surfaces, the underlying time-scale of vegetation changes involves centuries or even a few thousand years. In this form of succession, permanent changes by species evolution are to be expected, which will affect community formation.

A clearly permanent manifestation, such as drainage of a habitat, is truly a historical phenomenon. These historical phenomena cause departures from the original vegetation pattern and community structures because of a more or less fundamental change in habitat conditions. Enough time may have passed so that the vegetation is again in balance with the prevailing habitat factors. In that case, one can easily mistake the modified vegetation for the original natural vegetation, unless the history of the area is studied also.

A permanent shift in vegetation pattern through the effects of past perturbations has been documented for Central Europe (ELLENBERG 1963) and Southeast Europe (HORVAT, GLAVAČ and ELLENBERG 1974).Beginning with the glaciation which covered Central Europe to the Alps, a reinvasion of plant species occurred from refuge areas. The European beech (*Fagus sylvatica*) was among the last invaders that became reestablished only 3000 to 4000 years (or 30 to 80 tree generations) ago. Much of its undergrowth vegetation is believed to have been there earlier. Today, both the tree layer and undergrowth vegetation in so-called near-natural beech forests are considered more or less in balance with the prevailing habitat factors.

Ever since the retreat of the ice, the European vegetation has been subjected to the effects of man, either directly or indirectly. Hunters were the first to cause some indirect effects through cropping of native wildlife. Then agriculture became important; numerous farms were already established in the Neolithikum in 2000 to 800 BC. Cattle, sheep, goats, and horses browsed in the forests. Fire was used as a tool to gain more agricultural and grazing land. Then followed the period of tribe-migration (Völkerwanderung), which resulted in reinvasion of crop fields by woody plants. In the middle ages, settlements became permanent and forest destruction increased till about 1800. In addition to construction- and fire-wood a major initial forest-use was a potash industry which was sustained by progressive forest burning. Then followed organized forestry with a preference for conifer plantations. Except in mountainous terrain and on steep slopes, most of today's forest vegetation in Europe is on land that was at one time used for other purposes, ploughed up, or otherwise denuded or semidenuded. Therefore, the "near-natural" European beech forests are mostly secondary. In some ancient forest reserves, the surviving oak trees with their naturally looking gnarled trunks and limbs, were found to be distorted from branch and foliage harvesting, which was a widespread agricultural practice of obtaining cattle fodder. Pastures are a very old form of European grassland that originated from grazing domesticated animals in forests, while hay meadows are considered more recent (only about 1000 years old). They originated apparently from wetland pastures. Today they are widespread on well-drained soils and the narrow, caespitose grass-forms, and species compositions are believed to have evolved in response to a long-standing management practice of repeated regular mowing.

The detailed historical investigations based on pollen analysis, soil profiles with charcoal remains, early written records. inferences drawn from historic settlements, and studies of least disturbed remnant vegetations, allowed only for a generalized reconstruction of the original vegetation of Central Europe. In many areas, the modifications are so strong that a reconstruction is impossible.

Yet, in spite of these modifications, which range from slight to total departure of the original condition, most of Central Europe's plant communities are in balance with the prevailing habitat factors if one considers management as part of these. This means that recurring species combinations are frequent and can be quite reliably predicted from environmental gradients. This contrasts somewhat with the North American vegetation where, in many areas, perturbations from forest fires and other disturbances have caused a prevalence of secondary successional stages (DAUBENMIRE 1968). In such areas, the communities are in differing degrees of balance with the prevailing habitat factors, and species recombinations along environment gradients cannot be expected as regularly as in the European vegetation. The term "balance" is here interpreted in relation to the way in which the perturbating habitat factors operate: (a) as a very frequently and regularly repeated influence (i.e., intensive management by mowing, fertilization, grazing, or immediate replanting after logging, which is characteristic of Europe); or (b) as long-interval, more or less unpredictable perturbations (e.g., wild fires in North America). From another viewpoint, one can consider the European vegetation to be in a rather artificial balance because of man's strong influence and the North American vegetation to be in a more natural balance because of a less intensive form of management characteristic of most of North America's wild lands (outside the agriculturally used lands).

12.82   Evolutionary Stress Factors.   Of the five factors mentioned as basic causes of community formation (SECTION 12.2), all are intimately linked with evolution. Evolution may be defined as a process that shapes the genetic development of the biota of a given region into a specific direction over a long, but indefinite period of time. This direction depends on what one may call the "evolutionary stress factors," which differ from region to region. These are a group of ecological factors that have caused biological populations to adapt genetically to the regional physical and social environments in which they evolved. Here are the major evolutionary stress factors.

   1.   Climate (particularly temperature, rainfall, and light) and other habitat factors (such as soil water and nutrients).

   2.   Geographic-historic factors, such as geographic isolation which is related to access; and invasions or introductions (by man).

   3.   Species interaction in the form of competition and herbivory.

4.   Recurring perturbations, such as fire, storms, floods, volcanism, climatic extremes, etc.

We may recognize climate as a first-order stress factor that has caused the evolution of a specific combination of life forms in a given region. It is well known that similar climates in different parts of the world support a similar mix of life forms and community structures (CHAPTER 8). However, this life form similarity is not perfect, because certain life forms may be poorly represented by species or even be absent, such as, for example, native succulents in the half-deserts of Australia (WALTER 1971).

The geographic-historic factors are primarily responsible for the specific combinations of species available in an area. But in contrast to climate, which can be said to recur with close similarity in many areas of the earth, the geographic-historic factors are regionally unique. The size of the areas containing biotic within-similarity, or what may be called a naturalized flora, varies with degree of geographic isolation or access or with time of arrival of the taxa. Time of arrival is particularly important. The example of the European beech, given earlier, shows that this important community-structure forming tree became naturalized only 3000 to 4000 years ago. Similarly, species brought to the Hawaiian Islands by the early inhabitants 1000 to 2000 years ago, such as the tree *Aleurites moluccana* or the grass *Heteropogon contortus*, appear to have invaded all available habitats.

However, this is not true for certain recent introductions (of 50 to 100 years ago), such as the trees *Eugenia cuminii* and *Psidium cattleianum* in Hawaii. Such plant species on the move obscure the predictability of species combinations along environmental gradients. This means that similar recombinations of species or recurring communities are not as frequent in regions where the species distributions are not yet in balance with the prevailing environmental factors. The effect is similar in areas with successional disturbances. This phenomenon of species on the move is probably rare in continental areas, but common in island ecosystems. It cannot be ignored in causal community analysis.

12.83   Species Coexistence.   Species interaction by competition was explained before as being just as important as the habitat factors in the formation of communities. A continued competitive stress among the members of two species with similar requirements for a limited resource is believed by some zoologists to result in the extinction of one species by the other (GAUSE and WITT 1935). This is the so-called principle of competitive exclusion or the "VOLTERRA-GAUSE Principle." In the context of plant ecology at least, this principle needs some qualification. Competitive exclusion of a species from a given habitat or community is a common phenomenon in succession (see CHAP. 13). But competitive exclusion leading to extinction of one species by another is probably a rare phenomenon in nature. It can only occur in very uniform environments that allow one species to gain strength over its opponent. Such situations may occur on small islands. In most natural environments, one may better speak of competitive displacement. This merely involves a shift in the distribution range of one or both of the species competing for the same space and resource along an environmental gradient. Different patterns of competitive displacement were discussed in SECTION 12.61, and these are nothing more than forms of competitive exclusion along only a limited segment of an environmental gradient.

Competitive displacement presents an apparent conflict with the ecological group concept as presented in CHAPTER 11. As we have said there, ecological groups are defined as species of closely similar life forms that grow together in relation to a specific environmental factor. How is it possible that these species, which apparently are strong competitors for the same space, do coexist in nature in sympatric (i.e., closely parallel) distributions? It was even claimed that evolution may go in two directions: a development toward spatial overlap along environmental gradients, of which competitive displacement is the extreme form, and a development toward spatial synchronism, i.e., ecological groups. Evolution of ecological groups can be understood as a development of "specialists" that evolved to exploit a common niche. This is easily visualized for other biota, such as bark beetles or phyllosphere fungi. In both groups, we find several species that have become adapted to the same substrate. An increase in the number of species occupying the same general niche may have two possible explanations. There may be a further specialization on a finer level among the members of an ecological group, or, in the absence of further specialization, the members may have developed a greater tolerance to the presence of competitors. As pointed out in CHAPTER 11, the members of an ecological group do not coincide exactly in all ecological properties.

The first possibility, an increased specialization among plant species, has been interpreted as leading inevitably to a further restriction in spatial amplitude along environmental gradients and to further overlap in horizontal range (WHITTAKER 1970, 1972). It is just as conceivable that an increased specialization on the level of fine-structure and function may proceed also in vertical space (by root stratification) as well as in time. Integration of function in time has been stressed by POORE (1968) as an explanation of coexistence and sympatric distribution of so many species of trees of similar structure in the multispecies rain forest. These two alternatives in specialization, integration in vertical space, and timing of functions, do not require spatial overlap along environmental gradients as an explanation.

The second possibility, development of tolerance for the presence of close competitors, is certainly the mechanism that allows members of the same species to coexist in aggregation in nature. It is axiomatic (and stated already by DARWIN 1859) that the members of the same species are also the most severe competitors, because they have the most closely similar requirements. This is also indicated by their nearly identical structures. It is quite conceivable that such a mechanism for tolerance can evolve also on the next higher level of structural similarity, which we have called closely similar life forms. The cause for the development of such tolerance may be the fluctuation of the common resource sought by the members of an ecological group.

Therefore, evolution may go in two opposite directions, i.e., the development of species with dissimilar requirements and the development of species with similar requirements. On a precise level of similarity, however, all species may develop complementary adaptations. Close spatial correlation along environmental gradients among similar species may be explained by complementation in vertical space, time, and tolerance.

12.84. Species Diversity. The question of why communities differ in their number of species has been asked primarily by evolutionary biologists. However, this question is currently gaining importance also in sociological ecology. Data for elucidation of this question is as yet very meager. But evidence may be found when considering the following points.

Evolution is usually thought to progress toward increased specialization (WILSON 1969). Assuming this to be the dominant trend, one may expect wide-ranging species to become adapted genetically to more specific segments of their distribution range. This is abundantly demonstrated by research in genecology (TURESSON 1922-1931, BAKER 1952 and others) and the existence of physiologically different ecotypes (CLAUSEN, KECK and HIESEY 1948, HIESEY, NOBS and BJÖRKMAN 1971). It is, therefore, conceivable that community patterns, in areas with longer evolutionary histories, are controlled more by the long-standing habitat factors operative at specific community sites than by competitive displacement among closely similar life forms.

However, this deduction is not necessarily true. Specialization is favored by natural selection only as far as it is efficient (MILLER 1967). This in turn depends on the nature of the evolutionary stress factors. For example, certain plant species occur with more individuals per unit area when the resource is in low supply. When the resource is increased, competition increases among the same individuals because of their increase in size. Mortality occurs among the weaker individuals. The end result is fewer individuals per area with an increase in the quantity of the resource (MUELLER-DOMBOIS and SIMS 1966). Similarly, eutrophication can result in the elimination of the more tolerant species in an ecological group. Such eutrophication may be a component among regional evolutionary stress factors, for example, fertilization by volcanic ash or sedimentation from periodic floods. In that case, historical perturbations in the form of a fluctuating resource may prevent an increase in the number of species.

Therefore, regional species diversity does not appear to be merely a function of time and access of the taxa, but is equally a function of the kind and *modus operandi* of the regional evolutionary stress factors. Recurring historical perturbations, such as fire, storms, floods, volcanism, climatic extremes, etc., are expected to interfere with specialization (i.e., genetic adaptation to the continuously active stress factors, such as annual seasonality of climate or a uniform or narrowly oscillating browsing pressure). For example, HARPER (1969) reports that the number of herbaceous species in England had increased in areas where rabbit grazing had become a constant stress factor in grass communities. In this case the increase in diversity may be related to two factors, the small size of the plants and the continuity of the grazing pressure. Constant clipping has kept the surviving plants small. More individuals can be accommodated per unit area, if they remain small. The constant browsing pressure eliminated competition by taller growing plants and may have allowed for further specialization among species that can complete their life cycle within close distance of the ground.

In this case, browsing has become a continuously operative stress factor producing a stable environment. In contrast, herbivory in the form of long-term interruptions with periods of high pressure and periods of low pressure may act as a periodic disturbance. In this case, the survival value of plants would be decreased by specialization in response to either the more constantly operating environmental factors or the perturbation itself. The most efficient adaptation, or the one of highest survival value, would seem to be development of tolerance to the perturbation and thus limited specialization (or speciation) in response to the group of continuously operative factors.

We may conclude, therefore, that specialization, or adaptation by speciation, may be high in regions with stable environments, i.e., those without periodic perturbations. This conclusion is supported by historic data on species diversity in lakes by GOULDEN (1969). But likewise, speciation may be high in unstable environments, i.e., where perturbations are a periodic and regularly recurring phenomenon. However, speciation is expected to be low in regions where perturbations are of irregular and unpredictable occurrence. A similar conclusion was reached by SLOBODKIN and SANDERS (1969). Speciation can be expected to be low in such areas because here specialization offers no advantage in survival. Instead, tolerance to environmental changes and maintenance of a generalized adaptation as found in wide-ranging species is of greater survival value.

12.85 Relation of Structure and Function. If it were possible to analyze precisely the nature of the evolutionary stress factors operative in a habitat, speciation and community evolution would be closely predictable. However, such an analysis seems possible only on a regional or more generalized level. Gross-structural similarity between broad communities offers a first basis for the comparison of functions. Such gross-structural vegetations are the world formations or biomes recognized as evergreen broad-leaved forest, coniferous forest, deciduous forest, grassland, scrub, desert, etc. However, it should be clear from the foregoing discussion that it is not possible to predict, for example, from one grassland formation to another, more than the gross-functions that are related to climatic seasonality and perhaps, fire. WHITTAKER and WOODWELL (1972) characterized grassland as having evolved generally in response to fire. It is true that the evolution of grassland in response to climate may be of very restricted geographic occurrence (WALTER 1971) or that climate may play only a contributory role, for example by dry seasons that promote the occurrence of fires. However, grassland varies from one region to another by grass life forms and species.

Perhaps a more appropriate generalization would be that mechanical influences in general are more important than fire alone as evolutionary stress factors in the formation of grassland. Mechanical factors include fire, animal browsing, trampling, avalanches, sedimentation by floods, and storms. It seems plausible that the caespitose or bunchgrass life form evolved in relation to fire, while the stoloniferous and rhizomatous grass life form evolved in relation to constant grazing pressure, whereby the latter may be a special adaptation to high animal density and trampling. Similarly, the bamboo-grass life form may have evolved in response to periodically devastating storms. This may be the reason for the many endemic bamboo species found on Taiwan, an island subjected to annual typhoons. These mechanical stress factors vary from region to region. Similarly, the kinds and number of indigenous animals and their mode of feeding may have caused a specific response-mechanism in native grasslands that find limited duplication elsewhere.

For example, scalping or sod-removal in short-grassland in the dry zone of Ceylon by elephants was found to create only temporary disturbances (MUELLER-DOMBOIS and COORAY 1968). The grasses involved were mostly stoloniferous and rhizomatous perennials that recovered the exposed barren soil surfaces of several square meter size completely in less than three months. A similar removal of grasses in a different region with the same climate and soil may create a more lasting disturbance. However, evidence of this sort is still lacking and studies of structurally similar communities in similar climates but in different biogeographic regions have only recently begun under the International Biological Program (IBP), e.g., the Mediterranean Scrub and Desert Scrub programs in the US/IBP.

Therefore, in inferring function from community structure, it is necessary also to analyze the number, kinds, intensities, and frequencies of the regional stress factors as much as possible. However, because important stress factors may be historical rather than current, any thorough causal analysis must be based on experimentation.

12.9   CAUSAL ANALYSIS OF PLANT COMMUNITIES

The final aim of causal-analytical vegetation research is to explain the origin of the total species combination of certain plant communities. Considering the present status of our knowledge, we are as yet far removed from this goal. The multitude of environmental effects and the influences of the plants upon each other, which can hardly be assessed without experimentation, complicate the analysis considerably. However, they also make this task a very challenging and interesting one.

Two complementary approaches seem of particular importance.

1.   A historical analysis of at least the main groups of regional environmental factors and perturbations with regard to their spatial distribution, frequency, and intensity in time. Identification and evaluation of evolutionary stress factors depend on a knowledge of their permanency and pattern of operation.

2.   Determination of the physiological optimum or fundamental niche of key plants in the community and a comparative determination of their ecological optimum or realized niche. These analyses will show to what extent the realized niche or ecological optimum is the result of competition or direct adaptation. On an exact basis such comparisons are possible only by experimentation which involves growing species that occur in association in nature in parallel experiments—once in isolation and once in groups under competition.

Particularly difficult and time-consuming are experiments with long-lived plants, for example forest trees. One cannot, without reservation, assume the response of seedlings to be equivalent to that of mature trees. However, the principles operative in the interaction of different plants and the formation of plant communities, can in many cases also be explained from experiments with short-lived plants, i.e., therophytes. Hemicryptophytes, particularly grasses, make useful experimental material because they can be reproduced vegetatively with ease, which provides for elimination of genetic variability. Therefore, these more simple objects play a similar role in causal vegetation ecology as do certain short-lived experimental animals in genetics.

Succession, Climax, and Stability

13

13.1 THE CONCEPT OF SUCCESSION

Assume the formation of a new terrestrial surface. This occurs in nature through lava flows, deposits of volcanic ash, filling-in of lakes, formation of landscars, and similar geomorphological processes. The result, in all terrestrial climatic regions favorable for plant growth, is an invasion of plants, sooner or later.

Soon after the invasion of new plants, a pioneer plant community may become evident. Invasion of new plants usually continues and the already present species may reproduce and become more abundant. Thus, a species enrichment may be observed on the same habitat that is associated with an increase in number of individuals or an increase in cover of certain species. Arrival of different life forms further increases the structural complexity. Concurrently with species enrichment, increases in number of individuals, and structural complexity, some early invaders may disappear. This usually is the result of competitive replacement among the species. But, disappearance of species may not always be associated with replacement by new species. It can also result in certain species excluding others. In many cases the net effect is a decrease in the number of species after passing a certain stage of maximum taxonomic diversity.

The general process of vegetation change described here has been called succession. TANSLEY (1920) has defined succession as "the gradual change which occurs in vegetation of a given area of the earth's surface on which one population succeeds the other." CLEMENTS (1916) distinguished six subcomponents: (a) nudation, i.e., the exposure of new substrate; (b) migration, i.e., the arrival of disseminules; (c) excesis, which refers to germination, establishment, growth and reproduction; (d) competition, which may result in species replacement; (e) reaction, which involves a habitat change through the species; and (f) final stabilization, the climax.

CLEMENTS' outline of succession still has validity. However, one may emphasize other subprocesses, for example, numerical changes within populations, changes in life form integration, or changes in genetic adaptation of populations in the course of evolution. Evolutionary changes certainly enter into the concept of long-term succession, although they are usually treated as a separate topic (see SECTION 12.8).

Succession as a study orientation is concerned with all changes in vegetation that occur on the same habitat in the course of time. This allows for a wide scope, which aims at establishment of the patterns and causes of succession. (For a brief historical development of the succession concept, see SECTION 2.5)

13.2 KINDS OF SUCCESSIONS

Vegetation changes may be of several kinds. The above described general sequence of a species enrichment associated with increased structural complexity is known as *progressive* succession. However, vegetation change may also involve the loss of species and a decrease in structural complexity as a result of site degradation. This may result, for example, from the loss of soil nutrients. Such vegetation changes may be distinguished as *regressive* or *retrogressive* succession (LÜDI 1930). Basically similar subprocesses may take place during a regressive succession. For example, grasses and sedges may replace trees upon a change in the soil-water regime through accumulation of excess-water in a formerly well-drained habitat, or a dwarf-heath vegetation may gradually replace a forest community because herbivores may eliminate any tree regeneration. Thus, it does not seem useful to restrict the concept of succession only to progressive vegetation changes.

GAMS (1918) made the point that it was unnecessary to distinguish kinds of successions by whether they originated from natural or man-made disturbances. The vegetation changes are hardly different, when, for example, a forest is destroyed through clear cutting or by a devastating storm, or if new soil becomes exposed through a landslide or through artificial soil movement. Instead, GAMS emphasized that successions differed fundamentally by the pattern of timing of the perturbations. He distinguished three kinds of successional sequences on this basis:

1. "Normal" sequences, which originate from constantly or rapidly repeated influences. Examples are: regularly repeated mowing of grassland, grazing and browsing, regularly used resting places of animals, frequent trampling, and so on. Such sequences lead eventually to an equilibrium stage, which can be viewed as a perpetuating community. The actual succession, i.e., the change-over phase, is then only temporary and of relatively short duration.

2. "Rhythmic" sequences. These originate from more or less regularly repeated or cyclic influences that recur at intervals of several months or years. For example, the rotation from cereal-weed communities to root-crop weed communities on the same agricultural field may be called a "rhythmic" succession. In this case, the change in agricultural treatment is the perturbation.

3. "Catastrophic" sequences. These originate from sudden perturbations of nonrhythmical nature, such as volcanic eruptions, earthquakes, landslides, fires, windfalls, bombs, clearings, drainage of a lake, or formation of a new lake, etc.

GAMS' concept reflects an earlier outlook. This no longer appears very useful for the following three reasons:

1. The nature of a perturbation, whether natural or man-caused, may not be critical for a specific local succession, but the difference has an evolutionary significance. For example, where fire has been a long-standing natural perturbation in a region (this could be from lightning or early man), the vegetation becomes evolutionarily adapted to it. Such is the case on sandy substrates in SE Manitoba, where a

fire-adapted pine (*Pinus banksiana*) dominates (see FIGURE 11.2). This tree species has serotinous cones that are opened only by the intense heat of a fire. Also, many associated species, for example, the undergrowing *Vaccinium* species are fire-adapted. They readily resprout from rootstocks after any ground fire. In contrast, in areas where fire is a recent introduction of technological man, its effects can be much more damaging because the flora is not adapted to such violent disturbances.

2. GAMS' "catastrophic" sequences, when followed over a longer time, may also turn out to be rhythmical. For example, volcanic eruptions, fires or windfalls may occur as cyclic phenomena in a particular region. In these cases they also may have an important evolutionary significance.

3. GAMS' "catastrophic" sequences include perturbations that result in total destruction of existing ecosystems, for example, volcanic eruptions, land slides, or the formation of a new lake, but also others, such as fire or windfall, that result only in partial destruction of an existing ecosystem. The difference between a succession starting on totally new substrates or one beginning within an existing, partially disturbed, ecosystem is of considerable importance. This difference has led to the distinction between primary and secondary successions.

Moreover, it is difficult to fit into GAM's scheme long-term successions that develop in response to climatic changes.

It appears sufficient for clarification to distinguish three major kinds of vegetation change by nature and timing of the perturbations and by the vegetation changes themselves. These are phenological changes, secondary succession, and primary succession.

1. Phenological changes do not involve only the flowering, fruiting, leaf emergence, and leaf fall on trees or other perennials, but also the periodic appearance and disappearance of therophytes and geophytes in a community. Such changes in species and number of individuals may not only be of seasonal occurrence, but they may also vary from year to year. In some desert habitats, annuals and geophytes may appear only after a penetrating rain that may come once in several years (WALTER 1971). Such phenological vegetation changes are excluded from the concept of succession. Phenological aspects were discussed already under vegetation structure (CHAP. 8).

2. The concept of *secondary succession* refers to all non-phenological vegetation changes that occur in already established ecosystems. This includes GAMS' "normal," "rhythmical," and in parts his "catastrophic" sequences. A secondary succession originates only from a partial disturbance of an ecosystem. Perhaps the most extreme case, in degree of disturbance, of a secondary succession is an old-field suc-

cession that begins after abandonment of an agricultural cropfield. In this situation, the soil is the remnant component of a formerly established ecosystem because it was already developed, it was supporting a stand of plants and it may still have a fully established community of soil organisms. Since any secondary succession proceeds within the context of an established ecosystem, it has a head start and the vegetation changes are relatively fast. The time-scale from pioneer stage to a stage of relative equilibrium or stability may be measured in years or decades. Because of the relatively short time-span in secondary successions, the process can be viewed as temporary. This implies, for example, that evolutionary changes in the ecological properties of the species usually may be ignored. If the source of disturbance is a recurring perturbation, such as fire in some areas, secondary successions can even be viewed as cyclic phenomena. In some marginal cases, secondary succession may be difficult to distinguish from phenological changes. For example, a large tree may fall down in a mature forest creating a gap in the canopy. A secondary succession will follow in the gap. Such gap-successions may be found in various stages throughout the forest. In these cases, they can be interpreted as an expression of the dynamic equilibrium of the forest and form part of its normal structural pattern. Thus, size of area is also important in the succession concept, and such small-size phenomena can be called "microsuccessions" or "serules" (DAUBENMIRE 1968). However, such variations in size should not really present a conceptual problem, if one clarifies properly the nature of the process. In general, successions are considered *directional* rather than *cyclic* vegetation changes (KERSHAW 1964, DAUBENMIRE 1968). But secondary successions can be viewed as both cyclic and directional, depending on how one wishes to emphasize the underlying time-scale.

3.   *Primary succession,* in contrast to secondary succession, is the community formation process that begins on substrates that had never before supported any vegetation. This is the kind of succession that gave rise to the classical concept of CLEMENTS and others as explained above. The time-scale in primary succession may involve centuries or even thousands of years. This depends on what condition the investigator considers to be the equilibrium or climax stage (see SECTION 13.5). On such a long time-scale, evolutionary changes in species adaptation and speciation cannot be ignored. For example, some tropical areas have never had any known geomorphological nudation during angiosperm evolution. Therefore, more primitive or less specialized plant life forms and species may have occupied the same sites that are occupied today by a highly evolved multispecies rain forest. Moreover, the same species cannot be considered eco-

logically constant over hundreds of years. However, this does not affect the reality of primary succession. But it does affect some of the hypothetical constructs about primary succession.

In an evolutionary sense, a primary succession is never cyclic, but progresses continuously in time. Yet, the initiations for a primary succession can be cyclic or erratically repeated within a region. For example, in a volcanic region lava flows may occur repeatedly in a cyclic pattern, or landslides may occur repeatedly in a mountainous region. However, physiographic aging on the same site may occur indefinitely. The substrates giving rise to primary succession may emerge from water bodies by silting of lakes, by emergence of a flood plain along a stream, or they may originate from geomorphological exposure of rock material, either solid bedrock, coarsely broken rocks, or fine rock or sand particles.

In the first case, the ensuing succession is called after CLEMENTS, a hydrosere; in the second it is referred to as a xerosere. COOPER (1913) used the terms hydrarch and xerarch successions for the same phenomena. The opposing initial differences in substrate moisture regimes, supporting either hydrophytic or xerophytic pioneer communities, are assumed to disappear in the course of primary succession. The validity of this assumption will be discussed later (SECTION 13.5). It is of course a fact, that rock material weathers upon exposure and therefore the moisture regime becomes in time more favorable for plant growth on such a site, unless the fine particles are eroded. The latter is true in many sloping topographic positions. Similarly, the terrestrialization process in a water body causes fundamental changes in the substrate water regime which affect plant growth. Thus, the more pronounced changes in primary succession are associated with changes in the water relations of the soil.

Such changes can be considered irreversible because they are part of the physiographic or geomorphological aging process of a landscape. The plant succession associated with such substrate transformations has also been called *physiographic succession* (COWLES 1899). This is another important distinguishing characteristic of a primary succession. The causes for the vegetation change due to such site changes have been called *allogenic* causes by TANSLEY (1929, 1946). He distinguished these from *autogenic* causes of succession, which involve vegetation changes caused by the organisms themselves. An autogenic change would, for example, be the disappearance of a species due to competition, allelochemical interaction (WHITTAKER and FEENY 1971), or parasitism. However, the separation into auto- and allogenic causes of succession is somewhat artificial. For example, competition among plants is ultimately controlled by outside or abiotic environmental factors.

The opportunties for studying primary succession are much more limited than for the study of secondary succession. Moreover, an exact study of primary succession can only be made for certain phases, such as the invasion of plants on newly exposed substrates or the long-term change of vegetation from historical records and indices. Any other approaches in the derivation of primary succession sequences remain rather hypothetical. But they may, nevertheless, offer a useful theoretical construct. For example, the idea is commonly presented (e.g., ODUM 1971) that the succession from a pioneer stage towards a climax (or a steady equilibrium state) represents a more and more complete utilization of the resources of the environment by increased species diversity and structural complexity, and that stability results from complete utilization and recycling of all available resources, rather than their removal from the system. This idea, however, can be applied to any progressive succession, whether primary or secondary, and it is thus relative unless the process of increasing species diversity and structural complexity is more clearly defined. In the commonly understood successional sense increasing species diversity and structural complexity simply means increasing the number of species and life forms on the same area from the already available flora. In the evolutionary sense, the same process means speciation, i.e., the development of new taxa. Secondary succession involves only the first type of process, while primary succession may begin with the first, and then continue with the second, i.e., speciation. This adds a certain complexity to the concept of primary succession, which has led to considerable controversy. However, the controversy seems somewhat unnecessary.

Numerous secondary successions can occur within the course of a primary succession, while a primary succession remains an open-ended process. An end-stage to primary succession, however, can be defined on a relative basis as a point in time when vegetation changes are slow. This will be discussed in SECTION 13.5.

13.3   METHODS OF STUDYING SUCCESSION

There are a number of methods that can be used to study vegetation changes. They fall into two general categories, (a) studies on the same area and (b) side-by-side comparisons. As a rule, the first are more reliable than the second.

13.31   Studies on the Same Area.   These can be based on the following: (a) permanent plots, (b) studies of exclosures, (c) air photos taken at different times, (d) historical and file records, and (e) evidence of change found in the present community.

Permanent plot studies. In forest stands, trees can be labelled and periodically remeasured for diameter, height changes, and mortality. New seedlings can be marked.

Herbaceous vegetation may be assessed and periodically reevaluated in small chart-quadrats whose position is permanently fixed. Reevaluation may take the form of mapping the cover outline, of measuring cover by the point-frequency method or by counting individuals and by photographic records.

One may also reevaluate herbaceous vegetation in larger plots without paying attention to the exact location of each former sample by statistically adequate resampling of systematic (transects) or random locations (see SECTION 6.4). In general, the quantitative methods described in CHAPTERS 6 and 7 are all useful for periodic remeasurements of communities to assess successional changes.

Where the changes in species quantities are rapid, and particularly where the changes are qualitative, succession can also be studied in permanent relevés using the BRAUN-BLANQUET method (MUELLER-DOMBOIS and SPATZ 1972).

Studies of exclosures. This is a special type of succession study that relates to investigating the course of community development under exclusion of grazing animals. It is in principle similar to the permanent plot study of succession, but does not always require an initial analysis of the community at the time the fence is built around the study plot. An exclosure usually allows for an experimental comparison of arrested succession or even regressive succession outside with progressive succession as occurring in the exclosure.

Studies of air photographs taken at different times. Air photos of the same area taken at different times can be an excellent tool for studies of succession. However, the detail obtainable is limited by the scale, and usually only the more general structural changes, such as the development from grass cover to woody plant cover, or a change in dominance of certain tree species, may be evaluated by this technique.

Air photo comparisons are based on the same principle as the permanent plot method, but the two methods provide two quite different sets of information.

Historical and file records. Written descriptions of the vegetation cover of an area at earlier dates may reveal useful information about succession, particularly if supplied with earlier vegetation maps and photographs. The usefulness of such information depends entirely on the objectives of the study. Exact successional trends are not usually revealed from such records.

Evidence of change found in the present community. There are two general sources of evidence of change, (a) organismic remains in the soil and (b) the population structure of species currently present.

(a) *Organismic remains in the soil.* Under certain conditions, information on succession can be obtained from organismic remains such as charcoal, fossils and pollen, found preserved in the substrate. The presence of charcoal in grassland soil gives evidence of two facts—presence of a former woody vegetation and the occurrence, at one time, of a fire. Fossils are preserved only in substrates that were waterlogged, thus excluding oxygen, at the time of deposition. Pollen is preserved also in dry soils, for example, in raw humus. Pollen analysis for historical studies of vegetation development is the major research inquiry of palynology. Of course, another form of historical information in ordinary soils may be the remains of former human habitation, such as pottery, tools or bones. These provide information on former land use, not on vegetation directly.

(b) *The population structure of species currently present. Structural analysis of the woody species.* An analysis of the woody plant species of a community by size and/or age classes may reveal important trends of succession. A species with a large number of individuals in the small size classes and fewer in the large size classes shows either active reproduction and maintenance on the site, or invasion from a neighboring site. The interpretation, whether such a trend shows maintenance or invasion, has to come from other evidence, for example, an analysis of the surrounding stand. A species showing the opposite trend, or no reproduction at all, may be interpreted as disappearing from the community. However, it is important also to consider the longevity and the rate of growth of a species. Large-sized individuals may range considerably in age but little in size. Thus, few smaller-sized individuals may be enough to maintain the population. Or, a species with a very fast growth rate may maintain itself actively in spite of being represented with only few small-sized individuals and a greater number of large-sized ones.

Such analyses of stand structure are the most important tools for obtaining information on succession (BRAUN 1950), because permanent plot data are usually not available. Size class records are the next best information to repeated measurements in permanent plots (for the method see SECTION 7.31).

Current species composition. AICHINGER (1949, 1951) defined a number of undergrowth community types in scotspine stands in Austria as vegetation development types. He was able to evaluate each community in terms of incoming (increasing) and outgoing (decreasing) species that occurred together with actively reproducing and quantitatively balanced species. In this way, he could group communities occurring side-by-side into a successional or developmental order.

A similar approach was used by CURTIS and McINTOSH (1951), who arranged sample stands of Wisconsin forests in order of their amounts of sugar maple, *Acer saccharum* (see SECTION 10.71). The stand with the highest amount of sugar maple was considered the successionally most advanced stand. All others were ranked in relation to this stand, and the resulting order was interpreted as a chronosequence.

Such an interpretation is scientifically defendable only where the course of vegetation succession is well known from permanent plot studies.

Succession studies based on an interpretation of current species compositions are often as hypothetical as those discussed next.

13.32 Side-by-Side Comparisons. Succession trends, (changes in time) are usually judged from a study of contemporary communities occurring side-by-side, i.e., in geographically separate places. The reason most of our conclusions about succession are based on this side-by-side approach is that few investigators have an opportunity to follow the changes occurring on the same habitat for any length of time. Other evidence of change on the same habitat, from written records or organismic remains, are rare opportunities.

Commonly, successional information is even inferred from studies of different communities where the dates of disturbance or the starting time of successional development are unknown. In these cases, side-by-side communities are arranged into a time sequence on the basis of relative differences in community development. The validity of such treatments depends on a thorough knowledge of the ecology of the entire area studied, and often such treatments include a strong hypothetical element. The better studies of this sort are supplemented with data about current population structures.

Closely similar substrates with different known dates of formation provide a very useful basis in succession studies. Such substrates include differently aged lava flows, volcanic ash substrates, differently aged landscars, sand banks, or dunes. Similarly, different dates of partial ecosystem disturbances, such as logging, fire, or the time a form of management was discontinued, such as the elimination of agricultural use of pastures or crop fields, provide opportunities for studying succession through side-by-side observations.

However, one important criterion has to be strongly adhered to in all such studies. All other ecological factors, except the time of disturbance or successional start, must be held as constant as possible. Particularly the climate and the soil type must be the same in the area where the side-by-side comparison is made. One cannot infer a chronosequence of vegetation development from comparisons of communities in a rain forest climate with those occurring in a seasonal climate. Likewise, one cannot validly infer a chronosequence from community-structure studies on physically or chemically different substrates. An example of common misuse is drawing conclusions about succession from communities on soils with different depths overlying bedrock. WENDELBERGER (1953) found that such soils were in equilibrium with the climatic factors and that they had different depths of overlying fine soil material from the start. Therefore, they could not be presumed to form a time sequence.

Since climate and soil vary from place to place, it is precisely speaking not possible to keep all ecological factors, but time, constant. For this reason, it is of utmost importance that an area is well studied and known ecologically before different communities can be interpreted in the form of a chronosequence or succession scheme. This implies that all geographic points in the area under study can be placed into a generally applicable ecological land or habitat classification (SECTION 11.2).

13.4 EXAMPLES OF SUCCESSIONS

Successions have received so much attention in the ecological literature that an adequate review of this subject would fill a book by itself. Obviously, there is no room for this here. Instead, we will concentrate on a few examples that elucidate some generally applicable principles.

13.41 Primary Successions. The classical division of this subject is into xerarch and hydrarch successions. However, it seems adequate, for differentiating the facts from the hypotheses in primary succession studies, to separate this topic into (a) successions on stable substrates and (b) successions on instable substrates.

13.41.1 Successions on Stable Substrates. Strictly speaking, there are no permanently stable substrates. However, it is readily appreciated that dunes, alluvial flats with periodic siltations, certain steep slope positions constantly losing soil material, or slope bottom positions receiving periodic or continuous colluvial deposits are associated with instable substrates. What we mean here by stable substrates are those edaphic components in which a significant change in either topography or soil water regime (a *physiographic successsion*) cannot be discovered in permanent plots or in an investigator's life time. These include, for example, most of the rock or glacial till successions described in the literature. It is characteristic of these stable substrate sites that indications of succession on the same site are obtained from structural changes of species populations rather than from physiographic evidence. The latter has always been obtained only from side-by-side comparisons of different substrates. This has led to some rash conclusions.

COOPER's (1913) otherwise excellent ecological study of the climax forest of Isle Royal (Lake Superior) suffers in its successional interpretation from an overemphasis of dynamism. While there is no doubt that Isle Royal emerged as a barren rock island sometime after glaciation (about 10,000 years ago) and then probably became invaded in the form of a xerarch succession as described by COOPER, the climax forest may not also have developed from a hydrosere. COOPER claimed that the climax trees (*Abies balsamea, Betula alba* var. *papyrifera,* and *Picea canadensis*) had entered already into the later stages of his hydroseres. It is probable that he was dealing with spatial ecotones (transition zones) where upland species may overlap into somewhat boggy habitats. This situation can be described as a broad ecological amplitude of these climax species but not automatically as a lowering of a water table or an upward growth of the soil. COOPER's exaggeration is his interpretation of all spatial ecological series in terms of a time series without any evidence that his nonclimatic climax communities were in fact still giving way to the climatic climax community of the region. (For an explanation of the climax concept, see SECTION 13.5).

Primary succession on a new lava rock deposit has recently been studied in the same location over a period of nine years on the floor of Kilauea Iki Crater, Hawaii (SMATHERS and MUELLER-DOMBOIS 1972). The crater occurs in tropical montane rain forest climate at an elevation of 1200 m with an annual rainfall of about 2500 mm and a monthly rainfall in excess of 100 mm except in June, when rainfall may drop slightly below 100 mm. The mean annual temperature is near 16°C.

The sequence of life form invasions and their progression is recorded in TABLE 13.1. The first macroscopic plant life forms to appear, one year after the new lava floor had been deposited on the crater floor, were three: an alga (*Stigonema panniforme*), two mosses (*Campylopus densifolius* and *C. exasperatus*), and a fern (*Nephrolepis exaltata*). A fourth cryptogamic life form, a lichen (*Stereocaulon volcani*), appeared in the third year after the new substrate had formed. Seed plants

TABLE 13.1   *Progression of Plant Life from Crater Floor Edge toward Center in Kilauea Iki, Hawaii (After SMATHERS and MUELLER-DOMBOIS 1972).*

DIRECTION ON CRATER FLOOR	YEAR AFTER ERUPTION	1960 1	1961 2	1962 3	1963 4	1966 7	1968 9	TOTAL DISTANCE TO CENTER
NE	Transact 'a'[a]	11 m	11 m	31 m	22 m	150 m	250 m	320 m
NW	'b'	—	—	10 m	30 m	80 m	120 m	385 m
SW	'c'	—	—	30 m	40 m	60 m	120 m	340 m
SE	'd'	—	3 m	40 m	48 m	90 m	120 m	400 m
	Number of species	4	6	10	17	23	30	
	Life forms:							
	Algae	1	2	4	4	5	5	
	Mosses	2	2	3	3	4	5	
	Ferns	1	2	2	3	4	5	
	Lichens	—	—	1	2	2	2	
	Higher vascular plants	—	—	—	5	8	13	

[a] 1m wide belt transects.

appeared in the fourth year. These included five species. Four of these were native sclerophyllous woody species (a potential tree *Metrosideros collina,* two shrubs, *Vaccinium reticulatum* and *Dubautia scabra* and a subshrub or semiwoody herb, *Hedyotis centranthoides*), and one was an introduced semiwoody, microphyllous herb (*Lythrum maritimum*). The introduced herb disappeared in the following years and only the four native woody plant species became firmly established. After the fourth year, other seed plant life forms invaded the lava surface. These included perennial grasses and soft-leaved woody plants.

In the ninth year, three pioneer communities had become evident.

1.   A cryptogamic pioneer community consisting of algae, mosses, ferns and lichens. This community occurred all around the crater and had advanced furthest towards the center.

2.   A phanerogamic pioneer community consisting of the above life forms plus native seed plants. This community also occurred all around the crater, but had not yet advanced as far to the center as the pure cryptogamic community.

3. A more complex phanerogamic pioneer community, which included several exotic seed plants. The additional species were all soft-leaved perennials including two native shrubs (*Cyrtandra* sp. and *Coprosma ochracea*). This third community was more complex because it contained a fuller complement of life forms, potential trees, shrubs, perennial herbs and grasses.

Growing with the *Metrosideros* seedlings, was a second potential tree species, the exotic *Buddleja asiatica*. All shrubs were native, namely the four species mentioned above. The herbaceous plants, except for *Hedyotis centranthoides*, were introduced species: a scapose, tall herb *Anemone japonica*, a stoloniferous creeping herb *Fragaria vesca* var. *alba* and three perennial grasses, *Paspalum dilatatum*, *Andropogon virginicus* and *Setaria geniculata*. In the ninth year, this community had only entered the first few meters onto the moister side of the crater floor and where the neighboring vegetation was undisturbed.

The three pioneer communities, therefore, differed greatly in their progression on to the new substrate.

While they could be distinguished as separate entities by time of appearance, spatial distribution, and life form composition, they occurred merely as overlapping communities. That is, the first cryptogamic community persisted in the second phanerogamic community, and these two persisted within the third more complex phanerogamic community. Therefore, succession in terms of population replacement had not yet occurred within the first 9 years after the eruption had ceased. Succession occurred merely in the form of species invasion and enrichment. The number of macroscopic plant species increased rapidly from 4 in the first year to 30 in the ninth year (TABLE 13.1).

While plant life zone expanded rapidly from year to year, from the crater floor edge towards the center, the filling-in with plant life of the zone itself proceeded extremely slowly.

The information is conveyed in TABLE 13.2. The cumulative transect area occupied in meters was obtained by summing the yearly progression of plant life along the four one-meter wide belt-transects shown on TABLE 13.1. Thus, the plant life zone expanded from an initial 11 m in the first year to 610 m in the ninth year. This expanding life zone was determined by the progression of algae.

TABLE 13.2 shows how uniformly this expanding life zone was filled by the different life forms. The record is given in percent frequency out of the total transect area occupied by plant life. It is shown that algae filled the expanding life zone with 100 percent frequency in the first year. Except for the third year, their frequency or uniformity

TABLE 13.2   *Frequency of Plant Life on Crater Floor Area Occupied,*
Percent (After SMATHERS and MUELLER-DOMBOIS 1972).

Year after eruption	1960 1	1961 2	1962 3	1963 4	1966 7	1968 9
Cumulative transect area occupied[a]	11	14	111	140	380	610
Algae	100	93	43	87	97	97
Mosses	0.9	29	22	22	21	20
Lichens	—	—	26	51	92	77
Ferns	27	50	11	16	20	18
Seed plants	—	—	—	0.4	0.6	0.6

[a] Each area refers to 1m wide belt transects.

of distribution remained very high. The third year was an unusually
dry year with a mean rainfall 1000 mm below the normal. This resulted
in much less vapor steaming and probably was the cause for the thin-
ning-out of algae in that year. The ferns were similarly affected. It is
shown that the frequency of the other life forms was much lower than
that of the algae. Only lichens increased rapidly in uniformity of spread
after their appearance in the third year, and they reached a similar
near-uniform distribution in the life zone as the algae in the seventh
year. Ferns and mosses showed much less uniformity in distribution
with frequencies between 16 and 22 percent after the fourth year. The
seed plants were very scattered showing less than 1 percent frequency
after they appeared in the plant life zone in the fourth year.

Yet, while TABLE 13.2 gives an indication of the filling-in process
in the plant-occupied area, the lava flow surface in the ninth year after
the eruption still looked practically barren. This was due to the fact
that the plant cover was as yet practically insignificant. Wherever
present in a meter square quadrat, the species occupied only a small
fraction of the quadrat, usually less than 0.1 percent of the surface.

Moreover, the individual life forms occupied different microhabitats.
The ferns and seed plants grew only in the crevices and cracks of the
pahoehoe (i.e., smooth, pavement-like) lava. The ferns could grow with
their rhizomes surficially attached to the fractured rock in these cracks,
while the seed plants grew only in crevices where their roots were at
least somewhat imbedded in accumulated fines from windblown fractured rock flakes or ash. The algae, mosses and lichens could grow on

the smooth surface of the pahoehoe lava. But the algae colonies were found mostly imbedded in the rockpores. The mosses probably also started with their protonemata in the pores, while the developed moss-sporophytes grew right on the rock-surface, and also in the crevices. The lichen-initials also occurred in the surface-rockpores but soon extended over the exposed surface. While the algae and mosses thrived in microhabitats receiving moisture from steam condensation and run-off, the lichens were absent in those places.

The initial stages of primary succession on lava rock as reported in this study follow closely the sequence of evolution of plant life from the more primitive nonvascular cryptogamic life forms over the vascular cryptogams (ferns) to woody and then the herbaceous vascular flowering plants.

Gymnosperms (the more primitive vascular plants with naked seeds), though forming an important stage in plant evolution, did not appear in this succession. Gymnosperms were not available as seed sources and have never arrived by natural means on to the Hawaiian Islands. Another observation of perhaps evolutionary significance in ecosystem development shown in TABLE 13.1 is that two trends seem to occur almost simultaneously. One is the arrival of more than one species equipped to utilize the same general niche. For example, the number of macroscopic algae species equipped to grow on the smooth pavement surfaces rose to five in the seventh year. The other trend is the development of structural complexity with the arrival of different life forms which are equipped to utilize quite different portions of the same habitat. For example, algae, mosses and ferns arrived together in the first year. Each life form occupied different general niches, the algae took to the pores on the pavement surface, the mosses grew in some of the crevice bottoms, and the ferns stuck to certain crevice-walls with their rhizomes. In addition, there occurred overlap of niches, but no evidence of competition between different life forms. Yet, as is evident from the progression of the three pioneer communities, the increase in structural complexity lags somewhat behind the increase in species of the same structural group or life form.

A telescoped sequence of a simulation of plant evolution in primary succession can only be expected on raw rock surfaces and not necessarily also on new geological substrates that are already broken down into fines. These include ash substrates, glacial till, dunes, and so on. Here vascular plants may be expected among the first mascroscopic plants. For example, on the raw ash deposit of the 1960 Kapoho eruption, a sedge (*Cyperus polystachyos*) occurred among the first plant life forms. On sand dunes rhizomatous grasses are usually the first invaders (COWLES 1899, ELLENBERG 1963, DAUBENMIRE 1968).

Therefore, a generalized sequence of plant life forms in primary succession can only be stated with regard to the kind of substrate and the floristic offering in terms of available disseminules. Another important variable to be considered is macroclimate.

The volcanic rock succession described is, of course, only the beginning of a primary succession, which from here on may be said to go on indefinitely or to a climax stage. To obtain an idea of what the future trend may be, one must resort to comparing existing communities of different, successive ages. This requires a good knowledge of the ecology of an area.

In areas where the community ages are undeterminable, one can only use the maintenance trends of the major species, which one can obtain from structural analyses, periodic remeasurements of permanent plots, and a knowledge of the autecology of the major species, for example, in terms of their relative tolerance to shade. However, where community ages are known, one can attach a time-scale to the succession. This is of considerable interest for answering the question of just how dynamic is a succession. Of course, the community age in primary succession cannot be established simply from counting the growth rings of trees. Instead, one needs to know the time of origin of a community, which may be closely related to the age of the substrate.

Such a study was recently done by ATKINSON (1969, 1970) on dated Hawaiian lava flows. His study included the area in the humid montane climate near Kilauea Iki. He applied soil analyses involving the rate of calcium loss, titanium gain and pH decrease as a means of dating prehistoric lava flows, because lava flow dates have only been recorded for about the last 200 years. He recognized several stages from barren lava flow to rain forest, via a *Stereocaulon* lichen–*Nephrolepis* fern stage, a *Dicranopteris* matted fern–with scattered small-sized *Metrosideros* trees stage, to a closed *Metrosideros-Cibotium* (tree fern) forest. ATKINSON believed that the latter closed rain forest, in form of a currently self-perpetuating community, was reached in 400 years. This is perhaps the fastest known rate of a primary succession to a perpetuating forest community. This is understandable, because the area is in a humid climate with year-round favorable growth temperatures.

However, this so-called perpetuating rain forest occurs on only incipiently weathered lava rock. Therefore, no significant site change or physiographic succession has occurred in this primary succession. For this, one has to go to an older island of the Hawaiian chain, where lava rock has changed to soil. On such hydrolhumic latosols (Oxisols or Ultisols), we usually find that the tree ferns form only a minor component. Instead, a number of native small woody trees and shrubs occur. These soils are several millions of years old. Here even the tree species *Metrosideros collina* occurs with different varieties, which shows that evolutionary change enters as a variable into the concept of primary succession.

13.41.2 *Successions on Instable Substrates.* These are the kinds of primary successions in which the substrate of the site changes more or less concomitantly with the vegetation. When we say concomitantly here, we mean that the site changes were observed to occur so rapidly that the changes in communities appeared to be closely correlated with the site changes. For example, the classical dune succession of COWLES (1899) is such a succession; many observed hydroseres also fall into this caterogry.

The inland dune succession described by COWLES was derived from comparisons of dunes with their associated plant cover. Thus, COWLES' conclusions of a sequence in time were derived from a sequence that in reality occurred in space, i.e., from an ecological series. However, his side-by-side observations were done so carefully that the derived time sequence appeared very convincing. He recognized a system of dunes that extended from the beach (near the shore of Lake Michigan) to stationary embyonic dunes to active wandering dunes to passive or established dunes, and he clearly explained the interdependency of this dune system with the prevailing wind forces and the plant communities on the different dune types. He observed that the size of the stationary dunes was related to the plant life forms. The smallest dunes were occupied by grasses, such as *Ammophila arundianacea*. This grass has an elaborate rhizome system that fixes the wind-blown sand so that it becomes a laterally widespread mound. In contrast, other grass dunes with *Elymus canadensis* were rather narrow, because this grass does not form rhizomes. Dunes occupied with *Salix adenophylla* or *Prunus pumila* were larger than the grass dunes both in height and area, because *Salix* and *Prunus* show extensive horizontal and vertical root growth. *Populus monilifera* dunes were the highest since this species grows tall, but the dunes were not wide because *Populus* does not spread horizontally.

He explained the changeover from one dune type to another in the wandering dune complex as an incapacity of the respective dominant plant life form to grow with the ever-increasing buildup of sand beyond a certain threshold point. This threshold point was reached when the root system became disconnected from the capillary fringe of the water table. When finally the dunes outgrow all available life forms, they become wandering dunes. These are moving with the wind either till they become smaller, in which case they may repeat the plant life form associated sequence, or till they have moved to a location where the wind no longer accumulates sand beyond the vegetative sand-fixing capacity of the developing community. In the latter case, the dunes become passive or "established."

On these established dunes, COWLES observed low-stature *Tilia americana* thickets with associated river-bottom plants and next to these, tall *Tilia* forests, which he believed to indicate the next stage in development. *Acer saccharinum* and *Fagus ferruginea* forests occurred also on the same topographic positions, and these he believed to form the normal climax or finally stabilized type of the Lake Region. On the windward slopes of the established dunes he noted quite a different community with a dominance of evergreens, which seemed to develop from heath (i.e., here *Arctostaphylos uva-ursi, Juniperus communis*) to coniferous forests of pines (*Pinus banksiana, P. strobus, P. resinosa*). He noted that the pine forests on windward slopes of the established dunes can easily collapse through a slight change in physical condition of the substrate and that this may result in a regressive succession followed by a new dune and community life cycle. He also observed that *Quercus tinctoria* may follow the pine foerst in the south of the lake region, but he was very cautious in not stating this as a fact.

Thus, COWLES' conclusions of a chronosequence from the side-by-side occurrence of different dune and community types is highly reasonable, as he did not extrapolate beyond the finally stable topographic positions with their communities to a single end-stage. He carefully evaluated and stressed the edaphically more stable conditions, which retain a certain community mosaic in his dune succession. However, he did not provide a time-scale for the dune succession, and the reader of his paper may get the impression of a very rapidly changing environment. SNOW (1913) showed that dune successions are not really that fast, but require at least decades to show noticeable changes in vegetation types. OLSON (1958) was able to attach a time-scale to COWLES' dune succession. He found from radiocarbon dating that some of the established dunes are at least 12,000 years old. For the earlier stages, he determined dates of a few months for the smallest grass dunes, 4 to 8 years for the larger grass dunes, and 20 years for a foredune ridge occupied by jack pine (*Pinus banksiana*). The dates were obtained by counting annual rhizome elongation in the grass dunes and from growth ring counts on jack pine. Radiocarbon dates for the taller stationary dunes turned out to be 2500 to 3500 years. To the finally established or passive dunes, OLSON remarked that the blueberry-huckleberry community does not seem to begin to invade the black oak (*Quercus tinctoria*) dunes until after 6 to 10 centuries.

OLSON also noted from his comparative analysis of soil carbon, nitrogen, moisture equivalent, carbonates and cation exchange relations that most of the soil improvement in fertility and water holding capacity of the original barren dune sand occurs within about 1000 years after dune stabilization.

Thus, we may conclude that early development is fast, but then slows down considerably. The initiation of the climax vegetation stage associated with physiographic stability of the dunes, however, may take as much as 10,000 years.

TANSLEY (1939) summarized several studies of hydroseres done by PEARSALL in England. A factual account of a small part of hydrosere development is shown on the two maps (FIG. 13.1). It can be seen that the 1.8 m water-depth contour line extending through the south-center part on each map had contracted after 15 years. Thus, sediments had accumulated. The *Salix* stand in the north part of the map had extended further southward. The *Salix* stand is the most terrestrial of the communities shown. The *Carex* community south of the *Salix* stand had replaced much of the *Phragmites* community. The *Typha* community extending east from the stream mouth had spread by replacing much of the *Phragmites,* while *Phragmites* replaced much of the *Scirpus lacustris* community. The submerged *Fontinalis* moss community in the center of the 1914-1915 map became much smaller in 1929. Thus, definite changes in plant distribution are here associated with substrate changes in the form of soil accumulation indicating a typical allogenic succession in TANSLEY's terminology. However, this 15 year change only shows a minor shift or small step in the course of a hydrarch succession.

A more complete picture of a hydrosere is diagrammed in the form of a scheme on TABLE 13.3. Such a scheme is meaningful only when the patterns and ecological relationships in an area are well studied and when there is clear evidence for successional trends. The chronosequential relationships can then be presented as a simplified succession scheme.

TABLE 13.3 gives an example of a succession scheme of communities based on historic records and maps. The communities were found to form upon filling of eutrophic lakes in Northwest Germany. TÜXEN and PREISING (1942) distinguished three successional series in this area.

1.   A primary, progressive succession from open water to a black alder swamp forest as the final, natural community.

2.   A regressive succession that started through management. This included cutting of the black alder forest, subsequent mowing, artificial drainage operations, and fertilization, finally arriving at a fresh *Cirsium* meadow that is closely related to an *Arrhenatherum* meadow community.

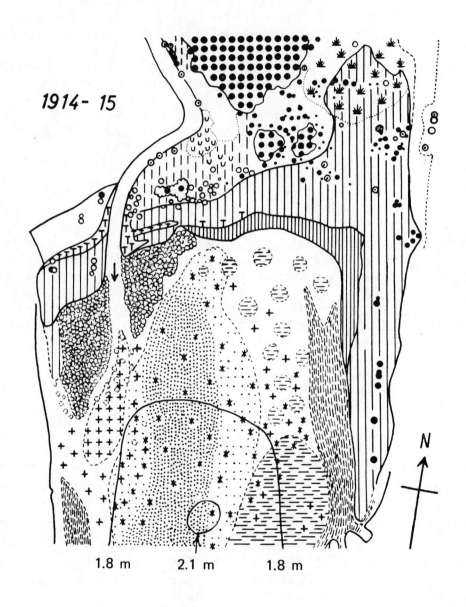

1914- 15

8

1.8 m   2.1 m   1.8 m

N

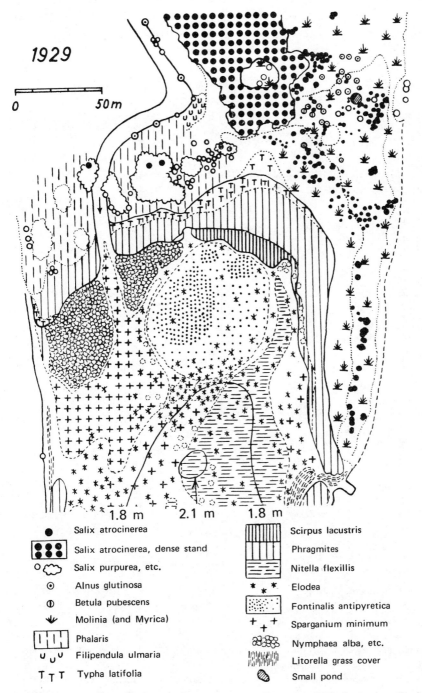

1929

0   50 m

1.8 m   2.1 m   1.8 m

●	Salix atrocinerea
	Salix atrocinerea, dense stand
○	Salix purpurea, etc.
⊙	Alnus glutinosa
⏀	Betula pubescens
↯	Molinia (and Myrica)
|I|	Phalaris
∪ ∪	Filipendula ulmaria
T T T	Typha latifolia

‖‖‖	Scirpus lacustris
	Phragmites
	Nitella flexillis
* *	Elodea
	Fontinalis antipyretica
+ +	Sparganium minimum
	Nymphaea alba, etc.
	Litorella grass cover
	Small pond

FIGURE 13.1. A site at the mouth of a stream in England showing a hydrosere development after a lapse period of 15 years of increased sedimentation. (After TANSLEY 1939:604–605.)

TABLE 13.3   Example of a Successsion Scheme Referring to the Terrestrialization of Eutrophic Waters in Northwest Germany (Simplified, After TÜXEN and PREISING 1942).

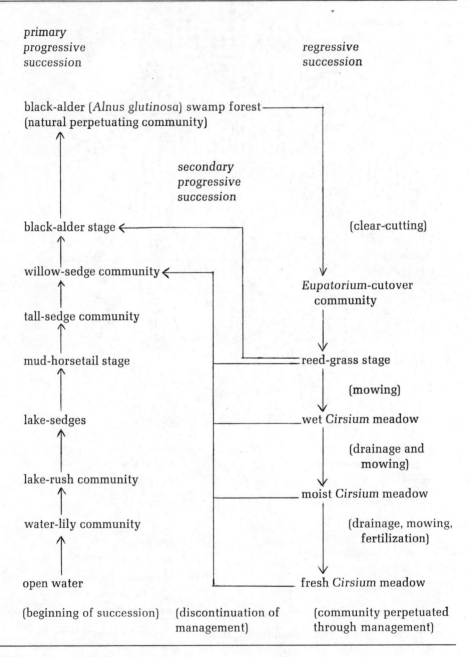

3. A secondary, progressive succession, which can begin at any stage of a regressive succession immediately upon discontinuation of the management practices.

Such a scheme provides for an elucidative arrangement of successionally related communities. It therefore forms a valuable addition to the classification scheme of BRAUN-BLANQUET, in which successionally related communities, such as these aquatic, swamp, and meadow communities, are never treated together because of their floristic dissimilarity. Of course, a similarly elucidative arrangement can be obtained by a simple ecological series in which successional relationships may not occur or may not have been discovered.

It may be noted that the primary progressive succession of this hydrosere ends in a black-alder swamp forest, which is not a climatic climax community, but rather an edaphic climax community (SECTION 13.5).

13.42 Secondary Successions. We will now discuss three kinds of secondary succession.

13.42.1 Abandoned Field Succession. The most complete disturbance leading to secondary successions is the conversion of forest to agricultural crop land. This involves cutting trees and removing stumps and subsequent seeding or planting of several generations of crop plants. In this process the soil microflora and fauna may be considerably altered. The soils may be depleted of much of their original nutrient potential. This is characteristic of the shifting agriculture still practiced today in many tropical countries. When such crop fields are abandoned, a succession begins.

Such a succession is secondary, because the same soil had already once supported a vegetation. An exception would be the entire removal of the former soil profile by accelerated erosion. In such cases, one can speak of a primary succession.

The typical succession on abandoned fields is much more rapid and involves a sequence of life forms different from the primary succession on volcanic rock discussed above. Abandoned field succession has received much attention in the southeastern United States (OOSTING 1942, KEEVER 1950, QUARTERMAN 1957, McCORMICK and BUELL 1957, ODUM 1960). A thorough review was compiled by HAUG (1970).

On abandoned fields, the first plants to appear are typically annual weeds, if available in the regional flora. Biennial weeds soon become associated with these (in the temperate zone) and then perennial herbaceous plants (temperate and tropical zone), particularly members of the grass family. In some areas, for example in central Oklahoma, a definite annual grass stage (of *Aristida oligantha*) may precede the occurrence of perennial grasses. This may be related to particularly low levels of nitrogen (RICE, PENFOUND and ROHRBAUGH 1960, ROUX and WARREN 1963).

Thereafter woody perennials are added to the perennial grasses. At first, these may be rapidly growing suffrutescent or thin-branched woody shrubs (chamaephytes). Arborescent shrubs and pioneer, fast-growing, shade-intolerant trees are next. Chamaephytes are usually found to form a successional stage only on acid, oligotrophic soils. On mesotrophic soils, phanerophytes may follow the herb stage right away (ELLENBERG 1963). When these grow to maturity and form an interlocking canopy, the shade-intolerant herbaceous plants, particularly the grasses, disappear. Shade-tolerant herbaceous plants or even mosses may become established in their places. At the same time, shade-tolerant tree species may arrive as seedlings. The shade-intolerant canopy trees do not regenerate anymore in sufficient numbers to replace the mature members of their populations. Instead, the shade-tolerant, usually slow-growing trees eventually reach into the canopy and may even overlap the shade-intolerant pioneer trees, thereby hastening their disappearance.

The last plant life form to arrive is usually vascular epiphytes. In the tropical rain forest in Hawaii, epiphytes seem to show also a succession from algae to mosses to ferns to herbaceous (*Astelia menziesii*, et al.) and woody seed plants (*Metrosideros collina, Cheirodendron trigynum*, et al.). However, the different epiphytic life forms persist side-by-side in different microhabitats and the succession is primarily only an arrival sequence and enrichment pattern in species.

Woody lianas are usually also among the last arrivals, but herbaceous vines are typically a seral plant life form that thrives in open forests. They may, as weeds, retard the closing of a tree canopy by suppressing the tree reproduction. But they may also act as nurse crops (e.g., the fern *Dicranopteris linearis*), allowing shade-tolerant tree species to become established.

In a study of primary productivity on an abandoned field in South Carolina, ODUM (1960) came to the conclusion that succession may involve a series of equilibrium stages associated with the prevalence of major life forms rather than a continuous change with species changes as is usually postulated.

While the life-form sequence as indicated here may be quite generally valid, the species compositions vary from region to region. The sequence of plant life forms to occur in succession depends also on the size of the disturbed area. Where the area is small, perennial, stoloniferous plants may invade the area from a neighboring undisturbed territory. Such mat-forming plants in Hawaii are, for example, the native fern *Dicranopteris linearis,* the introduced grass *Melinis minutiflora,* and the introduced composite herb *Eupatorium riparium.* These perennial herbs can quickly cover a denuded surface of less than a few hundred square meters in size. Annual herbs, under these conditions, may not have a chance to form a pioneer stage in secondary succession.

Secondary successions of this sort are repeated in this or similar forms with regional, habitat, and disturbance variations. Not all secondary successions end in forest. This depends on the regional climate and the periodicity of the disturbance. In high-alpine, desert, and tundra climates, forest does not form the end stage. Here successions are much simpler, because only a limited number of life forms can tolerate such extreme environments.

In a review paper on successional theory, McCORMICK (1968) re-emphasized EGLER's (1954) distinction of species arrival patterns. These are the so-called "relay floristics" and the "initial floristic composition" patterns. The idea of relay floristics relates to the popular notion that species groups arrive and then disappear as groups during the course of succession. The idea of initial floristic composition implies that the species participating in a successional sequence are already on the site at the start of the sequence. They are present in form of seeds or in form of vegetative propagules. EGLER's initial floristic composition concept suggests as mechanism that the annuals are merely the quickest to develop into dominance. Thereafter, the perennial herbs (e.g., perennial grasses) come into dominance, not because they arrive later on the site, but because they respond somewhat more slowly to the cessation of a perturbation. The shrubs respond even more slowly than the perennial herbs, and then the trees more slowly than the shrubs. In the latter two life forms, the slower response may be a combination of prolonged seed dormancy and slow initial growth rates as seedlings.

The initial floristic composition concept explains certain anomalies that have been observed in secondary succession sequences. For example, a denuded forest area may be exhausted of tree seeds. Succession may proceed relatively fast to the shrub stage to then stagnate for a very long time before trees become eventually established by relay floristics. There is little doubt that the effect of initial floristic composition (*sensu* EGLER) has been underestimated in the interpretation of successional phenomena and that the concept is significant in sharpening the prediction of successional sequences. However, both patterns, i.e. staggered *in situ* development and staggered arrival of disseminules (i.e. relay floristics) are undoubtedly applicable in most secondary successions. The first pattern of successional species development (i.e.

initial floristic composition) is of even more general importance in the next discussed type of secondary succession than on abandoned fields.

13.42.2 *Logging and Fire Succession.*   Secondary successions following logging and fire usually involve a less complete sequence of life forms in the Douglas-fir (*Pseudotsuga menziesii*) region in British Columbia (MUELLER-DOMBOIS 1959, 1965b). Cutting removes only the tree layer. Subsequent slash burning may contract the original undergrowth vegetation to a relict community, but only the shade-preferring mosses may disappear entirely. The shade-tolerant vascular undergrowth plants can persist and sometimes even expand their ground cover by lateral vegetative growth. A new flora of forest weeds becomes established on the burned over sites. This flora is often present in the form of seeds. Particularly prevalent after fire is the composite *Senecio sylvatica*. In the first year after the fire, the fireweed *(Epilobium angustifolium)*, appears first with a basal leaf rosette, thereafter with a tall flowering stalk. This species becomes then dominant in height (1.5 m tall on the average) masking by its omnipresence (on all cutover habitats) the presence of the original forest undergrowth plants. However, the fireweed takes advantage only of barren spots. Thereby it checks the expansion of the relict undergrowth plants, but it does not replace them in competition. The frequency of fireweed may be 100 percent in contiguous 1 m^2 quadrats, but its cover is rarely as high. Instead, there is only partial shading below a fireweed stand. This favors the persistence of the relict undergrowth plants.

Eventually, tree seedlings become established, pushing through the fireweed canopy. As soon as the fireweed individuals are in partial shade, they cease to reproduce. Soon after, they disappear in the shade of the developing tree canopy. The only persistent cutover element inside immature forest stands is the bracken fern (*Pteridium aquilinum*).

The recognition of stages in secondary succession is of great importance in the evaluation of a regional vegetation cover. Certainly, not all variation in the vegetation of a region can be attributed to intrinsic environmental variations between habitats. On the same topographic habitat type, one may find several different plant communities in the floristic region. Such differences are primarily caused by partial disturbances resulting in secondary successions.

13.42.3 *Exotic Species Succession.*   A still different phenomenon of secondary succession is found commonly on islands. An island vegetation may show different communities on the same habitat types that are not caused merely by different stages in secondary successions. Superimposed, one may find newly evolving floristic patterns that are caused by introductions of new species by man. Newly invading, dominant community-structure forming species may be in one part of an island, but may not have reached equivalent habitats in other parts. Because of such naturalizing species, secondary successions in island habitats may be changing from time to time and thus may not follow an essentially reversible floristic pattern as in floristically saturated continental areas.

13.5 THE CLIMAX CONCEPT AND STABILITY

A succession ends with a community in which the species perpetuate themselves through reproduction. Such a community has been called a climax* community. In a climax forest, the tree species should be present as seedlings, saplings, subcanopy, and canopy trees. Likewise, the undergrowth species (shrubs and herbs) should be present with all sizes from seedling stage to maturity. Whether or not a community is indeed a climax community can be ascertained through a structural analysis of its species populations (SECTION 7.31).

In a climax community the species composition should remain the same over a long period of time. Because of evolution, a constantly perpetuating, stable species composition is strictly speaking not possible. However, the climax concept has validity as a stage of relative stability in species composition, where species population turnover is unnoticeable. This does not mean that a climax community is static or without vegetation dynamics. In fact, overmature or diseased individuals of the member species in a climax community die any time or, periodically, new ones appear as seedlings and immature ones grow to maturity. This form of dynamic equilibrium was well documented by COOPER (1913) in a number of quadrats for the climax forest of Isle Royal. This pattern also involves local successions in places within the community, for example, where an old tree has broken down. If the tree were large, it would create a considerable gap or opening in the canopy, which would set the stage for the start of a small-scale secondary succession on that locality. But in some regions mature trees tend to fall down in groups, thereby creating somewhat larger successional communities that may form a mosaic of phases as indicated in FIGURE 13.2. The most common idea of a climax forest is represented by the mixed-age or mixed-size phase, which in the mapped sample of the mixed conifer-broad-leaf forest in FIGURE 13.2 is only 1 of 6 recognizable phases. The optimal phase is represented by a group of near-

---

* The Greek term "climax" means ladder. Thus, its original meaning was identical with succession. But it always has been interpreted as the "final step of the ladder."

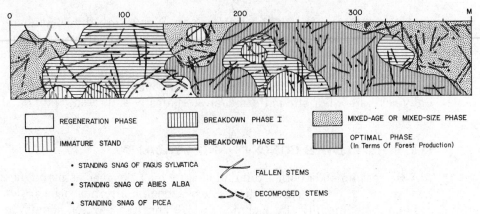

FIGURE 13.2. *Mosaic of successional phases mapped in a near-pristine conifer-broad-leaf forest in Yugoslavia. (After LEIBUNDGUT; see ELLENBERG 1963:272.)*

mature, relatively even-sized and even-aged trees that form a closed canopy. These trees probably grew up together after a windfall.

Therefore, a climax community may be described as being in equilibrium with the prevailing environmental factors of the habitat whereby the member species are in a dynamic balance with one another. This inter-species balance may be brought about by various forms of interaction. The more general among these are complementarity in the use of the resources available at the habitat and a balance in competition and other forms of interferences (e.g., herbivory).

13.51   The Monoclimax Concept.   As mentioned before, successional series starting from open water or semiaquatic conditions are called hydroseres after CLEMENTS (1916), TANSLEY (1929), and other Anglo-American ecologists, while successions from barren rock or soil to the final stage of vegetation development are called xeroseres. GAMS (1918) pointed out that it is often difficult to decide in individual situations, to which of the two groups a certain vegetation sequence belongs. CLEMENTS believed that all successions of a region lead eventually to the same, climatically controlled final stage, the so-called climatic climax association. This idea is known as the monoclimax hypothesis.

It is questionable in many cases whether such a final stage is ever reached and whether the series includes all stages from beginning to end. CLEMENTS' hypothesis—that the plant communities of an open water body and those establishing on bare rock are developing into a single climax community within a macroclimatically uniform region—appears rather convincing in its simplicity. However, the validity of this hypothesis has lost more and more ground. Already GAMS (1918), LÜDI (1930) and particularly DU RIETZ (1930, from p. 347 on) have criticized this hypothesis with good reasoning, citing unconforming examples from different continents and climates.

This hypothesis hardly fits mountainous areas, where several different communities always remain in existence side-by-side, because of the strong relief-energy and the distinct differences in local climate. But the primary differences between certain soils may not be eliminated over long periods of time even in areas with level landscapes and intermediate moisture relations. For example, DU RIETZ (1930:349) described the contrast in vegetation on sandstone and basalt-soils, where successional development was undisturbed since Tertiary. TÜXEN (1933) referred to a similar contrast between the natural forests on diluvial sand and loess-loam soils in Northwest Germany.

Moreover, it is very improbable that the terrestrialization of a water body with a stable water table will eventually give rise to a forest community that is equivalent to those on soils without ground water. TÜXEN and PREISING (1942) did not extend their succession scheme (TABLE 13.3) beyond that of an alder swamp forest, which they considered the final, natural stage in succession. Such a final community that is predominately controlled through soil factors is known also as an edaphic climax or "perpetuating or stable community" (Dauergesellschaft). Such a community should not be confused with a climatic climax community, although the former is, of course, also controlled to a certain extent through macroclimate.

The oak-beech mixed-wood forest could be considered the climatic climax association on sandy soils of the Northwest German lowland, whereby the proportion of beech depends to a certain extent on the quality of the soil. This community develops well only on soils with low ground water tables. However, an alder swamp forest can never develop into an oak-beech mixed-wood forest without artificial drainage or a lowering of the ground water table through a peculiar natural phenomenon. The upward growth of the peat soil from undecomposed organic matter which is the major reason of terrestrialization in this hydrosere, ceases when the ground water table fails to rise to the surface, or when it reaches the surface only at rare occasions.

The succession scheme of TÜXEN and PREISING is supported at all stages through direct observations or vegetation-historical investigations. However, an extension of the scheme to a climatic climax community could be done only on a strictly hypothetical basis. The sparse evidences, given by TANSLEY, for such an extrapolation are all not entirely convincing, as they refer to "partially drained" peat soils.

While initial phases of xerosere and hydrosere successions have been supported by data, the complete convergence of communities from these starting points to a climatic climax has never been observed on the same site. It seems plausible, however, that a climatic climax community can originate from new rock substrates. And indeed side-by-side observations of plant community stages on differently aged volcanic surfaces have shown that relatively stable, probably self-perpetuating communities may establish within 400 years on Hawaii in the rain forest climate (ATKINSON 1970). However, the substrate of these *Metrosideros-Cibotium* (tree fern) communities is still relatively undecomposed parent rock material augmented only with organic matter and some ash (which is also only incipiently weathered). Rain forest communities on mature latosols (Oxisols) in Hawaii show quite different species compositions, and it is not yet clear whether evolution of new species may not take place at a faster rate than soil evolution from raw rock to latosol. If species evolution is faster than or equivalent in rate to soil evolution, a climatic climax (*sensu* CLEMENTS) may never be formed, because new species evolve before the ecosystem, i.e., community plus soil are in complete equilibrium with the prevailing climate.

In some continental tropical areas, there is evidence that no major geological disturbance has occurred since the Mesozoic (125 million years ago). It can be assumed that the same substrate supported evolutionarily more primitive vegetation, such as the giant equisetum and fern forests, then, the primitive gymnosperm forests and, since early Tertiary (6 million years ago), the more modern angiosperm forests. Therefore, these present-day tropical rain forests did not originate from either a hydro- or a xerosere. Only their ancestral vegetation did. The evolutionary development of species and vegetation on essentially the same substrate makes it difficult to conceive vegetation development as being calibrated with soil development. It has also been suggested that the monoclimax hypothesis requires geomorphological equilibrium with the prevailing macroclimate. This would involve complete erosion and base-levelling (DOMIN 1923). Wherever this has occurred in the Hawaiian Island chain (i.e., the Leeward Island group), the climate changed from humid to summer-drought and the substrate from volcanic soil to coral sand. The resulting vegetation then is a strand vegetation and is no longer a rain forest. One may object that islands are too small to conform to patterns that are essentially continental. Rain forests in Hawaii result from orographic climatic phenomena, which disappear with base-levelling. However, this takes many millions of years.

13.52 The Polyclimax Concept. Another idea of ecosystem stability is presented in the polyclimax hypothesis (DOMIN 1923, TANSLEY 1929, DU RIETZ 1930, CAIN 1947). This postulates that there may be a number of different climax communities within a climatic region, but these can be in dynamic equilibrium with the local habitats and their controlling environmental factors. Thus, one may distinguish self-perpetuating communities as edaphic or topographic climaxes, if the controlling factor complex is associated more strongly with local site (such as extremes in soil water regime or local wind exposure, rainfall, insolation) than with the prevailing macroclimate, as in the case of the climatic climax. Thus, the polyclimax concept postulates that a "climax landscape" consists of a mosaic of edaphic, topographic, or ecoclimatically different communities. One of these is usually geographically dominant. Then the landscape can be named after this prevailing climax community ("climax landscape," "vegetation zone," "biogeoclimatic zone").

One may also distinguish a fire climax (see DAUBENMIRE 1968)— a vegetation that is periodically arrested in its development by fires and in which fire-resistent or fire-adapted species have become established. Or, a vegetation that developed under constant grazing pressure and is kept in a dynamic equilibrium through grazing may be called a grazing climax. This hypothesis is definitely useful as an orientation scheme, if not too rigidly interpreted.

The various climax stages are merely relatively stable communities with self-perpetuating species populations, which may form the endpoints of secondary successions. In fact, such relatively stable communities are slowly changing, because of species evolution, soil evolution, geomorphological evolution, and long-term macroclimatic changes.

These ideas can be briefly summarized in the form of a two-dimensional time-scale diagram (FIG. 13.3) that was once presented in similar form by V. J. KRAJINA in an ecology lecture. The diagram shows an ecological series from xerophytic to hydrophytic ecosystems on the abscissa. In the course of primary succession, moving the time-scale along the ordinate upwards, these ecosystems maintain their original environmental controls to a large degree. This allows for the recognition of a climatic climax (cc) and two or more edaphic climaxes (ec) in a biogeoclimatic zone. We may assume a slight convergence of the edaphic climaxes towards the zonal climatic climax over a very long period of time. This convergence is indicated by the arrows pointing from the edaphic climaxes (ec) to the climatic climax (cc). However, this convergence is never completed, it is open-ended, because of counterbalancing forces, such as erosion, deposition, climatic change, etc. A secondary succession can occur at any time within a primary succession in the

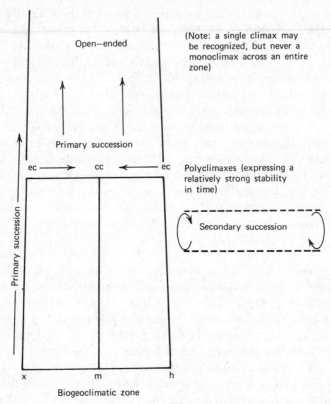

FIGURE 13.3. *The relationship of polyclimax and monoclimax hypotheses in primary succession. (Approximated after KRAJINA, unpublished). Legend: x=xerophytic; m=mesophytic; h=hydrophytic ecosystems; ec=edaphic climax; cc=climatic climax. Arrows from ec to cc indicate an extremely slow tendency of change toward more mesophytic conditions. Looking at the open-ended development, these mesophytic tendencies are never realized because of counteracting forces. Secondary successions may occur in a cyclic pattern within a primary succession at any point in time.*

form of a cyclic phenomenon. It may be arrested before reaching the edaphic or climatic climax because of repeated fire, grazing or other influences.

A somewhat different polyclimax concept is represented by the view that a macroclimatic region may include not only one climatic climax, but several that may be found on different mature soil types in the same region or zone (VON BÜLOW 1929, TÜXEN 1933, TÜXEN and DIEMONT 1937, ELLENBERG 1959). Thus, this idea may be distinguished as "polyclimatic climax concept." Many macroclimatic regions may contain not only one widespread basic textural soil type, but two or three. For example, on the profile diagram showing the forest habitat types in southeastern Manitoba (FIG. 11.2), the upland soils consist of either "pure" sand (those occupied by jack pine) or of soils with a significant clay component (those occupied by mixed-wood). These textural differences reflect basic parent material differences and there is no reason to believe that they will ever disappear. It would be very difficult to decide on a single climatic climax for this region. Since there are two widespread types of well-drained soils, namely habitat 2 on sand and habitat 7 on soils with clay lenses (bisequa grey wooded soils), one may also expect two climatic climaxes for this macroclimatic region instead of only one. The mature jack pine (*Pinus banksiana*) stands on sand are, of course, not nearly approaching a climatic climax forest because they are mostly one-generation stands that originated after wild fires. They represent instead a fire climax in their present condition. The regional coexistence of basic textural soil types that never converge is a common phenomenon, which limits the applicability of the monoclimax hypothesis. Of course, one may argue that this depends partly on the definition of a macroclimatic region. However, in southeastern Manitoba the two textural parent materials form a mosaic in the same macroclimatic region. Perhaps a more valid counter argument would be that very similar final species combinations could become established on the well-drained sites of the two differently textured soils. The latter possibility could occur in floristically poor areas where the community-structure forming species may have wider ecological amplitudes than in floristically rich areas.

This modified polyclimax concept is still more realistic as a thinking scheme than the one interpreting only one climatic climax in any macroclimatic region. For this modified scheme, the diagram in FIGURE 13.3 is still applicable, but a soil-moisture regime sequence as shown there on the abscissa may be applied to each major textural soil group. For example, for southeastern Manitoba, two such schemes would be required, one for the sandy soils and one for the finer textured soils. However, the complete soil-moisture regime range may not be realized on each major textural soil group and particularly at the hydrophytic extreme, soil textural differences may become insignificant. Yet, this does not interfere with the idea that one may recognize more than one climatic climax in a macroclimatic region.

Since the parent material, and not the climate, differs one may argue that these different climatic climaxes could just as well be named edaphic climaxes. However, this would bypass the distinction. The more widely known polyclimax concept is still an "offspring" of the monoclimax concept in so far as it recognizes only one climatic climax in a macroclimatic region (namely the natural or near-natural community on well-drained substrates) plus a number of other climax communities that are controlled primarily by topographic, edaphic, or other factor complexes. In contrast, the polyclimatic climax concept contains the idea that certain soil parent material differences will never disappear even when a landscape has totally aged physiographically (i.e., has attained the state of "base-levelling"), and that in such landscapes one can find more than one kind of well-drained soil with natural (or near-natural) communities that are in balance with the regional climate.

13.53   Difficulties with Both Climax Concepts.   In spite of accepting *cum grano salis* both the monoclimax and polyclimax hypotheses as reasonable thinking schemes, certain problems need further clarification. It appears that there are two major conceptual difficulties in both climax concepts. These may be stated as follows:

1.   A concept of relative stability in time (climax) has been applied to relationships in space without much further qualification. It is, therefore, scientifically more defendable to use a strictly geographic concept for spatial relations. This is provided in the Russian concept of zonal, azonal, and extra-zonal vegetation (SECTION 13.55).

2.   A concept referring to community and ecosystem stability (climax) has been equated with physiographic stability. This means that community stability was seen as equal to stability in soil development plus stability in geomorphological development. It has been shown that a perfect calibration of community stability with physiographic stability cannot be expected in most cases. The classical dune succession of COWLES (1899), in which the instability of the substrate was the primary cause of vegetation succession, is an exception. In the dune ecosystem, community stability is indeed calibrated with physiographic stability. This exception was interpreted subsequently as the rule. But, as WHITTAKER (1953) had already emphasized, the concept of community stability becomes clearer when defined independently of physiographic stability.

However, any community has a spatial as well as a temporal structure, and both these aspects have to be considered in the concept of stability.

Since the term climax has been used without explaining these two meanings, it has led to considerable controversy. It may thus be better not to use the term climax at all (WALTER 1937, EGLER 1947), or to use it only with a clarifying definition (WHITTAKER 1953).

13.54   Community Stability.   Community stability is considered here the same as ecosystem stability. The reason for this is that a balance among the species of a community can only be maintained if the environmental factors are in some form of balance also.

In addition to community structural aspects, ODUM (1969) discussed a number of functional community attributes that should be considered in the concept of succession and ecosystem stability. A major successional phenomenon pointed out by ODUM is a change in the ratio of gross production (P) to community respiration (R). At maturity of the ecosystem (i.e., in the stable stage) the P/R ratio should approach 1, because the community then has reached an optimal biomass that is presumed to be in balance with the habitat factors. Maintenance of this biomass at maturity requires only as much energy input, in form of photosynthates (i.e., gross production), as is lost through respiration. Thus, the community net production (i.e., the increment or yield of organic materials) would be low during the stable stage of the ecosystem, but high during its developmental or seral stages. ODUM also points out that food chains are characteristically web-like in the stable stage, while they tend to be linear and simplified during earlier seral stages.

A symposium was devoted to the problem of diversity and stability in ecological systems (BROOKHAVEN SYMPOSIUM 1969). A considerable number of interpretations of the stability concept were presented, and it became clear that the term stability has an even greater spectrum of meanings than the term climax. Similarly, it was shown that the term diversity is very ambiguous, or at least, subject to abuse.

PRESTON (1969) pointed out that to the physicist, stability means resistence against displacement; to the economist, it refers to the cycles of ups and downs; to the zoologist or population ecologist, it refers to constancy in population size.

LEWONTIN (1969) emphasized that stability is a dynamic concept. But he contrasted stability to constancy, by interpreting constancy as a static concept. In reality, the two concepts are very closely related. The opposite of constancy is variation or fluctuation and not stability. The opposite of stability is instability or fragility. Constancy over a period of time is no less a dynamic concept than variation over a period of time. The dynamic aspect of the stability concept lies in the idea that stability in living systems must include a certain fluctuation or variation in time. Thus, a community or ecosystem is considered stable even if it undergoes rhythmic changes. This is very important, because the living components of an ecosystem reproduce, undergo metabolism, grow, and die any time and there are always shifts and adjustments within the matrix of a community. The main problem in the stability concept is not that it must include dynamics, but rather how much and what kind of dynamics it may include.

MARGALEFF (1969) brought out that stability is an aspect of organization in a community or ecosystem, a measure of persistence in time. He says, "a system is stable if, when changed from a steady state, it develops forces that tend to restore it to its original condition." Thus, resiliency or "homeostasis" is an important quality of a stable ecosystem (see also FOSBERG 1965b). This interpretation brings out a close relationship of stability and regulatory forces. This idea was expressed also in the polyclimax concept, where the regulatory forces producing community stability are named, such as climate, topography, edaphic factors, animal influences (grazing climax), or fire.

WILSON (1969) pointed out that community stability relates to the concept of dynamic equilibrium, and he distinguishes four stages of species-equilibria:

1.   A "noninteractive species equilibrium." This applies primarily to pioneer communities, where the species invading a habitat occupy vacant niches. Such species combinations are considered noninteractive.

2.   An "interactive species equilibrium." This applies to communities composed of noninteractive plus interactive species. Here the interactive species are those with similar niche adaptation. Such species form ecological groups (SECTION 11.1).

3.   An "assortative species equilibrium." This is interpreted as a stage where interaction among species has resulted in long-lived combinations.

4.   An "evolutionary species equilibrium." This is considered the final stage of community development when the member species have become adapted genetically to one another and to the local environment.

WILSON's four species-equilibria are interesting interpretations of stages in primary succession, which are recognized through degrees of competitive interaction and structural complexity in community development. This interpretation also provides for the evolutionary processes of speciation as seen in the framework of the community and ecosystem.

The application of equilibrium or stability concepts within the course of primary succession is not new. For example, CLEMENTS considered all communities in a region, but the climatic climax, as seral stages in primary succession. The polyclimax concept implied further persistence or permanence to these stages by emphasizing that they are in equilibrium with certain regulatory factors of the environment.

To be meaningful, the concept of stability or dynamic equilibrium must be relatable to some scale of time. However, since the rates of change vary tremendously with kinds of communities and environments, an absolute time scale has no general meaning. Equilibrium stages thus must be interpreted as stages of relative stability (slow change) in relation to stages of relatively fast changes.

If seen in the context of *relative* rates of change, the concept of community stability as a stage of relatively slow changes assumes an important meaning.

To the vegetation ecologist, the most important measures of community stability appear to be the species composition and structure of the community (WHITTAKER 1953, DAUBENMIRE 1968, and others). If the species composition of a community remains relatively constant over a period of several decades, one may speak of a stable community. This concept of stability is less rigid than that of the population ecologist, because it allows for some fluctuation in density of individuals or in the quantities of each species comprising a community. Of course, changes in the quantities of the major structure-forming species of a community may indicate a shift in development leading to instability.

Therefore, it seems necessary to interpret community stability also in relation to persistence in life form composition and structure. Clearly, a naturally imposed change in structure (for example, a change from closed forest to open forest or woodland) should be interpreted as instability, even if not accompanied by a change in species composition. However, if such a structural change is reversed, as may be the case in the time following a storm, the structural change would not be interpreted as instability but rather as a resistance to change, i.e., stability.

Therefore, community stability is best interpreted as a stage of dynamic equilibrium that persists over at least several decades in terrestrial plant communities. It is recognized by a relatively constant species composition and structure over this time span. The two parameters—species composition and structure—are measurable, and predictions about stability can be made from structural analyses or population dynamics studies within communities. Certain compositional and structural variations within a community are a normal phenomenon in a stable community. But the nature of this variation must be such that it gives evidence of mere oscillation, fluctuation or reversible change rather than of a progressive, irreversible change. If evidence leads to the latter, the community loses its character and may be considered instable or fragile.

13.55 A Spatial Concept Involving General Stability. Community stability is determined by both the kind and degree of change over a period. But it is also related to the geographic size of the community.

For example, on a 400 m² sample-plot basis, a forest community may be considered unstable because a large tree may fall down, creating a gap that becomes invaded with weeds. This would result locally in a new species composition and structure. In one hectare of the same forest, several such gaps may be found, thus the weeds or other heliophytes occupying these gaps belong to the total species composition.

Therefore, geographic scale enters into the concept of stability. It is thus necessary to specify the community concept one has in mind when speaking of community stability. The stability of a formation is expected to be less than that of a vegetation region, while the stability of an association is expected to be less than that of a formation, and the stability of a synusia is expected to be less than that of an association. In reference to different sizes of ecosystems, one may speak of local or habitat stability, watershed or landscape unit stability, or of regional stability. In each case the stability relations would be expected to be quite different.

When referring to broad zones of natural vegetations that have not undergone major compositional and structural changes for some length of time, it seems advisable to avoid the term climax altogether and to use instead the terms zonal, azonal, and extrazonal or intrazonal, as orginally developed in Russian geobotany (see WALTER 1943, 1954, 1964). These have merely a geographic implication. The three terms relate exclusively to the natural plant cover. A zonal plant community corresponds more or less to a climatic climax community and an azonal community to an edaphic climax community controlled primarily by extreme soil conditions. The terms extrazonal and intrazonal have no corresponding meaning in the succession concept.

The term zonal applies to a vegetation unit of higher order, which reflects a close relation to the current climatic conditions of a larger region and which has developed without significant human interference on soils with nonextreme properties. For example, mesophilous mixed hardwood forests with *Fagus sylvatica* as the dominant component represent the zonal vegetation cover in the lowland of Southwest Germany.

The zonal forest vegetation cannot get established on habitats subject to excessive moisture or drought, or those subject to frequent flooding or extreme nutrient deficiency. Here we find azonal plant communities, such as sedge meadows, peatmoss communities, swamp forests, rock-vegetation, alluvial forests, or acid-tolerant oak forests. Such communities are found not only in one climatic vegetation zone, but they occur, in very similar compositions, also in other zones. The aquatic vegetation in particular shows very little dependence on zonal climates.

However, the delimitation between zonal and azonal vegetation is not abrupt. For example, forests on moist mineral soil have predominantly zonal character. They show, however, somewhat greater similarity to the forests on similar habitats in the neighboring zone than do the forests on less moist habitats. In general, the concept of zonal vegetation should not be implied too narrowly so that one does not equate the concept with certain subassociations, variants, or sociations, but rather with formations, subformations, alliances, or associations (in some cases). Otherwise, the definition will always remain debatable. A good example of such a broad classification is shown by HORVAT's (1954) vegetation map of Southeast Europe entitled *Klimatogene Vegetationseinheiten* (climatogenic vegetation units, see also HORVAT, GLAVAČ and ELLENBERG 1974).

The zonal vegetation of a relatively warm zone is also found frequently in cooler zones on southern exposures, where it is favored by a warmer local climate. Conversely, the zonal vegetation of cooler zones can reappear on northern exposures in the warmer neighboring zone. Such island-like vegetation units that reach into the neighboring zone are called extrazonal.

The term intrazonal refers to vegetation units that are found only in one vegetation zone where they occur on locally unique habitats, e.g., the *Corylus colurna-Fagus* mixed-woods on dry slopes within the *Fagion moesiacum* zone of Southeast Europe.

Vegetation terms of rank geographically equivalent to zonal, azonal, extrazonal and intrazonal but referring to strongly man-influenced vegetations would be: modified vegetation (if it is at least partly natural) and cultivated vegetation (if it is artificial, that is, established through cultural practices).

In a way, this geographic concept is similar to the polyclimax concept, but it involves stability only on the level of an ecologically broadly defined vegetation zone. Here the stability-controls are only in four general categories, (a) the zonal climate or macroclimate is understood as controlling the character of the zonal vegetation, (b) strongly macro-climate-overriding edaphic or physiographic controls are considered to shape an azonal vegetation, (c) local climates that do not conform to zonal macroclimate are considered to give rise to an extrazonal vegetation, and (d) man's influence, when pronounced, is considered to result in modified vegetation.

This concept appears less confusing or controversial for an understanding of general spatial relationships than the mono- or polyclimax concepts because it involves stability only at a broad geographic scale at which stability is relatively high. But the zonal concept is not a dynamic concept. Instead, it is useful in establishing a frame within which community dynamics and stability can be studied in detail with regard to the kind and severity of perturbations and the response of biological communities of differing sizes. It must be understood that all these concepts are idealistic schemes that cannot be expected to fit reality in all respects. However, as thinking schemes, they are extremely important. As such, they are all subject to revision in a process of successively closer approximation to reality as new significant information is derived.

Vegetation and Site Mapping

14

14.1 THE CONCEPT OF VEGETATION AND SITE MAPPING

Mapping for ecological purposes usually involves the graphic portrayal, in two dimensions, of the patterns or mosaics of plant communities or of the habitats or sites of a given area. For certain purposes, it may involve the mapping of species populations or of individuals. Direct mapping of existing vegetation must be clearly distinguished from mapping of potential vegetation which may never come into existence. Apart from the objective, this distinction is often related to the mapping scale, as will be explained below.

Furthermore, vegetation is often used as an aid in site or habitat

411

mapping. In this case, the investigator must know the ecology of an area in considerable detail before he can begin to map. This relates particularly to knowing the site indicator value of individual species, species groups (e.g., ecological groups) or communities, and to knowing the ecological effect of site factors, such as topography, seepage and water table relations, soil texture and other soil characteristics, and disturbance variations.

Site mapping is defined here as mapping of the site potential or productive capacity of habitats with the aid of vegetation. It is thus also an aspect of vegetation mapping, but may involve the vegetation only indirectly. We will not be concerned here with strictly environmental mapping, such as soil, hydrological, or topographical mapping. In some cases, site mapping can be achieved through direct mapping of plant communities. In these cases, we can still speak of vegetation mapping. But, in many situations, special groups of species or sometimes individual species have to be isolated with reference to site potential, and their reliability as indicators has to be tested and balanced against the habitat concept applied.

It is important to understand that site mapping in the sense here defined is a sort of "total site" or "combined" mapping, whereby total or combined mean that both vegetation and environmental parameters may purposely be combined in the mapping criteria. Site mapping is used particularly for applied purposes.

14.2 THE MOSAIC OF PLANT COMMUNITIES

14.21 Natural Community Complexes. A plant community is unimaginable without the space it occupies. Consequently, a plant community is a geographic phenomenon with spatial variation. These variations, however, are not haphazard. Instead they are correlated with recurring combinations of habitats that tend to form a "pattern" or "mosaic."

Plant communities, whose habitats occur in a mosaic pattern, are referred to as a community complex. DU RIETZ (1930:338) recognized four categories of community complexes according to geographic arrangement of habitats and degree of complexity:

1. Mosaic-complex, i.e., a certain natural community complex occupying a small space. For example, the mosaic complex of a developing bog with its hummocks and depressions, described by GRISEBACH.

2. Zonation-complex, i.e., a regularly recurring natural zonation

in a small area. This includes, for example, the zonation around a small snow-bed in the Alps, or the communities corresponding to the various zones at a lake shore or a river bank; or the concentrically arranged mosaic-complexes of a bog with its central area, the zone near the margin, the eroded marginal slope, and the surrounding fen.

3.   Vegetation region, i.e., a natural community complex that is usually quite variable and distributed over a large area, and in which one or a few communities are predominant. For example, we have the tundra complex, the region of the North European coniferous forests, the region of the Atlantic oak-birch forest with its intervening bogs and fens and inland dunes, or the region of the evergreen Mediterranean forests, etc. The vegetation region of DU RIETZ corresponds largely to SHELFORD's (1963) concept of "biome."

4.   Vegetation belt, i.e., a natural complex of vegetation regions, which coincide in their relative elevation; for example, the mountainous belt of the eastern slope of the humid-tropical Andes; or the Central European lowland belt, within which one can distinguish the region of the oak forests on acid soils and that of the pine forests in the continental climate; but in particular, the region of the beech-dominated mixed forests on loamy soils.

The four geographic concepts are usually sufficient for the description of natural vegetation complexes. The spatial arrangement of vegetation units has received special attention in vegetation geography (SCHMIT-HÜSEN 1959, 1968).

14.22   Community Complexes of the Cultivated Landscape.   We can distinguish two kinds of community ·complexes—replacement and contact.

14.22.1   Replacement Communities.   In proposing the above four categories DU RIETZ had primarily the natural mosaic habitats and communities in mind. However, what is more obvious to the eye in the present-day landscape, is the mosaic of managed vegetation, such as plantations, forests, meadows, heathland, agricultural fields, etc. Wherever these man-induced and managed communities are growing on essentially the same physical habitats as the original natural communities they are known as "replacement communities."

For example, replacement communities of the calcareous beech forests (*Fagetum elymetosum*) of the Central European Mountains are, when grazed, the semidry grasslands (*Mesobrometum*); when mowed and fertilized, they are *Arrhenatherum* meadows; and when used as cropland they form a *Delphinium* weed community. Through medieval coppice practice, the beech forest was changed into an oak-hornbeam forest (*Querceto–Carpinetum primuletosum*) and through reforesting of the *Mesobrometum,* a coniferous forest community originated with abundant *Brachypodium pinnatum* in the undergrowth. If the semidry grassland is left to itself or only grazed from time to time, it is invaded by a large number of shrub species. All of these and still other communities form the replacement community complex or the cultivated landscape (complex) of the calcareous beech forest. SCHWICKERATH (1954) named the replacement community complex a "community ring." Such rings or complexes play a role particularly in the geographical classification of the landscape and in the mapping of habitats.

The number of replacement communities on a given habitat is limited. For example, in place of the moist oak-birch forest (*Querco roboris–Betuletum molinietosum*) in the North German lowland, there can only occur the moist sand-heath (*Calluno–Genistetum molinietosum*) community upon grazing and extensive use, not however the crowberry-heath (*Calluno-Genistetum empetretosum*) or any other heath community. Completely impossible is the development of heaths of the Mesobrometum type, i.e., of dry grassland and semidry grassland on base-enriched soils. Upon proper range management, it is often possible to replace the moist oak-birch forest by a certain subassociation of the dog's tail grass meadows (*Lolio–Cynosuretum lotetosum*), while conversion into other meadow communities requires a much greater effort. Even when the land is used for crop production, only certain definite replacement communities are possible, i.e., a weed community with a number of acidity and moisture indicating species. However, upon heavy liming of the soil, the weed community becomes enriched with nitrate-preferring plants.

Since the choice of crop plants is limited by certain intrinsic habitat factors, the crops can likewise be considered as belonging to the regional complex of replacement communities of former forest vegetation. For example, alfalfa (*Medicago sativa*) does not grow on soils formerly occupied by the humid oak-beech forest, and barley, wheat, and sugar beet show poor yields. In contrast, rye and especially oats, but also red clover, fodder-beets, pulp-stem cabbage, and several field-vegetable crops are grown successfully.

The stable communities of a vegetation region together with their replacement communities form a vegetation complex which recurs frequently in a similar way under similar geologic, climatic, and management conditions. KRAUSE (1952) studied such a complex over its entire range of distribution from Central Europe to Asia, and he presented the more important plant communities in the form of a table. This table is constructed in a manner similar to the synthesis tables of individual plant communities or releves. In this table a characteristic combination of regularly recurring plant communities is differentiated from a series of only locally occurring plant communities, which are characteristic to the different regions. This may be sufficient for a general understanding, but for further details, one should consult the original work.

14.22.2 Contact Communities. Most replacement communities can occur in geographically neighboring positions. Where this is the case they are called "contact communities" after TÜXEN and PREISING (1942). This pattern, however, does not apply only to replacement communities, but, for example, also to rush- and tall-sedge communities or to the *Arrhenatherum-* and *Cirsium* meadows which frequently occur adjacent to one another as their habitats differ only in degree by one factor, namely soil moisture. Otherwise, they are rather similar (see FIG. 14.1). The surest way of studying community complexes is through the mapping of vegetation units.

14.3 MAPPING OBJECTIVES

Maps are helpful for an understanding of the spatial relations of plant communities or vegetation units. Here are some of the ways maps can be useful during different stages of an ecological investigation of an area.

1. *For providing a framework for research.* When beginning an ecological investigation of an area, it is useful first to subdivide the general vegetation cover into more homogeneous subunits. For this purpose, air photographs are the best tool. The more obvious patterns on the air photo can be outlined in the laboratory. By means of repeated field reconnaissance, the pretyped patterns can be investigated and corrected, where necessary. On this basis, one can usually quite readily establish structurally defined vegetation units or dominance communities (i.e., units defined by dominant species).

2. *For locating sample stands or relevés.* Maps established from air photo-field reconnaissance may be used as a basis for distributing sample plots. These may be used to describe the floristic content of the established subunits, for testing the validity of the subdivisions in various parts of the mapped area, for verifying the boundaries drawn on the map, or for other purposes (such as permanent plots for phenology studies, succession studies, studies of animal activity, etc.). The map provides the opportunity for a fair distribution of sampling in the sense of geographic distribution and vegetation variation. Air photos alone may serve the purpose of locating relevés in this way. Once the air photos are converted to a map, the first step of an investigation is

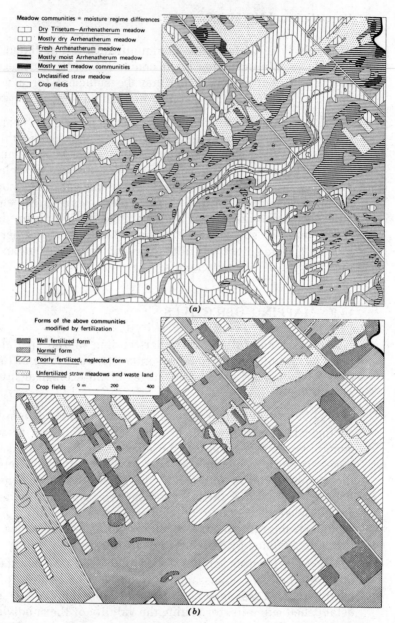

FIGURE 14.1. *Maps of grassland communities in the Danube valley south of Ulm (see TABLE 9.7). (a) Unranked communities as moisture indicators. (b) Vegetation forms (facies), which have developed following fertilization and management and not in response to differences in soil. The latter correspond in some respects to the sociations of DU RIETZ. Well-fertilized form = Arrhenatherum and other tall grasses; poorly fertilized form = predominantly short grasses and low herbs, such as Plantago lanceolata. (After ELLENBERG 1952.)*

already accomplished and the map may serve to generate a number of working hypotheses as briefly indicated above.

3. *To aid in classifying the vegetation.* Without mapping the vegetation units, a classification may look good on paper, but when tested in the field, may show gaps because certain transitions or ecotones have not been fully considered. Thus, a map can serve as a test of a classification because it forces the investigator to accommodate all variations in his scheme. In so doing, the mapping process may result in corrections and thereby aid in deriving a realistic classification.

4. *To give a detailed representation of the spatial structure of a vegetation pattern or mosaic.* For this purpose, large scale maps that show well-defined vegetation units, are indispensable. Although any map is only a telescoped abstraction of the real vegetation, the most realistic maps are those done correctly at larger scales. For example, maps portraying zonation complexes (FIG. 13.1) fall into this category.

5. *For succession studies.* All vegetation maps based strictly on the existing vegetation form a documentation "frozen" in time. They may thus serve to evaluate successional changes (as demonstrated by FIGURE 13.1). This is not automatically true for site maps based on plant indicators, as these are prepared to show the site potential, which is usually a more permanent property than the existing vegetation.

6. *To show the geographic distribution of a specific vegetation unit.* A detailed causal analysis of a specific vegetation unit (e.g., an ecological group or association) may be expanded into a chorological investigation. In this case, it is not necessary to portray the entire vegetation mosaic of a region. Instead, one can show the geographic distribution of this vegetation unit by areal shading or by dots on overview maps (small scale maps), or one can indicate the general distribution of the unit by outlining the area on a small-scale map.

7. *To aid in causal analytical research of plant communities.* Vegetation type maps based strictly on the currently existing vegetation can be compared to environmental maps. The latter must be based strictly on environmental parameters and preferably should be to the same scale. Geological maps, soil maps, maps showing the distribution of climatic factors, history maps, etc. allow the relating of the mapped vegetation types to various environmental factors. Ecosystems can be established on this basis, and indications for the causes of vegetation variation may also be obtained.

8. *For applied purposes.* Vegetation units that reflect the factors of their habitats closely can be used for the preparation of habitat maps. However, such maps may also be prepared in a more direct way. By

using the environmental indicator properties of species groups or communities, site maps can be prepared that do not show the vegetation itself. Instead, one can show, for example, the suitability of the area for certain cultivated plants or the habitat types whose boundaries are defined by indicator plants.

In this latter case, the map is not really a vegetation map, but a site map, i.e., a map showing the site potential that was evaluated through various aspects of the existing or former vegetation or through experimental work. For example, regeneration experiments through seeding or planting may have provided an answer to the relative success of reestablishing trees in a certain set of forest habitats. For mapping this regeneration chance over a wider area, one may use indicator plants and other characteristics, such as topography or soil texture, to define the habitats. This area covers a wide field in the practical application of vegetation ecology. For the manager it is often impossible, or at least very difficult, to translate results of vegetation research into practical uses. In this regard, the vegetation ecologist should deem it his task to help the land manager in agriculture, forestry, range management, landscape architecture, and park and wildlife management in interpreting the results of research. The best method for doing this type of research interpretation is often by the preparation of maps for specific use-purposes or the collaboration in an interdisciplinary project that aims at the preparation of such maps.

14.4 MAP CONTENTS AND SCALES

In both vegetation and site maps, the contents and scales depend primarily on three factors: (a) on the objectives for which a map is prepared, (b) on the detail and accuracy of the underlying information, and (c) often unfortunately, also on the funds available for the preparation and printing of the map.

An excellent bibliography of published vegetation maps has been compiled by KÜCHLER and McCORMICK (1965, KUCHLER 1966, 1968, 1970). In this the maps are arranged by country and publication date. The map legends are reproduced in the original text so that one can obtain a clear picture of their information content and applicability.

The following is an outline of map-information possibilities which vary with the scale:

1. *Small-scale maps for general overview.* This category includes maps of 1 : 1 million or smaller. On a 1 : 1 million map, 1 cm on the map represents 10 km in the field. Such maps can present as a rule

only the prevailing vegetation units of formation rank (see CHAP. 8) or generalized stable units (such as zonal, azonal, and prevailing modified vegetation units) or the natural climax communities forming the presumed final stages of primary succession (see CHAP. 13). Examples are KÜCHLER's (1964, 1965) 1 : 7.5 million map of the "Potential Natural Vegetation" of the United States; ROWE's (1957, 1959) 1 : 6.4 million map of the "Forest Classification" of Canada; KRAJINA's (1969) 1 : 5.5 million map of the "Biogeoclimatic Zones" of British Columbia; or GAUSSEN's et al. 1 : 1 million "International Vegetation Map," including one for Sri Lanka, Ceylon (1964). In a still smaller category, but of interest for its clarity and information content, is the world vegetation map of SCHMITHÜSEN (1968) at the scale of 1 : 25 million.

2. *Intermediate-scale maps for regional orientation.* These include maps in the range from 1 : 1 million to about 1 : 100,000 (1 cm on map = 1 km in the field). These maps may already permit the representation of floristically defined vegetation units such as alliances (see CHAP. 8) or dominance communities, or structurally defined communities. However, at this scale range the vegetation units are often generalized to show the "potential natural" vegetation (for definition, see next section) rather than actually existing vegetation boundaries. An example is GAUSSEN's 1 : 200,000 map of the "Vegetation of France," a section of which is shown by KÜCHLER (1967:259), or MUELLER-DOMBOIS' (1972) "Generalized Vegetation Map" of Ruhuna National Park, Ceylon, which was published at 1 : 140,000. KÜCHLER (1954) devoted a special discussion to vegetation maps at this intermediate scale range.

3. *Large-scale maps.* These include most maps in the scale range of 1 : 100,000 to 1 : 10,000 (1 cm on map = 100 m in field; or 1 cm^2 on map = 1 hectare). Such maps allow representation of many or nearly all of the vegetation units that may be defined through dominant species, or other floristic definitions, or through structural attributes (for explanation, see APPENDIX C). In this scale range the actual boundaries of existing vegetations can be mapped. Such maps require months or years of preparation and intensive field work. Thus, only small countries, as for example, Belgium, may consider a country-wide ecological mapping at this scale range. In most instances, vegetation or site maps at this scale are prepared only for certain areas as case examples. The usual topographic base maps are in this scale range. Examples are KÜCHLER and SAWYER's (1967) 1 : 30,000 map of the "Vegetation west of Maenam Ping in Thailand," MUELLER-DOMBOIS' (1969) 1 : 32,000 "Vegetation map of Ruhuna National Park, Ceylon" (which was the basis for the generalized vegetation map referred to above), or

the six 1 : 10,000 ecological maps prepared by different authors for the International Methods comparison in Switzerland (ELLENBERG 1967).

4. *Detail maps of very large scale.* Examples would be maps at scales of 1 : 5000 or 1 : 1000 (1 cm on map = 10 m in the field, or 1 ha = 1 dm^2 on the map). Such maps are prepared only for special purposes, for example, for the documentation of a nature reserve or a research plot. Maps at such large scales are of particular use in studies concerned with vegetation changes. The succession map in FIGURE 13.1 is in this scale range. Another example is the 1 : 2500 vegetation map of the Neeracher Riet in Switzerland (ELLENBERG and KLÖTZLI 1967).

5. *Chart-quadrats for mapping all important species.* Square meter quadrats can usually be mapped conveniently at a scale of 1 : 10 (10 cm on note sheet = 1 m on ground, or 1 cm^2 on paper = 1 dm^2 on ground). These are useful for herbaceous species and tree seedlings (see CHAP. 5). For trees, somewhat smaller scales are appropriate such as 1 : 100 (1 cm on note sheet = 1 m on ground). Thus, a belt-transect of 10 x 50 m requires a note sheet of 10 x 50 cm, which allows one to map individual positions of trees and shrubs and their crown outlines. Correct positions are best mapped by subgridding the belt transect with string, in 10 x 10 m subplots or smaller, if necessary. Such chart quadrats are useful for many purposes, such as a detailed stand structure description, the mapping of tree undergrowth patch communities, the mapping of dynamic phases (such as shown in FIG.13.2) and as a basic technique for permanent tree plot records. It is also often useful to represent the chart quadrat of a forest sample in the form of a profile diagram (see CHAP. 8, FIG. 8.3).

14.5   ACTUAL AND POTENTIAL VEGETATION

The "actual" vegetation, i.e., the currently existing vegetation mosaic of an area, can only be represented on maps with large scales. These are the map types 3, 4, and 5 discussed above, with scales from approximately 1 : 100,000 and larger. On such maps, most of the vegetation units can be shown by their actual areal extent. This is true for the *presently existing vegetation* as well as for the *historically real* vegetation, i.e., for the vegetation that was once present on a given area at an earlier time. However, the reconstruction of historically real vegetation is often rather inaccurate.

Strictly speaking, any vegetation mapping requires a generalization, abstraction, or typification of the really present vegetation pattern. For example, on a 1 : 5000 map one can represent associations, subassociations or even variants (*sensu* BRAUN-BLANQUET), but not facies, i.e., species or species groups that dominate only over several square meters, such as shown on the succession maps (FIGURE 13.1). Each "unit" designated by a symbol on a map is an abstraction or typification of a real plant mosaic.

Overview maps of small and intermediate scales (types 1 and 2 above) do not, as a rule, permit us to reproduce the vegetation mosaic of an area in its currently existing or historically existing real pattern. In preparing an overview map, one can resort to various means. This depends on the purpose of the map.

The simplest generalization is to represent only the prevailing or major existing vegetation types and to omit the others. The boundaries encircling small units will simply be omitted and the small units are then absorbed into the larger unit that contacted or included the smaller units. This form of generalization was applied in MUELLER-DOMBOIS' Ceylon map cited under point 2 above.

More difficult is the task of representing *types of vegetation complexes* within given limits of geographic scale, that is, small mosaic complexes, zonation complexes or vegetation regions or subregions as discussed in SECTION 14.2.

The typification of vegetation complexes becomes really problematic, when the area to be mapped is no longer covered with natural or near-natural vegetation, but instead has been converted into a cultivated landscape. Then, the side-by-side occurrence of replacement communities is often so arbitrary or without any similarity relationship that they cannot be grouped into types. Instead of attempting to map the real vegetation, many investigators prefer in these situations to project an image of the "natural" vegetation, i.e., the vegetation considered to be there without the interference by man.

"Natural" should not be confused with "original," i.e., with the kind of vegetation that was present in an area before man had any influence on it. In Europe, for example, the beginning of man's interaction with the vegetation lies so far back in history that at that time, there was a different prevailing climate and a different floristic assemblage on the continent.

"Natural vegetation" can be understood in the sense of "climax" vegetation or "zonal" vegetation (see CHAP. 13) or with TÜXEN (1956) as "potential natural" vegetation. In each case, the concept is highly hypothetical, especially when it comes to the drawing of boundaries on maps. It is impossible to predict with high accuracy what plant cover a given land segment will support in the course of time, if man and his domesticated animals were totally removed from the scene. Such a prediction becomes uncertain particularly in view of the differential rates at which a replacement community may develop into a relatively stable "natural" community. This rate is determined by the local climate, the soil, the current status of the vegetation and the competitive interactions of the species. In some cases, the development into a new steady state or dynamic equilibrium with the new set of factors may take so long that soil, climate, and the genetic structure of important species populations may have undergone significant changes also.

To deal with this dilemma, TÜXEN (1956) suggested basing the concept of "potential natural vegetation" on the presently existing vegetation and site mosaic. He defined the concept of "potential natural vegetation of today" as the vegetation structure that would become established if all successional sequences were completed without interference by man under the present climatic and edaphic conditions (including those created by man). With this magic formula, TÜXEN narrowed the problem to the core of vegetation-environmental relationships.

Maps representing potential natural vegetations of today (or briefly "potential vegetations") are basically nothing more than small-scale site maps that are prepared by using the present-day vegetation as an indicator. For their preparation, one also often uses soil investigations and soil maps, climatic data and historical information. (Example: Vegetation mapping at 1 : 200,000 of Northwest Germany as carried out by the Bundesanstalt für Vegetationskartierung, Hannover.)

Potential natural vegetation can only be constructed. Such vegetation has never really existed and it never will exist in the form projected on the map. This is because the potential natural vegetation is a conceptual abstraction that was established from a knowledge of the existing vegetation, its developmental tendencies, and its site relationships. Therefore, a potential natural vegetation map provides a mirror-image of the current state of knowledge with respect to the present vegetation potential of a region. If these restrictions are clear in one's mind, such maps can be used to advantage, either for practical purposes or as starting bases for other research.

The idea of potential natural vegetation and its complexities continues to receive much attention by continental-European vegetation ecologists. Anglo-American ecologists are often unaware of these concerns or they consider these efforts as something highly subjective and rather unimportant. These differences in concern become understandable, however, if one considers that the vegetation of Europe, particularly that of southern and western Europe has been modified by man for thousands of years. The question of what would grow naturally in such a landscape under the given climatic and soil conditions, must occupy the vegetation ecologists as well as land managers in such countries. An answer to that question can help substantially to clarify the ecological potential of such a country. It also may help to clarify its relationship to other landscapes which are covered with vegetations that are less strongly modified by man.

In contrast, in North America, the long-lasting effects of man on the vegetation became noticeable only a few hundred years ago. These effects are reasonably well known with regard to their distributions and intensities. Therefore, they rarely present themselves as research problems in vegetation ecology. Moreover, there are still so many remnants of the original vegetation, at least in the forested regions of North America, that their direct study is still possible without resorting to hypothetical situations.

Only at the margins of arid landscapes, i.e., along the prairie and semidesert borders does it seem necessary in North America to give some thought to the potential natural vegetation. It is possible that forest vegetation here (as found in the south European arid landscapes) was much more widespread than has been assumed so far.

14.6 MAP SYMBOLS AND COLORS

Questions relating to such technicalities as map symbols and colors have received much discussion and tentative agreements. KÜCHLER (1967) devoted a whole chapter to this. However, because of the multivariate character of the plant cover, it has not been possible so far to come to a generally satisfactory solution. Nevertheless, it seems of practical value to adhere to certain rules. Commonly, particular site properties are signified by the following colors:

> red = very dry (or warm)
>
> yellow = dry (or moderately warm)
>
> green = intermediate (with respect to moisture and heat)
>
> blue = wet (or cool)

Color intensity is often used to indicate differences in the structure of the vegetation. For example: strong color = forest vegetation; increasingly paler color with narrow to wider hatching = scrub, dwarf-scrub, grassland; and pale color or no color without hatching = sparse plant cover, desert or rockland, etc. This color system is essentially that of GAUSSEN.

In the choice of symbols on colored fields, the following rules have proven useful: none or only simple symbols = large area-covering vegetation type; conspicuous or complicated symbols = rare or small-area covering vegetation types.

Rules useful for color- and symbol-application on the finally drafted or printed map are not necessarily useful in the field-preparation process. At this stage, clear differences in color and symbols are more important than emphasis on similarity relationships of vegetation types. In the preparation of large-scale maps, one often uses only numbers or letters for the vegetation units. For easy correcting, these symbols are preferably entered initially in pencil. This is a good practice also on air photos.

14.7 FIELD MAPPING AIDS

14.71 Topographic Maps and Aerial Photographs. Field mapping of anything, other than chart-quadrats or belt-transects at very large scales, is best done with the aid of aerial photographs and topographic maps. Vegetation mapping has become very much easier during the last few decades with the great advances in aerial photography. Aerial photographical techniques are continually improving and there is no need to discuss details of these techniques in this book. However, a few major kinds of commonly available mapping aids will be mentioned.

1. *Topographic maps.* If possible, these should have only faint colors or no coloration at all and they should preferably be at a scale larger than the final vegetation or site map.

2. *Air photo mosaics.* These are specially prepared aerial photographs in which the distortion of scale has been removed by piecing the central undistorted parts of several aerial photographs together to result in one sheet true to scale. Their general disadvantage is a certain loss of resolution or clarity of the photographic images, because they are rephotographed from the originals.

3. *Black and white aerial photographs.* These are the most commonly used mapping aids. Their general advantage is greater clarity, and thus recognition of detail, at the same scale than seen on 2. Another major advantage is the possibility of using these standard air photos in pairs for three-dimensional viewing under a stereoscope. This permits recognition of minor variations in topography and often relative heights of trees, and, under certain conditions, even vertical stratification of open forests. The aerial image becomes much clearer and more detailed if viewed under a stereoscope.

4. *Infrared aerial photographs.* These are black and white photos taken with infrared filters that permit even clearer differentia-

tion of certain vegetation units. In addition, they may permit easy recognition of finer within-unit variation such as phenological differences or dying trees, etc. For increased differentiation, the black and white air photo pattern can also be converted into a highly contrasting color spectrum, particularly if the infrared photography is done with a scanner.

5. *Colored aerial photographs.* These offer many advantages over 3 and 4. Their present high cost, however, is often not in balance with the mapping objectives. Particularly useful are color photos taken simultaneously with different filters (e.g., GERRESHEIM 1971).

Also when mapping color photographs, stereoscopic viewing offers more contrast and detail than the viewing of single photographs.

The great advantage in the use of aerial photographs for vegetation mapping lies firstly in the relative ease of drawing boundaries around units and secondly, in the ease of maintaining the relative uniformity or equivalence of the units across the entire map area. The boundaries are best drawn directly on the air photographs.

The final map preparation from air photos of types 3, 4, and 5 requires, in most cases, elimination of the scale-distortion that is present in all single, vertically taken air photos towards their margins. This is usually achieved by hand-transfer of the air photo boundaries to a topographic map.

This can be done by preparing a tracing of the contour lines and major landmarks of the topographic map that is then matched in scale (by enlarging or reducing) to the central undistorted portion of the air photo. The vegetation type boundaries are then transferred onto either the same or a second transparent overlay as far as they can be matched with the contour lines and landmarks. The distorted portions are neglected and the boundary lines are subsequently connected with the tracing on the next overlapping air photo, from which the boundaries are transferred in the same manner. An advantage in mapping is to have a transparent overlay with contour lines prepared before field mapping. On this contour overlay, the distorted areas on each air photo can be indicated by a polygonal boundary that encloses only the scale-undistorted central part of each air photograph and discards the rest.

This aspect of transfer of air photo boundaries to topographic maps is a job that can be done by any skilled draftsman, but it is very time-consuming work. From this viewpoint, air photo mosaics are ideal as they do not require this transfer.

Vegetation units defined by their total floristic composition cannot always be clearly recognized on air photos because the structure visible on air photos is primarily conditioned by the dominant species and their developmental status. In floristically very rich areas, as in parts of the tropics, the air photo patterns may be relatable only to dominant life forms containing several species and lacking dominants. In any case, it is always necessary to accompany air photo mapping with intensive field work to achieve a "calibration" of the field variation or pattern of plant communities with the patterns seen on the air photos. Without field work and frequent ground checks an air photo-derived map may be worthless.

If one has only topographic maps for mapping, an exact orientation on the ground may cause some problems. This can be solved by measuring distances by passing from certain mapped landmarks, by use of stakes, pegs, and flagging tapes and by walking known distances along compass lines.

14.72   Map Keys.   Before starting with routine mapping, one should always prepare a map key that serves to identify the units without ambiguity. Such a key also forces the mapper into establishing an exact definition of what he wishes to portray. Moreover, a map key permits several workers to map in exactly the same way.

When using the BRAUN-BLANQUET method, a key is automatically obtained from the differential synthesis table which shows the key species in the form of the differential species (see CHAP. 9, TABLE 9.11). If one works with structurally defined vegetation units, it is advisable to define their salient properties in words. Such a key is shown as an example in APPENDIX C. This key was prepared for mapping structural vegetation types in the dry zone of Ceylon at the scale of 1 : 31,-680. Mapping was done on air photo mosaics, and the key represents a large-scale local adaptation of FOSBERG's system (CHAP. 8). The key gives four main categories: forest, scrub, herbaceous cover types and other areas. Each of these shows a number of quantitatively defined map units that can be mapped by any experienced person with a minimum of field work. The finished map (MUELLER-DOMBOIS 1969) was to serve as a framework for floristic detail-sampling and for animal activity surveys (MUELLER-DOMBOIS 1972) and for correlation with environmental maps of the same scale to establish the major ecosystems of the area. Another map key is shown in APPENDIX D. This key was prepared for mapping forest sites or habitat types in SE Manitoba to a scale of 1 : 15,840 (MUELLER-DOMBOIS 1965a). This is not a vegetation key, but one based on a combination of surface-soil, topographic and vegetation properties. This key was used to map the habitat types shown in FIGURE 10.15.

It would be a mistake to consider a map key as final or incorrectable during the mapping process. Even after thorough preparation, one often makes new observations which may demand a correction or modification. Species combinations or structural variations that cannot be grouped into the scheme established so far should not be forced into it. In this case, it is best to leave the questionable area unmapped for the time being. Instead, one may establish new relevés or other clarifying investigations, which may eventually lead to a correction. The gap can then be closed, but the area already mapped has to be rechecked to see that the earlier units are balanced properly with the corrected ones.

Aids for the construction of potential natural vegetation maps at small scales have already been briefly discussed in SECTION 14.5. Each landscape has certain unique problems that require special aids and adaptations. An excellent reference, for further details, is KÜCHLER (1967).

14.8 COMPARISON OF VEGETATION AND SITE MAPPING METHODS

Vegetation and site mapping for applied purposes, particularly in forestry, has received its greatest attention in Europe. At the 1954 World Forestry Congress in Dehra Dun, SUKACHEV suggested that his method of site mapping should be applied internationally. The chairman of the site-research section, De PHILLIPPIS, made a counter proposal that an international comparison of several methods should be made. Details were stipulated at the 1956 IUFRO Congress (International Union of Forest Research Organization) in Oxford and at a workshop in Warsaw in 1959. Sixteen institutes showed interest in participating. The criteria established at the Warsaw meeting were as follows:

1. By different methods, a forest area was to be fully evaluated with its site-spectrum in terms of climate, soil, vegetation, other biological factors, and human impact. The interaction of these factor-complexes was to be represented as clearly as possible. A site considered in its biological totality was referred to as an ecosystem, a biogeocoenosis, or a biochore, depending on the author's preference. This integrated approach to site evaluation was also aimed at an assessment of the actual and potential productivity.

2. As many as possible site-evaluation methods were to be compared. Coordinating institute was to be the Geobotanic Institute RÜBEL in Switzerland (ELLENBERG).

3. It was suggested to have the methods comparison in different regions with strong within-variation of sites. According to SUKACHEV, at least the following methods should be compared:

(a) The method of BRAUN-BLANQUET as an example of a phyto-sociological method working particularly with species combinations.

(b) The method of AICHINGER as an example of a phytosociological method, which emphasizes the dynamics of plant communities.

(c) The method of SUKACHEV that aims at an assessment of the biogeocoenosis, but that uses, for mapping in particular, the dominance-relations of plant species (SUKACHEV and DYLIS 1964).

(d) A combined site and phytosociological method that integrates both approaches, i.e., recognition and mapping of units through plant indicators and direct site factors.

(e) A method whose main emphasis lies on *direct* site evaluation without the use of plants as indicators.

(Method explanations are summarized below.)

4.   The suggested test areas were: one in Poland, one in the Soviet Union, and one in Switzerland. Each was to be a forest land area of about 500 ha. Choice of the specific area was left to the site-research institutes of the host countries. Field work was to begin in 1960-1961.

5.   Upon completion of the work, the information value of each method was to be evaluated by an international team of silvicultural experts. The following criteria were to be used in judging the methods:

(a) The information value for practical silviculture and other forestry purposes.

(b) The cost of application of each method.

(c) The flexibility of the method in terms of changing developments in silvicultural techniques.

(d) The information value of the method for related disciplines such as landscape planning, soil science, phytosociology, etc.

The outcome of these plans for the Swiss example appeared in a publication (with five maps and three fold-out tables) edited by ELLEN-BERG (1967).

A 420 ha area was selected in Switzerland on the northern foothills of the Alps (Forstkreis Zofingen, Kanton Aargau) ranging in elevation from 460 to 713 m above sea level. This area was studied and mapped at 1 : 10,000 by application of the following five methods:

1.   BRAUN-BLANQUET method; on a purely phytosociological basis (by FREHNER).

2.   AICHINGER method; phytosociological with emphasis of dynamics (by BOSSE-MARTIN under direction of AICHINGER).

3.   In place of the method of SUKACHEV, who could not participate personally (nor was he able to send his collaborators to Switzerland), the method of SCHMID was included in the comparison (by

SAXER, under direction of SCHMID). SCHMID's method evaluates the site spectrum of an area by means of "vegetation belts" and "biocoenoses." This will be further explained below.

4. A *combined* method using both vegetation and direct evaluation of site factors. The Eberswalder method of KOPP was applied by EBERHARDT (under direction of KOPP with participation of PASSARGE).

5. Instead of a direct site evaluation method without use of plants (for which unfortunately no person could be found), KOPPP's separate (transparent) map of "site forms" was used to provide an example.

The four work teams each supplied its complete data with a comprehensive interpretive text and a colored 1 : 10,000 map of the approximately 5 km long and 1 to 2 km wide area. These were then published in one volume with the reviewer's evaluations. Examples of these maps are shown in FIGURE 14.2 for identical area segments.

Since we considered these examples as very instructive, a summary of the procedures applied in each method is given below. The base maps used were 1 : 5000 topographic sheets.

1. *BRAUN-BLANQUET method.* A large number of relevés is placed into the tentatively recognized communities. From these, synthesis tables are prepared from which locally applicable differential species are isolated (procedure as in CHAP. 9). These are then used in certain combinations as key species for mapping locally valid vegetation units. These local units are later grouped by character species into the general hierarchical system of plant communities. The site factors of each vegetation unit, at least the soil of each, are also investigated. However, mapping is based strictly on floristic criteria. The map units at this large scale of 1 : 10,000 are mostly variants within subassociations and associations (for definition see CHAP. 8).

2. *Method according to SCHMID.* Only a few vegetation samples are analyzed in detail for orientation. These are then sorted as members of already previously defined "vegetation belts." These vegetation belts are defined as large vegetation units that are established on the basis of a number of species with similar geographic distributions. The vegetation belts are thus broadly conceived floristic-historic units whose boundaries intergrade or overlap to some extent. This continuum problem is overcome by the recognition of "isolation districts," which are ecotonal geographic-floristic subunits of a vegetation belt. The vegetation belts represent large-scale floristic provinces or areal types. Each vegetation belt, which may therefore be better understood as a floristic belt, contains a number of phytocoenoses. The allocation of a vegeta-

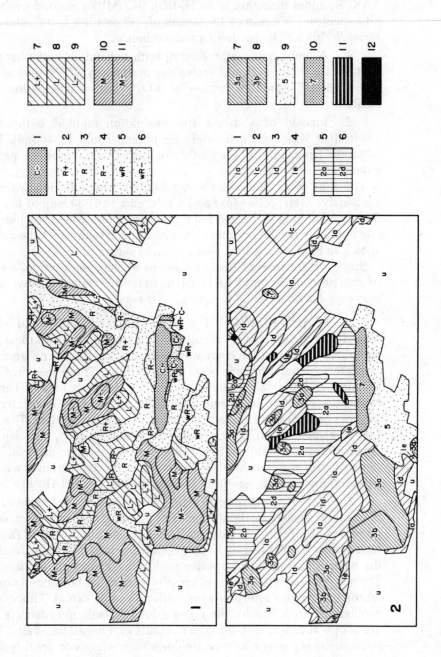

1 SITE-FORM GROUPS (COMBINED METHOD AFTER KOPP)

1 dry, calcareous sites
2 moist, nutritionally rich sites
3 fresh, nutritionally rich sites
4 dry, nutritionally rich sites
5 fresh, nutritionally rich sites in warm locations
6 dry, nutritionally rich sites in warm locations

7 moist, nutritionally less rich sites
8 fresh, nutritionally less rich sites
9 dry, nutritionally less rich sites
10 fresh, nutritionally medium sites
11 dry, nutritionally medium sites

2 NATURAL FOREST COMMUNITIES (AFTER BRAUN-BLANQUET)

1—4 *Melica nutans–Fagus sylvatica* Association (=*Melico–Fagetum*)

1 Subassoc. with *Asperula odorata* (=*asperuletosum*)
2 Subassoc. with *Blechnum spicant* (=*blechnetosum*)
3 Subassoc. with *Luzula sylvatica* (=*luzuletosum*)
4 Subassoc. with *Cornus sanguinea* (=*Cornetosum sanguineae*)

5—6 *Milium effusum–Fagus sylvatica* Association (=*Milio–Fagetum*)

5 Subassoc. with *Dryopteris disjuncta* (=*dryopteridetosum*)
6 Subassoc. with *Luzula sylvatica* (=*luzuletosum*)

7—8 *Melampyrum pratense–Fagus sylvatica* Association (=*Melampyro–Fagetum*)

7 Subassoc. typicum

8 Subassoc. with *Leucobryum glaucum* (=*leucobryetosum*)
9 *Pulmonario officinalis–Fagus sylvatica* Association (= *Pulmonario–Fagetum*)
10 *Carex spp.–Fagus sylvatica* Association (=*Carici–Fagetum*)
11 *Acer pseudoplatanus–Fraxinus excelsior* Association (= *Aceri–Fraxinetum*)
12 *Carex remota–Fraxinus excelsior* Association (=*Carici remotae–Fraxinetum*), Subassoc. with Chrysosplenium alternifolium (=*Chrysosplenietosum*)

Remark: Nos. 7, 8, and 10 indicate dry sites, No. 10 indicates moist sites, No. 11 indicates wet sites; all other communities indicate fresh (i.e., intermediate) sites. Nos. 2, 3, and 6—8 are acidity indicators, Nos. 4, 9, and 10 are limestone indicators.

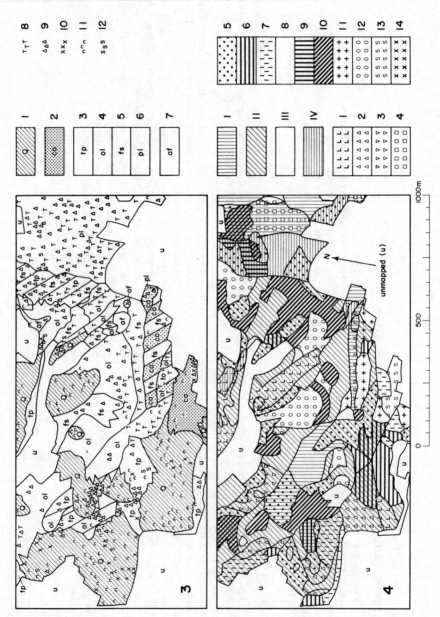

FIGURE 14.2. Examples of maps.

3 VEGETATION-BELTS AND BIOCOENOSES (AFTER SCHMID)

1 *Fagetum sylvaticae* mixed with *Quercus robur–Calluna*-belt-associates of the Europe–Asia Minor segment of the *Fagus–Abies alba*-belt in the Isolation-district of the Molasses-foothills of the Alps, on the most acid and degraded Molasses-soils

2 *Fagetum sylvaticae* mixed with *Quercus–Tilia–Acer*-belt-associates of the Europe–Asia Minor segment . . ., on sandstone with traces of limestone superimposed

3 *Fagetum sylvaticae typicum* of the Europe–Asia Minor segment . . ., on sandstone

4 *Fagetum sylvaticae* of the Europe–Asia Minor segment . . ., on oligotrophic Molasses-slopes

5 *Fagetum sylvaticae* of the Europe–Asia Minor segment . . ., on moist, gravelly slopes

6 *Fagetum sylvaticae* of the Europe–Asia Minor segment . . ., on nutritionally rich, moist, gravelly plateaus

7 *Acereto–Fraxinetum* of the Europe–Asia Minor segment. . . .

8 *Abies alba*

9 *Picea abies*

10 *Larix decidua*

11 *Pinus sylvestris*

12 *Pinus strobus*

4 VEGETATION-DEVELOPMENT-TYPES (AFTER AICHINGER)

Basic hatching:

I *agrum solum silicicolum* (former arable field)

II *semi-superirrigatum*

III *superirrigatum* (watered from above only)

IV *paludosum* (moist)

Superimposed symbols:

1 *Luzula luzuloides–Vaccinium*-type

2 *Luzula luz.–Luzula sylvat.–Vaccinium*-type

4 *Luzula–Oxalis*-type

5 *Luzula–Vaccinium–Oxalis–Asperula*-type

6 *Luzula–Vaccinium–Asperula*-type

7 *Luzula–Asperula*-type

8 *Luzula–Oxalis–Asperula*-type

9 *Vaccinium–Oxalis–Asperula*-type

10 *Oxalis–Asperula*-type

11 *Asperula*-type

12 *Rubus–Oxalis–Asperula*-type

13 *Rubus–Asperula*-type

14 *Rubus–Vaccinium–Oxalis–Asperula*-type

Remark: Nos. 1–3 are acidity indicators, No. 11 is a limestone indicator.

433

tion sample to one of the previously defined floristic belts is done on the basis of the proportion of species in the sample, whose geographic distribution coincides most closely to those of an established floristic or vegetation belt. The phytocoenoses are defined by species combinations, particularly by dominant trees and by the growth forms (life forms) of all species in the sample. Thus, dominance and fine-structure of species form the mapping criteria. For unit interpretation, site factors and human influences are evaluated as far as they can be easily ascertained in the field.

The basic unit is the vegetation belt (and its penetration into a specific area), which is always represented by a phytocoenosis. The latter are named after the dominant trees, further by belt-relationship, isolation district, and by soil characteristics.

3. *Method of AICHINGER.* A dense net of relevés (about one per hectare) is distributed throughout the study area. These are analyzed according to the BRAUN-BLANQUET method and are used for establishing undergrowth community types. These undergrowth types are used also for mapping, but the actual mapping is done indirectly from interpolation among the network points (relevés). On the basis of soil-surface configuration (topographic position) and other easily recognizable soil characteristics, the undergrowth types are combined into ecological indicator groups that are designated by Latin environmental names (e.g., *super-irrigatum* or *paludosum*). The development (i.e., progressive or regressive successional trends) of each vegetation type is then interpreted from general experience. Succession is therefore not used as a criterion for mapping.

The basic map unit is called vegetation development type, which is named after dominant tree species, dominant undergrowth species, and site characteristics.

4. *Eberswalder combined method after KOPP.* Vegetation analysis follows essentially the method of BRAUN-BLANQUET, except vegetation units are established only for the study area. These units are called site-vegetation types. They are identified by combinations of plant indicator groups ("ecological groups") which correspond roughly to differential species. Soil investigation begins, initially independently of vegetation analysis, with a very detailed investigation of numerous soil pits. But these are mostly in the same plots, in which the vegetation samples are analyzed. This is followed by establishment of locally applicable soil types (so-called "soil forms"). These are designated by geographic place names combined with the main textural class and the greater soil group to which the profiles belong (e.g., Stolten-loam-brown earth). A detailed laboratory analysis of soils is also carried through. Further habitat characterization is added by analysis of the general cli-

mate of the area and its local variations which are evaluated from topographic position, slope, exposure to sun, local wind influences, etc.

The knowledge gained from the study of vegetation, soil, and local climate is combined into so-called site-form groups, which are thus synthetic units. Only these are portrayed as map units for management purposes. All other characteristics that are not of practical value to management are omitted. The site-form groups are briefly circumscribed on the map by such phrases as "lime-rich sites with below-normal water supply," or, "warm, rich sites with normal water supply." Only the relatively stable site properties (such as the basic moisture and nutrient regimes) and no properties influenced by management (such as humus type, certain undergrowth plant groupings, etc.) find recognition in the concept of "site form." The latter are, however, supplied as additive characterization on overlay maps and as descriptive text. On the basis of these stable units, the area does not require remapping with changes in the silvicultural treatments.

5. *Soil survey method.* This method is based on a large number of soil pits and soil profile descriptions as already described (under 4) for the establishment of "soil forms." (A soil map of this kind is today considered insufficient for forest site evaluation at least in Western Europe. This method thus was not even included as a separate approach in this comparison.)

The four work-teams applied their methods independently without any discussion or contact during the field work or subsequent preparation of manuscripts. After completion of the printed maps and texts in the summer of 1966, invitations for judging the outcome were sent to twelve silviculture professors of different countries. The names of these experts were determined by vote through the silviculture section of UFRO. Four individuals accepted the invitation, OLA-BØRSET (Norway), KÖSTLER (West Germany), LEIBUNDGUT and ROTH (both from Switzerland). They received the maps and texts for review before the site meeting and thereafter met for a two-day field meeting in the study area and for a subsequent day of discussions indoors. Their final reports are published in the same volume in English (ELLENBERG 1967:283).

The judges rated the methods for their silvicultural information value as follows:

No. 1 The Eberswalder combined method of KOPP

No. 2 The BRAUN-BLANQUET method

No. 3 The method of SCHMID

No. 4 The method of AICHINGER

Some of the major reasons given for this rating order were:

Number 1 • Gave the most accurate and detailed silvicultural information.

• The method has its greatest advantage in areas variously disturbed by human or other activity because it stresses the stable site properties.

• It was the most time-consuming of all methods (4 ha/man-day), but the extra time was considered justified for areas with highly intensive silviculture.

Number 2 • In general, this strictly floristic mapping method showed very good correlation with silviculturally important environmental variations in this area, the time required was much less (13 ha/man-day) than for No. 1, but the judges cautioned against the site indicator value of this method.

• Two separate associations (*Querco–Abietum* and *Melico–Fagetum*) were mapped on one basic site, because part of this site had been disturbed by human activity. This modification was considered temporary and unimportant with regard to the basic potential of the site.

• One of the map units (*Milio–Fagetum*) reflected clearly the increasing influence of the mountainous climate in the southern part of the map area, while the others did not recognize this unit. Number 1 only made a passing reference to it.

Number 3 • The method of SCHMID aims at providing an interpretation of world vegetation. It thus presents information considered unnecessary for silvicultural purposes, while it lacks detail needed for the latter.

• However, it was felt that application of the detailed "growth form" (life form) classification for environmental interpretation held great promise that could be further exploited for applied purposes.

• The method took the least time (with 18 ha/man-day).

Number 4 • The map of AICHINGER was prepared in an office after evaluation of the network of about 500 relevés. The boundaries thus were more arbitrarily drawn than on all other maps.

• The dynamic interpretation of primarily human activity on the sites was appreciated, but it was criticized that no successional trends were indicated on the map.

• The judges considered some of the terminology for the main groups confusing. For example, the term "paludosum" was used for a mineral soil on a high plateau with a suspended water table.

Time used for the AICHINGER method (10 ha/man-day) differed little from that used for the BRAUN-BLANQUET method (13 ha/man-day).

It is clear from this comparison of methods that the best approach to forest site classification and mapping for practical forestry purposes is a method that combines phytosociological information with environmental information. A similar conclusion was reached in a recent review paper of forest site classification methods in Canada by BURGER (1972). All four methods do combine these two sets of information to some extent, but the first was the most consequent in this respect. One reason for the superiority of the combined approach is the recognition of certain vegetation patterns as merely temporary and others as more permanent. In this way, more reliance can be given to the permanent patterns, a point stressed also very much by DAUBENMIRE (1968). A second important reason is simply the much greater knowledge and information obtained on the local environmental variations and their significance when these are given equal study, a point also stressed particularly by KRAJINA (1960, 1969).

However, while an intensification of local environmental study is of unquestionable advantage for site evaluation, the approach to be taken depends entirely on the objectives. An example is the method of SCHMID, whose primary purpose is a geographic-floristic and historic evaluation of the vegetation of the world. His map of the study area showed only one major color (green) indicating that he considered this area merely as a part of a much wider geographic concept. From a scientific viewpoint, all interesting objectives are acceptable provided they do not violate the basic scientific method of inquiry, which is to check any emerging trend carefully against the evidence available.

With regard to additional objectives, all methods complemented each other—the BRAUN-BLANQUET method by its scope of extending the level of comparison of the local community-site spectrum to a much broader regional system; the method of SCHMID by adding a historic-floristic and evolutionary dimension to the local study; and AICHINGER's method by adding a dynamic interpretation with regard to degrading or improving silvicultural treatments. In spite of these complementary aspects, the general results of the four studies were in close agreement, which shows that the methods are basically oriented towards an environmental interpretation. However, the combined method is directly geared to the forest site use objectives that the investigation called for. For this reason, it gave the best results.

PART
V

Conclusions

Synthesis of Aims and Methods
in Vegetation Ecology

15

15.1 SCIENTIFIC AND PRACTICAL IMPORTANCE

Vegetation ecology has recently gained in scientific importance. It has also become more important in problem solving. This impetus is related to the general realization that the solving of environmental questions is so essential to life on this planet.

Today, vegetation still forms the immediate environment of man and his domesticated stock over large areas of the earth's surface. For maintenance of a quality human environment, vegetation and botanical organisms in general must remain geographically dominant.

Plant communities are indicators of the environment. They respond not only to one environmetal factor, but also to an interacting group of factors. The plant community integrates these influences and reacts sensitively to changes in the balance of environmental stresses.

Vegetation is usually the most readily recognized component of ecosystems. Plant communities are often used to identify and define the boundaries of ecosystems.

Plants are the primary producers in ecosystems. An exact knowledge of the structure and composition of plant communities is therefore important for an understanding of trophic relationships.

Disturbances of the biological balance through foreign plant or animal species or through direct interference by man are often readily recognized by changes in the physiognomy, structure, and species composition of the vegetation.

15.2 FUTURE DEVELOPMENT OF VEGETATION ECOLOGY

Future development in the scientific application of vegetation ecology should (and probably will) go in two major directions:

1. Laterally, i.e., by closing the gaps of our knowledge of the vegetation in little or as yet uninvestigated areas, for example, in parts of South America, Africa, and Asia.

2. Vertically, i.e., by deepening our understanding of the plant community mosaics at all levels of abstraction, particularly by:

(a) increasingly exact determinations of the *structure* of communities and species population patterns therein (CHAP. 5 through 10) as a prerequisite for:

(b) causal-analytical research (CHAP. 12) and for experimental modifications of plant communities as well as for:

(c) exact research of temporal phenomena, such as rhythmic events (of a phenological nature (SECTION 8.5)), as distinct from directional changes of succession (CHAP. 13) and the study of evolutionary trends (CHAP. 12). These together may provide the means for predicting future vegetation development, if:

(d) the spatial environmental variations of habitats and along gradients are fully taken into consideration (CHAP. 11).

In addition, vegetation ecology can contribute in the future greatly to:

3. Research into the relationships of biological diversity and stability of different landscapes and into the changes of these landscape attributes as influenced by technological man.

4. Ecosystem research in the sense indicated under 2 above by deepening our understanding particularly of the sociological environment of ecosystems.

The future problem-solving capacity of vegetation ecology depends not only on its advances made in addressing itself to scientific questions, but also on the involvement of vegetation ecologists in practical issues related to our environmental problems. With respect to current environmental problems, the tasks of vegetation ecology can be seen particularly in three areas:

5.   Classification of the earth's landscapes into ecologically equivalent habitats, zones, or regions as the basis for environmentally oriented planning (i.e., mapping of ecological landscape units with the aid of vegetation).

6.   Prediction of consequences of land management (from agriculture, forestry, engineering, etc.) from the knowledge gained through the study and mapping of ecological similarities and differences between and within areas. (The same managerial treatments will have the same success only in ecologically equivalent landscapes.)

7.   Systematic and detailed investigations of the effects of man's activity on specific vegetations, plant communities or ecosystems so that the predictive capacity (mentioned under point 6) becomes increasingly more reliable.

Therefore, vegetation ecology is intimately involved and must still solve a *broad range of many-sided general and specific tasks.* Vegetation ecology enters only now a phase in which it can contribute to a sound practical application of the findings gained through interdisciplinary ecosystem research. The latter is essentially designed to answer fundamental, scientific questions, and these are persued currently by integrated research teams in many nations, e.g., under the International Biological Program (IBP).

15.3   SYNTHESIS OF METHODS

For predicting the correct level of result-application of basic ecosystem research, it is necessary to arrive at a synthesis of those methods in vegetation ecology which have proven their validity. A particular need was to bridge the gap between the continental-European and Anglo-American schools of vegetation ecology, both of which have worked so far in too much isolation. As a consequence, they tended to forego the advantages inherent in the methods of other schools.

The continental-European approach to vegetation ecology, especially the releve technique for association analysis after the method of BRAUN-BLANQUET (CHAPS. 5 and 9) is particularly well suited to accomplish the tasks named under points 1, 3 and 5 in the preceding section. These tasks can be accomplished by this method in a rapid manner and on a global basis for comparisons within and among world formations (APPENDIX B). The BRAUN-BLANQUET method offers a solid foundation for vegetation mapping on large scales.

In continents, whose flora is as yet not sufficiently investigated, it is still necessary today to describe and map initially the formations (CHAP. 8), and often one has to be satisfied with small scale maps (such as 1:1 million or smaller). However, in floristically highly complex areas such as found generally in the tropics, large-scale maps (larger than 1 : 50,000) may also be developed initially on structural rather than floristic criteria (APPENDIX C). The floristic contents with an indication of species-quantities in communities comprise the fundamental ecological information. In more detailed (large-scale) structural units, these are also most readily evaluated by the releve´ method.

For exact analysis objectives in the sense of points 2 and 4 above, quantitative methods are essential such as those developed particularly in America and England, but also in Northern Europe. Quantitative field analysis methods and the computer processing methods of vegetation data are, however, really successful only where a meaningful, initially qualitative analysis precedes the quantification. For this, the BRAUN-BLANQUET method may serve as a recommendable basis and framework. It would be desirable that a proper familiarity with this method be achieved finally also among American investigators.

For an in-depth study of vegetation in those countries in which most information was derived from the BRAUN-BLANQUET method, we recommend in particular:

• The *point-frequency method* for obtaining reproducible values of cover and plant biomass in herbaceous communities (SECTION 6.54).

• The *BITTERLICH method* for rapid and accurate analyses of stem-cover (i.e., basal area) by species in semicomplex forest communities in which the undergrowth is primarily herbaceous (SECTION 7.5).

• The *point-centered quarter method* for tree density estimates, basal area and frequency analysis by species in mixed-species stands of closed forests (SECTION 7.63).

• The *count-plot method* along belt transects in subplots for the same three parameters as the point-centered quarter method in all forests that vary structurally in segments between closed and open (SECTION 7.3).

• The *line-intercept method* for assessing the cover of the more common species in all open vegetations, such as scattered bunchgrass, open scrub, or open forest vegetation (SECTION 6.55).

- The *local frequency method* in 100 subsquares of a square meter frame for permanent quadrat analysis of species abundance in pastures or other short herbaceous covers that are expected to show rapid changes in species composition (SECTION 6.32). This method is most profitably combined with the point-frequency method to obtain cover as a second important parameter.

For data processing of up to 50 to 200 relevés to develop keys for mapping floristically defined vegetation units, we recommend hand-processing by the BRAUN-BLANQUET table technique (CHAP. 9). Where more than about 200 vegetation samples are involved, computer processing may become more convenient. This depends primarily on the availability of and access to computer facilities.

For ordination purposes, we recommend the BRAY and CURTIS method for its high information value achieved with a relatively simple computation process. We believe that ordination of stands or species has a real advantage (a) where little is known about environmental correlations with any observed species combination, (b) in environmental gradient analysis to determine the position of each sample stand or species in relation to a known gradient, and (c) as an independent sorting method for subsequent classification.

Since plant communities are geographic phenomena, we suggest further that a vegetation study is not complete without a map and preferably with profile diagrams for map interpretation.

Since mapping requires decisions as to the limits of units, we also suggest that a vegetation study is not yet complete without a classification. Therefore, we consider classification an important part, whether an investigator includes an ordination of his samples or not.

Finally, we believe that quantitative field analysis, ordination, classification, and mapping are merely tools to aid in an understanding of vegetation. A complete study of vegetation includes not only an establishment of the patterns of vegetation and species combinations, but also an explanation of these through establishment of correlations and causes. Moreover, vegetation does not demand only an explanation of its spatial distribution and changes in time (succession), but also an explanation of its physiological processes. Yet, for reasons already explained the latter could not be covered in this work.

This book represents an attempt to build a bridge. The structural supports of this bridge are now considered as set, so to speak. The supports relate to the detailed discussion of the more important methods in vegetation ecology as well as to their advantages and disadvantages with respect to the defined aims. Therefore, we would be pleased if this book contributes to a positive understanding of the different orientations and schools in vegetation ecology.

Appendices

Appendix A

A Key to Raunkiaer Plant Life Forms with Revised Subdivisions (from ELLENBERG and MUELLER-DOMBOIS, 1967b)

I. KEY TO THE MAIN GROUPS OF PLANT LIFE FORMS

Aa   Autotrophic plants
  Ba   Kormophytes (= vascular plants)
   Ca   *Self-supporting plants*
    Da   Woody plants, or herbaceous evergreen perennials

> Plants that grow taller than 50 cm, or whose shoots do not die back periodically to that height limit*
>   **Phanerophytes** 1

---

* In particularly favorable environments (e.g., humid tropics and warm seepage water habitats) this height limit may be extended to 100 cm.

449

Plants whose mature branch or shoot system remains perennially within 50 cm above ground surface, or plants that grow taller than 50 cm, but whose shoots die back periodically to that height limit (see footnote on p. 449)

Chamaephytes 2

Db Perennial (including biennial) herbaceous plants with periodic shoot reduction

Periodic shoot reduction to a remnant shoot system that lies relatively flat on the ground surface

Hemicryptophytes 3

Periodic reduction of the complete shoot system to storage organs that are imbedded in the soil

Geophytes (Cryptophytes) 4

Dc Annuals. Plants whose shoot and root system dies after seed production and which complete their whole life cycle within one year   **Therophytes** 5

Cb *Plants that grow by supporting themselves on others*

Ea Plants that root in the ground

Plants that germinate on the ground and maintain their contact with the soil   **Lianas** (Eu-lianas) 6

Plants that germinate on other plants and then establish their roots in the ground, or plants that germinate on the ground, grow up the tree and disconnect their soil contact

Hemi-epiphytes (Pseudo-lianas) . 7

Eb Plants that germinate and root on other plants (these include dead standing plants, telegraph poles and wires, stumps and such like)   **Epiphytes** 8

Cc *Free-moving water plants* (=errants)

Errant Vascular Hydrophytes ... 9

Bb Thallophytes (=non vascular crytogams)

Fa *Plants attached to the ground surface* (here defined as any material making up the surface of the ground, such as mineral soil, rock, humus, litter, decaying wood or other solid media covering the surface)

GA Perennials

Cushion-formed or pulvinate mosses and liverworts and fruticose lichens   **Thallo-chamaephytes** 10

Flat-matted mosses and liverworts, foliose and crustose lichens, algae (including endolithic lichens and algae)

Thallo-hemicryptophytes 11

Gb Annuals   **Thallo-therophytes** 12

Fb *Plants attached to others*, either directly to the bark, leaves or such like, or indirectly to soil and humus pockets occurring in branch-forks, bark fissures, etc.

Thallo-epiphytes 13

Fc   *Free-moving autotrophic thallophytes* (=errants)

  Ha   Photosythesizers
    In water (salt, brackish or fresh)
    Errant Thallo-hydrophytes 14

    In snow and ice   **Kryophytes** 15

    At and near the soil surface (including humus and decaying wood)   **Edaphophytes** 16

    Chemosynthesizers   **Chemo-edaphophytes** 17

Ab   Semi-autotrophic plants
  Green plants growing attached to other living autotrophic plants
    Ia   Kormophytes   **Vascular Semi-parasites** 18
    Ib   Thallophytes   **Thallo-semi-parasites** 19

Ac   Heterotrophic plants

  Ka   Kormophytes
    Growing on living plants
    Vascular Parasites 20

    Growing on dead organic matter
    Vascular Saprophytes 21

  Kb   Thallophytes
    Growing on or in living plants
    Thallo-parasites 22

    Growing on dead organic matter
    Thallo-saprophytes 23

II. KEY TO THE SUBDIVISIONS OF THE MAIN GROUPS OF PLANT LIFE FORMS

Subdivisioning has been carried through the autrophic terrestrial plant life form groups 1–13. These are also the main producers of concern to the terrestrial plant ecologist. A satisfactory subdivision of life form groups 14–23 requires special knowledge and would go beyond the present scope.

The key employs the decimal system. The first digit designates the main life form group, e.g.,

  1. Phanerophytes
  2. Chamaephytes, etc.

The second digit denotes the next subgroup, e.g.,

  1.1 Phanerophytes with normal woody stems and branches
  1.2 Tuft trees
  1.3 Bottle trees, etc.

The third digit denotes an important characteristic in the phanerophytes:

1.01 Single-stemmed (scapose)=trees
1.02 Branched from near the base (caespitose)=shrubs

The fourth digit refers to height classes in the phanerophytes:

1.001   <2 m = Nanophanerophytes
1.002   2–5 m = Microphanerophytes
1.003   5–50 m = Mesophanerophytes
1.004   >50 m = Megaphanerophytes

The above digit places are reserved for these attributes in the phanerophytes. Other important attributes, such as evergreen versus deciduous, follow by added digits, and whatever attribute best characterizes the subgroups occupies further digit places. In the other life form groups, such as chamaephytes, hemicryptophytes, etc., the digits following the first one are also chosen for the next important characteristics, which are of course not always the same as in the phanerophytes. In the height classes, the smallest is always denoted with 1, the next higher one with 2, etc.

In places where a zero (0) occupies a certain digit place, the attribute is undetermined.

Abbreviations of life form names have been added only as far as they are already used in literature.

1. Phanerophytes (P)

1.1 Phanerophytes with normal woody stems and branches ..		P
1.11	Trees=single-stemmed phanerophytes with more or less numerous lateral branches (=scapose)	P scap
1.111	Dwarf trees=Nanophanerophytes <2 m	N P scap
1.112	Small trees=Microphanerophytes 2–5 m	Mi P scap
1.113	Large trees=Mesophanerophytes 5–50 m...........	Mes P scap
1.114	Giant trees=Megaphanerophytes >50 m	Meg P scap
1.12	Shrubs=Phanerophytes branched from near the base of the stem (=caespitose)	P caesp
1.121	Normal-sized shrubs=Nanophanerophytes <2 m	N P caesp
1.122	Tall shrubs=Microphanerophytes 2–5 m	Mi P caesp
1.123	Giant shrubs=Mesophanerophytes >5 m	Mes P caesp
1.13	"Krummholz"=creeping phanerophytes, whose stems or branches are bowed down, but whose height exceeds 50 cm vertically from the ground (=reptant habit)	rept
1.131	Typical "krummholz" <2 m	N P rept
1.132	Tall "krummholz" >2 m	Mi rept

The above height classes and the distinction between trees, shrubs and krummholz need to be applied to specified field conditions. For simplifying the mechanism of this classification the following separations are based on features that apply to almost all normally woody phanerophytes (P) in their appropriate size-class ranges, whether they are trees or shrubs.

1.100.1   Evergreen
            Broad-leaved

            Without bud protection, probably almost exclusively tropi-
            cal rain (ombro=o) forest species

1.100.11   Malacophyllous (=m; soft leaves that collapse immedi-
            ately when held over hot water vapor, e.g., *Macaranga*)   omP

1.100.12   Semi-sclerophyllous to sclerophyllous (=s, e.g., *Coffea*)   osP

            With bud protection

1.100.13   Malacophyllous (=m, e.g., *Hibiscus tiliaceus*) ....   mP

1.100.14   Sclerophyllous (=s, e.g., *Metrosideros collina*) ....   sP

1.100.15   Needle-leaved (belonido=b, e.g., *Pinus*) ....   bP

1.100.2   Summer-green or cold-deciduous (aestivo=a)

1.100.21   Broad-leaved (e.g., *Fagus*) ....   aP

1.100.22   Needle-leaved (e.g., *Larix*) ....   abP

1.100.3   Drought-deciduous (cheimo=c), mostly with strong
            bud protection during the dry season, e.g., *Erythrina* ..... cP

Each of these life forms can be further subdivided by:

            (a) Crown shape

1.100.001 With spherical crown (e.g., *Mangifera indica*)
    .002 With umbrella-like crown (e.g., *Samanea saman*)
    .003 With cylindrical crown (e.g., *Metrosideros collina* in ash-fallout
         areas on Hawaii)
    .004 With conical crown (e.g., many alpine temperate-zone conifers, but
         also for example young *Rhizophora mangle*)
    .005 With umbellate crown (e.g., *Albizia falcataria*)
    .006 With irregular crown or crown of indefinite shape

            (b) Crown extension

1.100.000.1 Crown restricted to the uppermost top of tree (most co-dominant
            trees of ombrophilous tropical lowland forest)
    .000.2 Crown restricted to upper ⅓ of tree height
    .000.3 Crown about ½ length of tree
    .000.4 Crown extending down to more than ½ of tree length
    .000.5 Crown extending to near the base of the tree

            (c) Leaf size (includes phyllodes)

    .000.01 Nanophyllous, usually less than 1 cm²
    .000.02 Microphyllous, usually less than 5 cm²
    .000.03 Mesophyllous
    .000.04 Macrophyllous, usually larger than 100 cm²
    .000.05 Megaphyllous (giant leaves), usually larger than 500 cm²

            (d) Leaf shape (includes phyllodes)

            Needle-shaped leaves, already accounted for as belonido; in-

Casuarina. The latter has narrow cylindrical phyllodes as photosynthetic organs, which appear needle-shaped.

.001 Scale-needles (e.g., *Thuja, Chamaecyparis*)
.002 Micro-needles, shorter than 1 cm
.003 Meso-needles, 1–5 cm
.004 Macro-needles, longer than 5 cm

Laminate leaves = broad-leaved, already accounted for; this group includes all but the following:
.005 Feathery leaves (many legume trees, e.g., *Albizia* spp.)

(e) Rooting features that are recognizable above the ground

1.100.000.000.1 Buttresses, board-roots (characteristic for many lowland tropical rain forest trees, e.g., *Ficus variegata, Shorea balanagaran* and other dipterocarps)
.2 Stilt-roots, regardless of function (e.g., *Pandanus, Rhizophora, Iriartea orbignyana*)
.3 Pneumatophores = asparagus- or knee-shaped episurface roots, e.g., *Avicennia, Sonneratia, Bruguiera*
.4 Aerial roots, suspended as adventitious roots from main stem or branches (e.g., *Eucalyptus robusta* in perhumid rain forest conditions, *Metrosideros,* several *Ficus* spp.)
.5 Xylopod = bulbous, water-storing, mostly subterranean stem base (e.g., *Capparis* spp.)

(f) Bark features

1.100.000.000.01 Green bark, mostly thin or moderately thick (e.g., *Commiphora*)
.02 Thin, smooth, nongreen bark (e.g., many humid tropical trees, e.g., *Albizia falcataria, Ficus religiosa*)
.03 Moderately thick, smooth, nongreen bark that remains relatively smooth at maturity (e.g., *Fagus sylvatica,* most *Abies* spp.)
.04 Moderately thick, smooth, nongreen bark becoming fissured at maturity (e.g., *Fraxinus excelsior,* Thuja)
.05 Thick, fissured bark (e.g., *Quercus robur, Pseudotsuga menziesii, Pinus ponderosa*)
.06 Thick, corky bark (e.g., *Quercus suber*)

(g) Thorns (thorns, spines, or prickles in the morphological sense)

1.100.000.000.001 Absent
.002 Very few, mostly on branches
.003 A few, mostly on stem
.004 A few, both on stem and branches
.005 Abundant
.006 Leaves or phyllodes reduced to thorns

(h) Position of inflorescence

1.100.000.000.000.1 Apically (eg., *Abies*)
.2 Laterally on branches or no definite position, i.e., throughout the crown (e.g., *Pheudotsuga*)

.3 On main stem or main branches, i.e., cauliflory (e.g., *Cercis siliquastrum*, some *Ficus* spp., *Theobroma cacao*, *Couroupita*)

1.2   Tuft trees. Phanerophytes with woody stems and large apical leaf-fronds or terminal, rosulate branches (=rosulate phanerophytes, e.g., palms and tree ferns) ....   P ros

1.201   Dwarf trees=Nanophanerophytes   <2 m ....   N P ros

1.202   Small trees=Microphanerophytes 2—5 m ....   Mi P ros

1.203   Large trees=Mesophanerophytes   5–50 m ....   Mes P ros

1.204   Giant trees=Megaphanerophytes   >50 m ....   Meg P ros

The above height classes should be applied to specific field conditions where they arise. Following are the more common forms of rosulate phanerophytes:

1.210   Unbranched

1.210.1   Simple leaf-fronds (e.g., some *Cecropia* spp.)

1.210.2   Feathery leaf-fronds (e.g., *Cocos*)

1.210.3   Fan-shaped leaf-fronds (e.g.,*Mauritia*)

1.220   Branched

1.220.1   Simple, linear leaf-fronds (e.g., *Xanthorroea*)

1.220.2   Feathery leaf-fronds (e.g., *Schizolobium excelsum*)

1.220.3   Fan-shaped leaf-fronds (e.g., *Hyphaene thebaica*)

1.230   Tufted, twin stems arising from common rootstock or rhizome

1.240   Hollow stem filled with roots (*Puya raimondii*)

1.200.4   Leaves with woolly hair cover (e.g., giant *Senecio* of high tropical mountains)

1.200.5   Semi-succulent leaves (e.g., *Aloë* spp.)

1.3   Bottle trees. Phanerophytes with markedly swollen, water-storing stem (phanerophyta dolaria=dol, e.g., *Adansonia*)..   P dol

1.300   Height and tree-shrub variations can be locally evaluated as before. Therefore the two zero digits.

1.300.1   Evergreen

1.300.11   Normal leaves (e.g., *Brachychiton*)

1.300.12   Leaf-fronds (palm, e.g., *Colpothrinax wrightii*)

1.300.13   Succulent leaves (e.g., *Aloë dichotoma*)

1.300.2   Drought-deciduous (including aphyllous forms)

1.4   Tall succulents, with succulent stem extending from base to apex or with upright growing succulent cladophylls ....   P succ

1.400   Height and tree-shrub variations as before, except megaphanerophytes, which are not present

1.410   Single-stemmed, but commonly branched

1.410.1   Cylindri-formed stem

1.410.2   Cladophyllous (e.g., *Opuntia macracantha*)

1.411.1   Nanophanerophytes (e.g., *Ferrocactus wislizenii*)

1.413.1   In height up to mesophanerophyte (e.g., *Carnegia gigantea*)

1.420.1   Caespitose, cylindri-formed, in height up to mesophanerophyte (e.g., *Pachycereus pringlei*)

1.5 Phanerophytes with herbaceous stem or variously lignified (but herbaceously derived) stem. This group includes all herbaceous and suffruticose (woody base with herbaceous branch-ends) perennials that become taller than about 50 cm and do not exhibit a periodic die-back to that height limit. In particularly favorable environments this height limit may be extended to 1 m, e.g., humid tropics, warm seepage water habitats.

1.500   Where applicable, height and scapose-caespitose variations can be evaluated as before. Therefore the two zero digits. A third variation for stoloniferous and rhizomatous forms appears useful for inclusion **here**

1.500.1   Phanerophytic grasses or graminoid phanerophytes ....   P gram

1.500.11   Lignified (e.g., various species of bamboo)

1.520.11   Caespitose

1.530.11   Reptant

1.500.12   Herbaceous (e.g., various species of sugar cane)

1.520.12   Caespitose

1.530.12   Reptant

1.500.2   Phanerophytic forbs (nongraminoid herbs) ....   P herb

1.500.21   Lignified (or suffruticose)

1.510.21   Scapose, with large leaf-fronds (e.g., *Musa* spp.)

1.520.21   Caespitose

1.520.211   With large leaf-fronds (some *Musa* spp.)

1.520.212   With normal branches and leaves (e.g., *Indigofera*)

1.530.21   ·Reptant

1.500.22   Herbaceous

1.510.22   Scapose

1.520.22   Caespitose

1.520.221   Centrally open, or with loose center

1.520.221.1   With large leaf-fronds (tall herbaceous ferns, e.g., *Hicriopteris*)

1.520.221.2   With normal branches and leaves (e.g., *Begonia* spp.)

1.520.222   Centrally dense, or with compact center, usually with flower stalk arising from center

1.520.222.1   Leaves relatively glabrous (nonwoolly, e.g., *Lobelia deckenii*, *Lupinus alopecurus*)

1.520.222.2   Leaves woolly

1.530.22   Reptant

2. Chamaephytes (Ch)

In addition to the features given in the key (p. 450) chamaephytes have typically a shoot-crowding habit. They are more or less broomy or bunchy from the ground up to 50 cm. This applies particularly to those with ascending shoots. If they become taller than 50 cm, branches or shoots thin out rapidly as a rule. This is the shoot portion that dies back periodically in the unfavorable season. In more favorable habitats this height limit may be extended to 100 cm, for classificatory reasons. Another typical chamaephyte habit is sprawling along the ground. Therefore, in contrast to phanerophytes, height differences are not as important. Instead, of major importance is the degree of lignification and the habit of the shoot system. But a height classification is given at the end:

2.1   Woody dwarf-shrubs. Woodiness completed into
branch-tips. Frutescent chamaephytes ....   Ch frut

2.11   Caespitose (most frequent, therefore "caesp" may be omitted)

2.111   Evergreen

2.111.01   Malacophyllous (m, e.g., *Daphne striata*) ....m Ch frut

2.111.02   Sclerophyllous (s, e.g., *Mahonia aquifolia*) ....s Ch frut

2.111.03   Aphyllous, phyllocladous (p, e.g., *Ephedra spp.*) ..p Ch frut

2.112   Cold-deciduous (aestivo=a, e.g., *Vaccinium myrtillus*)a Ch frut

2.12   Reptant (e.g., *Arctostaphylos uva-ursi*) ....   Ch frut rept

2.13   Pulvinate (cushion form, e.g., *Acantholimon spp.*) ....   Ch frut pulv

2.2   Semi-woody dwarf-shrubs. Woodiness restricted
to the base of the shoot system. Suffrutescent chamaephytes ....   Ch suff

2.21   Caespitose

2.211   Evergreen

2.211.01   Malacophyllous (e.g., *Helianthemum nummul.*) ...m Ch suff

2.211.02   Sclerophyllous (?) ....s Ch suff

2.211.03   Aphyllous, phyllocladous (?) ....p Ch suff

2.212   Cold or drought-deciduous (thero=t, e.g., *Vaccinium
parvifolium*) ....t Ch suff

2.22   Reptant (e.g., *Linnaea borealis*) ....   Ch suff rept

2.23   Pulvinate (?) ....   Ch suff pulv

2.24   Scapose (e.g., *Crotalaria mucronata*) ....   Ch suff scap

2.3   Herbaceous chamaephytes. Includes all nonwoody evergreen
perennial forbs, grasses and ferns that do not get much taller than 1 m
or die back periodically to a remnant shoot system that remains green
at least 25 cm above the ground surface. ....   Ch herb

2.31   Caespitose

2.311   Evergreen in the strict sense (e.g., *Dryopteris paleacea*)   Ch herb

2.312   Shoots dying back periodically, i.e., almost all at once (thero = t); transitory to hemicryptophytes (e.g.,t Ch herb
Andropogon virginicus) ....

2.32   Reptant

2.321   Evergreen (e.g., *Stellaria holostea*) .... Ch herb rept

2.322   Shoots with periodic die-back (t, *Stenotaphrum secundatum*) ....t Ch herb rept

2.33   Pulvinate .... Ch herb pulv

2.331   Globose (g., e.g., *Androsace helvetica*) ....g Ch herb pulv

2.332   Flat (f, e.g., *Silene acaulis*) ....f Ch herb pulv

2.34   Scapose .... Ch herb scap

2.4   Low succulents. These include all succulents below 50 cm height, except those that die back to a remnant portion at the soil surface (hemicryptophytes) or within the soil (geophytes), e.g., many succulents characteristic of the South African and American deserts ..   Ch succ

2.41   Stem-succulents (st, e.g., *Euphorbia mauretanica*) ......   Ch st succ

2.42   Leaf-succulents, some may be hemicryptophytes, which are here included (l, e.g., *Crassula* spp.) .... Ch l succ

2.43   Root-succulents, with subterranean storage organs (r, e.g., *Pachypodium bispinosum*) .... Ch r succ

2.5   Poikilohydrous chamaephytes. These are mostly ferns, as far as known, of arid climates, whose water balance changes with the humidity of the surrounding atmosphere. Their shoots survive the drought season in latent condition and become green immediately upon return of moister conditions (e.g.,*Cheilanthus hirta*) ....   Ch poik

Subdivisions as to height can be applied where necessary, for example:

2.000.1	Very low chamaephyte	< 3 cm
.2	Low chamaephyte	3—10 cm
.3	Typical chamaephyte	10—30 cm
.4	Tall chamaephyte	30–100 cm
.5	Very tall chamaephyte	> 100 cm

3. Hemicryptophytes (H)

The remnant shoot system, which during the unfavorable season lies relatively flat on the ground, is often protected by dead shoot remains. During the growing season the active shoots are always raised above the perennial ground-shoot. Hemicryptophytes are typically herbaceous throughout, but the maturing stem may show some secondary thickening (lignification), particularly when standing as a dead remnant, e.g., in many biennials.

3.10	Caespitose hemicryptophytes (bunched or circular shoot arrangement)	H caesp
3.101	Cold-deciduous shoot system (aestivo=a, e.g., *Dactylis*)	a H caesp
3.102	Drought-deciduous (cheimo=c, e.g., *Heteropogon contortus*)	c H caesp
3.103	Sparingly evergreen (e) during unfavorable season; transitory to chamaephytes (e.g., *Deschampsia flexuosa*) ...	e H caesp
3.20	Reptant hemicryptophytes (creeping or matted)	H rept
3.201	Cold-deciduous (e.g., *Agrostis stolonifera*)	a H rept
3.202	Drought-deciduous (e.g., *Tricholaena repens*)	c H rept
3.203	Sparingly evergreen (e.g., *Cynodon dactylon*)	e H rept
3.3	Scapose hemicryptophytes	
3.30	Without rosette	H scap
3.301	Cold-deciduous (e.g., *Scrophularia nodosa*)	a H scap
3.302	Drought-deciduous (e.g., *Chrysopogon acicularis*)	c H scap
3.31	Rosette	H ros
3.311	Cold-deciduous (e.g., *Bellis perennis*)	a H ros
3.312	Drought-deciduous (e.g., *Desmodium triflorum*)	c H ros
3.32	Semi-rosette (sem)	H sem
3.321	Cold-deciduous (e.g., *Ranunculus acer*)	a H sem
3.322	Drought-deciduous (e.g., *Erigeron canadensis* in summer-drought areas)	c H sem
3.4	Aquatic hemicryptophytes (hydrophyte=hyd) ...... hyd H	
3.41	Caespitose (e.g., *Isoetes*) .... hyd H caesp	
3.42	Reptant (e.g., *Pilularia*) .... hyd H rept	
3.43	Scapose (e.g., *Lobelia dortmanna*) .... hyd H scap	

In addition a breakdown into height classes can be applied, where required, as follows:

3.000.1	Very small hemicryptophyte	< 3 cm
.2	Small hemicryptophyte	3—10 cm
.3	Medium-sized hemicryptophyte	10—30 cm
.4	Tall hemicryptophyte	30–100 cm
.5	Very tall hemicryptophyte	> 100 cm

4. Geophytes (G. Cryptophytes)

These herbaceous plants with their survival organs well protected in the soil are typically found in climates with pronounced unfavorable seasons. However, they may occur also in less severe climates, where they may fill a temporary niche as complementary species in certain plant communities.

4.1	Root-budding geophytes (radicigemma = rad)	G rad
4.11	Spring-green (earizo = ear)	ear G rad
4.12	Summer-green (aestivo = a; e.g., *Cirsium arvense*)	a G rad
4.13	Rain-green (cheimo = c)	c G rad
4.2	Bulbous geophytes, arising from bulbs or corms	G bulb
4.21	Spring-green (e.g., *Leucoium vernum*)	ear G bulb
4.22	Summer-green (e.g., *Lilium martagon*)	a G bulb
4.23	Rain-green (e.g., *Stenomesson*)	c G bulb
4.3	Rhizome-geophytes, arising from rhizomes of various lengths	G rhiz
4.31	Spring-green (e.g., *Anemone nemorosa*)	ear G rhiz
4.32	Summer-green (e.g., *Agropyron repens*)	a G rhiz
4.33	Rain-green	c G rhiz
4.4	Aquatic geophytes (hydrophytic = hyd)	hyd G
4.41	Root-budding (?)	hyd G rad
4.42	Bulbous (?)	hyd G bulb
4.43	Rhizome (e.g., *Nymphaea*)	hyd G rhiz

Subdivisions as to caespitose, scapose or reptant growth habit can be applied where required:

4.001   Caespitose

4.002   Scapose

4.003   Reptant

Subdivisions for height differences can be applied as follows:

4.000.1	Very small geophyte	<3 cm
.2	Small geophyte	3—10 cm
.3	Medium-sized geophyte	10—30 cm
.4	Tall geophyte	30–100 cm
.5	Very tall geophyte	1—3 m
.6	Extremely tall geophyte	>3 m

5. Therophytes (T, Annuals)

As a rule therophytes live much less than a year and some complete their life cycle within a few weeks. However, exceptions are, for instance, the weeds in the winter-rye fields, which germinate in the fall and flower in the following vegetative period, or the succulent mesembrianthemums of African deserts that, because of their water-storing properties, may live longer than a year after a penetrating rain. Yet, they are therophytes in that they complete their life cycle within one favorable growing period and die after seed production. Not included are the "hapoxanthic" species (like *Agave, Argyroxiphium*, and others) that die after seed production, but grow for several years before reaching that state. They are true perennials and their life cycle does not depend on one favorable growing season.

5.10	Caespitose therophytes	T caesp
5.101	Spring-green (e.g., *Aira caryophyllea*)	ear T caesp
5.102	Summer-green (e.g., *Setaria viridis*)	a T caesp
5.103	Rain-green (e.g., *Chloris barbata*)	c T caesp
5.104	Winter-green, i.e., winter annuals germinating in fall and living till next summer or fall (metoporino=met, e.g., *Apera spica-venti*)	met T caesp
5.2	Reptant therophytes	T rept
5.201	Spring-green (e.g., *Veronica hederifolia*)	ear T rept
5.202	Summer-green (e.g., *Alopecurus geniculatus*)	a T rept
5.203	Rain-green (?)	c T rept
5.204	Winter-green, lasting through spring (e.g., *Stellaria media*)	met T rept
5.3	Scapose therophytes	
5.30	Without rosette	T scap
5.301	Spring-green (e.g., *Veronica triphyllos*)	ear T scap
5.302	Summer-green (e.g., *Chenopodium polyspermum*)	a T scap
5.303	Rain-green (e.g., *Eragrostis tenella*)	c T scap
5.304	Winter-green (e.g., *Ranunculus arvensis*)	met T scap
5.31	Ground-rosette, without leaves on the stalk	T ros
5.311	Spring-green (e.g.,*Erophila verna*)	ear T ros
5.312	Summer-green (?)	a T ros
5.313	Rain-green (?)	c T ros
5.314	Winter-green (e.g.,*Arnoseris minima*)	met T ros
5.32	Semi-rosette, with leaves on the stalk	T sem
5.321	Spring-green (e.g.,*Stenophragma thalianum*)	ear T sem
5.322	Summer-green (e.g., *Sonchus oleraceus*)	a T sem
5.323	Rain-green (?)	c T sem
5.324	Winter-green (e.g., *Capsella bursa-pastoris*)	met T sem
5.4	Aquatic therophytes (e.g., *Najas*)	hyd T
5.5	Succulent therophytes (e.g., *Portulaca oleracea*, in dry-season climates)	T succ

Subdivisions for height can be applied as follows:

5.000.1	Very small therophyte	<3 cm
.2	Small therophyte	3—10 cm

.3   Medium-sized therophyte   10—30 cm

.4   Tall therophyte   30–100 cm

.5   Very tall therophyte   1——3 m

.6   Extremely tall therophyte   >3 m

6. Lianas (L), Eu-lianas

Lianas, including vines, are treated as a special group, since they depend for their support on other, self-supporting plants or artificial props, which in turn determine also the height of the liana.

6.1   Phanerophytic and chamaephytic lianas, including all climbing plants that do not die back periodically to the ground ....   PL

6.11   Root climbers, closely attached to their support by modified adventitious roots (radici=r) ....   r PL

6.111   Woody (fruticose, e.g., *Hedera helix*) ....   r PL frut

6.112   Semi-woody (suffruticose, e.g., *Parthenocissus* spp.) ....   r PL suff

6.113   Herbaceous (?) ....   r PL herb

6.12   Winding climbers, encircling their support (strepano=st)   st PL

6.121   Woody (e.g., many tropical lianas) ....   st PL frut

6.122   Semi-woody (e.g., *Humulus lupulus*) ....   st PL suff

6.123   Herbaceous (e.g., *Polygonum dumetorum*) ....   st PL herb

6.13   Tendril climbers, attaching themselves by tendrils of different morphological origin (elitto=el) ....   el PL

6.131   Woody (e.g., *Bauhinia* spp.) ....   el PL frut

6.132   Semi-woody (some *Cissus* spp.) ....   el PL suff

6.133   Herbaceous (e.g.,*Passiflora* spp.) ....   el PL herb

6.14   Spread-climbers, propping their branches on other plants diateino=d) ....   d PL

6.141   Woody (e.g., *Chusquea* and other bamboos) ....   d PL frut

6.142   Semi-woody (e.g., many *Rubus* spp.) ....   d PL suff

6.143   Herbaceous (?) ....   d PL herb

Within each group, subdivisions are possible, e.g.,

6.000.01   Evergreen (no additional symbol)

6.000.02   Cold-deciduous, summer-green (aestivo=a) ....   a

6.000.03   Drought-deciduous, rain-green (cheimo=c)   c

Height classes can be applied in relation to the supporting life forms, or for a finer definition separate height classes may be devised by beginning with the fifth digit, i.e., 6.000.1, etc.

6.2   Hemicryptophytic lianas, dying back periodically to a remnant shoot system near the ground ....   HL

6.21   Root climbers (?) .... r HL
6.22   Winding climbers (?) .... st HL
6.23   Tendril climbers (e.g., *Vicia sepium*) .... el HL
6.24   Spread climbers (e.g.,*Galium mollugo*) .... d HL
6.3   Geophytic lianas, dying back periodically to subterranean
        storage organs .... GL
6.32   Winding climbers (e.g., *Convolvulus arvensis*) .... st GL
6.33   Tendril climbers (e.g.,*Lathyrus maritimus*) .... el GL
6.34   Spread climbers (e.g., *Corydalis claviculata*) .... d GL
6.4   Therophytic lianas .... TL
6.42   Winding climbers (e.g., *Polygonum convolvulus*) .... st TL
6.43   Tendril climbers (e.g., *Vicia hirsuta*) .... el TL
6.44   Spread climbers (e.g., *Galium aparine*) .... d TL

7. **Hemi-epiphytes,** Pseudo-lianas, Epiphytic Lianas (EL)

These plants may either be classified as lianas or as epiphytes depending on their developmental status during analysis. However, where their nature can be assessed with certainty the following classification may be applied:

7.1   Roots dying; plants that have climbed upwards and are losing or have lost their ground contact.

7.2   Roots winding around host plant or otherwise surrounding it (e.g., by ramifying roots); stranglers

7.3   Roots descending down without encircling host plant

7.31   Along trunk of host plant

7.32   Hanging free from branches of host plant

8. **Epiphytes** (E), Vascular Epiphytes

8.1   Epiphytes with normal root systems growing in soil or humus pockets (mostly facultative epiphytes)

8.11   Phanerophytes (e.g., *Cheirodendron trigynum*) .... PE

8.12   Chamaephytes .... ChE

8.121   Woody or suffruticose .... ChE frut

8.122   Herbaceous (e.g., *Psilotum nudum*) .... ChE herb

8.122.1 With leaves arranged in funnel shape (e.g., *Asplenium nidus*)

8.122.2 With leaves not showing any funnel arrangement, i.e., leaves as in normal soil-adapted herbs (e.g., many orchids)

8.2   Epiphytes with strongly modified root systems,

stems or leaves (or other unusual morphological modifications) adapted to growth on branches that have no soil or humus pocket-accumulations in branch forks, etc. (mostly obligative epiphytes)

8.21   Succulents (e.g., *Peperomia* spp. and many orchids) ...... ChE succ

8.22   Nonsucculent plants

8.221   With unmodified leaves (i.e., not in any way peculiar or unusual), but modified stems or roots

8.221.1   With strongly swollen stem-base (e.g., *Myrmecodia tuberosa*)

8.221.2   With green string-like roots (e.g., many orchids)

8.222   With strongly modified (i.e., unusual) leaves

8.222.1   With upright cup or funnel-shaped leaves (e.g., *Bromelia* spp.)

8.222.2   With leaves specially adapted to form humus layers (e.g., *Platycerium*)

8.223   With completely modified plant body, resembling bearded lichens (e.g., *Tillandsia usneoides*)

Two ecologically significant height strata can usually be evaluated:

8.000.01   Occupying sun-exposed positions in the upper tree canopy

8.000.02   Occupying lower canopy area and lower branches, growing in more shaded positions

9. **Errant Vascular Hydrophytes,** free-moving in water, not attached or rooted in the ground e.g., *Eichhornia, Salvinia, Lemna, Utricularia*)

Kormo-Hydrophyta natantia .... K Hyd Nat

Subdivisions possible

10. **Thallo-chamaephytes** (Th Ch)

10.1   Hummock-forming mosses (=sphagnoid bryophytes) ..... Br Ch sph

10.2   Heavy carpet-forming mosses (=reptant bryophytes, e.g., *Pleurozium schreberi*) ......·........................ Br Ch rept

10.3   Cushion-forming mosses (=pulvinate bryophytes, e.g., *Leucobryum glaucum*) .... Br Ch pulv

10.4   Cushion-forming or caespitose and fruticose lichens (=chamaephytic lichens, e.g., *Cladonia sylvatica*) .... Li Ch
Substrate differences may be recognized.

11. **Thallo-hemicryptophytes** (Th H)

11.1   Flat-lying mosses, bryo-hemicryptophytes (e.g., *Plagiothecium curvifolium*) .... Br H

11.2	Flat-lying liverworts, hepatic hemicryptophytes (e.g., *Marchantia*)	Hep H

11.3   Foliose lichens, lichen-hemicryptophytes (e.g., *Peltigera*)   Li H fol

11.4	Crustose lichens, crustaceous lichen-hemicryptophytes (e.g., *Lecidea*)	Li H crust

11.5   Endolithic lichens, living in stones near the surface ....   Li H end

11.6	Adnate algae, phyco-hemicryptophytes (e.g., some *Pleurococcus*)	Phyc H

11.7   Endolithic algae .... Phyc H end

12. **Thallo-therophytes** (Th T)

12.1   Short-living mosses, bryo-therophytes (e.g.,*Ephemerum*)   Br T

12.2   Short-living liverworts, hepatic therophytes (e.g., *Riccia*)   Hep T
       etc.

13. **Thallo-epiphytes** (T E), living on bark or leaves

13.1	Epiphytic thallo-chamaephytes	Th Ch E
13.11	Mosses (e.g., *Pseudisothecium*)	Br Ch E
13.12	Liverworts	Hep Ch E
13.13	Lichens (e.g.,*Alectoria*)	Li Ch E
13.2	Epiphytic thallo-hemicryptophytes	Th H E
13.21	Mosses (e.g., *Hypnum cupressiforme*)	Br H E
13.22	Liverworts	Hep H E
13.23	Lichens (e.g., *Parmelia*)	Li H E
13.24	Algae (e.g., *Pleurococcus*)	Phyc H E
13.3	Epiphytic thallo-therophytes	Th T E
13.31	Mosses	Br T E
13.32	Algae	Phyc T E

10.–13. Subdivisions incomplete,
14.–23. to be worked out later.

Appendix B
Tentative Physiognomic-Ecological Classification of Plant Formations of the Earth (revised from ELLENBERG and MUELLER-DOMBOIS, 1967a)*

This physiognomic classification has been elaborated as a basis for mapping world vegetation on a scale of 1:1 million or smaller. It will permit the

* Based on a discussion draft of the UNESCO working group on vegetation classification and mapping. A first list, prepared by SCHMITHÜSEN and ELLENBERG (1964) and revised by D. POORE and ELLENBERG (1965), has been discussed in Paris (Jan. 1966) by GAUSSEN (President), BUDOWSKI, ELLENBERG, FRÄNZLE, GERMAIN, KÜCHLER, LEBRUN, D. POORE, and SOCHAVA. The definitions of the terms have been added by the authors. For comments and corrections we thank the colleagues mentioned above, mainly KÜCHLER.

The authors will be grateful for any further comment. Please send it to H. Ellenberg. Untere Karspuele 2, D-34 Goettingen (West Germany).

worldwide comparison of ecological habitats indicated by equal plant life form combinations.*

Plant formations and other divisions in this context are conceived as combinations of plant life forms, i.e., as physiognomic units. For naming, ecological terms have been included for brevity; wherever possible, definitions are based on physiognomic criteria. In the following classification, units of unequal rank are distiguished by different symbols:

> **I, II, etc.** = **FORMATION CLASS**
> **A, B, etc.** = **Formation subclass**
> **1, 2, etc.** = **Formation group**
> **a, b, etc.** = **Formation**
> *(1), (2), etc.* = *Subformation*
> (a), (b), etc. = Further subdivisions

I. CLOSED FORESTS, formed by trees at least 5 m tall with their crowns interlocking.†

A. Mainly evergeen forests, i.e., the canopy is never without foliage, however, individual trees may shed their leaves.

1. Tropical ombrophilous forests. (Conventionally called tropical rain forests.) Consisting mainly of evergreen trees mostly with little or no bud protection, neither cold nor drought resistant. Truly evergreen, i.e., individual trees may stand leafless for a few weeks only and not at the same time as all the others. Leaves of many species with "drip tips."

a. Tropical ombrophilous lowland forest. Composed of numerous species of fast growing trees, some of them exceeding 50 m in height, generally with smooth bark, some with buttresses. Very sparse undergrowth and this composed mainly of tree reproduction. Palms and other tuft trees usually rare, lianas nearly absent except pseudo-lianas (i.e., plants originating on tree branches, subsequently rooting in the ground). Crustose lichens and green algae are the only constantly present epiphytic life forms; vascular epiphytes are less abundant than in b–d.‡

b. Tropical ombrophilous submontane forest. Tree growth largely the same as in a. In the undergrowth herbaceous life forms more frequent. The most important difference from a is the more frequent appearance of vascular epiphytes.

---

* Some formations are covering only small areas which cannot be represented on small scale maps. They have been mentioned here to facilitate a more general use of this physiognomic-ecological classification. A key to the plant life forms is given in Appendix A.

† In reproductive stage or as immature secondary growth temporarily less than 5 m tall, but individuals of scapose life form (i.e., real trees, not shrubs). In subpolar conditions, the limit may be only 3 m, in tropical ones 8 or 10 m.

‡ In lowlands vascular epiphytes are abundant only where fog frequently occurs, e.g., near the coast.

(IA1)

c. Tropical ombrophilous montane forest. (Corresponds most closely to textbook descriptions of the virgin tropical rain forest.) Abundant vascular and other epiphytes. Tree sizes markedly reduced (<50 m); crowns extending deeper down the stem than in a or b. Bark often more or less rough. Undergrowth abundant, often represented by rosulate nano-microphanerophytes (e.g., tree ferns or small palms); the ground layer rich in hygromorphous herbs and cryptogams.

(1) *Broad-leaved*, most common form

(2) *Needle-leaved* or microphyllous

(3) *Bamboo*, rich in tree-grasses replacing largely the tuft micro- or nanophanerophytes

d. Tropical ombrophilous "subalpine" forest. (Not including cloud forest or woodland. Considered unique by some investigators, but probably not important. Definition required.)

e. Tropical ombrophilous cloud forest. Tree crowns, branches and trunks as well as lianas burdened with epiphytes, mainly chamaephytic bryophytes. Also the ground covered with hygromorphic chamaephytes (e.g., *Selaginella* and herbaceous ferns). Trees often gnarled, with rough bark and rarely exceeding 20 m in height.

(1) *Broad-leaved*, most common form

(2) *Needle-leaved* or microphyllous

f. Tropical ombrophilous alluvial forest. Similar to b, but richer in palms and in undergrowth life forms, particularly tall forbs* (e.g., Musaceae); buttresses frequent.

(1) *Riparian* (on the lowest forested river banks, frequently flooded), mostly dominated by fast-growing trees; herbaceous undergrowth nearly absent, epiphytes extremely rare, poor in species

(2) *Occasionally flooded* (on relatively dry terraces accompanying active rivers), most common form of f; more epiphytes than in (1) and (3), many lianas

(3) *Seasonally water-logged* (along the lower river courses, where the water accumulates on large flats for several months); trees frequently with stilt roots; canopy density not uniform; as a rule poor in undergrowth, except for more open places

g. Tropical ombrophilous swamp forest. (Not along rivers, but on edaphically wet habitats, which may be supplied either with fresh or brackish water). Similar to f, but as a rule poorer in tree species. Many trees with buttresses or pneumatophores; mostly higher than 20 m.

(1) *Broad-leaved*, dominated by dicots

(2) *Dominated by palms*, but broad-leaved trees in the undergrowth

h. Tropical evergreen peat forest (with organic surface deposits).

---

* "Forb" (American) means "nongraminoid herb."

Poor in tree species, with lower canopy than g (as a rule not higher than 20 m). Trees have slow growth rates and thin diameters and are commonly equipped with pneumatophores or stilt roots.

(1) *Broad-leaved*, dominated by dicotylous plants

(2) *Dominated by palms*, which may be equipped with asparagus-shaped pneumatophores

2. **Tropical and subtropical evergeen seasonal forests.** Consisting mainly of evergreen trees with some bud protection. Foliage reduction during the dry season is noticeable, often as partial shedding. Transitional between **1** and **3**. Subdivisions a–c largely similar to those under **1**.

a. Tropical (or subtropical) evergreen seasonal lowland forest.

b. Tropical (or subtropical) evergreen seasonal submontane forest.

(1) *Broad-leaved*, most common form

(2) *Needle-leaved*

c. Tropical (or subtropical) evergreen seasonal montane forest. In contrast to **1** c no tree ferns; instead, evergreen shrubs are more frequent.

d. Tropical (or subtropical) evergreen dry "subalpine" forest. (Physiognomically resembling the winter-rain evergreen sclerophyllous dry forest (**8a**), usually occurring above the cloud forest (**1e**). Mostly evergreen sclerophyllous trees, smaller than 20 m, with little or no undergrowth (if not opened by human activity). Poor in lianas and epiphytes, except lichens.

3. **Tropical and subtropical semi-deciduous forests.** Most of the upper canopy trees drought-deciduous many of the understory trees and shrubs evergreen and more or less sclerophyllous. Nearly all trees with bud protection; leaves with "drip tips." Trees show rough bark, except some bottle trees, which may be present.

a. Tropical (or subtropical) semi-deciduous lowland forest. The taller trees are often bottle trees (e.g., *Ceiba*). Practically no epiphytes present. Undergrowth composed of tree reproduction and true woody shrubs. Succulents may be present (e.g., in form of thin-stemmed caespitose cacti). Both therophytic and hemicryptophytic lianas occur occasionally. A sparse herb layer may be present, mainly consisting of graminoid hemicryptophytes and forbs.

b. Tropical (or subtropical) semi-deciduous montane or cloud forest. Simila to a, but canopy lower and covered with xerophytic epiphytes (e.g., *Tillandsia usneoides*). Within group **3**, a submontane formation cannot be clearly distinguished.

4. **Subtropical ombrophilous forests.** (Present only locally and in small fragmentary stands, because the subtropical climate is typically a climate with a dry season. Where the subtropical ombrophilous forest occurs, e.g., Queensland/Australia and Taiwan, it usually grades rather

(IA4)

inconspicuously into the tropical ombrophilous forest. Its trees are less vigorous and allow some shrubs to grow in the understory. The subtropical ombrophilous forest should however not be confused with the tropical ombrophilous montane forest, which occurs in a climate with a similar mean annual temperature, but with less pronounced temperature differences between summer and winter.) Consequently, seasonal rhythms are more evident in all subtropical forests, even in the ombrophilous ones.

The subtropical ombrophilous forest is physiognomically more closely related to the tropical than to the temperate one. Therefore the subdivisions conform more or less to point **1** a–h.

5. **Mangrove forests.** (Occur only in the tidal range of the tropical and subtropical zones.) Composed almost entirely of evergreen sclerophyllous broad-leaved trees and shrubs with either stilt roots or pneumatophores. Epiphytes in general rare, except lichens on the branches and adnate algae on the lower parts of the trees.

(Subdivisions possible; transistions to **1** g exist).

6. **Temperate and subpolar evergreen ombrophilous forests.** (Occurring only in the extremely oceanic, nearly frostfree climate on the southern hemisphere, mainly in Chile). Consisting mostly of truly evergreen hemisclerophyllous trees and shrubs. Rich in thallo-epiphytes and in ground-rooted herbaceous ferns.

a. Temperate evergreen ombrophilous broad-leaved forest. Some vascular epiphytes and lianas present; height generally exceeds 10 m.

(1) *Without conifers*

(2) *With conifers admixed*

b. Temperate evergreen ombrophilous alluvial forest. Richer in herbaceous undergrowth than a. (Not yet sufficiently investigated).

c. Temperate evergreen ombrophilous swamp forest. (Perhaps existing, but not yet known.)

d. Subpolar evergreen ombrophilous forest. In contrast to a, vascular epiphytes lacking canopy height much reduced (in general less than 10 m).

7. **Temperate evergreen seasonal broad-leaved forests.** (With pronounced summer rainfall.*) Consisting mainly of hemi-sclerophyllous evergreen trees and shrubs. Rich in herbaceous chamaephytic and hemicryptophytic undergrowth. Very few or no vascular epiphytes and lianas.

° Correspond to the "lauraceous" forests of RUBEL, etc., which are often believed to be "winter-rain evergreen," but in reality cannot withstand much summer drought. Most frequent in East Asia.

Grades into subtropical or temperature ombrophilous forest. Probably includes subpolar types.

(Subdivisions similar to 2a–d possible.)

8. **Winter-rain evergreen broad-leaved sclerophyllous forests.** (Often understood as Mediterranean, but present also in southwest Australia, Chile, etc. Climate with pronounced summer drought.) Consisting mainly of sclerophyllous evergreen trees and shrubs, most of which show rough bark. Herbaceous undergrowth almost lacking. No vascular and only few cryptogamic epiphytes, but evergreen woody lianas present.

a. Winter-rain evergreen sclerophyllous lowland forest (including submontane). Corresponds largely to the description under **8.**

b. (Alluvial and swamp forests of this type perhaps existing, but not sufficiently known.)

9. **Temperature and subpolar evergreen coniferous forests.** Consisting mainly of needle-leaved or scale-leaved evergreen trees, but broad-leaved trees may be admixed. Vascular epiphytes and lianas practically lacking.

a. Evergreen giant conifer forest. Dominated by trees higher than 50–60 m (e.g., *Sequoia* and *Pseudotsuga* forest in the Pacific West of North America.

b. Evergreen (nongiant) conifer forest with rounded crowns. Dominated by trees 5–50 m high, with more or less broad, irregularly rounded crowns (e.g., *Pinus* spp.).

(1) *With evergreen sclerophyllous understory (Mediterranean)*

(2) *Without evergreen sclerophyllous understory*

c. Evergreen (nongiant) conifer forest with conical crowns. Dominated by trees 5–50 m high (only exceptionally higher), with more or less conical crowns (like most *Picea* and *Abies*).

d. Evergreen (nongiant) conifer forest with cylindrical crowns (boreal). Similar to c, but crowns with very short branches and therefore very narrow, cylindro-conical.

B. **Mainly deciduous forests.** Majority of trees shed their foliage simultaneously in connection with the unfavorable season.

1. **Drought-deciduous forests.** (Tropical and subtropical). Unfavorable season mainly characterized by drought, in most cases winter-drought. Foliage is shed regularly every year. Most trees with relatively thick, fissured bark.

a. Drought-deciduous lowland (and submontane) forest. Practically no evergreen plants in any stratum, except some succulents. Woody and herbaceous lianas present occasionally, also deciduous bottle-trees. Ground vegetation mainly herbaceous (hemicryptophytes, particularly grasses, geophytes and some therophytes), but sparse.

(1) *Broad-leaved*

(IB1)

    (2) *Microphyllous* (including feathery-leaved legume-trees)

    b. **Drought-deciduous montane (and cloud) forest.** Some evergreen species in the understory. Drought-resistant epiphytes present or abundant, often of the bearded form (e.g., *Usnea* or *Tillandsia usneoides*); canopy as in a. This formation is not frequent, but well developed, e.g., in northern Peru.

    (Further types of drought-deciduous forest may need recognition.)

    2. **Cold-deciduous forests with evergreen trees (or shrubs) admixed.** Unfavorable season mainly characterized by winter frost. Deciduous trees dominant, but evergreen species present as part of the main canopy or as understory. Climbers and vascular epiphytes scarce or absent.

    a. **Cold-deciduous forest with evergreen broad-leaved trees and climbers** (e.g., *Hedera helix* in Western Europe). Rich in cryptogamic epiphytes, including mosses. Even vascular epiphytes may be present at the base of tree stems.

    b. **Cold-deciduous forest with broad-leaved sclerophyllous understory** (e.g., sub-Mediterranean forest).

    c. **Cold-deciduous forest with evergreen needle-leaved trees.** (Further subdivisions possible).

    3. **Cold-deciduous forests without evergreen trees.** Deciduous trees absolutely dominant. Evergreen chamaephytes and some evergreen nanophanerophytes may be present. Climbers insignificant, vascular epiphytes absent (except occasionally at the lower base of the trees); thallo-epiphytes always present, particularly lichens.

    a. **Temperate lowland and submontane cold-deciduous forest.** Trees up to 50 m tall. Primarily algae and crustose lichens as epiphytes.

    b. **Montane or boreal cold-deciduous forest** (including lowland or submontane in topographic positions with high atmospheric humidity). Foliose and fruticose lichens, and bryophytes as epiphytes. Trees up to 50 m tall, but in montane or boreal forest normally not taller than 30 m.

    (1) *Mainly broad-leaved*

    (2) *Mainly deciduous coniferous* (e.g., *Larix*)

    (3) *Mixed broad-leaved and deciduous coniferous*

    c. **Subalpine or subpolar cold deciduous forest.** In contrast to a and b canopy height significantly reduced (not taller than 20 m). Tree trunks frequently gnarled. Epiphytes similar to b, but in general more abundant. Often grading into woodland (see **II**).

    (1) *With primarily hemicryptophytic undergrowth*

    (2) *With primarily chamaephytic undergrowth;* may merge with forests admixed with conifers (see **2c**)

    d. **Cold-deciduous alluvial forest.** (Flooded by rivers, therefore moister and richer in nutrients than a.) Trees and shrubs with high growth rates and vigorous herbaceous undergrowth.

(1) Occasionally flooded; physiognomically similar to a, with tall trees and abundant macrophyllous shrubby undergrowth

(2) Regularly flooded; trees not as tall and dense as in a, but herbaceous undergrowth abundant and tall (in Eurasia *Salix* or *Alnus* species frequently dominating)

e. Cold-deciduous swamp or peat forest. (Flooded until late spring or early summer, surface soil organic.) Relatively poor in tree species. Ground cover mostly continuous.

(Subdivisions like b.)

C. **Extremely xeromorphic forests.** Dense stands of xeromorphic phanerophytes, such as bottle trees, tuft trees with succulent leaves and stem succulents. Undergrowth with shrubs of similar xeromorphic adaptations, succulent chamaephytes and herbaceous hemicryptophytes, geophytes and therophytes. Often grading into woodlands (see **II**).

1. **Sclerophyllous-dominated extremely xeromorphic forests.** Life form combination as above, except for predominance of sclerophyllous trees, many of which have bulbose stem bases largely imbedded in the soil (xylopods).

(Subdivisions possible, but not yet sufficiently investigated.)

2. **Thorn-forests.** Species with thorny appendages predominate.
a. Mixed deciduous-evergreen thorn forest. Many merge with **1**.
b. Purely deciduous thorn forest. Most common form.

3. **Mainly succulent forests.** Tree-formed (scapose) and shrub-formed (caespitose) succulents very frequent, but the other xero-phanerophytes present as well.

II. **WOODLANDS.** (Open stands of trees). Formed by trees at least 5 m tall, with most of their crowns not touching each other, but covering at least 30% of the surface; grass cover sometimes present. This formation class does not include savannas or parklands.

A. **Mainly evergreen woodlands,** i.e., evergreen as defined in **IA.**

1. **Evergreen broad-leaved woodlands.** Mainly sclerophyllous trees and shrubs, no epiphytes.

(Subdivisions with regard to undergrowth variations possible.)

2. **Evergreen needle-leaved woodlands.** Mainly needle-or scale-leaved. Crowns of many trees extending to the base of the stem or at least very branchy.

a. Evergreen coniferous woodlands with rounded crowns (e.g., *Pinus*).

(1) With evergreen sclerophyllous understory (Mediterranean)

(IIA2)

(2) Without evergreen sclerophyllous understory (Mediterranean)
b. Evergreen coniferous woodland with conical crowns prevailing (mostly subalpine).
c. Evergreen coniferous woodland with very narrow cylindro-conical crowns (e.g., *Picea* in the boreal region).
(Further subdivisions according to undergrowth variations and frequency of epiphytes are possible.)

B. Mainly deciduous woodlands (see **IB**).

1. Drought-deciduous woodlands.
(Subdivisions more or less like forests.)

2. Cold-deciduous woodlands with evergreen trees (see **IB2**).

3. Cold-deciduous woodlands. (See **IB3**; most frequent in the subarctic region, elsewhere only on swamps or bogs). Without evergreen trees.
a. Broad-leaved deciduous woodland.
b. Needle-leaved deciduous woodland.
c. Mixed deciduous woodland (broad-leaved and needle-leaved).

C. Extremely xeromorphic woodlands. Similar to **IC**, the only difference being the more sparse stocking of individual trees.
(Subdivisions as under **IC**.)

III. SCRUB. (Shrublands or thickets.) Mainly composed of caespitose woody phanerophytes 0.5–5 m tall.* Each of the following subdivisions may be either of the following
shrubland = most of the individual shrubs not touching each other, often
         with a grass stratum
thicket   = individual shrubs interlocked

A. Mainly evergreen scrub. (Evergreen in the sense of **IA**.)

1. Evergreen broad-leaved shrublands. (Or thickets.)
a. Low bamboo thicket (or, less frequently, shrubland). Lignified creeping graminoid nano- or microphanerophytes.
b. Evergreen tuft tree shrubland (or thicket). Composed of small trees and wood shrubs (e.g., Mediterranean dwarf palm shrubland or Hawaiian tree fern thicket).
c. Evergreen broad-leaved hemi-sclerophyllous thicket (or shrubland). Caespitose, creeping or lodged nano- or microphanerophytes with relatively large and soft leaves (e.g., subalpine *Rhododendron* thickets, or *Hibiscus tiliaceus* matted thickets of Hawaii).

---

* Not to be confused with developing second growth forests, see footnote relating to I.
Sometimes, scrub may reach more than 5 m in height.

(Subdivisions possible.)

d. Evergreen broad-leaved sclerophyllous shrubland (or thicket). Dominated by broad-leaved sclerophyllous shrubs and immature trees (i.e., chaparral or macchia). May often merge with parkland, grassland or heath.

e. Evergreen suffruticose thicket (or shrubland). Stand of semi-lignified nanophanerophytes that in dry years may shed part of their shoot systems (e.g., *Cistus* heath.)*

(Additional units may be distinguished.)

2. **Evergreen needle-leaved and microphyllous shrublands** (or thickets).

a. Evergreen needle-leaved thicket (or shrubland). Composed mostly of creeping or lodged needle-leaved phanerophytes (e.g., *Pinus montana*, "krummholz").

b. Evergreen microphyllous shrubland (or thicket). Often ericoid shrubs (mostly in tropical subalpine belts).

(Further subdivisions possible.)

B. **Mainly deciduous scrub.** (Deciduous in the sense of **IB.**)

1.-3. (Subdivisions similar to **IIB1–3.**)

4. **Cold-deciduous shrublands.** (Or thickets).

a. Temperate deciduous thicket (or shrubland). More or less dense scrub without or with only little herbaceous undergrowth. Poor in cryptogams.

b. Subalpine or subpolar deciduous thicket (or shrubland). Upright or lodged caespitose nanophanerophytes with great vegetative regeneration capacity. As a rule completely covered by snow for at least half a year.

(1) With *primarily hemicryptophytic undergrowth*, mainly forbs (e.g., subalpine *Alnus viridis* thicket)

(2) With *primarily chamaephytic undergrowth*, mainly dwarf shrubs and fruticose lichens (e.g., *Betula tortuosa* shrubland at the polar tree line)

c. Deciduous alluvial shrubland (or thicket). Fast growing shrubs, occurring as pioneers on river banks or islands that are often vigorously flooded, therefore mostly with very sparse undergrowth.

(1) With *lanceolate* leaves (e.g., *Salix*, mostly in lowland or submontane regions)

(2) *Microphyllous* (e.g., *Tamarix*)

d. Deciduous peat shrubland (or thicket). Upright caespitose nanophanerophytes with *Sphagnum* and (or) other peat mosses.

(Subdivisions possible.)

C. **Extremely xeromorphic (subdesert) shrublands.** Very open stands of shrubs with various xerophytic adaptations, such as extremely sclero-

---

* Occasionally less than 50 cm tall, thereby grading to **IV A1 a.**

(IIIC)

morphic or strongly reduced leaves, green branches without leaves, or succulent stems, etc., some of them with thorns.

1. Mainly evergreen subdesert shrublands. In extremely dry years some leaves and shoot portions may be shed.

a. (Truly) evergreen subdesert shrubland.

(1) *Broad-leaved,* dominated by sclerophyllous nanophanerophytes, including some phyllocladous shrubs (e.g., mulga scrub in Australia)

(2) *Microphyllous,* or leafless, but with green stems (e.g., *Retama retam*)

(3) *Succulent,* dominated by variously branched stem and leaf succulents

b. Semi-deciduous subdesert shrubland. Either facultatively deciduous shrubs or a combination of evergreen and deciduous shrubs.

(1) *Facultatively deciduous* (e.g., *Atriplex-Kochia*-saltbush in Australia)

(2) *Mixed evergreen and deciduous,* transitional to **2**

2. Deciduous subdesert shrublands. Mainly deciduous shrubs, often with a few evergreens.

a. Deciduous subdesert shrubland without succulents.

b. Deciduous subdesert shrubland with succulents.

IV. DWARF-SCRUB AND RELATED COMMUNITIES, rarely exceeding 50 cm in height (sometimes called heaths or heath-like formations). According to the density of the dwarf-shrub cover are distinguished:

dwarf-shrub thicket	=branches interlocked
dwarf-shrubland	=individaul dwarf-shrubs more or less isolated or in clumps
cryptogamic formation with dwarf-shrubs	=surface densely covered with mosses or lichens (thallochamaephytes); dwarf-shrubs occurring in small clumps or individually; in the case of bogs locally dominating graminoid communities may be included

A. Mainly evergreen dwarf-scrub. Most dwarf-shrubs evergreen.

1. Evergreen dwarf-shrub thickets. Densely closed dwarf-shrub cover, dominating the landscape ("dwarf-shrub heath" in the proper sense).

a. Evergreen caespitose dwarf-shrub thicket. Most of the branches standing in upright position, often occupied by foliose lichens. On the ground pulvinate mosses, fruticose lichens or herbaceous life shoot systems (e.g., *Cistus* heath.)*

b. Evergreen creeping or matted dwarf-shrub thicket. Most branches creeping along the ground. Variously combined with thallochamaephytes in which the branches may be imbedded (e.g., *Loiseleuria* heath).

(Subdivision possible.)

2. **Evergreen dwarf-shrublands.** Open or more loose cover of dwarf-shrubs.

a. E v e r g r e e n c u s h i o n s h r u b l a n d. More or less isolated clumps of dwarf-shrubs forming dense cushions, often equipped with thorns (e.g., *Astragalus* and *Acantholimon* "porcupine"-heath of the East Mediterranean mountains).

b. E v e r g r e e n m o s a i c d w a r f - s h r u b l a n d. Colonies or clumps of dwarf-shrubs interrupted by other life forms, bare soil or rocks (e.g., *Erica tetralix* swamp heath. Transitions into **D** and **E** possible.

3. **Mixed evergreen dwarf-scrub and herbaceous formations.** More or less open stands of evergreen suffrutescent or herbaceous chamaephytes, various hemicryptophytes, geophytes, etc.

a. T r u l y e v e r g r e e n d w a r f - s c r u b a n d h e r b m i x e d f o r m a t i o n (e.g., *Nardus-Calluna*-heath).

b. P a r t i a l l y e v e r g r e e n d w a r f - s c r u b a n d h e r b m i x e d f o r m a t i o n. Many individuals shed parts of their shoot systems during the dry season (e.g., *Phrygana* in Greece).

B. **Mainly deciduous dwarf-scrub.** Similar to **A**, but mostly consisting of deciduous species.

1. **Facultatively drought deciduous dwarf-thickets.** (Or dwarf-shrublands.) Foliage is shed only in extreme years.

2. **(Obligatory) drought-deciduous dwarf-thickets.** (Or dwarf-shrublands.) Densely closed dwarf-scrub stands which lose all or at least part of their leaves in the dry season.

a. D r o u g h t - d e c i d u o u s c a e s p i t o s e d w a r f - t h i c k e t. Corresponding to **A1**a.

b. D r o u g h t - d e c i d u o u s c r e e p i n g o r m a t t e d d w a r f - t h i c k e t. Corresponding to **A1**b.

c. D r o u g h t - d e c i d u o u s c u s h i o n d w a r f - s h r u b l a n d. Corresponding to **A2**a.

d. Drought-deciduous mosaic (or mixed) dwarf-shrubland. Deciduous and evergreen dwarf-shrubs, caespitose hemicryptophytes, succulent chamaephytes and other life forms intermixed in various patterns.

3. **Mixed cold-deciduous and evergreen dwarf-thickets.** (Or dwarf-shrub lands.)

(Subdivisions similar to **2**.)

4. **Cold-deciduous dwarf-thickets.** (Or dwarf-shrublands.)

Physiognomically similar to **2**, but shedding the leaves at the beginning of a cold season. Usually richer in cryptogamic chamaephytes.

Subdivisions similar to **2**. Transitions into **D** and **E** possible. In **A** and **B**, further subdivisions are possible, e.g., subdivisions based on the

(IVB4)

distribution pattern and height of woody phanerophytes in the dwarf-scrub matrix, similar to **VA**.)

C. **Extremely xeromorphic dwarf-shrublands.** More or less open formations consisting of dwarf-shrubs, geophytes, therophytes and other life forms adapted to survive or to avoid a long dry season. Mostly subdesertic. (Subdivisions similar to **IIIC**.)

D. **Moss, lichen and dwarf-shrub tundras.** Slowly growing, low formations, consisting mainly of dwarf-shrubs and cryptogams, beyond the sub-polar tree line. (Except in boreal regions, dwarf-scrub formations above the mountain tree line should not be called tundras, because they are as a rule richer in dwarf-shrubs and grasses, and grow taller due to the greater radiation in lower latitudes.) Often showing plant patterns caused by freezing movements of the soil (cryoturbation).

1. **Mainly bryophyte tundras.** Dominated by mats or small cushions of chamaephytic mosses. Groups of dwarf-shrubs are as a rule scattered irregularly and are not very dense. General aspect more or less dark green, olive green or brownish.

a. Caespitose dwarf-scrub—moss tundra.

b. Creeping or matted dwarf-scrub—moss tundra.

2. **Mainly lichen tundras.** Mats of fruticose lichens dominating, giving the formation a more or less pronounced grey aspect. Dwarf-shrubs mostly evergreen, creeping or pulvinate.

a. Dwarf-scrub—lichen tundra.
(Other tundra types more or less rich in chamaephytes may be distinguished.)

E. **Mossy bog formations with dwarf-shrubs.** Oligotrophic peat accumulations formed mainly by *Sphagnum* or other mosses, which as a rule cover the surface as well. Dwarf-shrubs are concentrated on the relatively drier parts or are loosely scattered. To a certain extent they resemble dwarf-scrub formations on mineral soil. Graminoid hemicryptophytes, geophytes with rhizomes and other herbaceous life forms may dominate locally. Slowly growing trees and shrubs can grow as isolated individuals, in groups or in woodlands, which are marginal to the bog or may be replaced by open formations in a cyclic succession. The following subdivisions correspond to the classification of bog types adopted in Europe.

1. **Raised bogs.** By growth of *Sphagnum* species raised above the general ground-water table and having a ground-water table of their own. Therefore no more supplied by "mineral" water (i.e., water having been in touch with the inorganic soil), but only by rain water (truly ombrotrophic bogs).

a. Typical raised bog (suboceanic, lowland and submontane). Mosses dominating throughout, except on locally raised dry hummocks, which are dominated by dwarf-shrubs. Trees rare and, if present, concentrated on the marginal slopes of the convex peat accumulation. Mostly surrounded by a very wet, but less oligotrophic sedge swamp (Swedish "lagg," see **VD**).

b. Montane (or "subalpine") raised bog. Growing slower than the typical raised bog (or formed in an earlier period with a warmer climate and actually "dead" or being destroyed by erosion). Often covered with sedges or evergreen dwarf-shrubs. Micro- or nanophanerophytes (e.g., *Pinus montana*) locally dominating.

c. Subcontinental woodland bog. Temporarily covered by open wood of low productivity, which in a sequence of wetter years may be replaced by *Sphagnum* formations similar to a.

(Various subdivisions of a, b and c possible.)

2. **Nonraised bogs.** Not or not very markedly raised above the mineral-water table of the surrounding landscape. Therefore in general wetter and not as oligotrophic as **1**. Poorer in mosses than **1a**, to which various forms of transitions are possible.

a. Blanket bog (oceanic lowland, submontane or montane). The micro-surface of the bog is less undulating and less rich in actively growing mosses than in **1a**. Evergreen dwarf-shrubs are scattered as well as caespitose hemicryptophytes (sedges or grasses) and some rhizomatous geophytes.

b. String bog (Finnish "aapa" bog). Flat oligotrophic bog with strings in the boreal lowlands. The Finnish name indicates an open bog without or with only a few trees of very poor vigor, which grow on narrow and low elongated hummocks, the so-called strings. These peat strings are formed by pressure of the ice covering the more or less flooded bog from early fall to late spring. Only these strings are covered by dwarf-shrubs and are rich in *Sphagnum*. The main part of the bog is similar to a wet sedge swamp.

(Subdivisions of a and b possible.)

V. **TERRESTRIAL HERBACEOUS COMMUNITIES.** Grasses, graminoid and other herbaceous plants are predominant in the cover, but woody plants may be sparingly present (i.e., covering not more than 30%).

A. **Savannas and related grasslands** (tropical or subtropical grasslands and parklands). Trees or shrubs almost regularly present, often fire-scarred (fires frequent in the dry season).

1. **Tall-grass savannas.** Dominated by broad-leaved and tall grasses (corresponding to relatively humid conditions within the tropics and subtropics). Hemicryptophytic caespitose grasses most frequent, yellowing during the dry season. In general without cryptogams. Subdivisions are

(VA1)

based on the distribution pattern and height of woody phanerophytes in the grassy matrix.*

a. Woodland with patches of tall-grass savanna (woodland savanna). Tree groups and isolated trees which are interspersed by small patches of grassland.

(Subdivisions according to woodland formation, see **II**.)

b. Tall-grass savanna with isolated trees (tree savanna). Isolated trees dispersed more or less regularly over the grassy matrix.

(1) With evergreen broad-leaved trees

(2) With palms

(3) With deciduous trees

(4) With extremely xeromorphic trees or succulents

c. Tall-grass savanna with shrubs (scrub savanna). Stands of shrubs alternating in various patterns with grassland.
(Subdivisions similar to b.)

d. Tall-grass savanna (grass savanna). Practically without woody phanerophytes, in general due to anthropogenic influence. Normally called "tropical grassland," but the grass cover is physiognomically identical to above-mentioned units.

e. Flood savanna. Periodically inundated in various mosaic patterns, with either palms or groups of other trees on raised positions.

(1) With trees

(2) With scrub

(3) Without woody plants

2. Short-grass savannas. Dominated by narrow-leaved and more or less short grasses (indicating relatively drier conditions). In addition to perennial grasses annuals more frequent than in **1**, in some places even predominant.

(Subdivisions based on patterns formed by woody plants.)

a. Short-grass savanna with isolated trees.

(1) With evergreen trees

(2) With deciduous trees

(3) With xeromorphic trees except succulents

(4) With tree-succulents

b. Short-grass savanna with shrubs.

(Subdivisions as in a.)

---

* Studies have shown that these patterns are largely related to human influence, except in the case of the flood savanna. Savannas may form various mosaics with either forest, woodland or scrub, often fire-scarred.

c. Short-grass savanna (grass savanna). Without trees or shrubs (e.g., tropical montane grassland, like "puna" grassland of the Andes).

B. **Steppes and related grasslands** (e.g., North American "prairies," etc.; temperate, with late summer drought and winter frost season). Trees or shrubs absent as a rule, except on wetter sites,* e.g., along rivers, in ravines and in the forest border ecotone. Seasonal change of physiognomic-floristic aspects very pronounced. In spring therophytes and geophytes are most conspicuous, later hemicryptophytes dominate the aspect.

1. **Tall-grass steppes.** (Or prairies.) Caespitose grasses taller than 1 m dominate (indicating a more humid climate).

a. Tall-grass steppe with trees.

b. Tall-grass steppe with shrubs.

c. Tall-grass steppe without woody plants.

2. **Mid-grass steppes.** (Or prairies.) Intermediate between **1** and **3**. Medium-sized grasses locally frequent.

   (Subdivisions as in **1**.)

3. **Short-grass steppes.** (Or prairies.) Mostly composed of mat-forming, more or less low grasses.

   (Subdivisions a–c as in **1**.)

d. Short-grass steppe with suffrutescent plants.

4. **Forb-rich steppes.** Broad-leaved forbs, mostly hemicryptophytes, are frequent (in a climate transitional to forest climate, e.g., in Russia.) These "meadow steppes" resemble the meadows under **C**.
   (Subdivisions as in **1**.)

C. **Meadows, pastures or related grasslands.** (Temperate or subpolar grasslands in a forest climate with no marked dry season.) Mesophytic hemicryptophytes dominating as a rule. Below the snow cover many plants remain green at least partially during the whole winter.

1. **Meadows and pastures below tree line.** Anthropozoogenic formations in the forest climate belt (except avalanche grassland). As a result of different management distinguished as:

   meadow = grassland mainly used for hay production, growing rhythm and aspects largely determined by mowing once or several times per year

   pasture = grassland mainly used for grazing; no marked seasonal aspects; by selective grazing, the animals produce small scale mosaics of tall and low groups of plants

---

* Only in anthropogenic steppes, trees or shrubs and steppe-grassland may occur on the same physical habitat.

(VC1)

The following subdivisions are based on distribution patterns of woody phanerophytes:

a. Woodland pasture. Woodland opened up through grazing practice, consisting of isolated irregularly grouped trees.

b. Tree meadow (or pasture). Grassland with isolated trees.

c. Scrub pasture (or meadow). Shrub groups in grassland.

d. Grassy pasture without trees or shrubs. Frequently grazed. Height of grasses, legumes and other forbs varies with region and management.

(1) *Extensively grazed* (German "Triftweide," not fertilized); hard, thorny or other plants on which animals do not feed ("grazing weeds") are most frequent

(2) *Intensively grazed* (German "Standweide" or "Untriebsweide," fertilized); kept in closely cropped condition, forms a dense grass carpet; soft food plants dominating throughout; "grazing weeds" usually rare

e. Grassy meadow without trees or shrubs. Mowed and only exceptionally grazed. More or less tall caespitose grasses and tall, mostly scapose herbs dominating.

(1) *Litter meadow* (German "Streuwiese," mowed for obtaining straw for bedding animals in the stables; generally not fertilized and mowed only in the fall, after the shoots have dried); as a result the slowly developing bunch grasses become dominating

(2) *Hay meadow* (German "Futterwiese," generally fertilized; mowed for making hay to feed animals); rapidly developing in the early growing season, therefore rich in malacophyllous grasses and forbs, mowed several times a year; this management produces a very marked change in aspect

(a) Poor in spring geophytes (lowland or submontane)

(b) Rich in spring geophytes (montane or subalpine); snow cover disappears late in spring and prevents grasses from growing up earlier than geophytes (e.g., *Crocus* or *Narcissus*)

f. Sedge-rush meadow. More or less sclerophyllous graminoid herbs dominate, indicating periodically water-logged soil (transition to **D**).

g. Avalanche grassland. (The only nonanthropogenic meadow. Occurring as narrow strips of grassland between forests on steep slopes of high mountains, where avalanches, descending annually in spring, prevent forest establishment.) Composition of herb cover similar to d (1), but very variable. Major variations:

(1) *With shrubs* or damaged trees

(2) *Without shrubs*

2. **Pastures and meadows above mountain tree line.** (Or beyond northern tree line.) Only exceptionally with shrubs or gnarled trees. Covered with snow more than 6 months of the year.

a. Closed alpine (or subpolar) mat. Without snow cover at least 4–5 months.

(1) *Rich in graminoids*

(2) *Rich in forbs*

(3) *With dwarf-shrubs*

b. Alpine (or subnivean) mat-patches (upper alpine or subnivean). More or less open formation, covers the soil unevenly.

(Subdivisions similar to a.)

c. Snow bed formation. (Covered by snow more than 8–9 months yearly; water-logged by melting snow.) Open formation, rich in small forbs or forb-like dwarf shrubs (e.g., *Salix herbacea*).

(Subdivisions possible.)

D. Sedge swamps and flushes. Open formations on constantly or mostly water-logged ground, without or with only a few woody plants.

1. Sedge peat swamps and similar swamps. Dominated by sedges (i.e., graminoid hemicryptophyes or geophytes), seasonally flooded.

a. Tall-sedge swamp. (Frequently flooded, and commonly for long periods; as a rule natural.) Foliage taller than 30–40 cm, sedges dominating throughout; very few other life forms.

(1) *With creeping sedges,* forming large homogeneous stands (e.g., *Carex gracilis*)

(2) *With caespitose sedges,* forming tufts or hummocks (e.g., *Carex elata*)

b. Low-sedge swamp. (Flooded only little or only for short periods; mostly anthropogenic.) Small sedges (*Carex, Juncus, Scirpus*, etc., foliage not higher than 30 cm) of low productivity dominating, intermixed with many other herbaceous life forms.

c. Hard hummock swamp. Formed by very hard-leaved, mostly small sedges, therefore more compact and easier to walk on than a (2). Rare (e.g., in the tropical Andes).

2. Flushes (German "Quellfluren"). Evergeen herbaceous or cryptogamic vegetation growing on habitats where seepage water crops up at the surface. (Constantly wet, but rarely flooded.)*

a. Forb flush. Mostly dominated by small forbs.

(1) *Calcareous;* older parts of plants covered by a white or brownish crust of precipitated carbonate

(2) *Noncalcareous*

b. Moss flush. Dominated by mosses.

(Subdivisions like a.)

---

* Covering only very small areas, often forming mosaics with graminoid formations.

(V)

E. Herbaceous and half-woody salt swamps. Halophilous or salt-tolerant plants building more or less dense permanent formations. Most species are suffrutescent (half-woody). Shrubs or trees absent or only exceptionally present.

1. Halophytic half-woody shrub formations. Dominated by more or less succulent half-woody shrubs up to 1 m high, at least partially evergreen.

a. Marine half-woody salt marsh. (Growing near the sea coast on marine deposits. Flooded from time to time, but in any case for only a short period.) Rich in microscopic algae growing on the soil surface.

(1) *Succulent*; dominated by succulents (e.g., *Salicornia*)

(2) *Nonsucculent*; poor in truly succulent plants, frequently rich in more or less nitrophilous, quickly growing half-shrubs (e.g., *Obione* or *Artemisia* spp.); often forming narrow strips or bands along the rills on the seashore, where organic matter has been deposited

b. Inland half-woody salt marsh. Similar to a (but growing in continental depressions, flooded in wet seasons and drying out during the less rainy ones). In general more xeromorphic than a. Poor in algae.

(1) *Succulent*

(2) *Nonsucculent*

2. Salt meadows. Mainly herbaceous. More or less closed formations of hemicryptophytes, herbaceous chamaephytes and other nonwoody life forms, but poor in annuals.

a. Marine salt meadow. (Within the tidal range of temperate sea shores. but not flooded daily). More or less densely closed stands of quickly growing grasses and (or) succulent forbs.

(1) *Rich in succulents*

(2) *Poor in succulents*

b. Inland salt meadow. Similar to a (but growing in continental conditions). In general less vigorous and less dense.

(1) *Closed*; physiognomically similar to a (2), but more xeromorphic

(2) *Open*; most plants growing isolated or in tufts; soil with a more or less thick salt cover, at least in the dry season; transitional to **VI B**

F. Forb vegetation* and similar communities. More or less broad-leaved herbs dominating, normally mesophyllous and deciduous. Woody life forms only exceptionally present.

1. Mainly perennial forb communities. Dominated by nongraminoid hemicryptophytes and geophytes. Annuals sometimes present, but of little importance.

---

* In general covering small areas which cannot be represented on small scale maps. "Form" (American) means "nongraminoid herb" (German "Kraut").

a. Forest border herb formation. Between adjoining phanerogamic and herbaceous vegetation occurring as a narrow transitional band, consisting of hemicryptophytes, geophytes and therophytes. Growing more vigorously than the adjacent pasture or meadow.

b. Tall-forb formation (German "Hochstauden" formation). Dense stands of broad-leaved, mostly dicotyledonous herbs taller than 50 cm, mesophyllous, well provided with nitrogen and other nutrients.

c. Fern thicket (or heath). *Pteridium acquilinum* or *Dicranopteris* sp. dominating and forming nearly pure stands on pastured heathlands within forest regions. (The only forb formation covering large areas.)

d. Perennial forb formation on organic deposits at the flood lines. Consisting of broad-leaved herbs, growing abundantly on more or less decomposed organic deposits, which are often renewed by floods.

e. Perennial ruderal and clearing herb formation. More or less broad-leaved herbs (growing on debris, ruins and other places strongly influenced by man).

f. Mainly perennial weed formation on cultivated land. Mostly hemicryptophytic or geophytic weeds, growing more or less abundantly in the shade of cultivated perennial plant stands. Annual weeds are present, but not predominant (see **2**d). The significance of weed formations greatly diminished by use of herbicides.*

2. **Mainly ephemeral forb communities.** Therophytes more frequent than perennial herbs. Vegetation cover often not as dense as in **A**.

a. Tropical or subtropical ephemeral cloud desert forb formation. (Best developed on the coastal hills of Peru and northern Chile, where, from fall to spring, moving clouds moisten the vegetation and the soil by condensed water.) Dominated by annual broad-leaved herbs, which germinate at the beginning of the cloudy season and grow abundantly until the end of it, giving the landscape a fresh and green look. In the dry season the aspect is desert-like. Phanerophytes may be present as relics of natural cloud-woodland. Geophytes and cryptogamic hemicryptophytes or chamaephytes are constantly present and may become dominant locally.

b. Ephemeral halophytic formation. (Growing normally in more extreme conditions than salt meadows, see **VE2**.) More or less open formation of annual halophytes. Some permanent herbs and grasses may be present.

c. Ephemeral ruderal and clearing forb formation. Like **1**e, but dominated by annuals.

d. Mainly ephemeral weed formation on cultivated land. Similar to **1**f, but rich in annuals or species that germinate in fall and die after fructification during the next vegetation period.

3. **Episodical forb communities.** Very unstable ephemeral plant groupings

---

* On small-scale maps, crops rather than weed formations are to be shown.

(VF3)

appearing not regularly every year, but only when the growing conditions are favorable.

a. Episodical desert forb formation ("flowering desert"). Mostly broad-leaved, rapidly developing herbs with hardy seeds that germinate after episodical rain fall. Often concentrated in depressions (some of them hardly discernible) where the surface water accumulates. Sometimes this formation may fill the gaps between permanent subdesert plant, e.g., xeromorphic shrubs or succulents belonging to formations **III C** or **IV C.**

b. Episodical formation on pond muds and similar sites. (Developing, when the pond water has been drained—every year or after some or several years.) Dominated by forbs, whose seeds neither decay nor germinate while the bottom of the pond is inundated, but germinate and grow rapidly after the mud has emerged again.

c. Episodical forb formation on organic deposits at the flood lines. Similar to **1**d, but less permanent and mainly composed of annuals, whose seeds have been carried along together with organic deposits. Perennials may be present.

d. Episodical river bed formation. Ephemeral herbs, grasses or sedges developing in the dry parts of river beds during low water periods of more than 2 months. Depending upon seeds supplied by the river; therefore very unstable not only in density but also in species composition and pattern.

VI. DESERTS AND OTHER SCARCELY VEGETATED AREAS. (Subdeserts are included in the formation classes **III** to **V.**) Bare mineral soil determines the aspect more or less constantly. Plants are scattered or may be absent.

A. Scarcely vegetated rocks and screes.

1. Scarcely vegetated rocks.
a. Chasmophytic vegetation. Permanent plants rooting in fissures of rocks or walls.

(subdivisions according to life forms in different latitudes and altitudes.)

b. Adnate Bromeliaceae on rocks (only in the neotropics).

c. Cryptogamic mat on rocks.

(1) *Foliose lichens and mosses* dominant

(2) *Crustose lichens* dominant

(3) *Blue algae* dominant ("ink-strips," German "Tintenstriche"); dark strips on rocks caused by Cyanophyceae that grow actively when the water is trickling down

2. Scarcely vegetated screes. (More or less unstable, steep slopes of stones beneath weathering rocks.) Mostly permanent herbs or half-woody plants adapted to survive the movement of stones at the scree surface, sometimes even stopping them. Subdivisions mainly according to the length of the vegetation period:

a. Lowland and submontane scree formation.

b. Montane scree formation.

c. High mountain scree formation.

B. **Scarcely vegetated sand accumulations.** (Wood, scrub, grassland and other more or less closed formations on fixed dunes are treated in the formation classes listed above.) Moving quicksand with isolated plants that are contributing to its fixation; or bare shifting sand dunes. Vegetation covering not more than one third of the surface.

1. **Scarcely vegetated sand dunes.**

a. Tall-grass dune. (Coastal "white dune." Mostly rich in carbonates and nutrients.) Built up and partially covered by geophytic grasses or grass-like plants which are able to adapt their root and shoot system to new accumulations of sand that bury them in stormy periods.

(1) *Tropical and subtropical*

(2) *Temperate,* showing a marked annual growing rhythm

b. Short-grass dune. (Mostly continental, more or less acid and poor in nutrients). Low hemicryptophytic or geophytic grasses and sedges fixing the quicksand.

c. Forb dune (possibly existing).

2. **Bare sand dunes.** Only exceptionally with some isolated plants.

a. Shifting dunes in desert climate (natural).

b. Shifting dunes in forest climate (anthropozoogenic).

C. **True deserts.** Vegetation largely absent.

(Subdivisions possible only according to geological and morphological criteria. Desert valleys may be classified into one of the preceding formations.)

VII. **AQUATIC PLANT FORMATIONS** (except marine formations*). Composed of rooted and (or) floating plants that endure or need water covering the soil constantly or at most times of the year.

A. **Floating meadows.** Densely interwoven forbs and (or) mosses covering permanent fresh water accumulations. Most of the phanerogams being heliophytes, not true water plants.

1. **Mainly herbaceous floating meadows.** Dominated by sedges or herbs with rhizomes. Chamaephytes and even phanerophytes may be present.

a. Tropical and subtropical herbaceous floating meadow.

---

* To be worked out later.

(VIIA1)

b. Temperate and subpolar herbaceous floating meadow, with pronounced seasonal aspects.

2. Mainly mossy floating meadows. Mosses dominating throughout, but phanerogams may be present.

a. Mossy floating meadow (temperate or subpolar).

(Further formations possibly exist.)

B. Reed-swamps. Tall heliophytes rooting in the soil at the bottom of shallow lakes, slowly flowing rivers or similar waters.

1. Reed-swamp formations of fresh water lakes. Mostly broad-leaved plants which cannot endure high salt concentration. All shoots upright, only exceptionally floating in the water.

a. Tropical and subtropical fresh water reed-swamp. Seasonal aspects not pronounced.

b. Temperate and subpolar fresh water reed-swamp.
In winter time most plants yellow or dormant.

2. Reed-swamp formations of salt water lakes. More or less scleromorphous plants resisting high salt concentrations. Normally not as tall as **1**.

a. Tropical and subtropical salt water reed-swamp.

b. Temperate salt water reed-swamp.

3. Reed-swamp formations of flowing water. Shoots more flexible than in **1** and **2**, sometimes with floating leaves.

a. Tropical and subtropical reed-swamp on river banks.

b. Temperate reed-swamp on river banks.

C. Rooted floating-leaf communities.

(Subdivisions more or less similar to **B**.)

D. Rooted underwater communities. Comprised of aquatic plants that are structurally supported by water (i.e., non-self-supporting in contrast to aquatic heliophytes).

(Subdivisions possible.)

E. Free-floating (nonrooted) fresh water communities.

1. Broad-leaved, free-floating communities.

a. Tropical and subtropical broad-leaved, free-floating formation (e.g., *Pistia*, *Pontederia* and *Eichhornia*).

b. Temperate broad-leaved, free-floating formation. Disappearing in the cold season (e.g., *Stratiotes*).

2. Lemna-type free-floating communities.

(Subdivisions similar to **1**.)

3. Free-floating macroscopic algal communities.

(Subdivisions possible.)

Appendix C

A Key for Mapping Structural Vegetation Types in Southeast Ceylon on Air Photograph Mosaics at the Scale of 1:31,680* (from MUELLER-DOMBOIS, 1968)

The key is divided into four parts:

I   Forest cover types

II   Scrub cover types

III   Herbaceous cover types

IV   Other areas

Within each category, the map units are defined in such a way that they can be separated quantitatively from one another on the air photographs in the field. The method is a local adaptation of Fosberg's (1961) general key (Chap. 7).

---

* The key is not limited to this scale but can be applied to any larger scale and to a reduction in scale of about 50%.

489

I Forest cover types

1. *f*   =Low to medium stature forest with scattered emergents
Closed forest with interlocking or touching tree crowns covering an area of at least 6 ha. Low to medium stature refers to a height range of 5–15 m. The scattered emergents exceed 15 m in height. On the photographs this forest unit is usually shown by the darkest colored areas with a closely packed medium to coarse grained texture.

2. *af*   =High stature forest along almost permanent streams (alluvial forest)
Closed forest with interlocking or touching tree crowns covering an area of at least 6 ha. High stature implies that most of the canopy trees are taller than 15 m. On the photographs this unit is recognized by finely mottled light and dark tones and by dense coarse-grained texture resulting from large individual tree crowns. They form more or less distinct galleries along the almost permanent (year-round water-issuing) streams.

3. *df*   =Discontinuous low to medium stature forest with scattered emergents
Like *f* above, but with narrow, usually less than 70 m wide open strips and pathways interrupting in various patterns the continuity of the forest. These may be temporary drainage channels, erosion cuts and banks, grassy strips or rockoutcrop ridges.

4. *fs*   =Forest-scrub
A mosaic type composed of small-area (less than 6 ha) closed forests and closed scrub areas of the same size. Instead of small-area closed forests or in combination with them, the forest-scrub unit may also be characterized by scattered trees (at least 10–15/ ha) emerging from a closed scrub matrix. The latter can be called open forest with a closed scrub layer. However, a distinction between the two subunits (small-area forest and scrub, and open forest with closed scrub) is not practical at this map scale in this region.

5. *dfs*   =Discontinuous forest-scrub
The same as *fs* with discontinuity analog to *df*.

6. *ofs*   =Open forest-scrub
An open woody vegetation with about equal proportions of trees above 5 m and shrubs and trees below 5 m. Shrubs and trees may be nested but more commonly are individually dispersed. The herbaceous cover is also interrupted. Perennial grass mats surrounding nested woody plant groups or taller individual trees interchange with barren patches or annual herbaceous covers (depending on season). On the photograph the type appears light gray with dark dots showing the trees. Their number must be at

least 10–15/ha, otherwise this becomes the more widely spread open scrub with scattered trees = os(t) defined below.

7.   *(fs)*   =Forest-scrub islands
Scrub clumps or islands with tall trees at least 10 m in height, occurring close together over an area of at least 2 ha. The woody plant cover is more than 50%. The intervening openings may be grass cover, rock, barren sand or gravel-strewn surfaces. In contrast to *df*, the woody plant aggregations are small (usually less than ½ ha), whereas *df* shows larger continuous sections interrupted by strip-like openings.

II Scrub cover types

1.   *s*   =Scrub
Scrub with interlocking or touching shrub and/or tree (5 m or smaller) crowns resulting in a continuous canopy. As a rule, the scrub is never dense or thicket-like, but permits relatively easy penetration on account of the many animal trails beneath the canopy. These do not show up on the air photographs. The unit appears on the air photographs in a dark to medium gray tone with a fine-grained to smooth texture. Most scrub canopy heights are between 2.5 and 5 m.

2.   *s(t)*   =Scrub with scattered trees
The same as *s* above but with scattered trees emerging above 5 m (at least 5 trees/ha but less than 15 trees/ha). On the air photographs it is distinguished from *s* by the dark small spots showing the emergent trees. This unit is much more widespread than *s* and is one of the most important types in terms of area coverage.

3.   *ds*   =Discontinuous scrub
Like *s* above, but with narrow, usually less than 70 m wide open strips and pathways interrupting in various patterns the continuity of the scrub canopy. These may be drainage channels, erosion cuts and banks, elephant pathways, grassy or barren sand strips or rockoutcrop ridges. Tonal and textural patterns on the air photographs are the same as for *s* except for the crisscrossing white lines denoting the openings that disrupt the continuity of the shrub canopy. The continuous scrub sections cover areas of at least 2 ha.

4.   *ds(t)* =Discontinuous scrub with scattered trees
Like *ds* above, but with scattered emerging trees analog to *s(t)*. This unit is also more widespread than *ds* above and is nearly equally important in terms of area coverage as *s(t)*.

5.   *os*   =Open scrub
Individual shrubs and small trees (5 m or smaller) occur in loose formation, i.e., there are nested groups with interlocking crowns,

(II5)

but these and many individuals stand free and do not touch each other with their crowns. In this type much of the crown biomass is below 2 m in height. On the air photograph the unit appears in light gray tones with smooth texture. It is often hard to distinguish from grass cover.

6. *os(t)* =Open scrub with scattered trees
Like *os* above, but with scattered trees reaching above 5 m in height. The unit is intermediate in height of crown biomass between *os* and *ofs*, extending mainly to 5 m, but with some material above 5 m in form of the scattered trees, from 5–15 trees/ha. Among the 3 analog units (*os, os(t)* and *ofs*) this is currently the most important of the three in area coverage.

7. *(s)* =Scrub islands or clumped scrub
Like *(fs)*, but without the tall trees. However, the unit may contain trees of less than 10 m height. This unit is more common than *(fs)*. The scrub islands are usually less than ¼ ha in size.

III Herbaceous cover types

1. *g* =Short-grass cover
Matted and usually shortly cropped, closed grass cover. Suffrutescent and dwarf-shrubs may be present. Minimum size 2 ha.

2. *g(s)* =Short-grass cover with scrub islands
Like *g*, but with clumped scrub or shrub islands usually less than ¼ ha in size. In contrast to *(s)*, the scrub islands cover less than 50% of an area of at least 2 ha. The most common grass cover unit.

3. *g(fs)* =Short-grass cover with forest-scrub islands
Analog to *g(s)*, but with tall trees at least 10 m in height that are associated with the woody plant islands in the short-grass cover.

4. *g(t)* =Short-grass cover with scattered trees
Matted and usually closely cropped grass cover with scattered trees regardless of size. Typical savanna.

5. *g(s,t)* =Short-grass cover with scrub islands and scattered trees
Analog to *g(s)*, but with individual trees scattered between the scrub islands. Not common.

6. *g(w)* =Short-grass or graminoid cover with sections of sparse cover or barren areas near water
Similar to *g*, but with barren areas or sparse graminoid cover. These may have resulted from various causes, seasonal inundation, receding water lines, buffalo trampling, scalping by elephants or ploughing by pigs.

7. *r* =Sparse herbaceous and suffrutescent cover occurring more or less in pockets in rockoutcrop areas

Small, often fragmentary herbaceous and mixed low-shrub (semi-woody) communities that are interspersed in rockoutcrop areas resulting in sparse cover with little biomass on a hectare basis. Minimum size 2 ha.

8. *r(s)*   =Sparse herbaceous cover with scrub islands in rockoutcrop areas
Similar to *r* but with scattered scrub islands.

9. *r(fs)*   =Sparse herbaceous cover with forest-scrub islands in rockoutcrop areas
Similar to *r(s)*, but usually with larger (up to 1 ha) woody plant islands with associated trees or even small forest stands, defined as having more than 20% of its crown biomass above 5 m height.

10. *b*   =Sparse herbaceous and suffrutescent cover in beach and dune areas occurring usually in patches or zoned communities
This unit includes the barren surf-area, the *Spinifex* grass covers in the strand zone and all other patchy or sparse herbaceous and semi-woody low-scrub plant covers on coastal substrates.

11. *b(s)*   =The same as scrub islands
Aggregated shrubs or individually scattered shrubs up to 5 m height that cover in total less than 50% of the area. The rest like *b* above.

12. *b(fs)*   =Sparse herbaceous and suffrutescent cover with forest-scrub islands on coastal substrates
Similar to *b(s)*, but the islands are associated with trees or even small-area forests (up to ½ ha in size). Total woody plant cover is less than 50%. The areas are always larger than 2 ha.

IV   Other areas

x   =Recently strongly disturbed vegetation and cultivated areas (including town sites)
The disturbance of the vegetation was caused by man, primarily for 'chena.' But the latter not separated from cultivated field or town sites.

w   =Open water surfaces

Other information on the vegetation map includes important place names, names of the larger streams, roads and jeep tracks, and sample plot locations.

Appendix D

Key for Mapping Forest Habitat Types in Southeast Manitoba with the Help of 1:15,840 Standard Aerial Photographs (from MUELLER-DOMBOIS, 1965ₐ)

The key can also be used for mapping habitat types with the help of topographic base maps of larger scale, or simply for orientation in the field to determine the habitat in question. The key gives, at each step, two alternatives that lead to the next solution. Habitat type numbers refer to FIGURE 11.2 and TABLE 11.5. Symbols are explained in TABLE 11.5 (scientific names of species are listed following the key):

1. a Habitats on organic terrain 2
     (overlying organic matter > 6 in. deep)

1. b Habitats on mineral terrain 7
     (overlying organic matter < 6 in. deep)

2. a Habitats on firm, consolidated peat that does not
     move with fluctuations of the water table 3

2. b Habitats on "swimming" peat that moves up and down with periodic changes of water table; tree cover, if present, stunted, not taller than about 15–20 ft .... floating bog (not differentiated)

3. a Habitats covered dominantly with tamarack, swamp birch and tall sedges; surface is relatively smooth, and may be covered with low hummocks which may occupy up to 50% of the area .... BC type (No. 12)

3. b Habitats covered dominantly with black spruce (or black spruce stumps), occasionally with white cedar; surface with pronounced hummocks which occupy more than 50% of the area   4

4. a Habitats with shallow peat; up to 20 in. deep (measured from depressions between hummocks)=half bog types   5

4. b Habitats with peat deeper than 20 in., almost never covered with white cedar   6

5. a Habitats with level, muck-filled depressions (= sink holes), which may range from about 1 to 6 in diameter .... sk type (No.13)

5. b Habitats without muck-filled depressions; instead depressions covered with sphagnum, typically with upland herbs (e.g., palmate-leaved coltsfoot, dewberry, bunchberry) .... FS type (No. 5)

6. a Habitats with level, muck-filled depressions (=sink holes), which may range from about 1 to 6 ft in diameter .... skS subtype (No. 13)

6. b Habitats without muck-filled depressions; instead depressions covered with sphagnum; sphagnum and feathermoss peat forms a continuous hummocky sheet, herbaceous vegetation sparse .... S type (No. 14)

7. a Habitats covered dominantly with pine (or pine stumps)   8

7. b Habitats covered dominantly with hardwoods, spruces (white and occasionally black) or arborescent shrubs 14

8. a Habitats on ancient sand dunes on which green alder is absent from local depressions .... vd type (No. 1)

8. b Habitats on other sandy-textured landform types, including dunes with green alder in local depressions   9

9. a Habitats with bright, light gray mineral surface
horizons (Ae* >1 in.) on generally low lying
coarse sandy deposits   10

9. b Habitats without bright, light gray mineral surface horizons (Ae <1 in.) when on sandy soils;
on generally raised, level, undulating, or sloping positions; surface soil texture ranges from
coarse sand to sandy loam   11

10. a Habitats with rusty mottling immediately below
the bright, light gray mineral surface horizon;
with only sporadic occurrence of bearberry or
bearberry completely absent, with typically
Labrador tea, green and sometimes speckled
alder, and with moisture-indicating herbs (palmate-leaved coltsfoot, dewberry, bluebead-lily,
bunchberry and horsetail species) ....*om type* (No. 4)

10. b Habitats without rusty mottling immediately below the bright, light gray mineral surface horizon; some patches of twinflower among bearberry, which here is usually prevalent, often
with a few scattered bushes of green alder, and
scattered bunchberry; Labrador tea, palmate-
leaved coltsfoot, bluebead lily and horsetail
species usually absent; where recent ground
fires have occurred common scouring-rush is
usually abundant ....*of type* (No. 3)

11. a Habitats with dominantly low ericaceous shrubs
(bearberry, teaberry, blueberry) giving the impression of a "clean" forest floor with only
scattered larger shrubs present; bearberry forms
typically a carpet, which may or may not be
interrupted with barren needle-covered areas
and/or large patches of reindeer lichens; the
solum consists of uniform sand   12

11. b Habitats with dominantly nonericaceous, medium-sized to arborescent shrubs (particularly
snowberry and hazel) and abundant herbaceous vegetation (particularly grasses), pine
cover often mixed with birch and/or aspen; the
solum consists either of sands with Bt (colloid
enriched) horizons or entirely loamy sand or
sandy loam   13

12. a Habitats with abundant patches of reindeer
lichens, and numerous barren, needle-covered
spots (together occupying about 50% or more

° Ae = eluviated (i.e., leached) A horizon

of forest floor), and very sparse herbaceous cover ....*d+* subtype (No. 2)

12. b Habitats with nearly complete surface coverage of bearberry and only few barren spots if any and with less abundant patches of reindeer lichens ....*d* type (No. 2)

13. a Habitats with fairly dense arborescent shrub layer, in which hazel is prominent; other characteristic shrubs are green alder, raspberry, snowberry, downy arrow-wood, bush-honeysuckle; many herb species; characteristically present is wild sarsaparilla; bearberry sporadic, twinflower always present ....*mf* type (No. 7)

13. b Habitats with scrub layer usually less dense and snowberry typically dominant, hazel may be more sparsely represented; bearberry always present but usually patchy and interrupted by grasses and herbs; reindeer lichens typically absent or very sparse ....*mf* subtype (No. 7)

14. a Habitats near active streams with typically ash and elm as stand component and with dark mineral surface horizons of "garden soil" characteristics 15

14. b Habitats in other locations 16

15. a Habitats on alluvial bottomlands whose surfaces are only about 2–4 ft above the stream-level in mid-summer; typically with dominant ostrich-fern among the undergrowth ....*evm* type (No. 10)

15. b Habitats on higher alluvial terraces, whose surfaces are more than 6 or 8 ft above stream-level in mid-summer and on which ostrich-fern is absent or only very sporadic ....*evf* type (No. 9)

16. a Habitats in which the upper mineral soil profile is of sandy texture 17

16. b Habitats on which the upper mineral soil profile is of loamy sand, loam or finer texture 19

17. a Habitats in low, flat or depressional positions on flaring-out margins of beach deposits or on lacustrine sand with typically organic terrain in vicinity 18

17. b Habitats in raised positions on all sandy textured landforms except dunes 13

18. a Habitats with shallow black muck (up to 6 or 8
in. deep) overlying the mineral soil and with
tall sedges and clumps of marsh-marigold ....*ew type* (No. 11)
18. b Habitats with twin humus (typically root mor
over mull) without tall sedges and marsh-
marigold, often adjacent to and lower than *om*
type ....*mvm* type (No. 6)
19. a Habitats on low lying ground moraines with
grayish and dark-grayish colored upper mineral
soils, typically with faint yellowish mottling
near the surface, and usually with deep (4–8
in.) blackish humus cover   20
19. b Habitats on raised ground moraines with yellowish and yellowish brown loamy sand textured
surface soils and never with surface mottling,
and with brownish to dusky red humus cover
usually not exceeding 4 in. in depth ....*mf* type (No. 7)
20. a Habitats in local depressions with black muck
humus cover and tall sedges and clumps of
marsh-marigold ....*ew* type (No. 11)
20. b Habitats on broad flats without tall sedges and
marsh-marigold ....*mm* type (No. 8)

The following 36 plant species are used in the key: ash (*Fraxinus nigra*
Marsh, *F. pennsylvanica* Marsh), aspen (*Populus tremuloides* Michx.), bearberry (*Arctostaphylos uva-ursi* (L.) Spreng), birch (*Betula papyrifera* Marsh),
black spruce (*Picea mariana* Mill BSP), bluebead-lily (*Clintonia borealis* (Ait.)
Raf.), blueberry (*Vaccinium angustifolium* Ait.), bunchberry (*Cornus canadensis* L.), bush-honeysuckle (*Diervilla lonicerna* Mill.), common scouring-rush
(*Equisetum hyemale* L.), dewberry (*Rubus pubescens* Raf.), downy arrow-
wood (*Viburnum rafinesquianum Schultes*), elm (*Ulmus americana* L.), feathermoss (*Hylocomium splendens* Hedw.), green alder (*Alnus crispa* (Ait.) Pursh.),
hazel (*Corylus cornuta* Marsh.), horsetail species (*Equisetum pratense* Ehrh.
and *E. sylvaticum* L.) Labrador-tea (*Ledum groenlandicum* Oeder), marsh-
marigold (*Caltha palustris* L.), ostrich-fern (*Matteuccia struthiopteris* (L.)
Torado), palmate-leaved coltsfoot (*Petasites palmatus* Ait. Gray), pine (*Pinus
banksiana* Lamb.), raspberry (*Rubus idaens* L.), reindeer lichen (*Cladonia*
spp.), Schreber's moss (*Pleurozium schreberi* (Bird) M.H.), snowberry (*Sym-
phoricarpos albus* (L.) Blake and *S. occidentalis* Hook.), speckled alder (*Alnus
rugosa* (Du Roi) Spreng.), swamp birch (*Betula glandulosa* Michx.), tall sedges
(*Carex aquatilis* Wahlenb. and *C. lacustris* Willd.), tamarack (*Larix laricina*
(Du Roi) K. Koch), teaberry (*Gaultheria procumbens* L.), twinflower (*Linnaea
borealis* L.), wavy dicranum (*Dicranum rugosum* Brid.), white cedar (*Thuja
occidentalis* L.), white spruce (*Picea glauca* (Moench) Voss), wild sarsaparilla
(*Aralia nudicaulis* L.).

References

AGNEW, A. D. Q. 1961. The ecology of *Juncus effusus* L. in North Wales. J. Ecol. 49:83-102.

AICHINGER, E. 1949. Grundzüge der forstlichen Vegetationskunde. Ber. forstwirtsch. Arbeitsgem. Hochschule f. Bodenkulter, Vienna.

AICHINGER, E. 1951. Soziationen, Assoziationen und Waldentwicklungstypen. Angew. Pflanzensoziol. (Vienna) 1:21-68.

ALECHIN, W. W. 1926. Was ist eine Pflanzengesellschaft? Ihr Wesen und ihr Wert als Ausdruck des sozialen Lebens der Pflanzen. Repert. Spec. nov. 37:1-50.

ASHBY, E. 1948. Statistical ecology: II. A reassessment. Botan. Rev. 14:222-234.

ASHTON, P. S. 1965. Some problems arising in the sampling of mixed rain forest communities for floristic studies. p. 235-240. *In* Symposium on Ecological Research in Humid Tropics Vegetation (1963), sponsored by Gov. of Sarawak and UNESCO Science Coopor. Office for SE Asia. Tokyo Press Co. Ltd., Itabashi, Tokyo. 376 p.

ATKINSON, I. A. E. 1969. Ecosystem development on some Hawaiian lava flows. Ph. D. Dissertation. University of Hawaii, Honolulu, Hawaii. 191 p.

ATKINSON, I. A. E. 1970. Successional trends in the coastal and lowland forest of Mauna Loa and Kilauea Volcanoes, Hawaii. Pacific Science 24:387-400.

AUSTIN, M. P., and L. ORLOCI. 1966. Geometric models in ecology: II. An evaluation of some ordination techniques. J. Ecol. 54:217-222.

BAILEY, A. W., and C. E. POULTON. 1968. Plant communities and environmental relationships in a portion of the Tillamook burn, Northwestern Oregon. Ecology 49:1-13.

BAKER, H. G. 1952. The ecospecies—a prelude to a discussion. Evolution 6:61-68.

BARBOUR, M. G. 1970. Is any angiosperm an obligate halophyte? Am. Midland Naturalist 84:65-120.

BARBOUR, M. G., and R. T. LANGE. 1967. Seed populations in some natural Australian topsoils. Ecology 48:153-155.

BARTHOLOMEW, B. 1970. Bare zone between California shrub and grassland communities: The role of animals. Science 170:1210-1212.

BARTHOLOMEW, B. 1971. Role of animals in suppression of herbs by shrubs. Science 173-463.

BATCHELER, C. L. 1971. Estimation of density from a sample of joint point and nearest-neighbor distances. Ecology 52:703-709.

BEALS, E. 1960. Forest bird communities in the Apostle Islands of Wisconsin. Wilson Bull. 72:156-181.

BEARD, J. S. 1946. The Mora forests in Trinidad, British West Indies. J. Ecol. 33:173-192.

BECKING, R. W. 1957. The Zürich-Montpellier School of Phytosociology. Botan. Rev. 23:411-488.

BENNINGHOFF, W. S. 1966. The relevé method for describing vegetation. Michigan Botanist 5:109-114.

BENNINGHOFF, W. S., and K. J. CRAMER. 1963. Phytosociological analysis of aspen communities on three site classes for *Populus grandidentata* in western Cheboygan County, Michigan. Vegetatio 11:253-264.

BESCHEL, R. E., and P. J. WEBBER. 1962. Gradient analysis in swamp forests. Nature 194:207-209.

BILLINGS, W. D. 1965. Plants and the ecosystem. Fundamentals of Botany Series. Macmillan, London. 154 p.

BITTERLICH, W. 1948. Die Winkelzählprobe. Allg. Forst– u. Holzwirtsch. Ztg. 59:4-5.

BLISS, L. C. 1963. Alpine plant communities of the Presidential Range, New Hampshire. Ecology 44:678-697.

BONNER, J. 1950. The role of toxic substances in the interactions of higher plants. Botan. Rev. 16:51-65.

BORMANN, F. H. 1953. The statistical efficiency of sample plot size and shape in forest ecology. Ecology 34:474-487.

BORMANN, F. H., and M. F. BUELL. 1964. Old-age stand of hemlock-northern hardwood forest in central Vermont. Bull. Torrey Botan. Club 91:451-465.

BÖRNER, H. 1971. German research on allellopathy, p. 52-54. In Biochemical Interactions among Plants, National Academy of Sciences, Washington, D. C. 134 p.

BOURDEAU, P. F., and H. J. OOSTING. 1959. The maritime live oak forest in North Carolina. Ecology 40:148-152.

BOYSEN-JENSEN, P. 1949. Causal plant-geography. D. Kgl. Danske Vidensk. Selsk., Biol. Meddel 21:1-19.

BRANSON, F. A., R. F. MILLER, and I. S. McQUEEN. 1970. Plant communities and associated soil and water factors on shale-derived soils in northeastern Montana. Ecology 51:391-407.

BRAUN, E. LUCY. 1950. Deciduous forests of Eastern North America. Blakiston Co., Philadelphia. 596 p.

BRAUN, J. 1915. Les Cévennes méridionales (massif de l'Aigoual). Arch. Sci. phys. et nat. Genève, 4. Ser. 39/50. 207 p.

BRAUN-BLANQUET, J. 1928. Pflanzensoziologie. Springer-Verlag, 1st ed., Berlin, 1928, 2nd ed., Vienna, 1951. 631 p. 3rd ed., Vienna, New York, 1964. 865 p.

BRAUN-BLANQUET, J. 1932. Plant sociology; the study of plant communities. (Transl. by G. D. Fuller and H. S Conard.) Transl. of 1st ed. of Pflanzensoziologie (1928). McGraw-Hill, New York and London. 438 p.

BRAUN-BLANQUET, J. 1965. Plant sociology: The study of plant communities. (Transl. rev. and ed. by C. D. Fuller and H. S. Conard.) Hafner, London. 439 p.

BRAY, J. R., and J. T. CURTIS. 1957. An ordination of the upland forest communities of southern Wisconsin. Ecol. Monographs 27:325-349.

BROCKMANN-JEROSCH, H., and E. RÜBEL. 1912. Die Einteilung der Pflanzengesellschaften nach ökologisch-physiognomischen Gesichtspunkten. Wilhelm Engelmann, Leipzig. 68 p.

BROOKHAVEN SYMPOSIUM. 1969. Diversity and stability in ecological systems. Brookhaven Symposia in Biology No. 22. 264 p. (Reprod. by National Technical Inform. Service, Springfield, Va. 22151.)

BUELL, M. F., and J. E. CANTLON. 1950. A study of two communities of the New Jersey Pine Barrens and a comparison of methods. Ecology 31:567-586.

BÜLOW, K. von. 1929. Allgemeine Moorgeologie. Gebr. Borntraeger. 308 p.

BURGER, D. 1972. Forest site classification in Canada. Mitt. Ver. fur Forstl. Standortskunde u. Forstpflanzenzüchtung 21:20-36.

CAIN, S. A. 1938. The species-area curve. Am. Midland Naturalist 19:573-581.

CAIN, S. A. 1943. Sample plot technique applied to alpine vegetation in Wyoming. Am. J. Botany 30:240-247.

CAIN, S. A. 1947. Characteristics of natural areas and factors in their development. Ecol. Monographs 17:185-200.

CAIN, S. A., and G. M. DE O. CASTRO. 1959. Manual of vegetation analysis. Harper, New York. 325 p.

CAJANDER, A. K. 1909. Über Waldtypen. Acta Forest. Fenn. 1:1-175.

CAJANDER, A. K. 1925. Der gegenseitige Kampf in der Pflanzenwelt. Veröff. Geobot. Inst. Rübel, Zürich, 3. Festschr. Schröter:665-675.

CANFIELD, R. 1941. Application of the line interception method in sampling range vegetation. J. Forestry 39:388-394.

CAPLENOR, D. 1968. Forest composition on loessal and non-loessal soils in west-central Mississippi. Ecology 49:322-331.

CAPUTA, J. 1948. Untersuchungen über die Entwicklung einiger Gräser und Kleearten in Reinsaat und Mischung. Diss. ETH, Zürich. 137 p.

CATANA, H. J. 1963. The wandering quarter method of estimating population density. Ecology 44:349-360.

CATTELL, R. B. 1952. Factor analysis. Harper & Bros., New York. 462 p.

CESKA, A., and H. ROEMER. 1971. A computer program for identifying species-releve´ groups in vegetation studies. Vegetatio 23:255-277.

CLAPHAM, A. R. 1932. The form of the observational unit in quantitative ecology. J. Ecol. 20:192-197.

CLAUSEN, J., D. D. KECK and W. M. HIESEY. 1948. Experimental studies on the nature of species: III. Environmental responses of climatic races of *Achillea*. Carnegie Inst. Washington Publ. No. 581:1-129.

CLEMENTS, F. E. 1905. Research methods in ecology. University Publ. Co., Lincoln. 334 p.

CLEMENTS, F. E. 1916. Plant succession. An analysis of the development of vegetation. Carnegie Inst., Washington. 512 p.

CLEMENTS, F. E. 1928. Plant succession and indicators. H. W. Wilson Co., New York. 453 p.

CLEMENTS, F. E., J. E. WEAVER, and H. C. HANSON. 1929. Plant competition. An analysis of community functions. Carnegie Inst., Washington, Publication No. 398. 340 p.

COLE, L. C. 1949. The measurement of interspecific association. Ecology 30:411-424.

COLE, L. C. 1957. The measurement of partial interspecific association. Ecology 38:226-233.

CONWAY, V. M. 1962. The bogs of central Minnesota. Ecol. Monographs 19:173-206.

COOPER, W. S. 1913. The climax forest of Isle Royale, Lake Superior, and its development. Botan. Gaz. 55:1-235.

COTTAM, G. 1949. The phytosociology of an oak woods in southwestern Wisconsin. Ecology 30:271-287.

COTTAM, G. 1955. Correction for various exclusion angles in the random pairs method. Ecology 36:767.

COTTAM, G., and J. T. CURTIS. 1949. A method for making rapid surveys of woodlands by means of pairs of randomly selected trees. Ecology 30:101-104.

COTTAM, G., and J. T. CURTIS. 1956. The use of distance measures in phytosociological sampling. Ecology 37:451-460.

COTTAM, G., J. T. CURTIS, and B. W. HALE. 1953. Some sampling characteristics of a population of randomly dispersed individuals. Ecology 34:741-757.

COWLES, H. C. 1899. The ecological relations of the vegetation on the sand dunes of Lake Michigan. Botan. Gaz. 27:95-117, 167-202, 281-308, 361-391.

CURTIS, J. T. 1955. A prairie continuum in Wisconsin. Ecology 36:558-566.

CURTIS, J. T. 1959. The vegetation of Wisconsin. An ordination of plant communities. Univ. of Wisconsin Press, Madison. 657 p.

CURTIS, J. T., and H. C. GREENE. 1949. A study of relic Wisconsin prairies by the species-presence method. Ecology 30:83-92.

CURTIS, J. T., and R. P. McINTOSH. 1950. The interrelations of certain analytic and synthetic phytosociological characters. Ecology 31:434-455.

CURTIS, J. T., and R. P. McINTOSH. 1951. An upland forest continuum in the prairie-forest border region of Wisconsin. Ecology 32:476-496.

DAGET, P., and J. POISSONET. 1969. Contribution a l'etude des herbages des plateaux basaltiques de l'ouest du Cantal (Reconnaissance). Centre National de la Recherche Scientifique, Montpellier, France. Document No. 16. 117p.

DAGET, P., and J. POISSONET. 1971. Principes d'une technique d'analyse quantitative de la vegetation des formations herbacees. p. 85-100. *In* P. DAGET (ed.) Méthodes d'inventaire phyto-écologique et agronomique des prairies permantes. Centre National de la Recherche Scientifique, Montpellier, France. Document No. 56. 206 p.

DAGNELIE, P. 1960. Contribution à l'étude des communautes végétales par l'analyse factorielle. Bull. Serv. Carte phytogéogr. Sér. B, 5:7-71, 93-195.

DAHL, E. 1960. Some measures of uniformity in vegetation analysis. Ecology 41:805-808.

DANSEREAU, P. 1957. Biogeography, an ecological perspective. The Ronald Press, New York. 394 p.

DANSEREAU, P. 1963. The barefoot scientist. Colorado Quarterly 12:113.

DANSEREAU, P., et al. 1968. The continuum concept of vegetation: responses. Botan. Rev. 34:253-332.

DARWIN, C. R. 1859. On the origin of species by natural selection, or the preservation of favoured races in the struggle for life. John Murray, London. 251 p.

DAUBENMIRE, R. F. 1952. Forest vegetation of northern Idaho and adjacent Washington, and its bearing on concepts of vegetation classification. Ecol. Monographs 22:301-330.

DAUBENMIRE, R. F. 1959. Canopy coverage method of vegetation analysis. Northwest Sci. 33:43-64.

DAUBENMIRE, R. F. 1962. Plants and environment: A textbook of plant autecology. 2nd ed. John Wiley & Sons, New York. 422 p. 3rd ed. 1974.

DAUBENMIRE, R. F. 1966. Vegetation: identification of typical communities. Science 151:291-298.

DAUBENMIRE, R. F. 1968. Plant communities: A textbook of plant synecology. Harper & Row, New York. 300 p.

DIELS, L., and F. MATTICK. 1908. Pflanzengeographie. Samml. Göschen Nr. 389-389a. 1958: Walter de Gruyter, Berlin and Leipzig. 160 p.

DIERSCHKE, H. 1970. Zur Aufnahme und Darstellung phänologischer Erscheinungen in Pflanzengesellschaften. p. 291-311. In Tüxen (ed.) "Grundfragen und Methoden in der Pflanzensoziologie," Bericht über das Intern. Symposium des Intern. Vereins für Vegetationskunde. W. Junk, Den Haag. Revised edition, E. van der Maarel and R. Tüxen, eds., 1972.

DILWORTH, J. R., and J. F. BELL. 1972. Variable probability sampling—variable plot and three-P. A Pocket Book, O. S. U. Book Stores, Inc. Corvallis, Oregon. 129 p.

DOMIN, K. 1923. Is the evolution of the earth's vegetation tending towards a smaller number of climatic formations? Acta Botanica Bohemica 2:54-60.

DRUDE, O. 1896. Deutschlands Pflanzengeographie. Handb. Deut. Landes-u. Volkskunde 4:502.

DRUDE, O. 1902. Grundzüge der Pflanzenverbreitung. Vol. VI In A. Engler and O. Drude (eds.) Die Vegetation des Erde. Wilhelm Engelmann, Leipzig. 671 p.

DRUDE, O. 1913. Die Ökologie der Pflanzen. Die Wissenschaft, F. Vieweg, Braunschweig 50. 308 p.

DUCKER, S. C., W. T. WILLIAMS, and G. N. LANCE. 1965. Numerical classification of the Pacific forms of *Chlorodesmis* (Chlorophyta). Australian J. Botany 13:489-499.

DU RIETZ, G. E. 1921. Zur methodologischen Grundlage der modernen Pflanzensoziologie. Akadem. Abh. Wien. A. Holzhausen. 272 p.

DU RIETZ, G. E. 1930. Vegetationsforschung auf soziationsanalytischer Grundlage. Abderhalden, Handb. biol. Arbeitsmeth. 11:293-480.

EBER, W. 1972. Über das Lichtklima von Wäldern bei Göttingen und seinen Einfluss auf die Bodenvegetation. Scripta Geobotanica. 3:150.

EBERHARDT, E., D. KOPP, and H. PASSARGE. 1967. Standorte und Vegetation des Kirchleerauer Waldes im Schweizerischen Mittelland p. 13-134. *In* H. Ellenberg (ed.) Ecological and pedological methods of forest site mapping. Veröffentl. geobot. Inst. ETH, Stiftg. Rübel, Zürich, No. 39, 298 p.

EGLER, F. E. 1947. Arid southeast Oahu vegetation, Hawaii. Ecol. Monographs 17:383-435.

EGLER, F. E. 1954. Vegetation science concepts: I. Initial floristic composition, a factor in old-field vegetation development. Vegetatio 4:412-417.

ELLENBERG, H. 1939. Über die Zusammensetzung, Standort und Stoffproduktion bodenfeuchter Eichen– und Buchen–Mischwaldgesellschaften Nordwestdeutschlands. Mitt. Florist. Soziol. Arbeitsgemeinschaft Niedersachsen 5:1-135.

ELLENBERG, H. 1950. Unkraut–Gemeinschaften als Zeiger für Klima und Boden. Eugen Ulmer, Ludwigsburg. 141 p.

ELLENBERG, H. 1952. Wiesen und Weiden und ihre standörtliche Bewertung. Eugen Ulmer, Ludwigsburg. 143 p.

ELLENBERG, H. 1953. Physiologisches und ökologisches Verhalten derselben Pflanzenarten. Ber. Deut. Botan. Ges. 65:351-362.

ELLENBERG, H. 1954. Über einige Fortschritte der kausalen Vegetationskunde. Vegetatio 5-6:199-211.

ELLENBERG, H. 1956. Aufgaben und Methoden der Vegetationskunde. Eugen Ulmer, Stuttgart. 136 p.

ELLENBERG, H. 1959. Typen tropischer Urwälder in Peru. Schweiz. Z. Forstw. 110:169-187.

ELLENBERG, H. 1963. Vegetation Mitteleuropas mit den Alpen. Eugen Ulmer, Stuttgart. 943 p. 2nd ed. 1974.

ELLENBERG, H. 1967. Ecological and pedological methods of forest site mapping (in German, English summaries). Veröffentl. geobot. Inst. ETH, Stiftg. Rübel, Zürich, No. 39. 298 p. Including six maps at 1:10,000.

ELLENBERG, H. 1968*a*. Wege der Geobotanik zum Verstandnis der Pflanzendecke. Naturwissenschaften 55:462-470.

ELLENBERG, H. 1968*b*. Sichtlochkarten zur Ordnung, Klassifikation und Analyse pflanzensoziologischer Waldaufnahmen. p. 163-175. *In* R. Tüxen (ed.) Pflanzensoziologische Systematik. Bericht Intern. Symposion in Stolzenau (1964). W. Junk, Den Haag.

ELLENBERG, H. 1973. Die Ökosysteme der Erde: Versuch einer Klassifikation der Ökosysteme nach funktionalen Gesichtspunkten. p. 235-265. *In* H. Ellenberg (ed.) Ökosystemforschung. Springer-Verlag, Berlin, Heidelberg, New York. 280 p.

ELLENBERG, H., and G. CRISTOFOLINI. 1964. Sichtlochkarten als Hilfsmittel zur Ordnung und Auswertung von Vegetationsaufnahmen. Ber. geobot. Inst. ETH, Stiftg. Rübel, Zürich 35: 124-134.

ELLENBERG, H., and F. KLÖTZLI. 1967. Vegetation und Bewirtschaftung des Vogelreservates Neeracher Riet. Ber. geobot. Inst. ETH, Stiftg. Rübel, Zürich 37:88-103. With 1:2,500 map.

ELLENBERG, H., and F. KLÖTZLI. 1972. Waldgesellschaften und Waldstandorte der Schweiz. Mitt. Eidgen. Anstalt f.d. forstl. Versuchswesen (Zürich) 48:589-800.

ELLENBERG, H., and D. MUELLER-DOMBOIS. 1967*a*. Tentative physiognomic-ecological classification of plant formations of the earth. Ber. geobot. Inst. ETH, Stiftg. Rübel, Zürich, 37:21-55. (Republished 1969 as UNESCO report SC/WS/269 in slightly modified form, entitled " a framework for a classification of world vegetation," Paris. 26 p. Finalized UNESCO publication 1973. International classification and mapping of vegetation. Ecology and Conservation series No. 6, 93 p. and chart with map symbols.)

ELLENBERG, H., and D. MUELLER-DOMBOIS. 1967*b*. A key to Raunkiaer plant life forms with revised subdivisions. Ber. geobot. Inst. ETH, Stiftg. Rubel, Zurich, 37:56-73.

EVANS, F. C. 1956. Ecosystem as the basic unit in ecology. Science 123: 1127-1128.

EVANS, G. C., and D. E. COOMBE. 1959. Hemispherical and woodland canopy photography and the light climate. J. Ecol. 47:103-113.

EVENARI, M. 1961. Chemical influences of other plants (allelopathy). Handbuch der Pflanzenphysiologie 16:691-736.

FEKETE, G., and J. S. LACZA. 1970. A survey of the plant life-form systems and the respective research approaches: II. Annales Hist.-Natur. Musei Nation. Hungarici, Pars Botanica. 62:115-127. III (1971). 63:37-50. IV. (1972). 64:53-62.

FLAHAULT, C., and C. SCHRÖTER. 1910. Phytogeographische Nomenklatur. Proceed. Intern. Bot. Congress, Brussels. Extrait du compte-rendu:1-29.

FOSBERG, F. R. 1961. A classification of vegetation for general purposes. Trop. Ecol. 2:1-28. (Republished in modified form in G. F. Peterken. 1967. Guide to the check sheet for IBP areas. IBP Handbook No. 4. Blackwell Scientific Publications, Oxford and Edinburgh, p. 73-120.)

FOSBERG, F. R. 1965a. Vegetation as a geological agent in deltas, p. 227-233. In Scientific Problems of the Humid Tropical Zone Deltas and their Implications. Proceedings of the Dacca Symposium (1964). Published by UNESCO, Paris. 422 p.

FOSBERG, F. R. 1965b. The entropy concept in ecology. p. 157-163. In Proceedings of Symposium on Ecological Research in Humid Tropics Vegetation, Kuching, Sarawak, July 1963. Published by UNESCO Science Cooperation Office for Southeast Asia. Tokyo Press Co., Itabashi, Tokyo. 376 p.

GAMS, H. 1918. Prinzipienfragen der Vegetationsforschung. Ein Beitrag zur Begriffsklärung und Methodik der Biocoenologie. Vierteljahrsschr. Naturforsch. Ges. Zürich 63:293-493.

GARRISON, G. A. 1949. Uses and modifications for the "moosehorn" crown closure estimator. J. For. 47:733-735.

GATES, F. C. 1942. The bogs of northern lower Michigan. Ecol. Monographs 12:213-254.

GAUSE, G. F., and A. A. WITT. 1935. Behavior of mixed populations and the problem of natural selection. Am. Naturalist 69:596-609.

GAUSSEN, H., P. LEGRIS, M. VIART, and L. LABROUE. 1964. International map of the vegetation, Ceylon. Special sheet, Survey Dept. of Ceylon (Sri Lanka).

GERARD, R. W. 1965. Intelligence, information and education. Science 148:762-765.

GERRESHEIM, K. 1971. The Serengeti photo-ecological project. Serengeti Res. Hist. Public. No. 83. 10 p. (University of Dar Es Salaam, Tanzania).

GILMOUR, J. S. L., and J. HESLOP-HARRISON. 1954. The deme terminology and the units of micro-evolutionary change. Genetica 27:147-161.

GINZBERGER, A., and J. STADLMANN. 1939. Pflanzengeographisches Hilfsbuch. Springer-Verlag, Vienna. 272 p.

GLEASON, H. A. 1920. Some application of the quadrat method. Bull. Torrey Botan. Club 47:21-33.

GLEASON, H. A. 1926. The individualistic concept of the plant association. Bull. Torrey Botan. Club 53:7-26.

GLEASON, H. A. 1939. The individualistic concept of the plant association. Am. Midland Naturalist 21:92-110.

GOOD, R. D. 1964. A geography of flowering plants. 3rd ed. John Wiley & Sons, New York. 518 p.

GOODALL, D. W. 1952. Some considerations in the use of point quadrats for the analysis of vegetation. Australian J. Sci. Res., Series B 5:1-41.

GOODALL, D. W. 1953a. Objective methods for the classification of vegetation: I. The use of positive interspecific correlation. Australian J. Botany 1:39-62.

GOODALL, D. W. 1953b. Objective methods for the classification of vegetation: II. Fidelity and indicator value. Australian J. Botany 1:434-456.

GOODALL, D. W. 1953c. Point-quadrat methods for the analysis of vegetation. Australian J. Botany 1:457-461.

GOODALL, D. W. 1954a. Objective methods for the classification of vegetation: III. An essay in the use of factor analysis. Australian J. Botany 2:302-324.

GOODALL, D. W. 1954b. Vegetational classification and vegetational continua. Angew. Pflanzensoziologie (Vienna). Festschrift Aichinger 1:168-182.

GOODALL, D. W. 1970. Statistical plant ecology. Ann. Rev. of Ecol. & Systematics 1:99-124.

GOULDEN, C. E. 1969. Temporal changes in diversity. p. 96–102. In Diversity and stability in ecological systems. Brookhaven Symposia in Biology No. 22. 264 p. (Reproduced by Nat. Tech. Inform. Service, Springfield, Va. 22151.)

GOWER, J. C. 1967. A comparison of some methods of cluster analysis. Biometrics 23:623-637.

GRAEBNER, P. 1925. Die Heide Norddeutschlands. 2nd ed. Vol. V in the series: Die Vegetation der Erde, A. Engler and O. Drude (ed.). Wilhelm Engelmann, Leipzig. 277 p.

GRANDTNER, M. M. 1966. La Vegetation du Quebec Meridional. University of Laval Press, Quebec. 216 p.

GREIG-SMITH, P. 1964. Quantitative plant ecology. 2nd ed. Butterworths, London. 256 p. 1st ed. 1957.

GREIG-SMITH, P. 1965. Notes on the quantitative description of humid tropical forest. p. 227-234. In Symposium on ecological research in humid tropics vegetation, Kuching, Sarawak (1963). Sponsored by Gov. of Sarawak and UNESCO Science Cooperation Office for Southeast Asia. Tokyo Press Co., Itabashi, Tokyo. 376 p.

GRISEBACH, A. 1866. Der gegenwärtige Standpunkt der Geographie der Pflanzen. Behm's geogr. Jahrb. I:373-402. (Reprint 1880 in Ges. Abhandl. u. kleine Schriften:307-334. Wilhelm Engelmann, Leipzig.)

GRISEBACH, A. 1872. Die Vegetation der Erde nach ihrer klimatischen Anordnung. Wilhelm Engelmann, Leipzig, Vol. I, 603 p. Vol II, 635 p.

GRODZINSKIJ, A. M. 1973. Die Grundlagen der chemischen Wechselbeziehungen der Pflanzen (Ukrain). Akad. Wiss., Ukrain, SSR. 206 p.

GROSENBAUGH, L. R. 1952. Plotless timber estimates—new, fast, easy. J. For. 50:32-37.

GRUMMER, G. 1953. Die gegenseitige Beeinflussung höherer Pflanzen. Allelopathie. Gustav Fischer, Jena. 162 p.

HABEK, J. R. 1968. Forest succession in the Glacier Park cedar-hemlock forest. Ecology 49:872-880.

HANAWALT, R. B. 1971. Inhibition of annual plants by *Arctostaphylos*, p. 33-38. *In* Biochemical interactions among plants. National Academy of Sciences, Washington, D. C. 134 p.

HANSON, H. C., and E. D. CHURCHILL. 1961. The plant community. Reinhold Publishing Corp., New York. 218 p.

HARPER, J. L. 1969. The role of predation in vegetational diversity. p. 48-62. *In* Diversity and stability in ecological systems. Brookhaven Symposia in Biology No. 22. 264 p.

HARRIS, G. A. 1967. Some competitive relationships between *Agropyron spicatum* and *Bromus tectorum*. Ecol. Monographs 37:89-111.

HAUFF, R., G. SCHLENKER and G. A. KRAUSS. 1950. Zur Standortsgliederung im nördlichen Oberschwaben. Allg. Forst-und Jagdztg. 122:3-28.

HAUG, P. T. 1970. Succession on old fields. A review. M. Sc. Thesis. Colorado State University, Fort Collins, Colorado. 473 p.

HEIMANNS, J. 1954. L'accessibilité, terme nouveau en phytogéographie. Vegetatio 5-6:142-146.

HIESEY, W. M., M. A. NOBS, and O. BJÖRKMAN. 1971. Experimental studies on the nature of species. Carnegie Inst. of Washington Public. 628. 213 p.

HILLS, A. 1960. Regional site research. Forestry Chronicle 36:401-423.

HILLS, A. 1961. The ecological basis for land-use planning. Ontario Dept. of Lands and Forests. Tech. Series Research Rep. No. 46. 204 p.

HORVAT, J. 1954. Pflanzengeographische Gliederung Südosteuropas. Vegetatio 5-6:434-447.

HORVAT, J., V. GLAVAČ, and H. ELLENBERG. 1974. Vegetation Südosteuropas (with English summary). Gustav Fischer, Stuttgart. 864 p.

HUBER, H. 1955. Über Verbreitung und Standortsansprüche kalkfliehender Moose in der Umgebung Basels und ihre Bedeutung mit Hilfe statistischer Prüfverfahren. Ber. Schweiz. bot. Ges. 65:431-458.

HULT, R. 1881. Försok till analytisk behandling af växtformationerna. Meddel. Soc. Fauna Flora Fennica 8:1-155.

HUMBOLDT, A. von. 1805. Essay sur la Géographie des Plantes. Par Al. de Humboldt et A. Bonpland, redigé par Al. de Humboldt. Levrault, Schoell et Cie, Paris. 155 p.

HUMBOLDT, A. von. 1806. Ideen zu einer Physiognomik der Gewächse. Cotta, Stuttgart. 28 p.

JACCARD, P. 1901. Étude comparative de la distribution florale dans une portion des Alpes et du Jura. Bull. Soc. Vaud. Sc. Nat. 37:547-579.

JACCARD, P. 1912. The distribution of the flora of the alpine zone. New Phytol. 11:37-50.

JACCARD, P. 1928. Die statistisch-floristische Methode als Grundlage der Pflanzensoziologie. In Abderhalden, Handb. biol. Arbeitsmeth. 11:165-202.

JANZEN, D. H. 1971. Seed predation by animals. Ann. Rev. Ecol. & Systematics 2:465-492.

JENNY, H. 1941. Factors of soil formation. McGraw-Hill, New York. 281 p.

KEEVER, CATHERINE. 1950. Causes of succession on old fields of the Piedmont, North Carolina. Ecol. Monographs 20:229-250.

KELLER, B. A. 1923. Die Pflanzenwelt der russischen Steppen, Halbwüsten und Wüsten. Ökologische und phytosoziologische Studien. Russ. with summary in German (Woronesch). 835 p.

KERNER v. MARILAUN, A. 1863. Das Pflanzenleben der Donaulander. 2nd ed. by F. Vierhapper, 1929. Univ.-Verlag Wagner, Innsbruck. 452 p.

KERSHAW, K. A. 1964. Quantitative and dynamic ecology. Edward Arnold Publishing Co. Ltd., London. 183 p.

KERSHAW, K. A. 1968. A survey of the vegetation in Zaria Province, N. Nigeria. Vegetatio 15:244-268.

KLAPP, E. 1929. Thüringische Rhönhuten. Wiss. Arch. Landwirtsch. Abt. Pflanzenbau 2:704-786.

KLAPP, E. 1951. Leistung, Bewurzelung und Nachwuchs einer Grasnarbe unter verschieden häufiger Mahd und Beweidung. Zeitschr. f. Acker- und Pflanzenbau 93:269-286.

KNAPP, G., and R. KNAPP. 1954. Über Möglichkeiten der Durchsetzung und Ausbreitung von Pflanzenindividuen auf Grund verschiedener Wuchsformen. Ber. Deut. Botan. Ges. 67:410-419.

KNAPP, R. 1954. Experimentelle Soziologie der höheren Pflanzen. Eugen Ulmer, Stuttgart. 202 p. 2nd ed. 1967. Experimentelle Soziologie und gegenseitige Beeinflussung der Pflanzen. 266 p.

KNAPP, R. 1971. Einführung in die Pflanzensoziologie. Eugen Ulmer, Stuttgart. 388 p.

KNIGHT, D. H., and O. L. LOUCKS. 1969. A quantitative analysis of Wisconsin forest vegetation on the basis of plant function and gross morphology. Ecology 50:219-234.

KOPP, D., and H. HURTIG. 1965. Die forstliche Standortserkundung als Beitrag zu einer standörtlich-kartographischen Inventur der Kulturlandschaft, dargestellt am Beispiel des nordostdeutschen Tieflandes. Arch. Naturschutz u. Landschaftsforschung 5:3-25.

KÖPPEN, W. 1936. Das geographische System der Klimate. Vol. 1, Part C. In W. Koppen and R. Geiger, Handbuch der Klimatologie. Gebr. Borntraeger, Berlin.

KRAJINA, V. J. 1933. Die Pflanzengesellschaften des Mlynica-Tales in den Vysoke Tatry (Hohe Tatra). Mit besonderer Berücksichtigung der ökologischen Verhältnisse. Botan. Centralbl., Beih., Abt. II, 50:774-957; 51:1-224.

KRAJINA, V. J. 1960. Can we find a common platform for the different schools of forest-type classification? Silva Fennica (Helsinki) 105:50-59.

KRAJINA, V. J. 1965. Biogeoclimatic zones and biogeocoenoses of British Columbia. Ecology of Western North America (Published by Dept. of Botany, Univ. of British Columbia) 1:1-17.

KRAJINA, V. J. 1969. Ecology of forest trees in British Columbia. Ecology of Western North America (Published by the Dept. of Botany, Univ. of British Columbia) 2:1-147. Including map of "Biogeoclimatic Zones."

KRAJINA, V. J., L. ORLOCI, and R. C. BROOKE. 1962. Standard forms used for ecosystem studies, Appendix K:83-104. In Ecology of the forests of the Pacific Northwest. 1962 Progress Report of NRC Grant T-92, Dept. of Botany, Univ. of British Columbia. 104 p.

KRAUSE, W. 1952. Das Mosaik der Pflanzengesellschaften und seine Bedeutung für die Vegetationskunde. Planta 41:240-289.

KÜCHLER, A. W. 1954. Vegetation maps at the scale from 1:200,000 to 1:1,000,000. Paris. 8th Intern. Bot. Congress. Rapports et Communications, Section 7:107-122.

KÜCHLER, A. W. 1964. Potential natural vegetation of the conterminous United States. Manual to accompany the map. Am. Geogr. Soc., Special Publication No. 36. 116 p. Revised ed. of map, 1965.

KÜCHLER, A. W. 1966. International bibliography of vegetation maps. Vol. II: Europe. Univ. of Kansas Library Series. 584 p.

KÜCHLER, A. W. 1967. Vegetation mapping. The Ronald Press Co., New York. 472 p.

KÜCHLER, A. W. 1968. International bibliography of vegetation maps. Vol. III: USSR. Asia and Australia. Univ. of Kansas Library Series. 389 p.

KÜCHLER, A. W. 1970. International bibliography of vegetation maps. Vol. IV. Africa, South America and the World (General). Univ. of Kansas Library Series. 561 p.

KÜCHLER, A. W. 1972. On the structure of vegetation. Ecology 53:196-198.

KÜCHLER, A. W., and JACK McCORMICK. 1965. International bibliography of vegetation maps. Vol. I: North America. Univ. of Kansas Library Series. 453 p.

KÜCHLER, A. W., and J. O. SAWYER, JR. 1967. A study of the vegetation near Chiengmai, Thailand. Transact. Kansas Acad. Sci. 70:281-348. Including 1:30,000 map.

LACZA, J. S., and G. FEKETE. 1969. A survey of the plant life-form systems and the respective research approaches: I. Annales Hist.-Natur. Musei Nation. Hungarici, Pars Botanica. 61:129-139.

LAMBERT, J. M. 1948. A survey of the Rockland-Claxton Level, Norfolk. J. Ecol. 36:120-135.

LAMBERT, J. M., and W. T. WILLIAMS. 1962. Multivariate methods in plant ecology: IV. Nodal analysis. J. Ecol. 50:775-802.

LAUER, E. 1953. Über die Keimtemperatur von Ackerunkräutern und deren Einfluss auf die Zusammensetzung von Unkrautgesellschaften. Flora 140: 551–595.

LEVY, E. E., and E. A. MADDEN. 1933. The point method of pasture analysis. New Zealand Agric. J. 46:267-279.

LEWONTIN, R. C. 1969. The meaning of stability. p. 13-24. In Diversity and stability of ecological systems. Brookhaven Symposia in Biology No. 22. 264 p. (Reprod. by Nat. Tech. Inform. Service, Springfield, Va. 22151).

LIETH, H., and G. W. MOORE. 1971. Computerized clustering of species in phytosociological tables and its utilization for field work. I:403-422. In G. P. Patil, E. L. Pielou, and W. E. Waters (ed.) Spatial Patterns and Statistical Distributions. Penn. State Univ.: Press. 582 p.

LINDSEY, A. A. 1955. Testing the line strip method against full tallies in diverse forest types. Ecology 36:485-495.

LINDSEY, A. A., J. D. BARTON, and S. R. MILES. 1958. Field efficiencies of forest sampling methods. Ecology 39:428-444.

LIPPMAA, T. 1939. The unistratal concept of plant communities (the unions). Amer. Midland Naturalist 21:111-145.

LÜDI, W. 1930. Die Methoden der Suksessionsforschung in der Pflanzensoziologie. In Abderhalden, Handb. biol. Arbeitsmeth. 11:527-728.

MAAREL, E. van der. 1969. On the use of ordination models in phytosociology. Vegetatio 19:21-46.

McCORMICK, J. 1968. Succession. Student Publication of Grad. School of Fine Arts, Univ. of Pennsylvania. VIA 1:22-35, 131-132.

McCORMICK, J., and M. F. BUELL. 1957. Natural revegetation of a plowed field in the New Jersey Pine Barrens. Botan Gaz. 118:261-264.

McINTOSH, R. P. 1963. Ecosystem, evolution and rational patterns of living organisms. Am. Scientist 51:246-267.

McINTOSH, R. P. 1967a. The continuum concept of vegetation. Botan. Rev. 33:130-187.

McINTOSH, R. P. 1967b. An index of diversity and the relation of certain concepts to diversity. Ecology 48:392-404.

McINTOSH, R. P. 1970. Community, competition, and adaptation. Quart. Rev. Biol. 45:259-280.

McMILLAN, C. 1960. Ecotypes and community function. Am. Naturalist 94: 245-255.

McMILLAN, C. 1969. Ecotypes and ecosystem function. Bio-Science 19:131-134.

McNAUGHTON, S. J. 1968. Autotoxic feedback in relation to germination and seedling growth in *Typha latifolia.* Ecology 49:367-369.

McPHERSON, J. K., and C. H. MULLER. 1969. Allelopathic effects of *Adenostoma fasciculatum,* "chamise," in the California chaparral. Ecol. Monographs 39:177-198.

MAJOR, J. 1951. A functional, factorial approach to plant ecology. Ecology 32:392-412.

MAJOR, J. 1969. Historical development of the ecosystem concept. p. 9-22. *In* G. M. Van Dyne (ed.) The ecosystem concept in natural resource management. Academic Press, New York, London. 383 p.

MARGALEFF, R. 1969. Diversity and stability: a practical proposal and a model of interdependence. p. 25-37. *In* Diversity and stability in ecological systems. Brookhaven Symposia in Biology No. 22. 264 p. (Reprod. by Nat. Tech. Inform. Service, Springfield, Va. 22151).

MARR, J. W. 1967. Ecosystems of the east slope of the front range in Colorado. Institute of Arctic and Alpine Research, Univ. of Colorado, Contrib. No. 4. 134 p.

MILLER, R. S. 1967. Pattern and process in competition. Advan. Ecol. Res. 4:1-74.

MOORE, G. W., W. S. BENNINGHOFF, and P. S. DWYER, 1967. A computer method for the arrangement of phytosociological tables. p. 297-299. Proceedings, Ass. for Computing Machinery, National Meeting, 1967. Washington, D.C.

MOORE, J. J., S. J. FITSIMMONS, E. LAMBE, and J. WHITE. 1970. A comparison and evaluation of some phytosociological techniques. Vegetatio 20:1-20.

MORAVEC, J. 1971. A simple method for estimating homotoneity of sets of phytosociological releve's. Folia Geobot. Phytotax. Praha 6:147-170.

MORAVEC, J. 1973. The determination of the minimal area of phytocoenoses. Folia Geobot. Phytotax. Praha 8:23-47.

MORISITA, M. 1954. Estimation of population density by spacing method Mem. Fac. Sci. Kyushu University. E(Biol.) 1:187-197.

MORONEY, M. J. 1954. Facts from figures. Penguin Books Inc., Baltimore. 472 p.

MORRISON, R. G., and G. A. YARRANTON. 1970. An instrument for rapid and precise point sampling of vegetation. Can J. Botany 48:293-297.

MOTYKA, J., B. DOBRZANSKI, and S. ZAWADSKI. 1950. Wstepne badania nad lakami poludniowo-wschodneij Lubelszczyzny (Preliminary studies on meadows in the southeast of the province Lublin. Summary in English). Ann. Univ. M. Curie-Sklodowska, Sec. E. 5(13):367-447.

MUELLER-DOMBOIS, D. 1959. The Douglas-fir forest associations on Vancouver Island in their initial stages of secondary succession. Ph. D. Dissertation. University of British Columbia, Vancouver, B.C. 570 p.

MUELLER-DOMBOIS, D. 1964. The forest habitat types of southeastern Manitoba and their application to forest management. Canadian J. of Botany 42:1417-1444.

MUELLER-DOMBOIS, D. 1965a. Eco-geographic criteria for mapping forest habitats in southeastern Manitoba. Forestry Chronicle (Vancouver, B.C.) 41:188-206.

MUELLER-DOMBOIS, D. 1965b. Initial stages of secondary succession in the coastal Douglas-fir and Western Hemlock zones. Ecology of Western North America (Univ. of Brit. Col.) 1:38-41.

MUELLER-DOMBOIS, D. 1966. The vegetation map and vegetation profiles. p. 391-441. In M. S. Doty and D. Mueller-Dombois, Atlas for bioecology studies in Hawaii Volcanoes National Park. Univ. of Hawaii, Hawaii Botan. Science Paper No. 2. 507 p.

MUELLER-DOMBOIS, D. 1967. Ecological relations in the alpine and subalpine vegetation on Mauna Loa, Hawaii. J. Indian Botan. Soc. 46:403-411.

MUELLER-DOMBOIS, D. 1968. Vegetation cover types, Ruhuna National Park, Ceylon. Smithsonian-Ceylon Ecology Project Mimeo-Rep. No. 13. 8 p.

MUELLER-DOMBOIS, D. 1969. Vegetation map of Ruhuna National Park. 5 sheets at 1:31,680, Survey Dept. of Ceylon (Sri Lanka).

MUELLER-DOMBOIS, D. 1972. Crown distortion and elephant distribution in the woody vegetations of Ruhuna National Park, Ceylon. Ecology 53:208-226.

MUELLER-DOMBOIS, D., and R. G. COORAY. 1968. Effects of elephant feeding on the short-grass covers in Ruhuna National Park, Ceylon. Smithsonian-Ceylon Ecology Project Mimeo-Rep. No. 11. 12 p.

MUELLER-DOMBOIS, D., and C. H. LAMOUREUX. 1967. Soil-vegetation relationships in Hawaiian Kipukas. Pacific Science 21:286-299.

MUELLER-DOMBOIS, D., and H. P. SIMS. 1966. Response of three grasses to two soils and a water table depth gradient. Ecology 47:644-648.

MUELLER-DOMBOIS, D., and G. SPATZ. 1972. The influence of feral goats on the lowland vegetation in Hawaii Volcanoes National Park. Island Ecosystems IRP/IBP Hawaii (Univ. Hawaii). Tech. Rep. 13. 46 p.

MULLER, C. H. 1953. The association of desert annuals with shrubs. Am. J. Botany 40:53-60.

MULLER, C. H. 1966. The role of chemical inhibition (allelopathy) in vegetational composition. Bull. Torrey Botan. Club 93:332-351.

MULLER, C. H. 1969. Allelopathy as a factor in ecological process. Vegetatio 18:248-357.

MULLER, C. H. 1971. Phytotoxins as plant habitat variables. p. 64-72. In Biochemical interactions among plants. National Academy of Sciences, Washington, D.C. 134 p.

MULLER, C. H., and R. del MORAL. 1971. Role of animals in suppression of herbs by shrubs. Science 173:462-263.

MULLER, W. H., and C. H. MULLER. 1956. Association patterns involving desert plants that contain toxic products. Am. J. Botany 43:354-361.

NEWSOME, R. D., and R. L. DIX. 1968. The forests of the Cypress Hills, Alberta and Saskatchewan, Canada. Am. Midland Naturalist 80:118-185.

NICHOLS, G. E. 1917. The interpretation and application of certain terms and concepts in the ecological classification of plant communities. Plant World 20:305-319, 341-353.

ODUM, E. P. 1959. Fundamentals of ecology. W. B. Saunders Co., Philadelphia & London. 546 p. 3rd ed. 1971. 574 p.

ODUM, E. P. 1960. Organic production and turnover in old field successions. Ecology 41:34-49.

ODUM, E. P. 1969. The strategy of ecosystem development. Science 164: 262-270.

OLSEN, C. 1923. Studies in hydrogen-ion concentration of the soil and its significance to the vegetation, especially to the natural distribution of plants. Compt. rend. trav. Labor. Curlsberg 75:1-160.

OLSON, J. S. 1958. Rates of succession and soil changes on southern Lake Michigan sand dunes. Botan. Gaz. 119:125-170.

OOSTING, H. J. 1942. An ecological analysis of the plant communities of Piedmont, North Carolina. Am. Midland Naturalist 28:1-126.

OOSTING, H. J. 1956. The study of plant communities: An introduction to plant ecology. 2nd ed. W. H. Freeman and Co., San Francisco and London. 440 p.

ORLOCI, L. 1966. Geometric models in ecology: I. The theory and application of some ordination methods. J. Ecol. 54:193-215.

OVINGTON, J. D. 1962. Quantitative ecology and the woodland ecosystem concept. Adv. Ecol. Res. 1:103-192.

PAINE, R. T. 1971. The ecologist's Oedipus complex: community structure. Ecology 52:376-377.

PETERKEN, G. F. 1967. Guide to check sheet for IBP areas. IBP Handbook No. 4. Blackwell Scientific Publications, Oxford and Edinburgh. 133 p.

PETERSEN, A. 1927. Die Taxation der Wiesenländereien auf Grund des Pflanzenbestandes. Berlin.

PHILLIPS, E. A. 1964. Field ecology. Heath and Co., Boston. 100 p.

PIELOU, E. C. 1959. The use of point-to-plant distances in the study of the pattern of plant distributions. J. Ecol. 47:607-613.

PIELOU, E. C. 1969. An introduction to mathematical ecology. John Wiley & Sons, Inc., New York. 286 p.

PLATT, J. R. 1964. Strong inference. Science 146:347–353.

POISSONET, P. 1971. Comparaison des résultats obtenus par diverses methodes d'analyses de la végétation dans une prairie permanente. p. 39-71. In Ph. Daget (ed.) Méthodes d'inventair phytc-ecologique et agronomique des prairies permenentes. Centre National de la Recherche Scientifique, Montpellier, France. Documnt No. 56.

POISSONET, P. and J. POISSONET. 1969. Étude comparée de diverses methodes d'analyses de la végétation des formations herbacees denses et permanentes. Centre National de la Recherche Scientifique, Montpellier, France. Document No. 50. 120 p.

POORE, M. E. D. 1955. The use of phytosociological methods in ecological investigations. I. The Braun-Blanquet system. J. Ecol. 43:226-244. II. Practical issues involved in an attempt to apply the Braun-Blanquet system. J. Ecol. 43:245-269. III. Practical applications. J. Ecol. 43:606-651.

POORE, M. E. D. 1962. The method of successive approximation in descriptive ecology. Adv. Ecol. Res. 1:35-68.

POORE, M. E. D. 1964. Integration in the plant community. J. Ecol. 52 (Suppl.):213-226.

POORE, M. E. D. 1968. Studies in Malaysian rain forest: I. The forest on Triassic sediments in Jengka Forest Reserve. J. Ecol. 56:143-196.

PRESTON, F. W. 1969. Diversity and stability in the biological world. p. 1-12. *In* Diversity and stability of ecological systems. Brookhaven Symposia in Biology No. 22. 264 p. (Reprod. by Nat. Tech. Inform. Service, Springfield, Va. 22151).

QUARTERMAN, ELSIE. 1957. Early plant succession on abandoned cropland in the central basin of Tennessee. Ecology 38:300-309.

QUENOUILLE, M. H. 1950. Introductory statistics. Pergamon Press, London. 248 p.

QUENOUILLE, M. H. 1953. The design and analysis of experiment. C. Griffin & Co., London. 356 p.

RAABE, E. W. 1952. Über den "Affinitätswert" in der Pflanzensoziologie. Vegetatio 4:53-68.

RADEMACHER, B. 1959. Gegenseitige Beeinflussung höherer Pflanzen. Handbuch der Pflanzenphysiologie 11:655-706.

RAMENSKY, L. G. 1924. Die Grundgesetzmassigkeiten im Aufbau der Vegetationsdecke. Botan. Centralbl., N.F. 7:453-455.

RAMENSKY, L. G. 1930. Zur Methodik der vergleichenden Bearbeitung und Ordnung von Pflanzenlisten und anderen Objekten, die durch mehrere, verschiedenartig wirkende Faktoren bestimmt werden. Beitr. Biol. Pflanz. 18:269-304.

RAUNKIAER, C. 1913. Formationsstätistiske Undersøgelser paa Skagens Odde. Bot. Tidsskr. Kobenhavn 33:197-228.

RAUNKIAER, C. 1918. Recherches statistiques sur les formations végétales. Det. Kgl. Danske Vidensk. Selsk. Biol. Medd. 1:1-80.

RAUNKIAER, C. 1934. The life forms of plants and statistical plant geography; being the collected papers of C. Raunkiaer, translated into English by H. G. Carter, A. G. Tansley, and Miss Fausboll. Clarendon, Oxford. 632 p.

RAUNKIAER, C. 1937. Plant life forms. Clarendon, Oxford. 104 p.

REICHLE, D. E. (ed.). 1970. Analysis of temperate forest ecosystems. Springer, New York. 304 p.

RICE, E. L. 1964. Inhibition of nitrogen-fixing and nitrifying bacteria by seed plants. Ecology 45:824-837.

RICE, E. L. 1967. A statistical method of determining quadrat size and adequacy of sampling. Ecology 48:1047-1049.

RICE, E. L., and R. W. KELTING. 1955. The species-area curve. Ecology 36:7-11.

RICE, E. L., and W. T. PENFOUND. 1955. An evaluation of the variable-radius and paired-tree methods in the blackjack-post oak forest. Ecology 36:315-320.

RICE, E. L., and W. T. PENFOUND. 1959. The upland forest of Oklahoma. Ecology 40:593-608.

RICE, E. L., W. T. PENFOUND, and L. M. ROHRBAUGH. 1960. Seed dispersal and mineral nutrition in succession in abandoned fields in central Oklahoma. Ecology 41:224-228.

RICHMOND, T. de A., and D. MUELLER-DOMBOIS. 1972. Coastline ecosystems on Oahu, Hawaii. Vegetatio 25:367-400.

RISSER, P. G., and E. L. RICE. 1971. Phytosociological analysis of Oklahoma upland forest species. Ecology 52:940-945.

RISSER, P. G., and P. H. ZEDLER. 1968. An evaluation of the grassland quarter method. Ecology 49:1006-1009.

ROUX, E. R., and M. WARREN. 1963. Plant succession of abandoned fields in central Oklahoma and in the Transvaal Highveld. Ecology 44:576-579.

ROWE, J. S. 1959. Forest regions of Canada. Canada, Dept. of Northern Affairs and Nat. Resources, Forestry Branch Bull. 123. 71 p. Including 1957 map "Forest Classification of Canada." 3rd ed. 1972.

ROWE, J. S., P. G. HADDOCK, G. A. HILLS, V. J. KRAJINA, and A. LINTEAU. 1961. The ecosystem concept in forestry. p. 55-57. *In* V. J. Krajina (ed.) Ecology of the Forests of the Pacific Northwest. 1960 Progress Report. Univ. of British Columbia, Vancouver, B.C. 62 p.

RÜBEL, E. 1922. Geobotanische Untersuchungsmethoden. Gebr. Borntraeger, Berlin. 290 p.

RÜBEL, E. 1933. Geographie der Pflanzen 3. Soziologie, Handwörterbuch d. Naturwiss.Jena:1044-1071.

ST. JOHN, HAROLD. 1972. *Canavalia kauensis* (Leguminosae) a new plant species from the Island of Hawaii. Hawaiian Plant Studies 39. Pacific Science 26:409-414.

SCAMONI, A. 1954. Zur Frage der Charakterarten in der Vegetationskunde. Wiss. Z. Humboldt Univ. Berlin, Math. Nat. Reihe 3:339-343.

SCHIMPER, A. F. W. 1898. Pflanzengeographie auf ökologischer Grundlage. 3rd ed. 1935, by F. C. V. Faber, Vol. I:1-588, Vol. II:589-1612.

SCHLENKER, G. 1950. Forstliche Standortskartierung in Württemberg. Allg. Forstzeitschr. 40/41:1-4.

SCHLENKER, G. 1951. Regionalgesellschaft, Standortsgesellschaften und Bodenvegetationstypen. Mitt. Vereins für forstliche Standortskartierung 1:22-26.

SCHMELZ, D. 1969. Testing the quarter method against full tallies in old-growth forests. Indiana Academy of Science Proceedings 79:138.

SCHMID, E. 1954. Anleitung zu Vegetationsaufnahmen. Vierteljahrsschr. Naturforsch. Ges. Zürich 99, Beih. 1:1-37.

SCHMID, E. 1963. Die Erfassung der Vegetationseinheiten mit floristischen und epimorphologischen Analysen. Ber. Schweiz. Bot. Ges. 73:276-324.

SCHMITHÜSEN, J. 1959. Allgemeine Vegetationsgeographie. Walter de Gruyter & Co., Berlin. 261 p.

SCHMITHÜSEN, J. 1968. Vegetation maps at 1:25 million of Europe, North Asia, South Asia, SW Asia, Australia, N Africa, S Africa, N America, Central America, South America (north part), South America (south part). XVIII:321-346. In Grosses Duden-Lexikon (Bibliographisches Institut A. G., Mannheim).

SCHÖNHAR, S. 1954. Die Bodenvegetation als Standortsweiser. Ein Beitrag zur forstlichen Vegetationskunde Südwestdeutschlands. Allg. Forst- und Jagdztg. 125:259-265.

SCHOUW, J. F. 1823. Grundzüge einer allgemeinen Pflanzengeographie. Berlin. (In Danish, Copenhagen, 1822.)

SCHRÖTER, C. 1904. Das Pflanzenleben der Alpen. Eine Schilderung der Hochgebirgsflora. 2nd ed. 1923-1926. Albert Raustein, Zurich. Vol. I:1-833, Vol. II:845-1288.

SCHRÖTER, C. and O. KIRCHNER. 1896/1902. Die Vegetation des Bodensees, Kommissionsverlag der Schriften des Vereins für Geschichte des Bodensees u. seiner Umgebung (von J. T. Stettner). Lindau, I. 122 p. II. 865 p.

SCHWERDTFEGER, F. 1963. Autökologie. Paul Parey, Hamburg & Berlin. 461 p.

SCHWICKERATH, M. 1940. Die Artmächtigkeit. Repert. Spec. nov. 121:48-52.

SCHWICKERATH, M. 1954. Die Landschaft und ihre Wandlung. Dr. Rudolf Georgi, Aachen. 118 p.

SENDTNER, O. 1854. Vegetationsverhältnisse Südbayerns. Literarisch-artistische Anstalt, Munich. 910 p.

SHANKS, R. E. 1954. Plotless sampling trials in Appalachian forest types. Ecology 35:237-244.

SHELFORD, V. E. 1963. The ecology of North America. Univ. of Illinois Press, Urbana. 610 p.

SHIMWELL, D. W. 1972. The description and classification of vegetation. Univ. of Washington Press, Seattle. 322 p.

SILVA FENNICA. 1960. Forest types and forest ecosystems. Vol. 105, Helsinki. 142 p. (Containing the papers presented at the Symposium on Ecosystems held at the 9th Intern. Bot. Congress in Montreal, 1959).

SLOBODKIN, L. B., and H. L. SANDERS. 1969. On the contribution of environmental predictability to species diversity. p. 82-95. In Diversity and stability in ecological systems. Brookhaven Symposia in Biology No. 22, 264 p. (Reproduced by Nat. Tech. Inform. Service, Springfield, Va. 22151).

SMATHERS, G. A. and D. MUELLER-DOMBOIS. 1972. Invasion and recovery of vegetation after a volcanic eruption in Hawaii. Island Ecosystems IRP/IBP Hawaii (Univ. Hawaii). Tech. Rep. 10, 172 p.

SNOW, L. M. 1913. Progressive and retrogressive changes in the plant associations of the Delaware Coast. Botan. Gaz. 55:45-55.

SOKAL, R. R., and C. D. MICHENER. 1958. A statistical method for evaluating systematic relationships. Univ. Kansas Sci. Bull. 38:1409-1438.

SOKAL, R. R., and P. H. A. SNEATH. 1963. Principles of numerical taxonomy. Freeman, San Francisco. 359 p.

SØRENSEN, T. 1948. A method of establishing groups of equal amplitude in plant sociology based on similarity of species content. Det Kong. Danske Vidensk. Selsk. Biol. Skr. (Copenhagen) 5(4):1-34.

SPATZ, G. 1969. Elektronische Datenverarbeitung bei pflanzensoziologischer Tabellenarbeit. Naturwissenschaften (Heidelberg) 56:470-471.

SPATZ, G. 1970. Pflanzengesellschaften, Leistungen und Leistungspotential von Allgäuer Alpweiden in Abhängigkeit von Standort und Bewirtschaftung. Dissertation (Dr. agr.), Techn. Univ., Munich. 160 p. +20 looseleaf tables and a 1:5000 map.

SPATZ, G. 1972. Eine Möglichkeit zum Einsatz der elektronischen Datenverarbeitung bei der pflanzensoziologischen Tabellenarbeit. p. 251-261. In E. van der Maarel and R. Tuxen (eds.) Ber. Intern. Vereinigung f. Vegetationskunde 1970 (Basic problems and methods in phytosociology). W. Junk, The Hague.

SPATZ, G., and D. MUELLER-DOMBOIS. 1972. Succession patterns after pig digging in grassland communities on Mauna Loa, Hawaii. Island Ecosystems IRP/IBP Hawaii (Univ. Hawaii), Tech. Rep. 15. 44 p.

SPATZ, G. and D. MUELLER-DOMBOIS. 1973. The influence of goats on koa tree reproduction in Hawaii Volcanoes National Park. Ecology 54:870-876.

SPATZ, G., and J. SIEGMUND. 1973. Eine Methode zur tabellarischen Ordination, Klassifikation und ökologischen Auswertung von pflanzensoziologischen Bestandsaufnahmen. Vegetatio 28:1-17.

STEBLER, F. G., and C. SCHRÖTER. 1887. Einfluss des Beweidens auf die Zusammensetzung des Rasens. Landw. Jahrbuch d. Schweiz (Bern) 1:77-190.

STEBLER, F. G., and C. SCHRÖTER. 1892. Versuch einer Übersicht über die Wiesentypen der Schweiz. Landwirtsch. Jb. Schweiz 6:95.

STERN, W. L., and M. F. BUELL. 1951. Life-form spectra of New Jersey pine barrens forest and Minneosta jack pine forest. Bull. Torrey Botan. Club 78:61-65.

STEUBING, LORE. 1965. Pflanzenökologisches Praktikum. Paul Parey, Berlin, Hamburg. 262 p.

SUKACHEV, V. N. 1928. Principles of classification of the spruce communities of European Russia. J. Ecol. 16:1-18.

SUKACHEV, V. 1932. Die Untersuchung der Waldtypen des osteuropäischen Flachlandes, XI:191-250. *In* E. Abderhalden (ed.) Hanb. biol. Arbeitsmethoden. Urban & Schwarzenberg, Berlin & Vienna.

SUKACHEV, V. 1945. Biogeocoenology and phytocoenology. C. R. Acad. Sci. U.S.S.R. 47:429-431.

SUKACHEV, V., and N. DYLIS. 1964. Fundamentals of forest biogeocoenology. Transl. by J. M. MacLennan (1968). Oliver & Boyd. Edinburgh & London, 672 p.

SWAN, J. M. A., and R. L. DIX. 1966. The phytosociological structure of upland forest at Candle Lake, Saskatchewan. J. Ecol. 54:13-40.

SWAN, J. M. A., R. L. DIX, and C. F. WEHRHAN. 1969. An ordination technique based on the best possible stand-defined axes and its application to vegetational analysis. Ecology 50:206-212.

TANSLEY, A. G. 1920. The classification of vegetation and the concepts of development. J. Ecol. 8:118-149.

TANSLEY, A. G. 1929. Succession, the concept and its values. Proceed. Intern. Congress of Plant Sci., Ithaca 1926:677-686.

TANSLEY, A. G. 1935. The use and abuse of vegetational concepts and terms. Ecology 16:284-307.

TANSLEY A. G. 1939. The British Islands and their vegetation. 2nd ed. 1953. Univ. Press, Cambridge, Vol. I:1-484, Vol. II:487-930.

TANSLEY, A. G. 1946. Introduction to plant ecology. 2nd ed. 1949. Unwin Bros. Ltd, London. 260 p.

TANSLEY, A. G., and T. F. CHIPP. 1926. Aims and methods in the study of vegetation. The British Empire Vegetation Committee. Whitefriars Press, London. 383 p. (Library of Congress K51.T3)

TURESSON, G. 1922. The genotypical response of the plant species to the habitat. Hereditas 3:211-350.

TURESSON, G. 1923. The scope and import of genecology. Hereditas 4:171-176.

TURESSON, G. 1925. The plant species in relation to habitat and climate. Hereditas 6:147-236.

TURESSON, G. 1927. Habitat and genotypic changes, a reply. Hereditas 8:156-206.

TURESSON, G. 1930. The selective effect of climate upon the plant species. Hereditas 14:99-152.

TURESSON, G. 1931. The geographical distribution of the alpine ecotype of some Eurasiatic plants. Hereditas 15:329-346.

TÜXEN, R. 1933. Klimaxprobleme des nordwesteuropäischen Festlandes. Nederl. Kriudkund. Arch. 43:293-309.

TÜXEN, R. 1937. Die Pflanzengesellschaften Nordwestdeutschlands. Mitt. Florist. Soziol. Arbeitsgem. Niedersachsen 3:1-170.

TÜXEN, R. 1947. Der pflanzensoziologische Garten in Hannover und seine bisherige Entwicklung. 94th-98th Jb. Naturhistor. Ges. Hannover (1942/3-1946/7):113-288.

TÜXEN, R. 1950a. Grundriss einer Systematik der nitrophilen Unkrautgesellschaften in der Eurosibirischen Region Europas. Mitt. Florist. Soziol. Arbeitsgem. N.F. 2:94-175.

TÜXEN, R. 1950b. Pflanzensoziologie als unentbehrliche Grundlage der Landeswirtschaft. Studium Generale 3:396-404.

TÜXEN, R. 1956. Die heutige potentielle natürliche Vegetation als Gegenstand der Vegetationskartierung. Angew. Pflanzensoziol. (Stolzenau, Weser) 13:5-42.

TÜXEN, R. (ed.). 1970. Gesellschaftsmorphologie (Strukturforschung). Bericht uber das Internationale Symposium der Internationalen Vereinigung für Vegetationskunde in Rinteln, 4-7 April 1966. W. Junk, The Hague. 360 p.

TÜXEN, R., and H. DIEMONT. 1937. Klimaxgruppe und Klimaxschwarm. Jahresber. naturhist. Ges. Hannover 88/89:73-87.

TÜXEN, R., and H. ELLENBERG. 1937. Der systematische und der ökologische Gruppenwert. Ein Beitrag zur Begriffsbildung und Methodik in der Pflanzensoziologie. Mitt. Florist. Soziol. Arbeitsgem. 3:171-184.

TÜXEN, R. and E. PREISING. 1942. Grundbegriffe und Methoden zum Studium der Wasser- und Sumpfpflanzen-Gesellschaften. Deut. Wasserwirtsch. 37:10-17, 57-69.

UNGER, F. 1836. Über den Einfluss des Bodens auf die Vertheilung der Gewächse, nachgewiesen in der Vegetation des nördöstlichen Tirols. Rohrmann & Schweigert, Vienna. 368 p.

VRIES, D. M. de. 1949. Method and survey of the characterization of Dutch grasslands. Vegetatio 1:51-58.

VRIES, D. M. de. 1952. Objective combinations of species. Acta Bot. Neerlandica 1:497-499.

VRIES, D. M. de. 1954. Ecological results obtained by the use of interspecific correlation. European Grassland Conference, p. 32-36. O.E.E.C., Paris.

VRIES, O. de. 1934. Unkräuter und Säuregrad. Z. f. Pflanzenernähr., Düngung u. Bodenkunde 13:356-360.

WACKER, F. W. 1943. Vergleichende Prüfung von landwirtschaftlich brauchbaren Verfahren der Grünlandbestandesuntersuchung. Pflanzenbau 19:328-363.

WALKER, B. H., and C. F. WEHRHAN. 1971. Relationships between derived vegetation gradients and measured environmental variables in Saskatchewan wetlands. Ecology 52:85-95.

WALTER, H. 1937. Pflanzensoziologie und Sukzessionslehre. Z. f. Botanik 31:545-559.

WALTER H. 1943. Die Vegetation Osteuropas. Paul Parey, Berlin. 180 p.

WALTER, H. 1954. Klimax und zonale Vegetation. Angew. Pflanzensoziol. Vienna, Festschr. Aichinger 1:144-150.

WALTER, H. 1960. Grundlagen der Pflanzenverbreitung: I. Teil. Standortslehre. 2nd ed. Eugen Ulmer. Stuttgart. 566 p.

WALTER, H. 1964. Die Vegetation der Erde in öko-physiologischer Betrachtung. Band I: Die tropischen und subtropischen Zonen. 2nd edition, VEB Fischer, Jena. 592 p.

WALTER, H. 1971. Ecology of tropical and subtropical vegetation. (Transl. by D. Mueller-Dombois, ed. by J. H. Burnett). Oliver & Boyd, Edinburgh. 539 p.

WALTER H., and H. STRAKA. 1970. Arealkunde. Floristisch-historische Geobotanik. 2nd ed. Eugen Ulmer, Stuttgart. 478 p.

WALTER, H., and E. WALTER. 1953. Einige allgemeine Ergebnisse unserer Forschungsreise nach Südwestafrika 1952/53: Das Gesetz der relativen Standortskonstanz; das Wesen der Pflanzengemeinschaften. Ber. Deut. Botan. Ges. 66:227-235.

WARMING, E. 1909. Oecology of plants. An introduction to the study of plant communities. Oxford University Press, London. 422 p. (Modified English edition of original Danish publication: Plantesumfund. 1895). Second Impression. 1925.

WEBB, L. J. 1968. Environmental relationships of the structural types of Australian rain forest vegetation. Ecology 49:296-311.

WEBB, L. J., J. G. TRACEY, and K. P. HAYDOCK. 1967. A factor toxic to seedlings of the same species associated with living roots of the non-gregarious subtropical rain forest tree Grevillea robusta. J. Appl. Ecol. 4:13-25.

WEBB, L. J., J. G. TRACEY, W. T. WILLIAMS, and G. N. LANCE. 1970. Studies in the numerical analysis of complex rain forest communities. V. A comparison of the properties of floristic and physiognomic-structural data. J. Ecol. 58:203-232.

WEHSARG, O. 1954. Ackerunkräuter. Akademie-Verlag. Berlin. 294 p.

WENDELBERGER, G. 1953. Über einige hochalpine Pioniergesellschaften aus der Glockner-und Muntanitzgruppe in den Hohen Tauern. Verh. Zool.-bot. Ges., Vienna 93:100-109.

WENT, F. W. 1942. The dependence of certain annual plants on shrubs in southern Californian deserts. Bull. Torrey Botan. Club 69:100-114.

WHITTAKER, R. H. 1951. A criticism of the plant association and climax concepts. Northwest Sci. 25: 17-31.

WHITTAKER, R. H. 1953. A consideration of climax theory: the climax as a population and pattern. Ecol. Monographs 23:41-78.

WHITTAKER, R. H. 1954. Plant populations and the basis of plant indication. Angew. Pflanzensoziol. (Vienna) Festschrift Aichinger. I:183-206.

WHITTAKER, R. H. 1956. Vegetation of the Great Smoky Mountains. Ecol. Monographs 26:1-80.

WHITTAKER, R. H. 1957. Recent evolution of ecological concepts in relation to eastern forests of North America. Am. J. Botany 44:197-206.

WHITTAKER, R. H. 1960. Vegetation of the Siskiyou Mountains, Oregon and California. Ecol. Monographs 30:279-338.

WHITTAKER, R. H. 1962. Classification of natural communities. Botan. Rev. 28:1-239.

WHITTAKER, R. H. 1967. Gradient analysis of vegetation. Biol. Rev. 42 (misprinted as Vol. 49):207-264.

WHITTAKER, R. H. 1970. Communities and ecosystems. Macmillan Co., Collier-Macmillan Ltd., London. 162 p.

WHITTAKER, R. H. 1972. Evolution and measurement of species diversity. Taxon 21:213-251.

WHITTAKER, R. H., and P. P. FEENY. 1971. Allelochemics: chemical interactions between species. Science 171:757-770.

WHITTAKER, R. H., and G. M. WOODWELL. 1972. Evolution of natural communities, pp. 137-159. In J. W. Wiens (ed.) Ecosystem Structure and Function. Oregon State Univ. Press.

WILLIAMS, W. T., and J. M. LAMBERT. 1959. Multivariate methods in plant ecology: I. Association analysis in plant communities. J. Ecol. 47:83-101.

WILLIAMS, W. T., and J. M. LAMBERT. 1960. Multivariate methods in plant ecology: II. The use of an electronic digital computer for association analysis. J. Ecol. 48:689-710.

WILLIAMS, W. T., and J. M. LAMBERT. 1961. Multivariate methods in plant ecology: III. Inverse association analysis. J. Ecol. 49:717-730.

WILSON, E. O. 1969. The species equilibrium. p. 38-47. In Diversity and stability in ecological systems. Brookhaven Symposia in Biology No. 22. 264 p. (Reproduced by Nat. Tech. Inform. Service, Springfield, Va. 22151).

WIMBUSH, D. J., M. D. BARROW, and A. B. COSTIN. 1967. Color stereophotography for the measurement of vegetation. Ecology 48:150-152.

WINKWORTH, R. E., and D. W. GOODALL. 1962. A crosswire sighting tube for point quadrat analysis. Ecology 43:342-343.

WOLTERECK, R. 1928. Über die Spezifität des Lebensraumes, der Nahrung und der Körperformen bei pelagischen Cladoceren und über "Ökologische Gestalt-Systeme." Biol. Zentralblatt 48:521-551.

ZAHNER, R., and N. A. CRAWFORD. 1965. The clonal concept in aspen site relations, Chap. 18:229-243. *In* Forest-Soil relationships in North America. Second North Am. Forest Soils Conference. Oregon State Univ. Press. 532 p.

ZOLLER, H. 1954. Die Typen der *Bromus erectus*-Wiesen des Schweizer Juras. Beitr. z. Geobot. Landesaufnahme der Schweiz 33. 309 p.

Author Index

Subject Index